Merging Optimization and Control in Power Systems

IEEE Press
445 Hoes Lane
Piscataway, NJ 08854

Merging Optimization and Control in Power Systems

Physical and Cyber Restrictions in Distributed Frequency Control and Beyond

Feng Liu
Tsinghua University

Zhaojian Wang
Shanghai Jiao Tong University

Changhong Zhao
The Chinese University of Hong Kong

Peng Yang
Tsinghua University

IEEE Press Series on Control Systems Theory and Applications
Maria Domenica Di Benedetto, Series Editor

IEEE PRESS
WILEY

Published by John Wiley & Sons, Inc., Hoboken, New Jersey.
Published simultaneously in Canada.

Library of Congress Cataloging-in-Publication Data applied for:

ISBN: 9781119827924

Cover Design: Wiley
Cover Images: © metamorworks/Shutterstock; © Pobytov/Getty Images

Set in 9.5/12.5pt STIXTwoText by Straive, Chennai, India

To our families

Books in the IEEE Press Series on Control Systems Theory and Applications

Series Editor: Maria Domenica DiBenedetto, University of l'Aquila, Italy

The series publishes monographs, edited volumes, and textbooks which are geared for control scientists and engineers, as well as those working in various areas of applied mathematics such as optimization, game theory, and operations.

1. *Autonomous Road Vehicle Path Planning and Tracking Control*
   Levent Güvenç, Bilin Aksun-Güvenç, Sheng Zhu, Şükrü Yaren Gelbal
2. *Embedded Control for Mobile Robotic Applications*
   Leena Vachhani, Pranjal Vyas, and Arunkumar G. K.
3. *Merging Optimization and Control in Power Systems: Physical and Cyber Restrictions in Distributed Frequency Control and Beyond*
   Feng Liu, Zhaojian Wang, Changhong Zhao, and Peng Yang.

Contents

Foreword

With the ever-increasing penetration of distributed energy resources (DERs), such as distributed generations (DGs), electric vehicles (EVs), and customer energy storages (CESs), the centralized, hierarchical control architecture developed for the traditional power system faces unprecedented challenges arising from massive but fragmented controllable resources. On the one hand, it might be unwise to regulate millions of individual DERs dispersed across a vast land. On the other hand, such an architecture heavily depends on intensive communication and usually slows down the control response, which may not cope with the fast variation of renewable generations. These issues bring potential risks to the stability and economy of power system operation. In this regard, distributed control and distributed optimization approaches are expected to supplement the existing control structure to relieve the dependence on the control center by leveraging neighboring communications among controllable (aggregate) DERs.

Conventionally, the two approaches are investigated separately due to different goals and time scales. Specifically, the distributed control focuses on fast-time-scale stability problems like primary and secondary control, while the distributed optimization addresses the slow-time-scale optimization problems like economic dispatch and energy trading. However, our power system desires a coalescence of the two approaches to effectively organize and utilize the massive but fragmented controllable resources to facilitate stable and economic operation.

This is the first book to systematically explore the integration of optimization and control in power systems toward a distributed paradigm. Rooted in the idea of "forward and reverse engineering," the book suggests the methodologies of "optimization-guided control" and "feedback-based optimization." It envisions the distributed optimal control of power systems with appealing design frameworks, particularly when considering complicated constraints enforced by physical and cyber restrictions. This book also gives rigorous proofs with in-depth insights and rich demonstrative examples from practical applications, which may benefit both theoretical research and industrial deployment.

The authors are all professional and reputational young researchers who have accomplished impactful works in theories and applications of distributed control and optimization in power systems, ranging from transmission and distribution grids to microgrids. I am therefore delighted to see that they summarize their recent inspiring works and shed new light on the control design of power systems at the cutting edge of transition. I believe it is an outstanding reference for graduate students, researchers, and engineers interested in power system control and optimization.

Chengshan Wang
Director of School of Electrical and
Information Engineering, Tianjin University
Academician of Chinese Academy of Engineering

Preface

As a worldwide trend, our power system is undergoing a historical transition from the traditional to a new landscape with high penetration of renewable energies and power electronics. As P. W. Anderson said about the structure of science in 1972, "more is different." Not surprisingly, the shift of our power system follows the same principle.

With the explosion of renewable generation, millions of small-capacity, volatile, and heterogeneous renewable generators, such as wind-turbine generators and photovoltaics, have been encroaching on the dominion of a few large-capacity traditional generators. Moreover, the number is rapidly ever growing every day. With the new era looming, one natural but critical question arises: Can the current, top-down, central-dominant control architecture established for the traditional power system still cope with the transition? Despite a broad spectrum of opinions, at least one general agreement could be reached from disagreement. That is, the future power system calls for reshaping the control paradigm to make it more scalable to the vast number of devices, more compatible with ever-increasing heterogeneity, more efficient in fast-varying operation environments, more adaptable to diverse operation modes, and more robust against unexpected perturbations and even failures.

Recent years have witnessed remarkable progress in distributed control theory, convex optimization, measurement and communication technologies, and advanced power electronics engineering. The cross-fertilization among these terrific successes further creates potential opportunities to meet the challenges mentioned above. The two most common might be the following: (i) relieving the strong dependence on a central coordinator and (ii) reshaping the top-down architecture by breaking the original control hierarchy. The former advocates a distributed control paradigm to endow the power system with higher scalability, compatibility, and robustness, while the latter suggests emerging the slow-time-scale optimization and fast-time-scale control to achieve stronger adaptability and faster response. As a consequence, many fruitful and innovative researches have been conducted to date. This book is not an account of all the leading-edge achievements but the personal efforts of the authors alongside, focusing on how to address the physical and cyber restrictions in distributed control design for meeting the requirement of future power systems.

Particularly, in this book, we concentrate the attention in frequency control as the ever-increasing renewables and power electronics lead to a dramatic drop of system inertia, raising great concerns with frequency stability and system security. However, this book also stretches out to part of voltage control and beyond, involving different levels/sizes of power systems, from high-voltage transmission systems and middle/low-voltage distribution systems to microgrids. Advanced mathematical tools, design methodologies, and illustrative examples with numerical/experimental simulations are systematically organized, aiming to provide the reader with a vision on how to emerge optimization and control in power systems under the distributed control paradigm with special considerations posed by diverse physical and cyber restrictions.

This book is organized into four parts. Part I (Chapters 1–2) presents the introduction and necessary preliminaries in optimization theory, control theory, and power system modeling. Part II (Chapters 3–6) demonstrates how to bridge optimization and control to devise optimization-guided distributed frequency control from both "forward engineering" and "reverse engineering" perspectives. Physical restrictions arising from operational constraints and objective functions are addressed with rigorous proof. Part III (Chapters 7–8) is dedicated to designing distributed frequency and voltage control with different cyber restrictions due to imperfect communication, where the feedback-based optimization is mostly involved in. Lastly, Part IV (Chapters 9–11) extends to the scope of robustness and adaptability of distributed optimal control, such as the robustness against unknown disturbances, flexibility to partial control coverage, and adaptability to heterogeneous control configurations.

Hence, this book is intended to be of value to graduate students and researchers who work in theory and applications of distributed control and optimization, especially their applications in power systems. This also can be used as a reference by electrical engineers to improve industry practices in power system control.

October, 2021

Feng Liu, Beijing
Zhaojian Wang, Shanghai
Changhong Zhao, Hong Kong
Peng Yang, Beijing

Acknowledgments

This book summarizes our recent research about distributed optimal control of power systems. Most works were jointly carried out in the Department of Electrical Engineering at Tsinghua University and the Department of Electrical Engineering at the California Institute of Technology.

The authors would like to thank many people for their contributions to this book sincerely. Moreover, the authors are particularly indebted to Prof. Steven H. Low from the California Institute of Technology, Prof. Shengwei Mei and Prof. Qiang Lu from Tsinghua University, Prof. Felix F. Wu from the University of California at Berkeley, and Prof. Xinping Guan from Shanghai Jiao Tong University for their long-lasting inspiration, encouragement, and support to our research. We also thank Mr. Yifan Su, Mr. Sicheng Deng, Prof. Laijun Chen, and Prof. Ying Chen from Tsinghua University, Dr. John ZF Pang from California Institute of Technology, Prof. Peng Yi from Tongji University, and Prof. Ming Cao from the University of Groningen, who have contributed materials to this book. In addition, we acknowledge Sandra Grayson, Kimberly Monroe-Hill, Viniprammia Premkumar, and other staffs at IEEE Wiley for their assistance and help in preparing this book.

The authors appreciate the support from the Major Smart Grid Joint Project of the National Natural Science Foundation of China and State Grid (U1766206), the Natural Science Foundation of China (51677100, 62103265), the "Chenguang Program" supported by Shanghai Education Development Foundation and Shanghai Municipal Education Commission of China (20CG11), the Direct Grant (4055128) of the Chinese University of Hong Kong, and the Hong Kong Research Grants Council through Early Career Award (24210220).

1

Introduction

Modern power grids rely on hierarchical control architectures to achieve stable, secure, and economic operation, which involves various kinds of advanced measurement, communication, and control techniques [1, 2]. Under the pressures of global climate change and energy shortage, power systems have been undergoing fundamental changes. The past decade has witnessed the leaping penetration of renewable energy resources and distributed generations, the proliferation of electric vehicles, and the active participation of customers, all of which are devoted to recreating a more reliable, flexible, sustainable, and affordable power grid.

On the generation side, fossil fuels are gradually giving place to environment-friendly renewable generations. By the end of 2020, the total installed capacity of global renewable energies reached 2802.004 GW, including 1332.885 GW of hydropower, 732.41 GW of wind energy, and 716.152 GW of solar energy [3]. The rapid growth of renewable energy shows signs of speeding up in the near future. On the consumption side, many new forms of loads have emerged and started participating in system operation with unprecedented enthusiasm. These include, just to name a few, electric vehicles, active participation of load demand [4, 5], energy storages [6, 7], and microgrids [8].

Despite the tremendous environmental and economical benefits, the ongoing changing trends on both generation and consumption sides are gravely challenging the traditional power system control technologies. Renewable energies such as photovoltaics (PVs) and wind power generations (WTGs) are intrinsically uncertain [9], leading to volatile operating conditions. Besides, most PVs and WTGs are integrated via power electronic devices with low/zero inertia and may operate in various control modes. The load-side diversity and participation also complicate the system control problem, which requires a careful design of the interaction protocol to achieve the smart grid vision.

We have to thoroughly and carefully address all these issues to facilitate the grand ambition of the ongoing system revolution. This extremely challenging task calls for advanced future power system control technologies to handle volatile

Merging Optimization and Control in Power Systems: Physical and Cyber Restrictions in Distributed Frequency Control and Beyond, First Edition. Feng Liu, Zhaojian Wang, Changhong Zhao, and Peng Yang.

operating conditions and massive controllable participants. Unfortunately, the existing power system control paradigm that features a centralized hierarchical structure may fail to achieve this goal. Here, we envision a new paradigm that reshapes the hierarchy and merges optimization with control, which may provide a new opportunity to tackle the task. This chapter will first introduce the traditional control methods in power systems and then introduce some state-of-the-art methods.

1.1 Traditional Hierarchical Control Structure

The functional operation of a power system depends on its control systems. As one of the largest and most complicated man-made systems, the modern power system must keep the frequency strictly synchronized and the voltages around their nominal values among thousands of generators and loads spanning over continents and interconnected through tens of thousands of miles of electric wires and cables. Therefore, an appropriate control architecture appears to be highly crucial to a reliable and efficient operation of power systems, if not the most. As a matter of fact, during the past one hundred years, power system control has evolved to be a layered/hierarchical structure encompassing diverse time scales and control objectives, ranging from millisecond to years. Figure 1.1 illustrates the time scales of power system controls with different control objectives. Generally, a slow time-scale layer is more concerned with the economy of operation during a long-time period, while a fast time-scale layer focuses on stability and security during a short-time dynamic process.

The time-scale decomposition and the hierarchical control structure in traditional power systems greatly simplify the control synthesis problem. For example, in most cases, detailed fast time-scale dynamics can be neglected in slow time-scale control problems and vice versa. Such decomposition works well when the time scales of the two layers are significantly different. Even if the difference is less noticeable, e.g. between seconds and minutes, it is still acceptable. Nonetheless, the recent transition of our power system shows that it might be more suitable to combine layers in different time scales, say, merging slow optimization and fast control. This idea sets the first motivation for us to write the book.

In the rest of this section, we shortly introduce the hierarchies of traditional frequency and voltage controls.

1.1.1 Hierarchical Frequency Control

In an alternating current (AC) power system, frequency reflects the active power balance across the overall system. The frequency goes down when generation is

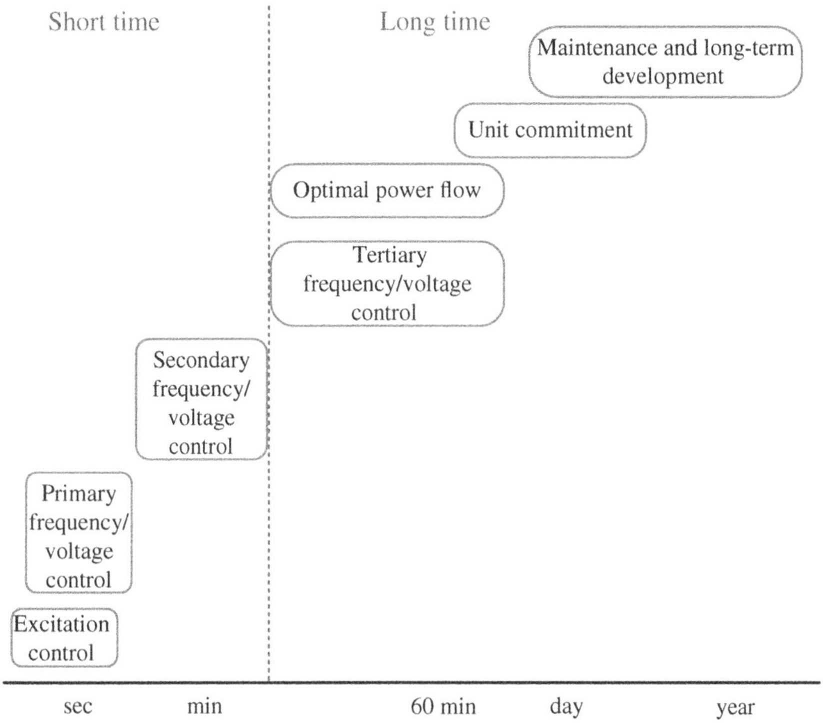

Figure 1.1 Time-scale decomposition of controls in a traditional power system.

less than load and vice versa. Therefore, a power system must adopt frequency control to maintain its frequency within a small neighborhood of the nominal value, such as 50 or 60 Hz.

In traditional power systems, most electric power is supplied by large-capacity synchronous generators. The huge rotating inertia of generators can serve as a buffer of kinetic energy to mitigate moderate power imbalance, limiting frequency changes instantaneously. For example, a sudden load demand increase will be naturally supported by extracting the kinetic energy from synchronous generators. Consequently, the frequency will drop. However, the kinetic energy stored in the inertia is quite limited, which is inadequate to cope with large or long-term frequency deviation. Therefore, intentional frequency control becomes a must to maintain system frequency more effectively and flexibly.

In accord with the control hierarchy mentioned above, frequency control includes three layers with respect to three different time scales, i.e. the primary control with a typical time scale in tens of seconds, the secondary control in several minutes, and the tertiary control in several minutes to tens of minutes, as shown in Figure 1.2. The first two layers act in a fast time scale that involves

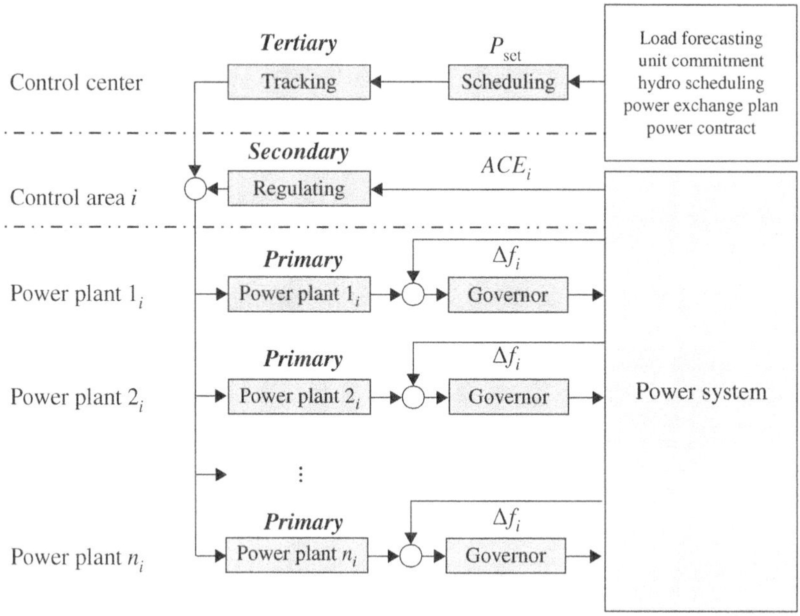

Figure 1.2 Diagram of hierarchical frequency control.

physical dynamics such as excitation control and governor control, while the last layer in a slow time scale involves operational or market dynamics such as economic dispatch (ED) and electricity market clearing. It is obvious from Figure 1.2 that the hierarchy of frequency control heavily relies on a control center.

1.1.1.1 Primary Frequency Control

Primary frequency control is designed to limit frequency deviation within an acceptable range. It is usually fulfilled via automatic governor regulation of generators. Denote by P_i^g the mechanical power of generator i and Δf_i the frequency deviation from the nominal value at bus i. Let the power compensation ΔP_i^g be $\Delta P_i^g = -k_i^p \Delta f$ with $k_i^p > 0$, and send it to change the valve opening of the prime mover, such that the system frequency regains a state of operating equilibrium.

Obviously, the primary frequency control is indeed a proportional feedback control, or droop control. It responds fairly fast to frequency deviation since only local frequency measurement is required. However, it may not restore system frequency to the nominal value due to the proportional control strategy.

In practice, not all generators in the system need to be equipped with a governor control. Those generators, however, usually are competent to provide fast power support. Hence one can still categorize them into primary frequency control when needed.

1.1.1.2 Secondary Frequency Control

Secondary frequency control is designed to eliminate frequency deviation. It is usually fulfilled via automatic generation control (AGC). In a multi-area power system, the area control error (ACE) is a linear combination of the deviations of system frequency and the tie-line powers delivered to or received from its neighboring areas [10]. For the ith area, the ACE is defined as

$$\mathrm{ACE}_i = K_i^s \Delta f_i + \sum_{j \in \mathcal{N}_i} \Delta P_{ij}$$

where $K_i^s > 0$ is a constant that stands for the responsibility of this area in response to the frequency deviation, which is referred to as area frequency response coefficient (AFRC). ΔP_{ij} is the power deviation of the tie-line connecting areas i and j. As a matter of fact, when the ACEs of all control areas converge to zero, the system frequency restores to the nominal value. Therefore, traditional AGC uses ACEs as the feedback signals to compute the control command that reflects the total required power compensation in the area. The obtained control command is then distributed to individual generators in proportion to their participation factors.

AGC typically adopts a proportional–integral (PI) control to drive ACEs to zero. However, as a power system always works in a time-varying environment, the ACEs do not converge to zero but rather fluctuate around zero. So does the system frequency.

1.1.1.3 Tertiary Frequency Control

Although the secondary frequency control can eliminate the frequency deviation, it is not responsible for achieving an optimal power generation allocation. Instead, this task is fulfilled by the tertiary frequency control that aims to minimize the operation cost of the system (e.g. generation cost or network loss) by reallocating the power production among generators. Thus, it is also called ED, usually performed every 5–15 min. This problem can be mathematically formulated as a constrained optimization problem, where the constraints include power balance, generation limits, line flow limits, etc. Traditionally, the ED problem is solved to generate the control commands in the control center, and then the control commands are sent to the dispatchable generators as the set points. In a deregulated power system, or power market, the ED problem is replaced by a market clearing problem. Pricing issues and strategic behaviors of participants need to be carefully considered as well.

1.1.2 Hierarchical Voltage Control

While frequency is determined by the active power balance of the overall system, voltage is more closely relevant to local reactive power supply. When the reactive power supply is inadequate, the voltage will be lower than the nominal value,

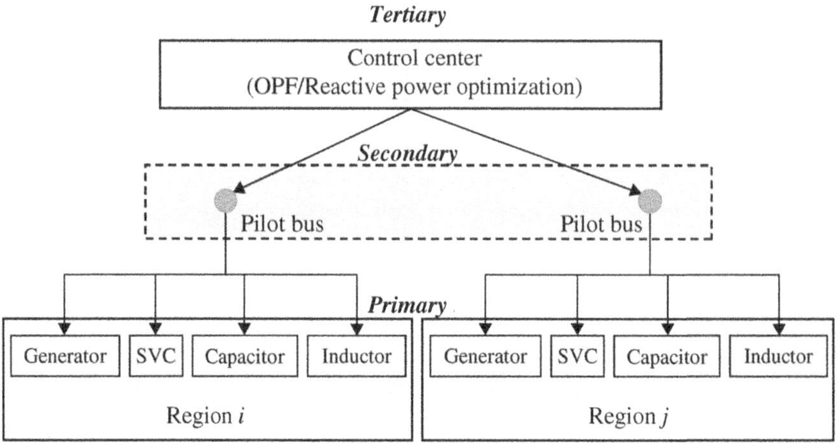

Figure 1.3 Diagram of hierarchical voltage control.

and vice versa. Therefore, voltage control is usually realized by regulating reactive power generation or consumption of generators, transformers, reactive power compensators, loads, etc. Usually, the load-based voltage regulation is adopted in an emergency only. Except for reactive power sources/loads, flexible AC transmission systems based on power electronics provide an additional option to voltage control by changing the equivalent line impedance.

Unlike frequency, voltage is more like a local indicator for reactive power adequacy than a system-wide indicator for active power balance. In addition, since reactive power is not suitable for long-distance transmission (for avoiding unexpected transmission loss and voltage instability), a voltage control problem is usually considered and implemented in local regions. For coordinating voltage controls in different regions to enhance system-wide performance, a hierarchical voltage structure emerges. Similar to frequency control, voltage control also has a central-dominant hierarchy that comprises three layers: the primary voltage control with a typical time scale in tens of seconds, the secondary voltage control in several minutes, and the tertiary voltage control in several minutes to tens of minutes. Figure 1.3 illustrates the diagram of hierarchical voltage control.

1.1.2.1 Primary Voltage Control

As the fastest control layer in the hierarchy, primary voltage control aims to stabilize the voltage rapidly when voltage deviation from the preset value exceeds a certain threshold. It is fulfilled by using local feedback control of various reactive power sources including synchronous generators, capacitors, series inductors, static Var compensator (SVC), static synchronous compensator (STATCOM), etc. For example, one can use automatic voltage regulator (AVR) to flexibly control the terminal voltage of generators (via the excitation system), often within a range of $\pm5\% \sim \pm10\%$ with respect to its rated value. In practice, PI controllers are extensively employed; otherwise, droop controllers or proportional controllers appear to be more helpful for a stable operation.

1.1.2.2 Secondary Voltage Control

As mentioned above, primary voltage control is built on fully local voltage deviation. Hence a controllable reactive power source or load in a subregion would provide little support to other subregions even if it has plenty of reactive power reserve. In this regard, secondary voltage control is developed. It is implemented in a control center of one region, within a time scale from tens of seconds to several minutes, or even longer up to the time constant of controlled devices. Secondary voltage control aims to coordinate the region-wide voltage control by utilizing the reference voltage of a pilot bus that represents the voltage situation or reactive adequacy in a local region. In operation, the reference voltage deviation of the pilot bus is sent to local primary voltage controllers, attached to the primary control signals, to coordinate all the controllable voltage/Var sources and loads within this region.

Except for reactive power sources and loads, onload tap changers can also be used to facilitate secondary voltage control by changing the taps of transformers. However, it should be aware that adjusting transformer taps only changes reactive power distribution rather than generating reactive power. Therefore, we only enable it when the reactive power is sufficient in the region. Otherwise, it may worsen the reactive power shortage, even resulting in disastrous voltage collapse.

1.1.2.3 Tertiary Voltage Control

Reactive power distribution has a remarkable influence on the power loss of power transmission. In this regard, tertiary voltage control is responsible for optimizing the reactive power distribution across the overall power grid. This target is achieved by changing the reference voltages of pilot buses in each control region. In this sense, tertiary voltage control can also be regarded as a particular type of optimal power flow (OPF) problem, where only reactive power is optimized. Therefore its time scale follows the pace of ED or slower, varying from several minutes to several hours.

1.2 Transitions and Challenges

The central-dominant hierarchical control architecture has effectively supported the operation of traditional power systems for decades. As power systems evolve into a new era, a critical question arises: Can the traditional control paradigm still apply? The key features in future power systems that may hinder the classical control paradigm include the following:

- **Massive entities**: A huge amount of dynamic devices are integrated into the power system from different voltage levels, including WTGs, PVs, electric vehicles, and energy storage, to name a few. The number of controllable and uncontrollable devices increases by orders of magnitudes, calling for a more scalable control scheme with an effective coordination.
- **Heterogeneous dynamics**: Due to the increasing diversity of control modes and physical natures of electrical devices, the dynamics in future power systems are extensively heterogeneous. It brings intricate interaction patterns and significantly complicates system-level analysis and control design, calling for a more compatible control scheme.
- **Uncertain and fast-changing environments**: The high penetration of renewable energy resources and complex loads significantly increases the uncertainties in power systems. In addition, distributed energies usually belong to individual owners, which could be switched on or off frequently. The increasing uncertainty leads to volatile operating conditions, which requires much faster control responses to retain power balance and economical operation in fast-changing environments, calling for a more adaptive and robust control scheme.

In light of these transitions, the classical center-dominant hierarchical control paradigm may fail to meet what the future power systems demand, and reshaping it to fit the future becomes a must. Ideally, we envision that the future paradigm should be scalable to the massive amounts of devices, compatible with ever-increasing heterogeneity, efficient in fast-changing operation environments, flexible to diverse operation modes, and robust against unexpected perturbations and even failures. Toward this ambition, fruitful progress has been made to combine advanced control and optimization theories with power system engineering. Two of the most promising topics among them are (i) relieving the dependence on central coordinators and (ii) reshaping the original control hierarchy. The former advocates a distributed control paradigm that endows the power system with higher scalability, compatibility, and robustness. The latter suggests merging the slow time-scale optimization and fast time-scale control to achieve stronger adaptability and faster response. The following two sections will briefly introduce the state of the art of these two innovative topics.

1.3 Removing Central Coordinators: Distributed Coordination

In terms of the way of coordinating, three approaches may be involved: centralized control, decentralized control, and distributed control.

Centralized control features a control center that collects information of all agents, performs a central computation to get control commands, and sends them back to each agent (Figure 1.4a). The control center can solve a complex centralized optimization problem to increase the economic efficiency of the whole system. However, the system may break down if the control center fails, which is the notorious single-point-of-failure issue. In addition, privacy becomes a big problem because the control center requires information from all agents. Another problem is that this approach is not scalable as it heavily relies on detailed information of all agents in the system. Moreover, collecting the information is time-consuming in large systems, which remarkably slows down the response.

Decentralized control needs no control center and adopts purely local control strategies to compute control commands, i.e. no communication between agents (Figure 1.4b). Thus, decentralized control usually has a rapid response, which has been widely adopted, particularly in the primary control. However, agents adopting decentralized control may conflict with each other since there is a lack of coordination.

The distributed control also needs no control center but requires communication among agents such that each agent can compute control commands locally[1] (Figure 1.4c). Roughly speaking, the complexity of distributed control is between those of centralized and decentralized ones. Similar to decentralized control, the structure of distributed control is also simple and easy to apply. However, distributed control can enable coordination among agents to facilitate global objectives as the agents can exchange information. Compared with centralized control, distributed control has several potential advantages. First, each agent only needs to share limited information with a subset of the other agents, i.e. its neighbors on the communication graph. This feature consequently improves cybersecurity, better protects privacy, and reduces the expense of communication networks. Second, distributed control avoids the single-point-of-failure issue and hence is more robust. Third, distributed control may be computationally superior to the centralized rival by decomposing a large-scale global problem into a set of small-scale local problems. Finally, distributed control can adapt to volatile operation conditions, such as fast variation of renewable generations and topology changes.

Since the distributed control structure inherits advantages both from the centralized and decentralized ones, it has been widely recognized as a promising solution to the aforementioned challenges and has inspired plenty of achievements in this field. Interestingly, it has many overlaps with another topic, distributed optimization, in both problem formulations and algorithms. The two closely related topics, sometimes, may lead to confusion. In this regard, this book adopts the following terminology. When considering distributed control, we design a controller and implement it to physical systems in a distributed manner, and hence the controller's dynamics directly couple with the physical dynamics. On the other hand,

1 The topology of a physical system is not necessarily identical to its communication graph.

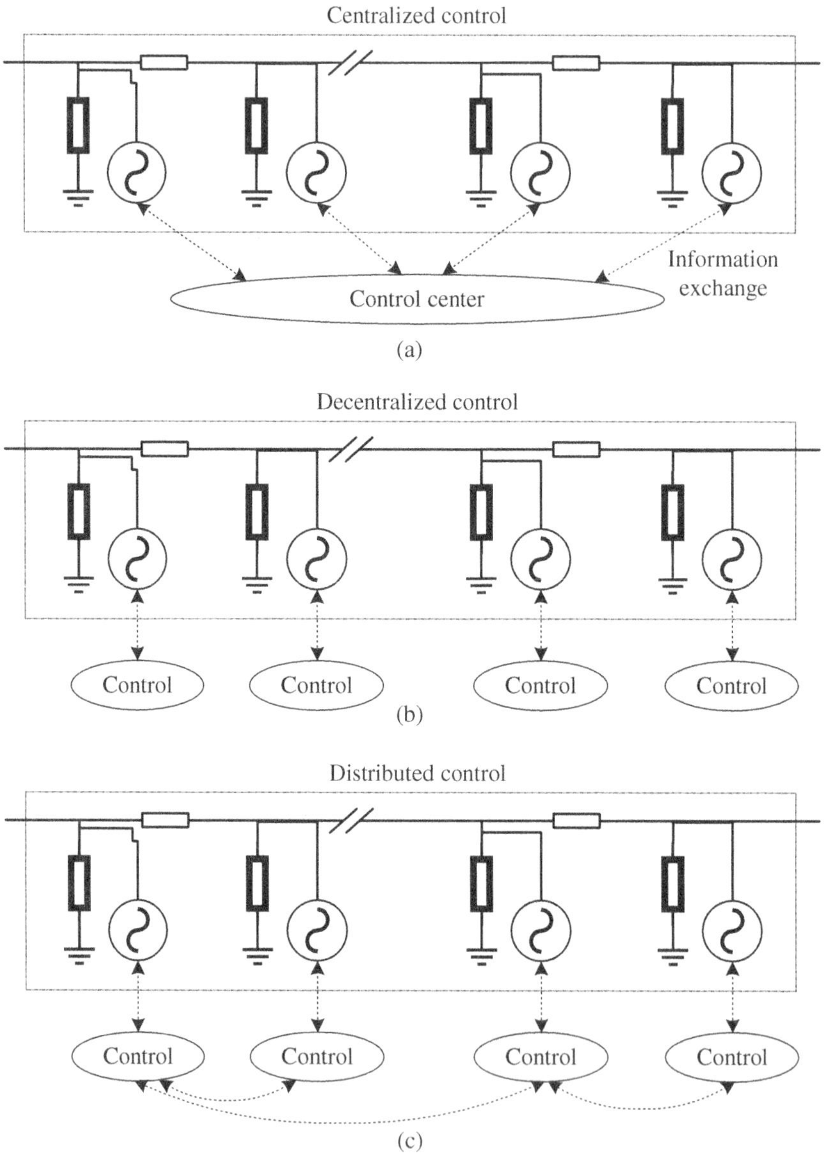

Figure 1.4 Diagram of centralized (a), decentralized (b), and distributed (c) control.

when considering distributed optimization, we construct a distributed algorithm to solve optimization problems subject to some snapshots of the physical systems while the solving process does not directly couple with the physical dynamics. In this sense, the primary and secondary frequency/voltage regulations are categorized as a control problem, while the tertiary frequency/voltage control, i.e. ED, is categorized as an optimization problem. However, we use distributed optimal control when both optimization and control are taken into account.

1.3.1 Distributed Control

Distributed control considers a group of dynamic agents as follows:

$$\dot{x}_i = f_i(x_i, x_k, u_i), \quad k \in \mathcal{N}_i \tag{1.1a}$$

$$y_i = g_i(x_i) \tag{1.1b}$$

where $x_i \in \mathbb{R}^{n_i}$ is the state variable of agent i, $y_i \in \mathbb{R}^{m_i}$ is the output, and $\mathcal{N}_i$ is the set of the neighbors of dynamic agent i on the physical network. Agent i can exchange information with its neighbors on the communication network, denoted by $\mathcal{N}_i^c$. Note that generally $\mathcal{N}_i^c \neq \mathcal{N}_i$.

We expect to design the following distributed controller that depends on the local output:

$$\dot{u}_i = h_i(y_i, y_j), \quad j \in \mathcal{N}_i^c \tag{1.2}$$

One of the most extensively used approaches is to adopt the consensus-based distributed control [11], which takes the following form:

$$\dot{u}_i(t) = \sum_{j \in \mathcal{N}_i^c} a_{ij} \left(y_j(t) - y_i(t) \right) \tag{1.3}$$

where a_{ij} is the weight between buses i and j. According to the consensus algorithm, $y_i(t) = y_j(t)$ holds for all i, j in the steady state. In this sense, the output y_j essentially serves as a global variable to coordinate all agents in the system. Individual agents estimate the variable locally and approach the consensus through iterative computation and information exchange. Therefore, by appropriately choosing the output y_j, various kinds of distributed controllers can be devised.

Specifically, in power systems, the global coordination variable could be generation ratio and marginal cost. The former one is the ratio between actual generation and the maximal capacity, which implies that all generators supply the load fairly up to their maximal capability [12–17]. References [12, 13] apply a consensus algorithm to the active power control of PVs for achieving a unified utilization ratio. This method can also apply in reactive power control [14–16] and harmonic control [17]. The latter implies that all generators share the same marginal cost and hence reach the economical optimum [18, 19]. Reference [18] shows that individual distributed generators in a microgrid can maintain identical marginal costs by using a consensus-based distributed controller while restoring the nominal frequency. This idea is further extended to direct current (DC) microgrids in [19].

Consensus-based distributed control has a simple structure, which can achieve a *fair* or *economical* operation. The fairness is realized by maintaining equal generation ratios, while the economy is realized by holding an equal marginal generation cost. On the other side, this simple structure conversely restricts the applicability of consensus-based distributed control when considering complicated optimization objectives or constraints such as line flow limits, which are very common in power systems. In such circumstances, we need more sophisticated designs, which will be discussed in detail in the remaining chapters of this book.

1.3.2 Distributed Optimization

Distributed optimization considers a group of agents that cooperatively solve the following separable (convex) optimization problem with constraints:

$$\min_{x_i,\ i=1,\ldots,N} \quad \sum_{i=1}^{N} C_i\left(x_i\right) \tag{1.4a}$$

$$s.t. \quad \sum_{i=1}^{N} f_i(x_i) \leq 0, \tag{1.4b}$$

$$x_i \in X_i, \ i = 1, 2, \ldots, N \tag{1.4c}$$

where $C_i\left(x_i\right), f_i\left(x_i\right)$ are convex functions and X_i is a convex set.

In distributed optimization, each agent carries out local computation on a subproblem and exchanges information with neighbors. Convergent computation gives an optimal solution to the original optimization problem. Some commonly used distributed optimization algorithms include consensus-based algorithms, dual decomposition, alternating direction method of multipliers (ADMM), and gradient-based algorithms, to name a few. An introduction to these algorithms can be found in Appendix A. Here, we briefly introduce the applications of the most popular consensus-based algorithms and ADMM in power systems.

Consensus-based algorithms are especially relevant in ED problems, where marginal costs or prices of generators serve as the global coordination variables for consensus [21–24]. In [22], an average consensus method is presented to solve the ED problem in a distributed fashion, where two stopping criteria are derived based on sign consensus. Reference [21] extends the consensus method to solve the ED problem with transmission losses. In [23] an incremental cost-consensus algorithm is suggested with a convergence proof. In addition, literature [24] shows that a consensus-based algorithm can track the optimal solution to the active power ED problem with generation capacity constraints. However, as mentioned in the previous part, this appears to be restrictive in dealing with complicated constraints.

Compared with consensus-based algorithms, ADMM has been more extensively employed to deal with complicated constraints, such as OPF problems. In terms of power flow models, the appliances of ADMM roughly fall into four categories: (i) AC OPF problems [25–27], (ii) DC OPF problems [28–30], (iii) distribution flow with the second-order cone relaxation [31–33], and (iv) linearized distribution flow [34–36]. Reference [25] proposes an asynchronous ADMM algorithm of the AC OPF problem to cope with the communication delay by extending the synchronous ADMM algorithm [26, 27]. In [28], a fully distributed accelerated ADMM algorithm is presented, where a consensus-based push-sum method is derived to improve the convergence. In [29], a consensus-based ADMM approach is employed to solve DC OPF problems. Reference [30] utilizes machine learning to predict the optimal dual variable under different realizations of system loads, remarkably accelerating the convergence of ADMM. The distribution flow (DistFlow) model is considered in references [31–36]. To cope with the nonconvexity of the DistFlow model, references [31–33] apply a second-cone relaxation while references [34–36] choose to linearize the DistFlow model by ignoring the line loss.

Note that ADMM algorithms are usually concerned with optimization problems only by assuming the dynamics of physical power systems are fast enough and negligible. One consequence is that the algorithms are not applicable in a fast-changing environment. Otherwise, the influence of physical power system dynamics will turn to be non-ignorable, making the assumption invalid.

1.4 Merging Optimization and Control

As analyzed above, distributed control and optimization have their own advantages and deficiencies and are applied in different layers. So a natural question is whether they could be combined to address both optimization and control issues in complicated environments. This question inspires the idea of merging primary and secondary controls in a fast time scale with ED or OPF in a slow time scale [10, 37, 38]. This leads to a cross-layer design approach of distributed optimal control that bridges the gap between optimization and control in different time scales. As a result, we expect the merged distributed optimal control could achieve optimality automatically and rapidly while guaranteeing system stability, even under complicated operational constraints and changing environments.

Generally, this idea has two sides: *optimization-guided control* and *feedback-based optimization*. On the one side, we intend to design a feedback controller for power systems, which could drive the system states to the optimal solution of an optimization problem in the steady state, such as ED. At the same time, the system should be asymptotically stable. On the other side, we hope to solve optimization problems by making use of feedback control. In this case, some state variables are obtained by measuring instead of solving complicated system equations, such as power flow equations. In this section, we explain them, respectively.

1.4.1 Optimization-Guided Control

The main idea of optimization-guided control is to design a (dynamic) feedback controller for power systems, which drives the system to the optimal solution to an optimization problem, such as ED or OPF. The framework is illustrated in Figure 1.5.

In this framework, the lower layer is the fast-time-scale dynamics of a power system with control inputs. We formulate it into a general control system:

$$\begin{cases} \dot{x} &= g_d\left(x, u_d'\right) \\ y_d &= h_d(x) \end{cases} \tag{1.5}$$

where x is the vector of state variables, u_d' is the vector of control inputs, and y_d is the vector of system outputs.

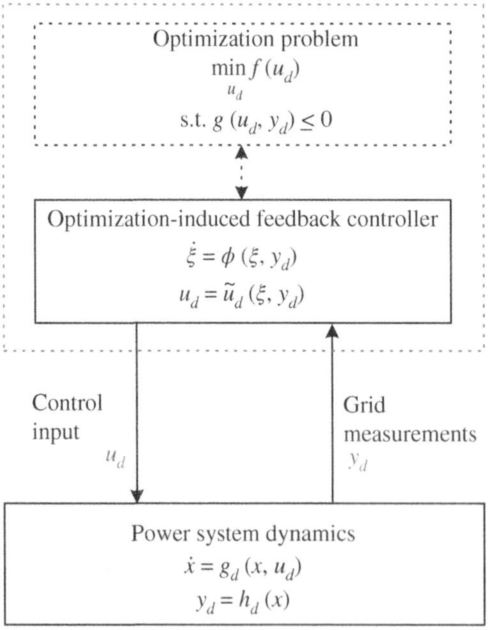

Figure 1.5 Conceptual diagram of optimization-guided feedback control.

The upper layer is the slow-time-scale ED or OPF process of the power system. We formulate it into a general constrained (convex) optimization problem with adjustable parameters:

$$\min_{u_d} \quad f\left(u_d\right) \tag{1.6a}$$

$$\text{s.t.} \quad g\left(u_d, y'_d\right) \leq 0 \tag{1.6b}$$

where u_d is the vector of decision variables and y'_d is the vector of adjustable parameters.

Our goal is to design a dynamic controller:

$$\begin{cases} \dot{\xi} &= \phi(\xi, y'_d) \\ u_d &= \tilde{u}_d(\xi, y'_d) \end{cases}, \tag{1.7}$$

such that the output of (1.7) at the steady state is equal to the optimal solution to (1.6). This equivalent relation can be established by noting that the solving process of optimization problem (1.6) essentially defines a dynamic process, e.g. the primal–dual gradient dynamics.

To merge the two layers above, we interconnect the controller (1.7) and the system (1.5) by letting $u_d = u'_d$ and $y_d = y'_d$,[2] as shown in Figure 1.5. Then we need to appropriately design the controller (1.7) such that the closed-loop power system guarantees the asymptotic stability with the equilibrium point identical to the solution to the optimization problem (1.6). This approach is closely related to the primal–dual gradient dynamics (or saddle-point dynamics) [39, 40] by noting that one can construct the controller (1.7) from the primal–dual gradient dynamics of the optimization problem (1.6). That is why this is called optimization-guided control.

The idea of optimization-guided control can date back to [41] that presents a methodology to regulate a nonlinear dynamical system to an optimal operation point, i.e. a solution to a given constrained convex optimization problem in terms of the steady-state operation. To this end, it constructs a dynamic extension of the Karush–Kuhn–Tucker (KKT) optimality conditions for the corresponding optimization problem. A similar method is applied in power systems [20], which exploits the pricing interpretation of the Lagrange multipliers to guarantee economically optimal operation at the steady state. This idea is further generalized as the notion of *reverse and forward engineering* methodology for designing optimal controllers, particularly in optimal frequency and voltage control of power systems [10, 37, 42, 43] Following and extending this idea, we will design frequency controllers considering various physical restrictions in this book.

---

2 Without confusion, we will not distinguish u_d/u'_d and y_d/y'_d in the rest of this chapter.

1.4.2 Feedback-Based Optimization

Feedback-based optimization provides an alternative way to merge optimization and control. One of the primary motivations for feedback-based optimization is to fast respond to power fluctuations due to time-varying loads and volatile renewable generations, since optimized strategies based on fixed operational conditions may not be applicable in this situation. Therefore, unlike optimization-guided control, feedback-based optimization is usually more concerned with the operational optimality than the stability of a power system. As a result, (quasi-) steady-state models of a power system other than its dynamical model are adopted.

The original idea of feedback-based optimization might come from straightforward intuition. A physical power system itself can be regarded as a *computer* that calculates and outputs exact information of power flow. Hence, one can accelerate the computation by directly measuring the state of the power system other than calculating it based on a mathematical model. From a control perspective, the use of measurements constitutes a feedback loop in optimization computing. That is why this approach is referred to as feedback-based optimization.

Figure 1.6 illustrates the conceptual framework of feedback-based optimization. In this framework, the lower level is the (quasi) steady-state power system described by the power flow equations:

$$0 = g_s(x, u_s)$$

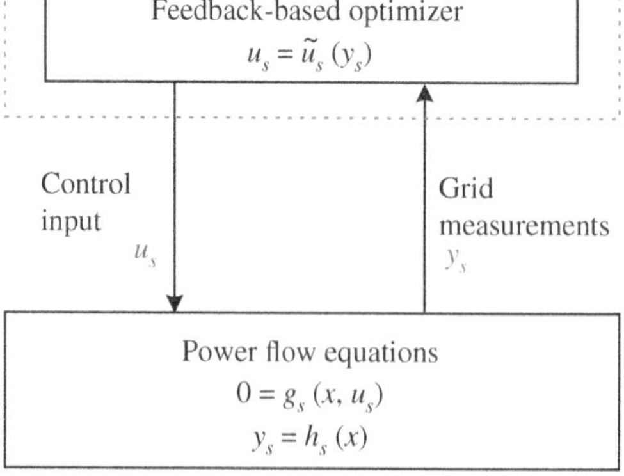

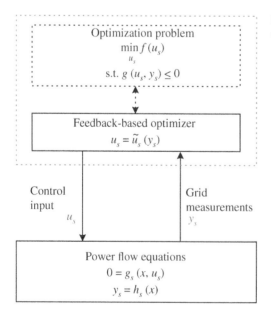

Figure 1.6 Conceptual diagram of feedback-based optimization.

where x is the system state, u_s is the controllable parameters such as generation set points, and y_s is the system outputs. The designer hopes to optimize the steady state x of the power system by adjusting u_s, rendering the following optimization problem:

$$\min_{u_s} \quad f(u_s) \tag{1.8a}$$

$$\text{s.t.} \quad g(u_s, y_s) \leq 0 \tag{1.8b}$$

In the above optimization problem, the power flow equations serve as part of the equality constraints. By measuring y_s from the physical power system, this constraint is eliminated. The optimal solution, u_s, to the problem (1.8) with respect to y_s is then sent to the physical power system to drive it to the desired working point.

The introduction of feedback in optimization has two salient advantages. First, it remarkably reduces the computational complexity by avoiding solving the high-dimensional nonlinear power flow equations. Second, it endows the optimization algorithms with the adaptability to time-varying working condition by feedback.

Similar to the optimization-guided control, the feedback-based optimization also involves the philosophy of forward and reverse engineering. However, the feedback-based optimization focuses on a slower time scale considering (quasi-)steady states, while the optimization-guided control needs to consider the dynamics of physical systems in a faster time scale. Nevertheless, in practical control system design, one sometimes needs to combine the two approaches to achieve satisfactory control performance.

1.5 Overview of the Book

This book includes four parts: (i) essential preliminaries of optimization and control theory and power system models (Chapter 2), (ii) optimization-guided controllers considering physical restrictions (Chapters 3–6), (iii) feedback-based optimization algorithms considering cyber restrictions (Chapters 7 and 8), and (iv) robustness and adaptability of the distributed control (Chapters 9–11). Related state-of-the-art technologies will be introduced and illustrated in detail with hands-on examples and real-world systems for the readers to learn about how to use them to address specific problems.

Part 1: Introduction and Preliminaries

After a general introduction in Chapter 1, Chapter 2 summarizes some essential preliminaries of distributed optimal controller design, including graph theory, convex analysis, operator theory, stability theory, power flow equations, and power system dynamics. It also provides a good reference to readers who are interested in related fields.

Part 2: Physical Restrictions

In this part, we design optimization-guided distributed frequency control of power systems considering physical restrictions.

Chapter 3 first introduces the notion of *reverse and forward engineering*. Furthermore, it presents a systematic method to design an optimization-guided feedback controller by taking the primary frequency regulation as an example. Particularly, we demonstrate how to bridge optimization and control by exploiting the structure of (partial) primal-dual dynamics.

Chapters 4 and 5 extend the methodology in Chapter 3 to incorporate operational constraints such as regulation capacity constraints and line flow limits into distributed optimal frequency control. Chapter 4 focuses on the per-node power balance, where the power mismatch is economically balanced within each area. A completely decentralized strategy is derived without communication. We address the input saturation issue by constructing a Lyapunov function using the projection technique, originally developed for variational inequality problems.

Chapter 5 focuses on the network power balance case, where power mismatch may be balanced by all generators and controllable loads across the overall system. We pay special attention to dealing with global inequality constraints imposed by line flow limits by employing the invariance principle for nonpathological Lyapunov functions.

Chapter 6 further considers nonsmooth objective functions, which widely exist in power systems. However, it is challenging to design distributed controllers in the presence of nonsmooth objective functions, as the traditional gradient descent approach does not apply. Here, we show how to leverage the Clark generalized gradient and differential inclusion to deal with such nonsmoothness issues.

Part 3: Cyber Restrictions

In this part, we design feedback-based optimization algorithms for both frequency and voltage control in power systems with cyber restrictions.

Chapter 7 investigates the distributed power control of microgrids with different kinds of asynchrony. We show how to fit different kinds of asynchrony into a unified framework and employ the operator splitting technique to prove the convergence. Specifically, frequency and voltage measurements are utilized as feedbacks to simplify the optimization algorithm.

Chapter 8 investigates the distributed optimal voltage control in distribution networks with different kinds of asynchrony. We combine the partial primal–dual gradient algorithm and the operator splitting techniques to handle complicated operational constraints and asynchrony. Moreover, we show how to utilize local measurements to facilitate the online implementation that is adjustable to time-varying environments.

Part 4: Robustness and Adaptability

This part further investigates the robustness and adaptability of the distributed optimal controllers.

Chapter 9 addresses the robustness of distributed frequency controller to unknown disturbances due to volatile renewable generations and loads. We show how to decompose the imbalanced power into subproblems in different time scales and resolve them by combining the primal–dual gradient, internal mode control, and L_2-gain theory.

Chapter 10 focuses on the adaptability of distributed optimal controller to partial control coverage, i.e. only a subset of generators is controllable. We show how to leverage incremental passivity conditions for both active and reactive power controls, which guarantee the closed-loop stability of the overall system even under large disturbances.

Chapter 11 addresses the adaptability of distributed optimal controllers to the heterogeneity of existing controls. We design a primal-dual gradient based controller of microgrids to unify different types of original controls. We show it is adaptable to the diversity of microgrids and plug-and-play operations.

This book also includes two appendices for the convenience of the readers. Appendix A provides a short review on typical distributed optimization algorithms, some of which are used in the previous chapters. Moreover, Appendix B presents the OPF theory of DC networks, which is omitted in Chapter 11 for clarity and simplicity.

Bibliography

1 A. Ipakchi and F. Albuyeh, "Grid of the future," *IEEE Power and Energy Magazine*, vol. 7, no. 2, pp. 52–62, 2009.

2 H. Bevrani, *Robust Power System Frequency Control*. Springer, 2009.

3 IRENA, *Renewable Energy Statistics 2021*. https://www.irena.org/publications/2021/Aug/Renewable-energy-statistics. 2021.

4 P. Palensky and D. Dietrich, "Demand side management: demand response, intelligent energy systems, and smart loads," *IEEE Transactions on Industrial Informatics*, vol. 7, no. 3, pp. 381–388, 2011.

5 P. Siano, "Demand response and smart grids–a survey," *Renewable and Sustainable Energy Reviews*, vol. 30, pp. 461–478, 2014.

6 S. J. Chiang, K. T. Chang, and C. Y. Yen, "Residential photovoltaic energy storage system," *IEEE Transactions on Industrial Electronics*, vol. 45, no. 3, pp. 385–394, 1998.

7 H. Zhou, T. Bhattacharya, D. Tran, T. S. T. Siew, and A. M. Khambadkone, "Composite energy storage system involving battery and ultracapacitor with dynamic energy management in microgrid applications," *IEEE Transactions on Power Electronics*, vol. 26, no. 3, pp. 923–930, 2011.

8 R. H. Lasseter, "Microgrids," in *IEEE Power Engineering Society Winter Meeting, 2002*, (New York, USA), pp. 305–308, IEEE, 2002.

9 J. Marcos, L. Marroyo, E. Lorenzo, D. Alvira, and E. Izco, "Power output fluctuations in large scale PV plants: one year observations with one second resolution and a derived analytic model," *Progress in Photovoltaics: Research and Applications*, vol. 19, no. 2, pp. 218–227, 2011.

10 N. Li, C. Zhao, and L. Chen, "Connecting automatic generation control and economic dispatch from an optimization view," *IEEE Transactions on Control of Network Systems*, vol. 3, no. 3, pp. 254–264, 2016.

11 R. Olfati-Saber, J. A. Fax, and R. M. Murray, "Consensus and cooperation in networked multi-agent systems," *Proceedings of the IEEE*, vol. 95, no. 1, pp. 215–233, 2007.

12 H. Xin, Z. Lu, Y. Liu, and D. Gan, "A center-free control strategy for the coordination of multiple photovoltaic generators," *IEEE Transactions on Smart Grid*, vol. 5, no. 3, pp. 1262–1269, 2014.

13 H. Xin, Y. Liu, Z. Qu, and D. Gan, "Distributed control and generation estimation method for integrating high-density photovoltaic systems," *IEEE Transactions on Energy Conversion*, vol. 29, no. 4, pp. 988–996, 2014.

14 F. Guo, C. Wen, J. Mao, and Y. D. Song, "Distributed secondary voltage and frequency restoration control of droop-controlled inverter-based microgrids," *IEEE Transactions on Industrial Electronics*, vol. 62, no. 7, pp. 4355–4364, 2015.

15 J. W. Simpson-Porco, Q. Shafiee, F. Dörfler, J. C. Vasquez, J. M. Guerrero, and F. Bullo, "Secondary frequency and voltage control of islanded microgrids via distributed averaging," *IEEE Transactions on Industrial Electronics*, vol. 62, no. 11, pp. 7025–7038, 2015.

16 N. M. Dehkordi, H. R. Baghaee, N. Sadati, and J. M. Guerrero, "Distributed noise-resilient secondary voltage and frequency control for islanded microgrids," *IEEE Transactions on Smart Grid*, vol. 10, no. 4, pp. 3780–3790, 2019.

17 J. Zhou, S. Kim, H. Zhang, Q. Sun, and R. Han, "Consensus-based distributed control for accurate reactive, harmonic, and imbalance power sharing in microgrids," *IEEE Transactions on Smart Grid*, vol. 9, no. 4, pp. 2453–2467, 2016.

18 X. Wu, C. Shen, and R. Iravani, "A distributed, cooperative frequency and voltage control for microgrids," *IEEE Transactions on Smart Grid*, vol. 9, no. 4, pp. 2764–2776, 2018.

19 Z. Wang, W. Wu, and B. Zhang, "A distributed control method with minimum generation cost for DC microgrids," *IEEE Transactions on Energy Conversion*, vol. 31, no. 4, pp. 1462–1470, 2016.

20 A. Jokić, M. Lazar, and P. P. van den Bosch, "Real-time control of power systems using nodal prices," *International Journal of Electrical Power & Energy Systems*, vol. 31, no. 9, pp. 522–530, 2009.

21 G. Binetti, A. Davoudi, F. L. Lewis, D. Naso, and B. Turchiano, "Distributed consensus-based economic dispatch with transmission losses," *IEEE Transactions on Power Systems*, vol. 29, no. 4, pp. 1711–1720, 2014.

22 H. Xing, Y. Mou, M. Fu, and Z. Lin, "Distributed bisection method for economic power dispatch in smart grid," *IEEE Transactions on Power Systems*, vol. 30, no. 6, pp. 3024–3035, 2014.

23 Z. Zhang and M.-Y. Chow, "Convergence analysis of the incremental cost consensus algorithm under different communication network topologies in a smart grid," *IEEE Transactions on Power Systems*, vol. 27, no. 4, pp. 1761–1768, 2012.

24 G. Chen, F. L. Lewis, E. N. Feng, and Y. Song, "Distributed optimal active power control of multiple generation systems," *IEEE Transactions on Industrial Electronics*, vol. 62, no. 11, pp. 7079–7090, 2015.

25 J. Guo, G. Hug, and O. Tonguz, "Impact of communication delay on asynchronous distributed optimal power flow using ADMM," in *2017 IEEE International Conference on Smart Grid Communications (SmartGridComm)*, pp. 177–182, IEEE, 2017.

26 X. Zhang, A. Papachristodoulou, and N. Li, "Distributed control for reaching optimal steady state in network systems: an optimization approach," *IEEE Transactions on Automatic Control*, vol. 63, no. 3, pp. 864–871, 2018.

27 S. Mhanna, G. Verbič, and A. C. Chapman, "Adaptive ADMM for distributed AC optimal power flow," *IEEE Transactions on Power Systems*, vol. 34, no. 3, pp. 2025–2035, 2018.

28 Y. Matsuda, Y. Wakasa, and E. Masuda, "Fully-distributed accelerated ADMM for DC optimal power flow problems with demand response," in *2019 12th Asian Control Conference (ASCC)*, pp. 740–745, IEEE, 2019.

29 M. Javadi, A. E. Nezhad, M. Gough, M. Lotfi, and J. P. Catal ao, "Implementation of consensus-ADMM approach for fast DC-OPF studies," in *2019 International Conference on Smart Energy Systems and Technologies (SEST)*, pp. 1–5, IEEE, 2019.

30 D. Biagioni, P. Graf, X. Zhang, A. S. Zamzam, K. Baker, and J. King, "Learning-accelerated ADMM for distributed DC optimal power flow," *IEEE Control Systems Letters*, vol. 6, pp. 2475–1456, 2020.

31 T. Xu and W. Wu, "Accelerated ADMM-based fully distributed inverter-based volt/var control strategy for active distribution networks," *IEEE Transactions on Industrial Informatics*, vol. 16, no. 12, pp. 7532–7543, 2020.

32 X. Kou, F. Li, J. Dong, M. Starke, J. Munk, Y. Xue, M. Olama, and H. Zandi, "A scalable and distributed algorithm for managing residential demand response programs using alternating direction method of multipliers

(ADMM)," *IEEE Transactions on Smart Grid*, vol. 11, no. 6, pp. 4871–4882, 2020.

33 A. Rajaei, S. Fattaheian-Dehkordi, M. Fotuhi-Firuzabad, and M. Moeini-Aghtaie, "Decentralized transactive energy management of multi-microgrid distribution systems based on ADMM," *International Journal of Electrical Power & Energy Systems*, vol. 132, p. 107126, 2021.

34 P. S˘ ulc, S. Backhaus, and M. Chertkov, "Optimal distributed control of reactive power via the alternating direction method of multipliers," *IEEE Transactions on Energy Conversion*, vol. 29, no. 4, pp. 968–977, 2014.

35 J. Chen, W. Zhang, Y. Zhang, and G. Bao, "Day-ahead scheduling of distribution level integrated electricity and natural gas system based on fast-ADMM with restart algorithm," *IEEE Access*, vol. 6, pp. 17557–17569, 2018.

36 W. Liao, P. Li, Q. Wu, S. Huang, G. Wu, and F. Rong, "Distributed optimal active and reactive power control for wind farms based on ADMM," *International Journal of Electrical Power & Energy Systems*, vol. 129, p. 106799, 2021.

37 C. Zhao, U. Topcu, N. Li, and S. H. Low, "Design and stability of load-side primary frequency control in power systems," *IEEE Transactions on Automatic Control*, vol. 59, no. 5, pp. 1177–1189, 2014.

38 F. Dorfler, J. W. Simpson-Porco, and F. Bullo, "Breaking the hierarchy: distributed control and economic optimality in microgrids," *IEEE Transactions on Control of Network Systems*, vol. 3, no. 3, pp. 241–253, 2016.

39 T. Kose, "Solutions of saddle value problems by differential equations," *Econometrica, Journal of the Econometric Society*, vol. 24, no. 1, pp. 59–70, 1956.

40 K. J. Arrow, L. Hurwicz, H. Uzawa, and H. B. Chenery, *Studies in Linear and Non-Linear Programming*. Redwood City, CA, USA: Stanford University Press, 1958.

41 A. Jokic, M. Lazar, and P. P. van den Bosch, "On constrained steady-state regulation: dynamic KKT controllers," *IEEE Transactions on Automatic Control*, vol. 54, no. 9, pp. 2250–2254, 2009.

42 L. Chen and S. You, "Reverse and forward engineering of frequency control in power networks," *IEEE Transactions on Automatic Control*, vol. 62, no. 9, pp. 4631–4638, 2016.

43 X. Zhou, M. Farivar, Z. Liu, L. Chen, and S. H. Low, "Reverse and forward engineering of local voltage control in distribution networks," *IEEE Transactions on Automatic Control*, vol. 66, no. 3, pp. 1116–1128, 2020.

2

Preliminaries

In this chapter, some fundamental preliminaries are briefly introduced, mainly including graph theory, convex analysis, projection operator, stability theory, passivity/dissipativity theory, power flow equations, and power system dynamics. In the following chapters, some of these preliminaries will be presented and further explained in need. However, we deliberately collect them here for clarity.

2.1 Norm

2.1.1 Vector Norm

Given a vector space X over a subfield of the complex space $\mathbb{C}^n$, a norm on X is a real-valued function $||x|| : X \rightarrow \mathbb{R}$, which satisfies the following properties:

- Subadditivity/triangle inequality: $||x + y|| \leq ||x|| + ||y||, \forall x, y \in X$.
- Absolute homogeneity: $||kx|| = |k|\,||x||, \forall k \in \mathbb{C}, x \in X$.
- Nonnegativity: $||x|| \geq 0, \forall x \in X$, $||x|| = 0$, if and only if $x = 0$.

The ith component of x is denoted by x_i. Some commonly used norms are listed below:

- 1-norm

$$||x||_1 = \sum_{i=1}^{n} |x_i| \tag{2.1}$$

  The 1-norm is simply the sum of the absolute values of the components, which is also called taxicab norm.
- 2-norm

$$||x||_2 = (x^*x)^{1/2} = \left(\sum_{i=1}^{n} |x_i|^2\right)^{1/2} \tag{2.2}$$

Merging Optimization and Control in Power Systems: Physical and Cyber Restrictions in Distributed Frequency Control and Beyond, First Edition. Feng Liu, Zhaojian Wang, Changhong Zhao, and Peng Yang.

The 2-norm is the well-known Euclidean norm. It is usually denoted by $||x||$ if there is no confusion.

- ∞-norm

$$||x||_\infty = \max_{1\le i\le n} |x_i| \tag{2.3}$$

- p-norm

$$||x||_p = \left(\sum_{i=1}^{n} |x_i|^p\right)^{1/p}, \quad 1 \le p < \infty \tag{2.4}$$

For $p = 1$, we get the 1-norm; for $p = 2$, we get the Euclidean norm; and as $p \to \infty$, the p-norm approaches the ∞-norm.

For the vector norms, the following inequalities hold:

- Reverse triangle inequality:

$$||x \pm y|| \ge |\,||x|| - ||y||\,|, \forall x, y \in X \tag{2.5}$$

- Hölder's inequality:

$$|\langle x, y\rangle| \le ||x||_p ||y||_q, \ \frac{1}{p} + \frac{1}{q} = 1, \ p, q > 0, \ \forall x, y \in X \tag{2.6}$$

The equality holds if and only if x, y are linearly dependent.

- Cauchy–Schwarz inequality: it is a special case of Hölder's inequality by merely taking $p = 2$, i.e.

$$|\langle x, y\rangle| \le ||x||_2 ||y||_2, \forall x, y \in X \tag{2.7}$$

In Euclidean space $\mathbb{R}^n$ with the standard inner product, the Cauchy–Schwarz inequality has the following form:

$$\left(\sum_{i=1}^{n} x_i y_i\right)^2 \le \left(\sum_{i=1}^{n} x_i^2\right)\left(\sum_{i=1}^{n} y_i^2\right) \tag{2.8}$$

In the rest of the book, $||x||$ represents the 2-norm of x if there is no confusion.

2.1.2 Matrix Norm

A real-valued function $||A|| : \mathbb{C}^{m\times n} \to \mathbb{R}$ is a matrix norm on $\mathbb{C}^{m\times n}$ provided that the following properties are satisfied [1]:

- Subadditivity/triangle inequality: $||A + B|| \le ||A|| + ||B||, \forall A, B \in \mathbb{C}^{m\times n}$.
- Absolute homogeneity: $||kA|| = |k|\,||A||, \forall k \in \mathbb{C}, A \in \mathbb{C}^{m\times n}$.
- Nonnegativity: $||A|| \ge 0, \forall A \in \mathbb{C}^{m\times n}$, $||A|| = 0$, if and only if $A = 0$.
- Submultiplicativity: $||AB|| \le ||A||\,||B||, \forall A^{m\times r}, B^{r\times n}$.

Denote $A := [a_{ij}] \in \mathbb{C}^{m\times n}$. Some commonly used matrix norms have the following form:

- 1-norm

$$\|A\|_1 = \max_{1\le j\le n} \sum_{i=1}^{m} |a_{ij}| \tag{2.9}$$

The 1-norm is simply the maximum absolute column sum of the matrix.

- 2-norm

$$\|A\|_2 = \sigma_{\max}(A) \tag{2.10}$$

where $\sigma_{\max}(A)$ is the largest singular value of matrix A[1].

- ∞-norm

$$\|A\|_\infty = \max_{1\le i\le m} \sum_{j=1}^{n} |a_{ij}|, \tag{2.11}$$

which is simply the maximum absolute row sum of the matrix.

- Frobenius-norm

$$\|A\|_F = \sqrt{\sum_{i=1}^{m}\sum_{j=1}^{n} |a_{ij}|^2} = \sqrt{\mathrm{trace}\,(A^*A)} = \sqrt{\sum_{i=1}^{\min\{m,n\}} \sigma_i^2(A)} \tag{2.12}$$

where $\sigma_i(A)$ are the singular values of A.

From definitions of 2-norm and Frobenius-norm, we have

$$\|A\|_2 \le \|A\|_F \tag{2.13}$$

The equality holds if and only if rank$(A) = 1$ or A is a zero matrix.

Example 2.1 *Take a matrix A as*

$$A = \begin{bmatrix} 1 & 2 & 0 & 2 \\ 3 & 2 & 1 & 3 \\ 5 & 0 & 6 & 3 \\ 2 & 4 & -9 & 7 \end{bmatrix}$$

Its different types of norm are

$$\|\boldsymbol{A}\|_1 = 16,\ \|\boldsymbol{A}\|_\infty = 22,\ \|\boldsymbol{A}\|_2 = 12.73,\ \|\boldsymbol{A}\|_F = 15.87$$

---

1 For a matrix A, its singular value is calculated by $\sigma(A) = \sqrt{\lambda\,(AA^*)} = \sqrt{\lambda\,(A^*A)}$, where $\lambda\,(A)$ is the eigenvalue of A and A^* denotes the conjugate transpose of A.

2.2 Graph Theory

2.2.1 Basic Concepts

An graph is an ordered pair $\mathcal{G} = (\mathcal{V}, \mathcal{E})$, where $\mathcal{V} = \{1,2, \dots, n\}$ is a set of vertices. $\mathcal{E} \subseteq \mathcal{N} \times \mathcal{N}$ is a set of edges, which are unordered pairs of vertices (i.e. an edge is associated with two distinct vertices). If for all $(i,j) \in \mathcal{E}$ also $(j,i) \in \mathcal{E}$ holds, we call $\mathcal{G}$ undirected; otherwise it is called a directed graph. The set of neighbors of the vertex i is denoted by $N_i = \{j \in \mathcal{V} : (i,j) \in \mathcal{E}\}$.

For an undirected graph $\mathcal{G}$, the degree of a given vertex i, denoted by $d(U_i)$, is the cardinality of the set N_i, i.e. $d(U_i) = |N_i|$. The degree matrix of $\mathcal{G}$, denoted by $\Delta(\mathcal{G})$, is a diagonal matrix, containing the vertex degrees of $\mathcal{G}$ on the diagonal, $\Delta(\mathcal{G}) = \text{diag}(d(U_i))$. The adjacency matrix $A(\mathcal{G})$ is a square matrix, where the elements of the matrix indicate whether a pair of vertices are adjacent or not in the graph $\mathcal{G}$. The definition is

$$[A(\mathcal{G})]_{ij} := \begin{cases} 1, & \text{if } (i,j) \in \mathcal{E}, \\ 0, & \text{otherwise.} \end{cases} \tag{2.14}$$

If the edges have been arbitrarily oriented, the $n \times m$ incidence matrix $C(\mathcal{G})$ is defined as

$$[C(\mathcal{G})]_{ie} := \begin{cases} -1, & \text{if } i \text{ is the tail of the edge } e = (j,i) \text{ for some vertex } j, \\ 1, & \text{if } i \text{ is the head of the edge } e = (i,j) \text{ for some vertex } j, \\ 0, & \text{otherwise.} \end{cases} \tag{2.15}$$

For the incidence matrix $C(\mathcal{G})$, it has $0 = \mathbf{1}^{\mathrm{T}} C(\mathcal{G})$, where $\mathbf{1}$ is a column vector with all elements being 1.

For a directed graph, the adjacency matrix is defined as

$$[A(\mathcal{G})]_{ij} := \begin{cases} 1, & \text{if } (j,i) \in \mathcal{E}, \\ 0, & \text{otherwise.} \end{cases} \tag{2.16}$$

The degree matrix $\Delta(\mathcal{G}) = \text{diag}(d_{in}(U_i))$, where $d_{in}(U_i) = \sum_{j:(j,i)\in\mathcal{E}} 1$ is the in-degree of vertex i. The incidence matrix for a directed graph can be defined similarly to (2.15) by skipping the preorientation.

2.2.2 Laplacian Matrix

Another important matrix representation of a graph $\mathcal{G}$ is the Laplacian matrix $L(\mathcal{G})$, which can be defined in different ways, resulting in the same object. The most straightforward definition (for both undirected and directed graphs) is

$$L(\mathcal{G}) := \Delta(\mathcal{G}) - A(\mathcal{G}) \tag{2.17}$$

It also can be obtained by

$$L(\mathcal{G}) = C(\mathcal{G}) \cdot C(\mathcal{G})^{\mathrm{T}} \tag{2.18}$$

For an (undirected) connected graph $\mathcal{G}$ and its Laplacian matrix L with eigenvalues $\lambda_1 < \lambda_2 \leq \cdots \leq \lambda_n$, there are the following properties:

- L is symmetric and diagonally dominant.
- L is positive semi-definite, i.e. $\lambda_1 = 0$. The eigenvector with respect to λ_1 is 1, i.e. $L1 = \mathbf{0}$.
- Every row sum and column sum of L is zero.
- The smallest nonzero eigenvalue of L, i.e. λ_2, is the so-called spectral gap.
- The trace of L is equal to $2m$, with m as the number of edges of $\mathcal{G}$.
- There exists a positive definite matrix $\Gamma \in \mathbb{R}^{n\times n}$ such that $\Gamma L = L\Gamma = \Pi$, where $\Pi = I_n - \frac{1}{n}\mathbf{1}_n\mathbf{1}_n^{\mathrm{T}}$. Moreover, the eigenvalues of Γ are $\lambda_\Gamma, \frac{1}{\lambda_2}, \cdots, \frac{1}{\lambda_n}$, where $\lambda_\Gamma > 0$ can be any positive constant [2].

The second smallest eigenvalue of L is the algebraic connectivity of $\mathcal{G}$. If $\mathcal{G}$ is disconnected, its algebraic connectivity is zero. The number of connected subgraphs in $\mathcal{G}$ is the dimension of the nullspace of L and the algebraic multiplicity of the zero eigenvalue. For a graph with multiple connected subgraphs, L is a block diagonal matrix, where each block is the respective Laplacian matrix for each subgraph. Sometimes, we also use the notion of "two-hop neighbors," i.e. neighbors of the neighbor. This topological relation could be represented by the matrix $L^2 = L \times L$, where $[L^2]_{ij} = 0$ implies i, j are not two-hop neighbors.

We will explain these matrices by an example.

Example 2.2 *For the undirected graph with five nodes in Figure 2.1, its degree matrix, adjacency matrix, and Laplacian matrix are*

$$\Delta = \begin{bmatrix} 2 & 0 & 0 & 0 & 0 \\ 0 & 3 & 0 & 0 & 0 \\ 0 & 0 & 2 & 0 & 0 \\ 0 & 0 & 0 & 3 & 0 \\ 0 & 0 & 0 & 0 & 2 \end{bmatrix}, \quad A = \begin{bmatrix} 0 & 1 & 0 & 0 & 1 \\ 1 & 0 & 1 & 1 & 0 \\ 0 & 1 & 0 & 1 & 0 \\ 0 & 1 & 1 & 0 & 1 \\ 1 & 0 & 0 & 1 & 0 \end{bmatrix},$$

$$L = \begin{bmatrix} 2 & -1 & 0 & 0 & -1 \\ -1 & 3 & -1 & -1 & 0 \\ 0 & -1 & 2 & -1 & 0 \\ 0 & -1 & -1 & 3 & -1 \\ -1 & 0 & 0 & -1 & 2 \end{bmatrix}$$

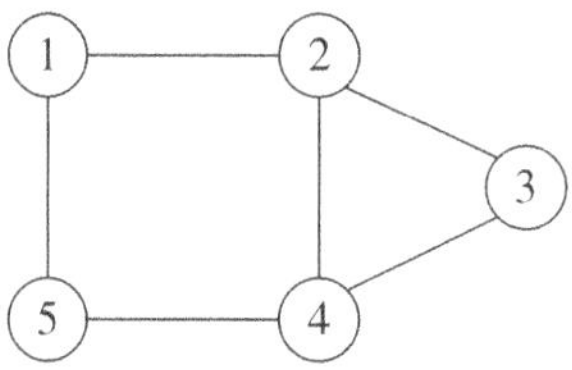

Figure 2.1 An illustrative graph with five nodes.

If the graph is oriented such that the head of an edge is the smaller number and the tail is the larger one, then the incidence matrix is

$$C = \begin{bmatrix} 1 & 1 & 0 & 0 & 0 & 0 \\ -1 & 0 & 1 & 1 & 0 & 0 \\ 0 & 0 & -1 & 0 & 1 & 0 \\ 0 & 0 & 0 & -1 & -1 & 1 \\ 0 & -1 & 0 & 0 & 0 & -1 \end{bmatrix}$$

It is easy to verify that $L = C \cdot C^T$.

The Laplacian matrix has the distributed structure naturally, i.e. if two nodes are disconnected, the corresponding element is zero. Then, the consensus can be written compactly as follows:

$$z_i = \sum_{j \in \mathcal{N}_i^c} (y_j - y_i), \forall i \iff z = -Ly$$

where z, y are vectors composed of z_i, y_i. The incidence matrix C has the similar feature, which also reveals the topology properties of the graph. It is useful in the distributed algorithm in this book (see Chapter 4).

2.3 Convex Optimization

In this section, we introduce fundamental concepts of convex optimization, including convex set, convex function, convex programming, duality, saddle point, and Karush–Kuhn–Tucker (KKT) conditions. These concepts set the cornerstone for mathematical optimization.

2.3.1 Convex Set

2.3.1.1 Basic Concepts

Definition 2.1 ***(Convex set)*** A set X is convex if for any $x_1 \in X, x_2 \in X$ and any α with $0 \leq \alpha \leq 1$, the point $\alpha x_1 + (1 - \alpha)x_2 \in X$.

Figure 2.2 illustrates a simple convex set and a non-convex set. Roughly speaking, a set is convex if, for any two of its points, the entire segment joining

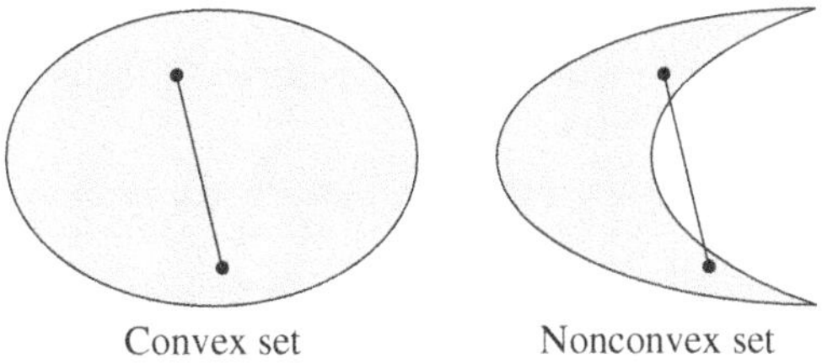

Figure 2.2 Illustration of (non)convex sets.

these points is contained in the set. Given basic convex sets we can create other convex sets through simple convexity-preserving operations. Let $\mathbb{X}$ and $\mathbb{Y}$ be linear subspaces. For example $\mathbb{X} := \mathbb{R}^n$ and $\mathbb{Y} := \mathbb{R}^m$:

- **Intersection**: if X_1 and X_2 are convex, then $X_1 \cap X_2$ is convex. This property can be extended to the intersection of an infinite number of convex sets, i.e. the intersection of an arbitrary collection of convex sets is convex.
- **Cartesian product**: if X_1 and X_2 are convex, then $X_1 \times X_2 := \{(x_1, x_2) | x_1 \in X_1, x_2 \in X_2\}$ is convex. This property can be extended to the Cartesian product of an infinite number of convex sets, i.e. the Cartesian product of an arbitrary collection of convex sets is convex.
- **Finite sum**: let X_1, X_2 be convex. Then $X_1 + X_2 := \{x_1 + x_2 | x_1 \in X_1, x_2 \in X_2\}$ is convex. Therefore, the sum of any finite number of convex sets is convex.

A point x is called a convex combination of points $x_1, \ldots, x_n$ if there exist $\alpha_1 \geq 0, \ldots, \alpha_n \geq 0$ and $\alpha_1 + \cdots + \alpha_n = 1$ such that $x = \alpha_1 x_1 + \cdots + \alpha_n x_n$. Then, we give the definition of the convex hull.

Definition 2.2 ***(Convex hull)*** The convex hull of a set X, denoted by $\mathbf{conv}\, X$, is the set of all convex combinations of points in X.

The convex hull $\mathbf{conv}\, X$ is the smallest convex set containing X. Figure 2.3 illustrates the convex hulls of two sets.

The set of all affine combinations of points in some set $C \subseteq \mathbb{R}^n$ is called the affine hull of C and denoted by $\mathbf{aff}\ C$:

$$\mathbf{aff}\ C = \left\{\theta_1 x_1 + \cdots + \theta_k x_k | x_1, \ldots, x_k \in C, \theta_1 + \cdots + \theta_k = 1\right\} \tag{2.19}$$

The affine hull is the smallest affine set that contains C in the following sense: if S is any affine set with $C \subseteq S$, then $\mathbf{aff}\ C \subseteq S$.

If the affine dimension of a set $C \subseteq \mathbb{R}^n$ is less than n, then the set lies in the affine set $\mathbf{aff}\ C \neq \mathbb{R}^n$. We define the relative interior of the set C, denoted by $\mathbf{relint}\ C$, as its interior relative to $\mathbf{aff}\ C$:

$$\mathbf{relint}\ C = \{x \in C | B(x, r) \cap \mathbf{aff} C \subseteq C \text{ for some } r > 0\},$$

where $B(x, r) = \{y | \|y - x\| \leq r\}$, the ball of radius r and center x in the norm $\| \cdot \|$. Here $\| \cdot \|$ is any norm; all norms define the same relative interior. We can then

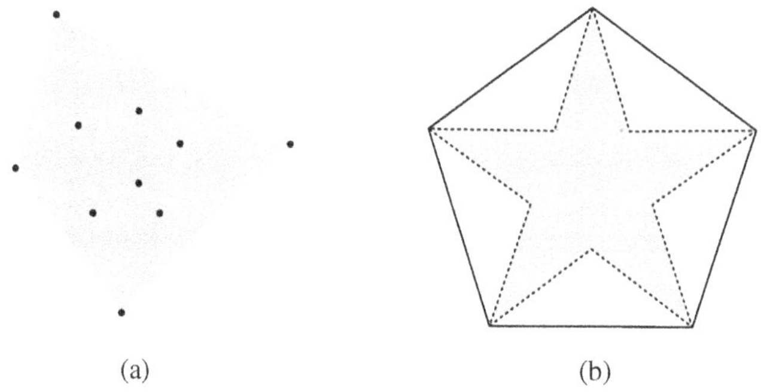

Figure 2.3 (a) The convex hull of 10 points. (b) The convex hull of a star.

define the relative boundary of a set C as **cl** $C\backslash$**relint** C, where **cl** C is the closure of C.

2.3.1.2 Cone

Convex cones are a particular class of convex sets, which play a significant role in optimization theory. Then, we will introduce some commonly used convex cones.

Definition 2.3 ***(Cone)*** A set $X \in \mathbb{R}^n$ is called a cone if, for every $x \in X$ and all $\alpha > 0$, the point $\alpha x \in X$.

A cone X is a convex cone, if, for any $x_1, x_2 \in X$ and $\alpha_1, \alpha_2 \geq 0$, there is $\alpha_1 x_1 + \alpha_2 x_2 \in X$.

Definition 2.4 ***(Polar cone)*** Let X be a cone in $\mathbb{R}^n$. The set $X^\circ = \{y \in \mathbb{R}^n : \langle y, x\rangle \leq 0, \forall x \in X\}$ is called the polar cone of X.

For every convex cone X, the polar cone X° is convex and closed.

Definition 2.5 ***(Normal cone)*** Consider a convex closed set $X \subset \mathbb{R}^n$ and a point $x \in X$. The set $N_X(x) = \{v | \langle v, y - x\rangle \leq 0, \forall y \in X\}$ is called the normal cone to X at x.

The normal cone is illustrated in Figure 2.4. The same as a polar cone, the normal cone is closed and convex.

Definition 2.6 ***(Tangent cone)*** Consider a convex closed set $X \subset \mathbb{R}^n$ and a point $x \in X$; the set

$$T_X(x) := \left\{ \lim_{k \to +\infty} \frac{x_k - x}{\tau_k} | x_k \in X, x_k \to x, \tau_k > 0, \tau_k \to 0 \right\}$$

is called the tangent cone to X at x.

Figure 2.4 Illustration of the normal cone.

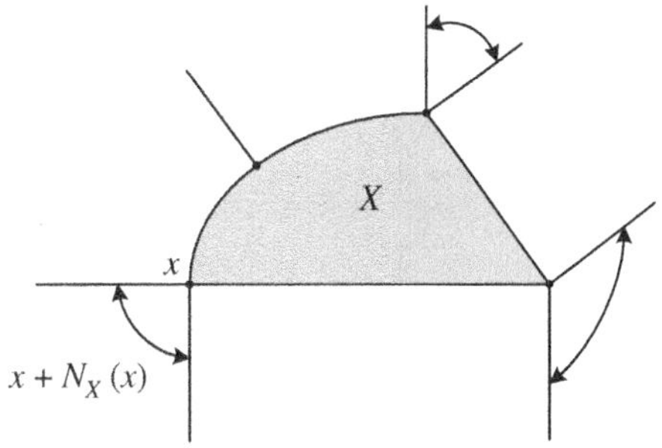

Figure 2.5 Illustration of the tangent cone.

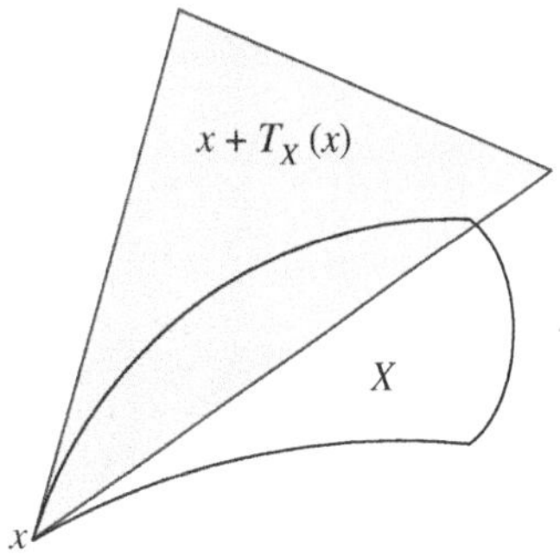

The notion of the tangent cone is illustrated in Figure 2.5. The tangent cone is the polar of the corresponding normal cone [3], which implies

$$T_\Omega(x) = \{y \in R^n | \langle s, y \rangle \leq 0, \forall s \in N_\Omega(x)\}$$

Definition 2.7 ***(Second-order cone)*** The second-order cone is defined as $C = \left\{x \in \mathbb{R}^n, \alpha \geq 0 \mid \|x\|_2 \leq \alpha\right\}$.

The notion of the second-order cone is illustrated in Figure 2.6.

2.3.2 Convex Function

2.3.2.1 Basic Concepts

Use **dom** $f(x)$ to denote the domain of function $f(x)$. Then, we give the definition of the convex function.

Definition 2.8 ***(Convex function)*** A function $f(x) : \mathbb{R}^n \to \mathbb{R}$ is convex (Figure 2.7) if **dom** $f(x)$ is a convex set and if, for all $x_1, x_2 \in$ **dom** $f(x)$, and α with $0 \leq \alpha \leq 1$, the following inequality holds:

$$f(\alpha x_1 + (1 - \alpha)x_2) \leq \alpha f(x_1) + (1 - \alpha)f(x_2) \tag{2.20}$$

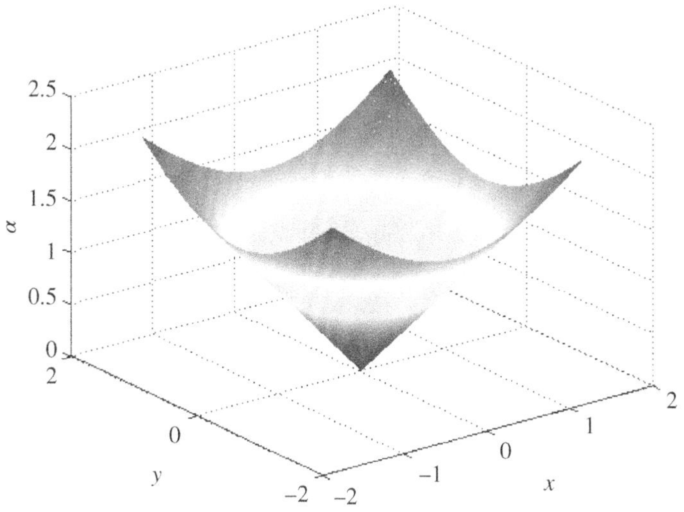

Figure 2.6 Illustration of the second-order cone. The cone is $\sqrt{x^2 + y^2} \leq \alpha$.

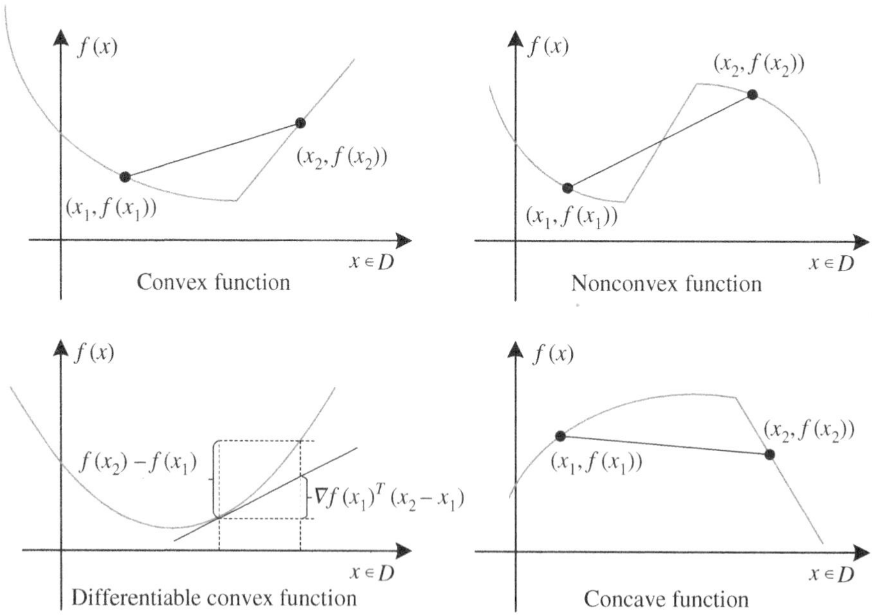

Figure 2.7 Illustration of a convex function.

A function $f(x)$ is strictly convex if inequalities in (2.20) strictly hold, whenever $x_1 \neq x_2$ and $0 < \alpha < 1$. A function $f(x)$ is called (strictly) concave if $-f(x)$ is (strictly) convex.

For a continuously differentiable function $f(x)$, i.e. its gradient $\nabla f(x)$ exists at each point in **dom** $f(x)$, we have the following first-order conditions [[4], Theorem 2.67].

Theorem 2.1 *Suppose $f(x)$ is continuously differentiable:*

- *$f(x)$ is convex if and only if for all $x_1, x_2 \in$ **dom** $f(x)$*

$$f(x_2) \geq f(x_1) + \nabla^T f(x_1)(x_2 - x_1) \tag{2.21}$$

- *$f(x)$ is strictly convex if and only if for all $x_1, x_2 \in$ **dom** $f(x)$ and $x_1 \neq x_2$*

$$f(x_2) > f(x_1) + \nabla^T f(x_1)(x_2 - x_1) \tag{2.22}$$

Moreover, if $f(x)$ is twice continuously differentiable, i.e. its Hessian $\nabla^2 f(x)$ exists and is continuous at each point in **dom** $f(x)$, we have the following second-order conditions [[4], Theorem 2.69].

Theorem 2.2 *Suppose f is twice continuously differentiable:*

- *$f(x)$ is convex if and only if for all $x \in$ **dom** $f(x)$, its Hessian $\nabla^2 f(x)$ is positive semi-definite.*
- *if the Hessian $\nabla^2 f(x)$ is positive definite for all $x \in$ **dom** $f(x)$, then $f(x)$ is strictly convex.*

Theorem 2.2 introduces an exact characterization for convexity, which, however, is not for strict convexity. The converse of the second assertion may not hold. For example, $f(x) = 2x^4$ is strictly convex but $f''(x) = 0$ at $x = 0$.

It should be noted the second assertion in Theorem 2.2 is different from the following condition:

$$x^T \nabla^2 f(x) x \geq 0 \quad \text{for all } x \in \textbf{dom} f(x).$$

Instead, to justify $\nabla^2 f(x) \succeq 0,\ x \in domf(x)$, we need [5]

$$y^T \nabla^2 f(x) y \geq 0 \quad \text{for all } y \in \mathbb{R}^n$$

no matter what $domf(x)$ is. Here, we use an example to explain this.

Example 2.3 *Consider the function*

$$f(x_1, x_2) = x_1 x_2,\ x_1 \geq 0, x_2 \geq 0$$

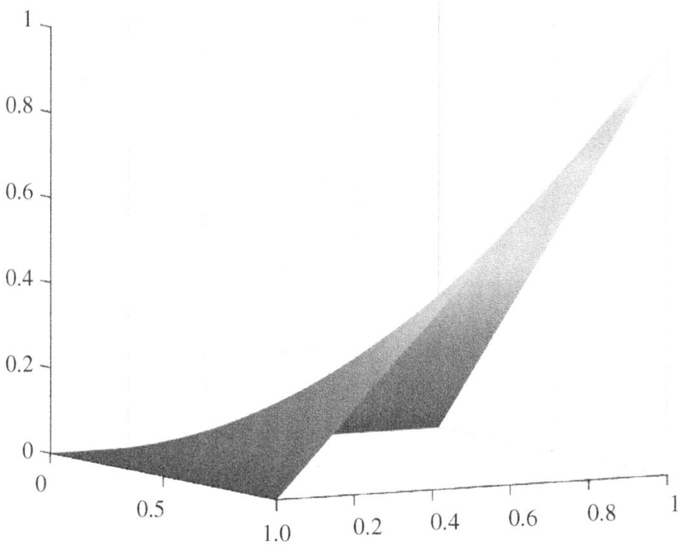

Figure 2.8 Illustration of function $f(x_1, x_2) = x_1x_2,\ x_1 \geq 0, x_2 \geq 0$.

The Hessian of $f(x)$ is

$$\nabla^2 f(x) = \begin{bmatrix} 0 & 1 \\ 1 & 0 \end{bmatrix},$$

Its eigenvalues are -1 and 1, implying that $f(x)$ is neither convex nor concave, as illustrated in Figure 2.8.

However, if we only check $x^T\nabla^2 f(x)x$, it is

$$x^T\nabla^2 f(x)x = 2x_1x_2$$

Obviously, we have $x^T\nabla^2 f(x)x \geq 0, \forall x_1 \geq 0, x_2 \geq 0$, which does not imply the convexity of $f(x)$.

Now, we introduce the concept of strong convexity for a twice continuously differentiable function.

Definition 2.9 ***(Strongly convex function)*** A twice continuously differentiable function $f : \mathbb{R}^n \to \mathbb{R}$ is strongly convex if, for all $x \in \mathbf{dom}\, f$, there exists a number $m > 0$ such that $\nabla^2 f(x) \geq mI$.

For a strongly convex function f, we have the following property [[6], Chapter 9.1.2].

Theorem 2.3 *Suppose f is strongly convex. Then there exists a number $m > 0$ such that $\forall x_1, x_2 \in \mathbf{dom}\, f$, the following inequality holds:*

$$f(x_2) \geq f(x_1) + \nabla^T f(x_1)(x_2 - x_1) + \frac{m}{2}\|x_2 - x_1\|_2^2 \tag{2.23}$$

In summary, if $f(x)$ is twice continuously differentiable, the following results hold:

- $f(x)$ is convex if and only if $\nabla^2 f(x) \geq 0$ for all x.
- $f(x)$ is strictly convex if $\nabla^2 f(x) > 0$ for all x (it is sufficient but not necessary).
- $f(x)$ is strongly convex if and only if $\nabla^2 f(x) \geq mI > 0$ for all x.

2.3.2.2 Jensen's Inequality

Equation (2.20) is called Jensen's inequality. It is easily generalized to convex combinations of more than two points: for a convex $f(x)$ with $x_1, \ldots, x_k \in \mathbf{dom}\, f$, and $\alpha_1, \ldots, \alpha_k \geq 0$, and $\alpha_1 + \cdots + \alpha_k = 1$, there is

$$f\left(\alpha_1 x_1 + \cdots + \alpha_k x_k\right) \leq \alpha_1 f\left(x_1\right) + \cdots + \alpha_k f\left(x_k\right)$$

This result can be extended to integrals and expectations with convex functions. For a convex function $f(x)$, if $p(x) \geq 0$ on $S \subseteq \mathbf{dom}\, f$, $\int_S p(x)\mathrm{d}x = 1$, then

$$f\left(\int_S p(x)x\mathrm{d}x\right) \leq \int_S f(x)p(x)\mathrm{d}x$$

provided the integrals exist.

For a convex function $f(x)$, if x is a random variable such that $x \in \mathbf{dom}\, f$ with probability one, then we have

$$f(\mathbb{E}x) \leq \mathbb{E}f(x)$$

provided the expectation exists.

Based on Jensen's inequality, we can obtain some useful results. For instance, the function $-\log x$ is convex, and we have

$$-\log(\alpha x + (1-\alpha)y) \leq -\alpha \log x - (1-\alpha)\log y = -(\log(x^\alpha y^{1-\alpha}))$$

with $0 \leq \alpha \leq 1$. Taking the exponential of both sides yields

$$x^\alpha y^{1-\alpha} \leq \alpha x + (1-\alpha)y \qquad (x, y > 0,\ 0 \leq \alpha \leq 1) \tag{2.24}$$

which is the general arithmetic–geometric mean inequality.

2.3.3 Convex Programming

A convex optimization problem is one of the form

$$\begin{aligned} \text{minimize} \quad & f_0(x) \\ \text{s.t.} \quad & f_i(x) \leq 0, \quad i = 1, \ldots, m \\ & h_j(x) = 0, \quad j = 1, \ldots, p \end{aligned} \tag{2.25}$$

where $f_0, \dots, f_m$ are convex functions and $h_1, \dots, h_p$ are affine functions. For a convex optimization problem, it has three requirements:

- The objective function must be convex.
- The inequality constraint functions must be convex.
- The equality constraint functions must be affine.

With a slight abuse of notation, the following problem is also called as a convex optimization problem:

$$\begin{aligned} \text{maximize} \quad & -f_0(x) \\ \text{s.t.} \quad & f_i(x) \leq 0, \quad i = 1, \dots, m \\ & h_j(x) = 0, \quad j = 1, \dots, p \end{aligned} \tag{2.26}$$

where $f_0, \dots, f_m$ are convex functions. It can be found that problems (2.25) and (2.26) have the same optimal solutions.

The feasible set of the optimization problem (2.25) is defined as

$$\Omega := \{x | f_i(x) \leq 0, \ i = 1, \dots, m, \quad h_j(x) = 0, \ j = 1, \dots, p\} \tag{2.27}$$

An optimal solution to the problem (2.25) may not exist, or may not be unique. We have the following theorem.

Theorem 2.4 *[5] Consider the problem (2.25):*

- *An optimal solution x^* exists if Ω is nonempty and compact and f is continuous.*
- *The optimal solution x^* is unique if f is strictly convex.*

This theorem also implies, for a convex optimization problem, any locally optimal point is also globally optimal [6].

2.3.4 Duality

In this subsection, we introduce the Lagrangian duality. The Lagrangian $L : \mathbb{R}^n \times \mathbb{R}_+^m \times \mathbb{R}^p \to \mathbb{R}$ associated with the problem (2.25) is defined as

$$L(x, \lambda, \nu) = f_0(x) + \sum_{i=1}^{m} \lambda_i f_i(x) + \sum_{j=1}^{p} \nu_j h_j(x) \tag{2.28}$$

with **dom** $L = \mathbb{R}^n \times \mathbb{R}_+^m \times \mathbb{R}^p$. λ_i is referred to as the Lagrange multiplier associated with the inequality constraint $f_i(x) \leq 0$. Similarly, ν_i is referred to as the Lagrange multiplier associated with the equality constraint $h_i(x) = 0$. The vectors λ and ν are called the dual variables or Lagrange multiplier vectors associated with the problem (2.25).

The dual function is defined as $g : \mathbb{R}^m \times \mathbb{R}^p \to \mathbb{R}$, which is the minimum of the Lagrangian over x. That is, for $\lambda \in \mathbb{R}^m_+$, $\nu \in \mathbb{R}^p$,

$$g(\lambda, \nu) = \inf_{x\in\mathbb{R}^n} L(x, \lambda, \nu) = \inf_{x\in\mathbb{R}^n} \left(f_0(x) + \sum_{i=1}^{m} \lambda_i f_i(x) + \sum_{j=1}^{p} \nu_j h_j(x) \right)$$

Since the dual function is the point-wise infimum of the affine functions of (λ, ν), it is concave, even when the problem (2.25) is not convex.

Then, the dual problem associated with the problem (2.25) is defined as

$$\begin{aligned} \text{maximize} \quad & g(\lambda, \nu) \\ \text{s.t.} \quad & \lambda \geq 0 \end{aligned} \tag{2.29}$$

In this context, the original problem (2.25) is sometimes called the primal problem. The Lagrange dual problem (2.29) is a convex optimization problem, since the objective to be maximized is concave and the constraint is convex. This is true whether the primal problem (2.25) is convex or not.

The optimal solution to the problem (2.29) is denoted by (λ^*, ν^*), which is also called the dual optimal or optimal Lagrange multipliers. The optimal value of the primal problem (2.25) is denoted by p^*. The optimal value of the dual problem (2.29) is denoted by d^*.

The dual function gives lower bounds on p^*, i.e.

$$g(\lambda, \nu) \leq p^*, \ \forall \lambda \geq 0 \tag{2.30}$$

This property is easy to verify. For any feasible points of the problems (2.25) and (2.29), $(\tilde{x}, (\lambda, \nu))$, $f_i(\tilde{x}) \leq 0$ and $h_i(\tilde{x}) = 0$ and $\lambda \geq 0$. Then, we have

$$\sum_{i=1}^{m} \lambda_i f_i(\tilde{x}) + \sum_{j=1}^{p} \nu_j h_j(\tilde{x}) = \sum_{i=1}^{m} \lambda_i f_i(\tilde{x}) \leq 0$$

Therefore,

$$\begin{aligned} g(\lambda, \nu) &= \inf_{x\in\Omega} L(x, \lambda, \nu) \\ &\leq L(\tilde{x}, \lambda, \nu) = f_0(\tilde{x} + \sum_{i=1}^{m} \lambda_i f_i(\tilde{x}) + \sum_{j=1}^{p} \nu_j h_j(\tilde{x}) \\ &\leq f_0(\tilde{x}) \end{aligned}$$

When $\tilde{x}$ takes one of optimal solutions, the inequality (2.30) follows. It gives a lower bound on p^*.

As a simple application of (2.30), we have the following result.

Theorem 2.5 ***(Weak duality)*** *[6] The lower bound on p^* can be obtained from the Lagrange dual function. We have the inequality*

$$d^* \leq p^* \tag{2.31}$$

The weak duality inequality (2.31) holds even when d^* and p^* are infinite. For example, if the primal problem is unbounded below, i.e. $p^* = -\infty$, we must have $d^* = -\infty$, implying the Lagrange dual problem is infeasible. Conversely, if the dual problem is unbounded from above, i.e. $d^* = +\infty$, we must have $p^* = +\infty$ as well, implying the primal problem is infeasible.

The difference $p^* - d^*$ gives the gap between the optimal value of the primal problem and the best lower bound on it that can be obtained from the Lagrange dual function. It is referred to as the optimal duality gap of the original problem, which is always nonnegative.

If the equality

$$d^* = p^* \tag{2.32}$$

holds, i.e. the optimal duality gap is zero, then we say that *strong duality* holds. In this situation, the optimal value can be obtained by solving the dual problem. However, strong duality does not, in general, hold. Then, we will introduce the condition to guarantee the strong duality.

Definition 2.10 ***(Slater's condition)*** For the problem (2.25), there exists an $x \in \mathbb{R}^n$ such that

$$\begin{aligned} f_i(x) < 0, \quad & i = 1, \dots, m, \\ h_j(x) = 0, \quad & j = 1, \dots, p \end{aligned}$$

Such a point is sometimes called strictly feasible, since the inequality constraints hold with strict inequalities.

Slater's condition can be relaxed when some of the inequality constraint functions f_i are affine. If the first k constraint functions $f_1, \dots, f_k$ are affine, then the Slater's condition is relaxed as there exists an $x \in \mathbb{R}^n$ with

$$\begin{aligned} f_i(x) &\leq 0, \quad i = 1, \dots, k, \\ f_i(x) &< 0, \quad i = k+1, \dots, m, \\ h_j(x) &= 0, \quad j = 1, \dots, p \end{aligned}$$

If all constraints are affine, Slater's condition degenerates to the feasibility of the optimization problem.

Then, we have the following result.

Theorem 2.6 ***(Strong duality)*** *[6] For the convex problem (2.25), if Slater's condition is satisfied, then the strong duality holds, i.e.* $d^* = p^*$.

It should be noted that Theorem 2.6 holds only for convex problems.

2.3.5 Saddle Point

In convex optimization problems, saddle point is a central concept of the duality theory. In this subsection, we will introduce the basic notion and some fundamental properties of saddle points.

Recall the Lagrangian (2.28); its saddle point is defined as below.

Definition 2.11 ***(Saddle point)*** A point $(\tilde{x}, (\tilde{\lambda}, \tilde{\nu})) \in X_0 \times \Lambda_0$ is called a saddle point of the Lagrangian (2.28) if for $\forall x \in X_0$ and $\forall(\lambda, \nu) \in \Lambda_0$ the following inequalities hold:

$$L(\tilde{x}, \lambda, \nu) \leq L(\tilde{x}, \tilde{\lambda}, \tilde{\nu}) \leq L(x, \tilde{\lambda}, \tilde{\nu})$$

In other words, a saddle point is such a point at which the maximum of the Lagrangian with respect to $(\lambda, \nu) \in \Lambda_0$ and the minimum with respect to $x \in X_0$ are attained, i.e.

$$\max_{(\lambda,\nu)\in A_0} L(\tilde{x}, \lambda, \nu) = L(\tilde{x}, \tilde{\lambda}, \tilde{\nu}) = \min_{x\in X_0} L(x, \tilde{\lambda}, \tilde{\nu})$$

With this definition, we have the following theorem.

Theorem 2.7 *If the Lagrangian has a saddle point $(\tilde{x}, \tilde{\lambda}, \tilde{\nu})$, then $\tilde{x}$ is a solution to the primal problem, and $(\tilde{\lambda}, \tilde{\mu})$ is a solution to the dual problem. Moreover, the strong duality holds at the saddle point.*

Conversely, assume that the strong duality holds at $(\tilde{x}, \tilde{\lambda}, \tilde{\nu})$. Then, $(\tilde{x}, \tilde{\lambda}, \tilde{\nu})$ is a saddle point of the Lagrangian (2.28).

2.3.6 KKT Conditions

We further assume that the functions $f_0, \dots, f_m, h_1, \dots, h_p$ in (2.25) are differentiable. Let x^* and (λ^*, ν^*) be any primal and dual optimal points with zero duality gap. We have the following condition.:

$$0 = \nabla f_0(x^*) + \sum_{i=1}^{m} \lambda_i^* \nabla f_i(x^*) + \sum_{j=1}^{p} \nu_j^* \nabla h_j(x^*) \tag{2.33a}$$

$$f_i(x^*) \leq 0, \quad i = 1, \dots, m \tag{2.33b}$$

$$h_j(x^*) = 0, \quad j = 1, \dots, p \tag{2.33c}$$

$$\lambda_i^* \geq 0, \quad i = 1, \dots, m \tag{2.33d}$$

$$\lambda_i^* f_i(x^*) = 0, \quad i = 1, \dots, m \tag{2.33e}$$

Equation (2.33) is called the *KKT* condition. The first Eq. (2.33a) states that the gradient of Lagrangian function with respect to x^* vanishes. Equations (2.33b) and (2.33c) state that x^* is primal feasible. Equation (2.33d) implies the dual feasibility. The last equation (2.33e) is due to the complementary slackness.

If a convex optimization problem with differentiable objective and constraint functions satisfies Slater's condition, then the KKT condition provides necessary and sufficient conditions for the optimality: Slater's condition implies that the optimal duality gap is zero and the dual optimum is attained, so x is optimal if and only if there are (λ, ν) that, together with x, satisfy the KKT condition.

Consider a convex optimization problem with domain constraints as follows:

$$\begin{aligned} \text{minimize} \quad & f_0(x) \\ \text{s.t.} \quad & f_i(x) \le 0, \quad i = 1, \dots, m \\ & h_j(x) = 0, \quad j = 1, \dots, p \\ & x \in X \end{aligned} \tag{2.34}$$

where $f_0, \dots, f_m$ are convex functions, $h_1, \dots, h_p$ are affine functions, and X is a convex set. The difference between (2.34) and (2.25) is that $x \in X$ is considered. Then, we have the following result.

Theorem 2.8 ***([4], Theorem 3.34)*** *Assume that x^* is the minimum of problem* (2.34), *the functions $f_0, \dots, f_m, h_1, \dots, h_p$ are differentiable, and Slater's condition holds. Then there exist $\lambda^* \in \mathbb{R}_+^m$ and $\nu^* \in \mathbb{R}^p$ such that*

$$0 \in \nabla f(x^*) + \sum_{i=1}^{m} \lambda_i^* \nabla f_i(x^*) + \sum_{j=1}^{p} \nu_j^* \nabla h_j(x^*) + N_X(x^*) \tag{2.35}$$

and

$$\lambda_i^* f_i(x^*) = 0, \quad i = 1, \dots, m \tag{2.36}$$

Conversely, if for some feasible point x^* of (2.34) *and some $\lambda^* \in \mathbb{R}_+^m$ and $\nu^* \in \mathbb{R}^p$ conditions* (2.35) *and* (2.36) *are satisfied, then x^* is the optimal solution to the problem* (2.34), *and (λ^*, ν^*) is the optimal solution to the dual problem of* (2.34).

From Theorem 2.8, we can get the KKT condition for the problem (2.34):

$$0 \in \nabla f(x^*) + \sum_{i=1}^{m} \lambda_i^* \nabla f_i(x^*) + \sum_{j=1}^{p} \nu_j^* \nabla h_j(x^*) + N_X(x^*) \tag{2.37a}$$

$$f_i(x^*) \le 0, \quad i = 1, \dots, m \tag{2.37b}$$

$$h_j(x^*) = 0, \quad j = 1, \dots, p \tag{2.37c}$$

$$\lambda_i^* \ge 0, \quad i = 1, \dots, m \tag{2.37d}$$

$$\lambda_i^* f_i(x^*) = 0, \quad i = 1, \dots, m \tag{2.37e}$$

The KKT condition (2.37) is a generalization of (2.33). It is very useful when the domain set is involved in an optimization problem. Obviously, if the domain set X is not considered ($X = \mathbb{R}^n$), the normal cone is $\{0\}$. Hence, the KKT condition (2.37) naturally degenerates to (2.33).

2.4 Projection Operator

2.4.1 Basic Concepts

For a set-valued operator $\mathcal{U} : \mathbb{R}^n \to \mathbb{R}^n$, its domain is $\text{dom}\mathcal{U} := \{x \in \mathbb{R}^n | \mathcal{U}x \neq \emptyset\}$. Then, the graph of $\mathcal{U}$ is given as follows [7].

Definition 2.12 The graph of $\mathcal{U}$ is

$$\text{gra}\,\mathcal{U} := \{(x, u) \in \mathbb{R}^n \times \mathbb{R}^n | u \in \mathcal{U}x\} \tag{2.38}$$

The monotonicity of $\mathcal{U}$ is defined as below.

Definition 2.13 For an operator $\mathcal{U}$ with $\forall(x, u), \forall(y, v) \in \text{gra}\,\mathcal{U}$, it is as follows:

- Monotone, if

$$\langle x - y, u - v\rangle \geq 0 \tag{2.39}$$

- Strictly monotone, if

$$\langle x - y, u - v\rangle > 0 \tag{2.40}$$

- Strongly monotone, if

$$\langle x - y, u - v\rangle \geq \alpha\|x - y\|^2, \quad \alpha > 0 \tag{2.41}$$

- Maximally monotone, if $\text{gra}\,\mathcal{U}$ is not strictly contained in the graph of any other monotone operator, i.e.

$$(x, u) \in \text{gra}\,\mathcal{U} \quad \Leftrightarrow \quad (\forall(y, v) \in \text{gra}\,\mathcal{U}),\ \langle x - y, u - v\rangle \geq 0 \tag{2.42}$$

For a nonexpansive operator, we have the following definition.

Definition 2.14 A single-valued operator $\mathcal{T} : \Omega \subset \mathbb{R}^n \to \mathbb{R}^n$ is as follows:

- Nonexpansive, if

$$\|\mathcal{T}(x) - \mathcal{T}(y)\| \leq \|x - y\|, \forall x, y \in \Omega \tag{2.43}$$

In other words, it is Lipschitz continuous with the constant 1.

- Firmly nonexpansive, if

$$\|\mathcal{T}(x) - \mathcal{T}(y)\|^2 + \|(\mathrm{Id} - \mathcal{T})(x) - (\mathrm{Id} - \mathcal{T})(y)\|^2 \leq \|x - y\|^2, \forall x, y \in \Omega \tag{2.44}$$

A point $x^* \in \Omega$ is a fixed point of $\mathcal{T}$ if $\mathcal{T}(x^*) \equiv x^*$. The set of fixed points of $\mathcal{T}$ is denoted by $\mathrm{Fix}(\mathcal{T})$.

Definition 2.15 For $\alpha \in (0,1)$, $\mathcal{T}$ is called α-averaged if there exists a nonexpansive operator $\mathcal{R}$ such that

$$\mathcal{T} = (1 - \alpha)\mathrm{Id} + \alpha\mathcal{R}$$

The α-averaged operator is denoted by $\mathcal{A}(\alpha)$.

Definition 2.16 For $\beta > 0$, $\mathcal{T}$ is called β-cocoercive if

$$\beta\mathcal{T} \in \mathcal{A}\left(\frac{1}{2}\right)$$

Moreover, there is $\beta\mathcal{T} \in \mathcal{A}\left(\frac{1}{2}\right)$ if and only if

$$\beta\|\mathcal{T}(x) - \mathcal{T}(y)\|^2 \leq \langle x - y, \mathcal{T}(x) - \mathcal{T}(y)\rangle, \forall x, y \in \Omega \tag{2.45}$$

More useful results about maximal monotonicity can be found in [7].

Example 2.4

1) *Let C be a nonempty closed convex subset. Then, the normal cone N_C is maximally monotone.*
2) *Let A be such that $A^* = -A$. Then A is maximally monotone.*
3) *Let f be lower semicontinuous convex. Then, its gradient ∂f is maximally monotone.*
4) *Let $\mathcal{T}$ be nonexpansive and let $\alpha \in [-1,1]$. Then $Id + \alpha\mathcal{T}$ is maximally monotone.*
5) *Let $\mathcal{T}$ be β-cocoercive, with $\beta > 0$. Then $\mathcal{T}$ is maximally monotone.*

2.4.2 Projection Operator

Consider a convex closed set $\Omega \subset \mathbb{R}^n$ and a point $x \in \mathbb{R}^n$. The point in Ω closest to x is called the projection of x on Ω, denoted by $\mathcal{P}_\Omega(x)$, which is illustrated in Figure 2.9. By the definition, $\mathcal{P}_\Omega(x)$ is the unique solution to the convex

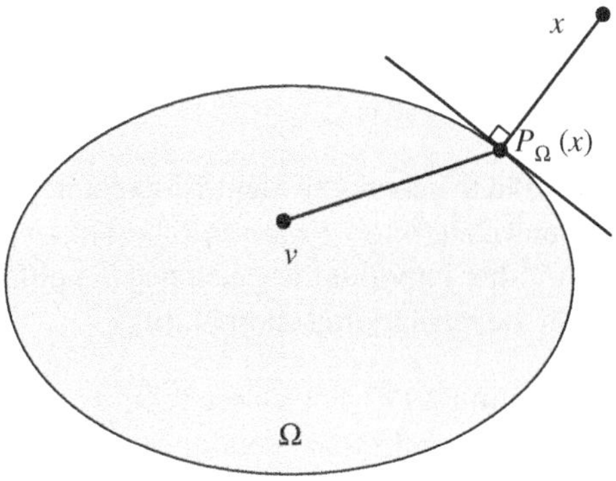

Figure 2.9 The projection of a point onto a closed convex set.

minimization problem in the variable y, where x is fixed:

$$\begin{aligned} \min \quad & \frac{1}{2}(y-x)^{\mathrm{T}}(y-x) \\ \text{s.t.} \quad & y \in \Omega \end{aligned}$$

The projection operator is always well defined by noting that $x = \mathcal{P}_\Omega(x)$ holds for every $x \in \Omega$.

Theorem 2.9 ***([4], Theorem 2.10)*** *If the set Ω is nonempty, convex, and closed, then for every $x \in \mathbb{R}^n$, there exists exactly one point $z \in \Omega$ that is the closest to x.*

Theorem 2.10 ***([4], Lemma 2.11)*** *Assume that $\Omega \in \mathcal{R}^n$ is a closed convex set. The following statements are valid:*

- *For all $x \in \mathcal{R}^n$, there is $z = \mathcal{P}_\Omega(x)$ if and only if $z \in \Omega$ and*

$$\langle v - z, x - z \rangle \leq 0, \ \forall v \in \Omega \tag{2.46}$$

- *For all $x_1 \in \mathcal{R}^n$, $x_2 \in \mathcal{R}^n$, we have*

$$\langle \mathcal{P}_\Omega(x_1) - \mathcal{P}_\Omega(x_2), x_1 - x_2 \rangle \geq \|\mathcal{P}_\Omega(x_2) - \mathcal{P}_\Omega(x_1)\|^2, \ \forall v \in \Omega \tag{2.47}$$

Figure 2.9 illustrates the projection of a point onto a closed convex set and the property (2.46) of the projection.

The projection operator is nonexpansive, and we have the following result.

Theorem 2.11 ***([4], Theorem 2.13)*** *Assume that $\Omega \in \mathcal{R}^n$ is a closed convex set. Then for all $x_1 \in \mathcal{R}^n$, $x_2 \in \mathcal{R}^n$, we have*

$$\|\mathcal{P}_\Omega(x_2) - \mathcal{P}_\Omega(x_1)\| \leq \|x_2 - x_1\| \tag{2.48}$$

For the projection in Euclidean space, we also have

Theorem 2.12 *([8], **Theorem 1.5.5**) The squared distance function $\rho(x) = \frac{1}{2}\|x - \mathcal{P}_\Omega(x)\|_2^2$ is continuously differentiable in x, and the gradient $\nabla\rho(x) = x - \mathcal{P}_\Omega(x)$.*

Use Id to denote the identity operator, i.e. $\mathrm{Id}(x) = x, \forall x$. Recall the definition of normal cone, and we have $\mathcal{P}_\Omega(x) = (\mathrm{Id} + N_\Omega)^{-1}(x)$ [[7], Chapter 23.1].

Another important projection is the differentiated projection, which is derived from the regular projection [9, 10].

Definition 2.17 For a closed convex set Ω, point $x \in \Omega$, and direction v, the differentiated projection operator is

$$\Pi_\Omega(x, v) = \lim_{\tau \to 0} \frac{\mathcal{P}_\Omega(x + \tau v) - x}{\tau} \tag{2.49}$$

The differentiated projection operator have the following properties.

Theorem 2.13 *[9] For the differentiated projection operator $\Pi_\Omega(x, v)$:*

- *If $x \in int(\Omega)$, then*

$$\Pi_\Omega(x, v) = v \tag{2.50}$$

- *If $x \in \partial\Omega$, then*

$$\Pi_\Omega(x, v) = v + \beta(x) n^*(x) \tag{2.51}$$

where

$$n^*(x) = \arg \max_{n \in N_\Omega(x)} \langle v^T, -n \rangle$$

$$\beta(x) = \max \left\{0, \langle v^T, -n^*(x) \rangle \right\}$$

The operator $\Pi_\Omega(x, v)$ in (6) is equivalent to the projection of v onto the tangent cone $T_\Omega(x)$, i.e. $\Pi_\Omega(x, v) = \mathcal{P}_{T_\Omega(x)}(v)$.

2.5 Stability Theory

In this section, we introduce the fundamentals of stability theory, including Lyapunov stability, invariance principles, input–output stability, and passivity and dissipativity theories. These concepts and theories lay a solid foundation for stability analysis and control system design.

2.5.1 Lyapunov Stability

Consider a time-invariant dynamical system:

$$\dot{x} = f(x(t)), \quad t \geq 0, \quad x(0) = x_0 \tag{2.52}$$

where $f : D \rightarrow \mathbb{R}^n$ and $D \subset \mathbb{R}^n$ is a domain. x_0 is the initial state. A point $x^* \in D$ is an equilibrium of (2.52) if $f(x^*) = 0$. Assume D contains an equilibrium point x^*, and we have the following stability definition.

Definition 2.18 An equilibrium $x^* \in D$ of (2.52) is as follows:

- **Stable** if $\forall \epsilon > 0, \exists \delta = \delta(\epsilon) > 0$ such that

$$\|x_0 - x^*\| < \delta \Rightarrow \|x(t) - x^*\| < \epsilon \quad \forall t \geq 0 \tag{2.53}$$

- **Asymptotically stable** if it is stable and δ can be chosen such that

$$\|x_0 - x^*\| < \delta \Rightarrow \lim_{t\to\infty} x(t) = x^* \tag{2.54}$$

- **Globally asymptotically stable** if $D = \mathbb{R}^n, x^*$ is stable and given any initial point $x_0 \in D$, $\lim_{t\to\infty} x(t) = x^*$.

A general method to prove the stability of an equilibrium point x^* of (2.52) is to find a so-called Lyapunov function $V(x)$, which is positive definite, and monotonically decreases along the solution trajectory of (2.52).

Let $V : D \rightarrow \mathbb{R}$ where $D \subset \mathbb{R}^n$ be a continuously differentiable function. The derivative of V along the trajectory of (2.52), denoted by $\dot{V}(x)$, is given by

$$\dot{V}(x) = \sum_{i=1}^{n} \frac{\partial V}{\partial x_i} \dot{x}_i = \frac{\partial V}{\partial x} f(x) \tag{2.55}$$

Then, we have the following Lyapunov's stability theorem.

Theorem 2.14 ***([11], Theorem 4.1)*** *Let $x = x^*$ be an equilibrium point of (2.52) and $D \subset R^n$ be a domain containing $x = x^*$. Let $V : D \rightarrow R$ be a continuously differentiable function such that*

$$V(x^*) = 0 \;\; and \;\; V(x) > 0 \; in \; D - \{x^*\} \tag{2.56}$$

$$\dot{V}(x) \leq 0 \; in \; D \tag{2.57}$$

Then, $x = 0$ is stable. Moreover, if

$$\dot{V}(x) < 0 \; in \; D - \{x^*\} \tag{2.58}$$

then $x = x^*$ is asymptotically stable.

This theorem only characterizes *local* stability of the system. When considering global stability, extra conditions should be needed. The following theorem is useful in justifying whether an equilibrium point x^* is globally asymptotically stable.

Theorem 2.15 ***([11], Theorem 4.2)*** *Let $x = x^*$ be an equilibrium point of (2.52). Let $V : R^n \to R$ be a continuously differentiable function such that*

$$V(x^*) = 0 \text{ and } V(x) > 0 \text{ in } D - \{x^*\} \tag{2.59}$$

$$||x|| \to \infty \Rightarrow V(x) \to \infty \tag{2.60}$$

$$\dot{V}(x) < 0 \text{ in } D - \{x^*\} \tag{2.61}$$

then $x = x^*$ is globally asymptotically stable.

A function satisfying (2.60) is called *radially unbounded.*

2.5.2 Invariance Principle

The condition of the Lyapunov stability is too harsh. Sometimes, it is difficult to find a function to satisfy all the conditions in Theorem 2.14. LaSalle's invariance principle is very useful in this situation. We first introduce some definitions. Let $x(t)$ be a solution of (2.52). A point p is a positive limit point of $x(t)$ if there is a sequence $\{t_n\}$, with $t_n \to \infty$ as $n \to \infty$, such that $x(t_n) \to p$ as $n \to \infty$. The set of all positive limit points of $x(t)$ is called the positive limit set of $x(t)$. A set M is said to be an invariant set with respect to (2.52) if

$$x(0) \in M \Rightarrow x(t) \in M, \quad \forall t \in \mathbb{R}$$

That is, if a solution belongs to M at some time instant, then it belongs to M for all future and past time. A set M is said to be a positively invariant set if

$$x(0) \in M \Rightarrow x(t) \in M, \quad \forall t \geq 0$$

Use **dist**(p, M) to denote the distance between a point p and a set M. If for each $\varepsilon > 0$ there is $T > 0$ such that $\text{dist}(x(t), M) < \varepsilon, \forall t > T$, we say that $x(t)$ approaches a set M as t approaches infinity.

A fundamental property of limit sets is stated in the following result.

Lemma 2.1 ***([11], Lemma 4.1)*** *If a solution $x(t)$ of (2.52) is bounded and belongs to D for $t \geq 0$, then its positive limit set L^+ is a nonempty, compact, invariant set. Moreover, $x(t)$ approaches L^+ as $t \to \infty$.*

The standard LaSalle's invariance principle is stated as follows.

Theorem 2.16 ***([11], Theorem 4.4)*** *Let $\Omega \subset D$ be a compact set that is positively invariant with respect to (4.1). Let $V : D \to R$ be a continuously differentiable function and $\dot{V}(x) \leq 0$ in Ω, E be the set of all points in Ω where $\dot{V}(x) = 0$, and M be the largest invariant set in E. Then every solution starting in Ω approaches M as $t \to \infty$ (Figure 2.10).*

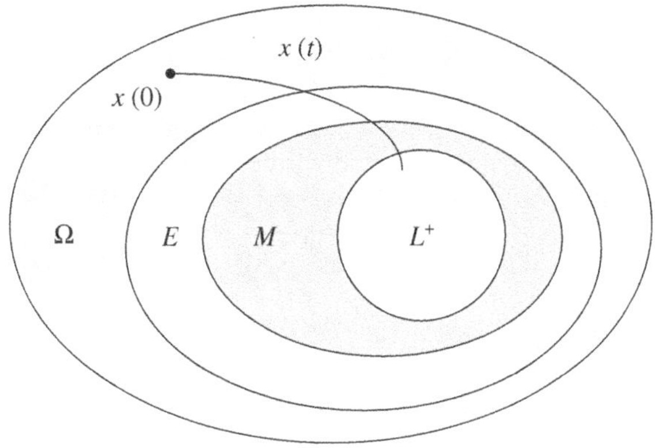

Figure 2.10 The illustration of the invariance principle.

Moreover, if the equilibrium is a unique point, then we have the following asymptotic and globally asymptotic stability results [[11], Corollary 4.1, 4.2].

Lemma 2.2 *Let* $x = x^*$ *be an equilibrium point for* (2.52). *Let* $V : D \to R$ *be a continuously differentiable positive definite function on a domain* D *containing the equilibrium* x^*, *such that* $\dot{V}(x) \leq 0$ *in* D. *Let* $S = \{x \in D | \dot{V}(x) = 0\}$, *and suppose that no solution can stay identically in* S, *other than the trivial solution* $x(t) \equiv x^*$. *Then, the equilibrium* x^* *is asymptotically stable.*

Lemma 2.3 *Let* $x = x^*$ *be an equilibrium point for* (2.52). *Let* $V : R^n \to R$ *be a continuously differentiable, radially unbounded, positive definite function such that* $\dot{V}(x) \leq 0$ *for all* $x \in R^n$. *Let* $S = \{x \in R^n | \dot{V}(x) = 0\}$, *and suppose that no solution can stay identically in* S, *other than the trivial solution* $x(t) \equiv x^*$. *Then the equilibrium* x^* *is globally asymptotically stable.*

Unlike Lyapunov's theorem, the LaSalle's invariance principle does not require the function $V(x)$ to be positive definite. When using LaSalle's invariance principle, the compact set Ω should be constructed at first, which does not have to be tied in with $V(x)$. In many cases, the construction of $V(x)$ will itself guarantee the existence of Ω. In addition, the LaSalle's theorem can be used in cases where the system has an equilibrium set, rather than an isolated equilibrium point.

2.5.3 Input–Output Stability

In some cases, the detailed system model is difficult to obtain except the input and output terminals. In this case, the concept of stability in the input–output sense is very useful. Here, we first introduce the norm for a continuous-time signal and then give the definition of $\mathcal{L}$ stability briefly.

The norm of a continuous-time signal is defined as

$$\|x\|_{\mathcal{L}_p} = \left(\int_0^{\infty} \|x(\tau)\|^p \mathrm{d}\tau\right)^{1/p}, \quad 1 \leq p < +\infty \tag{2.62}$$

Then, two commonly used norms $\mathcal{L}_2$-norm and $\mathcal{L}_\infty$-norm are given below:

$$\|x\|_{\mathcal{L}_2} = \left(\int_0^{\infty} \|x(\tau)\|^2 \mathrm{d}\tau\right)^{1/2}$$

$$\|x\|_{\mathcal{L}_\infty} = \sup_{t\geq 0}\|x(t)\|$$

Give the following dynamic system

$$\dot{x} = f(x, u) \tag{2.63a}$$

$$y = h(x, u) \tag{2.63b}$$

where $f : \mathbb{R}^n \times \mathbb{R}^m \to \mathbb{R}^n$ is locally Lipschitz, $h : \mathbb{R}^n \times \mathbb{R}^m \to \mathbb{R}^p$ is continuous, $f(0,0) = 0$, and $h(0,0) = 0$. For every initial condition x_0, (2.63) defines an input–output map $y = Hu, H : \mathcal{L}_e^m \to \mathcal{L}_e^p$ is called the input–output mapping. The $\mathcal{L}$ stability of (2.63) is defined as follows.

Definition 2.19 [11] The mapping H is $\mathcal{L}$ stable if there exists a class $\mathcal{K}$ function α, defined on $[0, \infty)$, and a nonnegative constant β such that

$$\|(Hu)_\tau\|_{\mathcal{L}} \leq \alpha\left(\|u_\tau\|_{\mathcal{L}}\right) + \beta$$

for all $u \in \mathcal{L}_e^m$ and $\tau \in [0, \infty)$. It is said to be finite-gain $\mathcal{L}$ stable if there exist nonnegative constants γ and β such that

$$\|(Hu)_\tau\|_{\mathcal{L}} \leq \gamma\|u_\tau\|_{\mathcal{L}} + \beta \tag{2.64}$$

for all $u \in \mathcal{L}_e^m$ and $\tau \in [0, \infty)$.

Another important concept is the $\mathcal{L}_2$-gain.

Definition 2.20 ***$\mathcal{L}_2$-gain*** The $\mathcal{L}_2$-gain of the system (2.63), denoted by g_2, is defined as

$$g_2 = \sup_{T\in[0,\infty)} \sup_{u\in\mathcal{L}_e^m} \frac{\left(\int_0^{\mathrm{T}} \|y(\tau)\|^2 \mathrm{d}\tau\right)^{1/2}}{\left(\int_0^{\mathrm{T}} \|u(\tau)\|^2 \mathrm{d}\tau\right)^{1/2}} \tag{2.65}$$

The $\mathcal{L}_2$-gain implies the comprehensive magnification of the output with respect to the input. This is the key of the robust control if we take the external disturbance as the input. Then, we introduce some results about $\mathcal{L}_2$-gain of both linear and nonlinear systems. We start with linear systems.

Theorem 2.17 *([11], Theorem 5.4) Consider the linear time-invariant system*

$$\begin{aligned} \dot{x} &= Ax + Bu \\ y &= Cx + Du \end{aligned} \tag{2.66}$$

where A is Hurwitz. Let $G(s) = C(sI - A)^{-1}B + D$. Then, the $\mathcal{L}_2$ gain of the system is $sup_{\omega \in \mathbb{R}} \|G(j\omega)\|_2$.

In linear time-invariant systems, we can find the exact value of $\mathcal{L}_2$-gain. In nonlinear systems, we can only find an upper bound, explained in the next theorem.

Theorem 2.18 *([11], Theorem 5.5) Consider the time-invariant nonlinear system*

$$\begin{aligned} \dot{x} &= f(x) + G(x)u, \quad x(0) = x_0 \\ y &= h(x) \end{aligned} \tag{2.67}$$

where $f(x)$ is locally Lipschitz and $G(x), h(x)$ are continuous over $\mathbb{R}^n$ with $f(0) = 0, h(0) = 0$. Suppose there is a continuously differentiable, positive semi-definite function $V(x)$ such that

$$\mathcal{H}(V, f, G, h, \gamma) := \frac{\partial V}{\partial x} f(x) + \frac{1}{2\gamma^2} \frac{\partial V}{\partial x} G(x) G^T(x) \left(\frac{\partial V}{\partial x} \right)^T + \frac{1}{2} h^T(x) h(x) \leq 0 \tag{2.68}$$

for all $x \in \mathbb{R}^n$ and some $\gamma > 0$. Then, for each $x_0 \in R^n$, the system (2.67) is finite-gain $\mathcal{L}_2$ stable, and its $\mathcal{L}_2$ gain is less than or equal to γ.

(2.68) is known as the Hamilton–Jacobi inequality. If we succeed in finding $V(x)$ satisfying (2.68), a finite-gain $\mathcal{L}_2$ stability result is obtained accordingly. Nevertheless, unfortunately, it is generally a very challenging task to find such a function $V(x)$.

2.6 Passivity and Dissipativity Theory

Passivity and dissipativity provide us with useful tools for the stability analysis of dynamic systems, usually accompanied by clear physical interpretations. In this section, we introduce basic notions and fundamental properties of passivity and dissipativity theory.

2.6.1 Passivity

The standard definition of passivity is given below.

Definition 2.21 The system (2.63) is said to be passive if there exists a positive semi-definite function $V(x)$ such that

$$u^{\mathrm{T}}y \geq \dot{V}(x) = \frac{\partial V}{\partial x}f(x,u), \quad \forall (x,u) \in \mathbb{R}^n \times \mathbb{R}^m \tag{2.69}$$

Moreover, it is:

- Strictly passive if $u^{\mathrm{T}}y \geq \dot{V}(x) + \psi(x)$ and $\psi(x) > 0, \forall x \neq 0, \psi(0) = 0$.
- Lossless if $u^{\mathrm{T}}y = \dot{V}(x)$.
- Input feedforward passive if $u^{\mathrm{T}}y \geq \dot{V}(x) + \phi(u)$ for some $\phi(u)$ with $\phi(0) = 0$.
- Input strictly passive if $u^{\mathrm{T}}y \geq \dot{V}(x) + \phi(u)$ and $\phi(u) > 0, \forall u \neq 0, \phi(0) = 0$.
- Output feedback passive if $u^{\mathrm{T}}y \geq \dot{V}(x) + \rho(y)$ for some $\rho(y)$ with $\rho(0) = 0$.
- Output strictly passive if $u^{\mathrm{T}}y \geq \dot{V}(x) + \rho(y)$ and $\rho(y) > 0, \forall y \neq 0, \rho(0) = 0$.

In all cases, the inequality should hold for all (x, u).

$V(x)$ is also called the storage function. Sometimes, the passivity inequality (2.69) is also defined in the integral form as

$$V(x(T)) - V(x(0)) \leq \int_0^T u^{\mathrm{T}}(\tau)y(\tau)\mathrm{d}\tau, \ \forall (x,u) \in \mathbb{R}^n \times \mathbb{R}^m, \forall T \geq 0 \tag{2.70}$$

Other definitions can be obtained similarly.

Lemma 2.4 *If the system* (2.63) *is output strictly passive with* $u^Ty \geq \dot{V} + \delta y^T y$, *and* $\delta > 0$, *then it is finite-gain* $\mathcal{L}_2$ *stable, and the* $\mathcal{L}_2$ *gain is less than or equal to* $\frac{1}{\delta}$.

For the system (2.63), we define its observability property.

Definition 2.22 ***(Observability)*** The system (2.63) is zero-state observable if $u(t) = 0, y(t) = 0, \forall t \geq 0$ implies $x(t) = 0, \forall t \geq 0$.

For a passive system, we have following stability results.

Theorem 2.19 *[11] For the system* (2.63), *the origin of* $\dot{x} = f(x, 0)$ *is as follows:*

- *Stable if the system* (2.63) *is passive with* $V(0) = 0$ *and* $V(x) > 0, \forall x \neq 0$.
- *Asymptotically stable if the system is strictly passive.*
- *Globally asymptotically stable if the system is output strictly passive and zero-state observable.*

Under the condition of asymptotic stability, if $V(x)$ *is proper, the origin will be globally asymptotically stable.*

Sometimes the equilibrium point is not zero, or even varying. In this situation, the concept of incremental passivity is introduced.

Definition 2.23 ***(Incremental passivity)*** The system (2.63) is said to be incrementally passive if there exists a positive semi-definite function $V(x)$ such that

$$\begin{aligned} & V\left(x_1(T), x_2(T)\right) - V\left(x_1(0), x_2(0)\right) \\ & \leq \int_0^T \left(u_1(\tau) - u_2(\tau)\right)^{\mathrm{T}} \left(y_1(\tau) - y_2(\tau)\right) \mathrm{d}\tau, \\ & \qquad \forall T \geq 0, \forall (x_1, u_1), (x_2, u_2) \in \mathbb{R}^n \times \mathbb{R}^m \end{aligned}$$

The concept of incremental passivity is useful in interconnected systems, particularly when considering the variation of equilibrium.

2.6.2 Dissipativity

In the definition of passivity, $u^{\mathrm{T}}y$ is called supply rate. If we use more general supply rate $s(u, y)$ instead, we have the following dissipative inequality:

$$V(x(T)) - V(x(0)) \leq \int_0^T s(u(\tau), y(\tau))\mathrm{d}\tau, \ \forall (x, u) \in \mathbb{R}^n \times \mathbb{R}^m, \forall T \geq 0 \quad (2.71)$$

With this inequality, we can give the definition of dissipative systems.

Definition 2.24 ***(Dissipativity)*** The system (2.63) is said to be dissipative with respect to the supply rate $s(u, y)$ if there exists a positive semi-definite function $V(x)$ such that the dissipative inequality (2.71) holds.

The system (2.63) is said to be cyclo-dissipative with respect to $s(u, y)$ if (2.71) holds but $V(x)$ is not necessarily nonnegative.

Roughly speaking, passivity is a special case of dissipativity, which can be understood by noting that, if the system (2.63) is dissipative, then it is:

- Passive if $s(u, y) = u^{\mathrm{T}}y$.
- Input strictly passive if $s(u, y) = u^{\mathrm{T}}y - \delta\|u\|^2$ with $\delta > 0$.
- Output strictly passive if $s(u, y) = u^{\mathrm{T}}y - \epsilon\|y\|^2$ with $\epsilon > 0$.

Another important type of supply rate is

$$s(u, y) = \frac{1}{2}\gamma^2 \left(\|u\|^2 - \|y\|^2\right), \quad \gamma \geq 0 \quad (2.72)$$

For the supply rate in (2.72), we have the following result.

Theorem 2.20 *[12] Suppose the system (2.63) is dissipative with respect to the supply rate in* (2.72)*. Then it has* $\mathcal{L}_2$ *gain equal to or less than* γ.

The statement above shows that dissipativity is a very generic concept. Notably, the abstract of the functions of supply rate makes it adaptable to different concepts, such as passivity and $\mathcal{L}_2$-gain. It also provides extra flexibility to deal with different inputs and outputs, which are mostly restricted by the physics of systems.

To relate dissipativity to stability, we introduce the concept of detectability, which is a weaker property than observability. The underlying logic is as follows: since dissipativity is essentially an input–output property of systems, it implies stability, provided that the convergence of output to a steady state can imply the convergence of all states to the equilibrium. This geometric property has been widely used in stability analysis in input–output interconnected systems.

Definition 2.25 ***(Detectability)*** The system (2.63) is zero-state detectable if $u(t) = 0, y(t) = 0, \forall t \geq 0$ implies $\lim_{t \to \infty} x(t) = 0$.

Then, we have the following stability results.

Theorem 2.21 *[12] Let V be continuously differentiable with $V(0) = 0$ and $V(x) > 0, \forall x \neq 0$. For a supply rate s satisfying $s(0, y) \leq 0$ and such that $s(0, y) = 0$ implies $y = 0$, where $h(0) = 0$. Suppose that the system (2.63) is zero-state detectable. Then $x = 0$ is asymptotically stable.*

If additionally V is proper, then $x = 0$ is globally asymptotically stable.

2.7 Power Flow Model

In this section, we introduce the power flow model, including the generic nonlinear power flow models, including the bus injection model and the branch flow model, and widely used linear power flow models, including the DC power flow model for high-voltage transmission networks and the linearized branch flow model for middle- or low-voltage distribution networks.

Consider an n-bus power system, and the set of buses is denoted by $\mathcal{N} := \{1, \dots, n\}$. Let $\mathcal{E} \subseteq \mathcal{N} \times \mathcal{N}$ be the set of lines, where $(i,j) \in \mathcal{E}$ if buses i and j are connected directly. The impedance of line (i,j) is $r_{ij} + \mathbf{k}x_{ij}$, where $\mathbf{k} := \sqrt{-1}$. The network admittance matrix containing the electrical parameters and topology information is denoted by $\mathbf{Y} := \mathbf{G} + \mathbf{k}\mathbf{B}$. The voltage at bus i is denoted by $\boldsymbol{U}_i := U_i e^{\mathbf{k}\theta_i}$, where U_i is the amplitude and θ_i is the angle. The current over line (i,j) is denoted by $\boldsymbol{I}_{ij} := I_{ij} e^{\mathbf{k}\theta_{ij}}$, where I_{ij} is the amplitude and $\theta_{ij} := \theta_i - \theta_j$ is the angle. Note that $\boldsymbol{I}_{ij} = \boldsymbol{I}_{ji}$ generally does not hold, e.g. for a lossy line. We usually denote active power by P, reactive power by Q, and apparent power by S, obeying Pythagoras' theorem, i.e. $S^2 = P^2 + Q^2$, or $S = P + \mathbf{k}Q$ in the form of complex number.

2.7.1 Nonlinear Power Flow

2.7.1.1 Bus Injection Model (BIM)

The power generation injected to bus i is denoted by $S_{Gi} = P_{Gi} + \mathbf{k}Q_{Gi}$. The load demand absorbed from bus i is denoted by $S_{Di} = P_{Di} + \mathbf{k}Q_{Di}$. Then, the power injection at bus i is $S_i := P_i + \mathbf{k}Q_i = S_{Gi} - S_{Di}$. Similarly, the current is $I_i = \sum_{j\in N_i} I_{ij}$, where N_i means the set of all the buses directly connected to the bus i. The power injection at the ith bus can be computed by

$$\begin{aligned} S_i &= \boldsymbol{U}_i \boldsymbol{I}_i^* \\ &= \boldsymbol{U}_i \left(Y_{ii}\boldsymbol{U}_i + \sum_{j\in N_i} Y_{ij}\boldsymbol{U}_j \right)^* \\ &= Y_{ii}^* U_i^2 + \boldsymbol{U}_i \sum_{j\in N_i} Y_{ij}^* \boldsymbol{U}_j^* \end{aligned} \tag{2.73}$$

Recalling that $\boldsymbol{U}_i := U_i e^{\mathbf{k}\theta_i}$ and $Y_{ij} = G_{ij} + \mathbf{k}B_{ij}$, (2.73) becomes

$$\begin{aligned} S_i &= Y_{ii}^* U_i^2 + \sum_{j\in N_i} U_i e^{\mathbf{k}\theta_i} U_j e^{-\mathbf{k}\theta_j} \left(G_{ij} - \mathbf{k}B_{ij} \right) \\ &= \left(G_{ii} - \mathbf{k}B_{ii} \right) U_i^2 + \sum_{j\in N_i} U_i U_j e^{\mathbf{k}\theta_{ij}} \left(G_{ij} - \mathbf{k}B_{ij} \right) \\ &= \left(G_{ii} - \mathbf{k}B_{ii} \right) U_i^2 + \sum_{j\in N_i} U_i U_j \left(\cos\theta_{ij} + \mathbf{k}\sin\theta_{ij} \right) \left(G_{ij} - \mathbf{k}B_{ij} \right) \\ &= G_{ii} U_i^2 + \sum_{j\in N_i} U_i U_j \left(G_{ij}\cos\theta_{ij} + B_{ij}\sin\theta_{ij} \right) \\ &\quad + \mathbf{k} \left[\sum_{j\in N_i} U_i U_j \left(G_{ij}\sin\theta_{ij} - B_{ij}\cos\theta_{ij} \right) - B_{ii} U_i^2 \right] \end{aligned} \tag{2.74}$$

Dividing (2.74) into real and imaginary parts, we have

$$P_i = G_{ii} U_i^2 + \sum_{j\in N_i} U_i U_j \left(G_{ij}\cos\theta_{ij} + B_{ij}\sin\theta_{ij} \right) \tag{2.75a}$$

$$Q_i = \sum_{j\in N_i} U_i U_j \left(G_{ij}\sin\theta_{ij} - B_{ij}\cos\theta_{ij} \right) - B_{ii} U_i^2 \tag{2.75b}$$

The corresponding active and reactive powers P_{ij}, Q_{ij} from bus i to bus j are

$$P_{ij} = U_i U_j \left(G_{ij}\cos\theta_{ij} + B_{ij}\sin\theta_{ij} \right) - G_{ij} U_i^2 \tag{2.76a}$$

$$Q_{ij} = U_i U_j \left(G_{ij}\sin\theta_{ij} - B_{ij}\cos\theta_{ij} \right) + B_{ij} U_i^2 \tag{2.76b}$$

The power flow model (2.75) is formulated under the polar coordinate system. There is also a power flow model under Cartesian coordinate system. Please refer to [13] for details.

2.7.1.2 Branch Flow Model (BFM)

As an alternative to the bus injection model, balanced radial distribution networks can be represented using the branch flow model [14]. We will use (i,j) and $i \to j$ interchangeably to denote the directed line from bus i to bus j. Define the active and reactive sending-end power flows on the line from bus i to bus j as P_{ij} and Q_{ij}, respectively. Denote by I_{ij} the current flow from bus i to bus j and by $\ell_{ij} = I_{ij}^2$ the squared magnitude of the current flow from bus i to bus j. For line (i, j), we have the Ohm's law:

$$U_i - U_j = z_{ij} I_{ij}, \quad \forall (i,j) \in E \tag{2.77}$$

The definition of branch power flow is

$$S_{ij} = U_i I_{ij}^*, \quad \forall (i,j) \in E \tag{2.78}$$

The power balance at each bus is

$$\sum_{k:j\to k} S_{jk} - \sum_{i:i\to j} \left(S_{ij} - z_{ij}\left|I_{ij}\right|^2\right) = S_j \tag{2.79}$$

where $\sum_{k:j\to k} S_{jk}$ is the total power supplied from bus j, $\sum_{i:i\to j} \left(S_{ij} - z_{ij}\left|I_{ij}\right|^2\right)$ is the total power transferred to bus j with $z_{ij}\left|I_{ij}\right|^2$ as the power loss on the line (i,j), and S_j is the power injected to bus j.

Substituting (2.78) into (2.77), we have

$$U_j = U_i - z_{ij}\frac{S_{ij}^*}{U_i^*}$$

Taking the magnitude squared, we have

$$U_j^2 = U_i^2 + \left|z_{ij}\right|^2 \ell_{ij} - \left(z_{ij} S_{ij}^* + z_{ij}^* S_{ij}\right) \tag{2.80}$$

Using (2.78) and (2.79) and resolving the result into real and imaginary parts, we have the branch flow equations [14, 15]

$$P_{ij} + P_j = \sum_{k \in N_j} P_{jk} + r_{ij}\ell_{ij} \tag{2.81a}$$

$$Q_{ij} + Q_j = \sum_{k \in N_j} Q_{jk} + x_{ij}\ell_{ij} \tag{2.81b}$$

$$U_i^2 - U_j^2 = 2(r_{ij}P_{ij} + x_{ij}Q_{ij}) - \left(r_{ij}^2 + x_{ij}^2\right)\ell_{ij} \tag{2.81c}$$

$$\ell_{ij} U_i^2 = P_{ij}^2 + Q_{ij}^2 \tag{2.81d}$$

Remark 2.1 ***(Comparison of BIM and BFM)*** The branch flow model (2.81) fully represents the power flows for a balanced radial network, which is also called DistFlow model. However, (2.81) is a relaxation for mesh networks due to the lack of a constraint ensuring consistency in the voltage angles [16]. Indeed, as explained in [14, 15], if the cyclic condition is added to (2.81), the resulting model is equivalent to the bus injection model for mesh networks. In most cases, the bus injection model is more convenient for analysis, while the branch flow model exhibits better numerical stability in computation.

2.7.2 Linear Power Flow

Both the power flow models (2.74) and (2.81) are nonlinear, resulting in nonconvex optimization problems that may cause difficulty in solving. Therefore, many works have focused on linear approximations of the power flow equations. This subsection introduces two most commonly used linear power flow models: the DC power flow and the linearized branch flow models.

2.7.2.1 DC Power Flow

In high-voltage transmission networks, there are three observations:

- The resistance of transmission lines is significantly less than the reactance, i.e. $r_{ij} \ll x_{ij}$.
- For most typical operating conditions, the difference in angles of the voltage phasors at two buses i and j connected by a line is very small.
- In the per-unit system, the numerical values of voltage magnitudes U_i and U_j are very close to 1 per unit.

The series admittance of line (i, j) is computed by

$$y_{ij} = \frac{1}{z_{ij}} = \frac{1}{r_{ij} + \mathbf{k}x_{ij}} = \frac{r_{ij}}{r_{ij}^2 + x_{ij}^2} - \frac{\mathbf{k}x_{ij}}{r_{ij}^2 + x_{ij}^2} = -Y_{ij} = -G_{ij} - \mathbf{k}B_{ij} \tag{2.82}$$

Since $r_{ij} \ll x_{ij}$, G_{ij} is much smaller than B_{ij}. Therefore, it is reasonable to use the approximation

$$G_{ij} \approx 0, \ B_{ij} = \frac{1}{x_{ij}}$$

Recalling (2.76), the active power flow can simplified to be

$$P_{ij} = U_i U_j B_{ij} \sin \theta_{ij} \tag{2.83}$$

From the second observation, the approximation $\sin(\theta_i - \theta_j) \approx \theta_i - \theta_j$ is valid. Then, (2.83) can be further simplified to be

$$P_{ij} = U_i U_j B_{ij}(\theta_i - \theta_j) \tag{2.84}$$

From the last observation, we have

$$P_{ij} = B_{ij}(\theta_i - \theta_j) \tag{2.85}$$

which is the so-called the DC power flow.

The DC power flow model is widely used in high-voltage transmission networks. The reactive power over the line is omitted in the model since it is very small compared with the active power. Nevertheless, in distribution networks, the resistance and reactance are comparable, so G_{ij} cannot be omitted. Therefore, the DC power flow may not be applicable in distribution networks, which motivates the development of linearized branch flow models.

2.7.2.2 Linearized Branch Flow

By dividing two sides of (2.81d) by U_i^2, replacing ℓ_{ij} in (2.81a), (2.81b), and (2.81c) with the resulting equations, we obtain

$$P_{ij} + P_j = \sum_{k \in N_j} P_{jk} + r_{ij} \frac{P_{ij}^2 + Q_{ij}^2}{U_i^2} \tag{2.86a}$$

$$Q_{ij} + Q_j = \sum_{k \in N_j} Q_{jk} + x_{ij} \frac{P_{ij}^2 + Q_{ij}^2}{U_i^2} \tag{2.86b}$$

$$U_i^2 - U_j^2 = 2\left(r_{ij}P_{ij} + x_{ij}Q_{ij}\right) - \left(r_{ij}^2 + x_{ij}^2\right) \frac{P_{ij}^2 + Q_{ij}^2}{U_i^2} \tag{2.86c}$$

In (2.86), the items $r_{ij}\frac{P_{ij}^2+Q_{ij}^2}{U_i^2}$ and $x_{ij}\frac{P_{ij}^2+Q_{ij}^2}{U_i^2}$ are active and reactive power losses over the line (i,j), respectively, which are usually very small compared with line power flows and hence omitted here. Denoting a new variable $V_i := U_i^2$, we can get the linearized branch flow model:

$$P_{ij} + P_j = \sum_{k \in N_j} P_{jk} \tag{2.87a}$$

$$Q_{ij} + Q_j = \sum_{k \in N_j} Q_{jk} \tag{2.87b}$$

$$V_i - V_j = 2(r_{ij}P_{ij} + x_{ij}Q_{ij}) \tag{2.87c}$$

As reported in [14, 15], the approximation error is usually at the order of 1%. The linearized model (2.87) has been extensively used in optimization and control of distribution networks due to its simplicity.

2.8 Power System Dynamics

Power systems are composed of many generators and loads integrated into different buses and interconnected by power lines, forming a dynamic power network. The buses can be divided into two types: generator buses and pure load buses. Those buses without generators and loads can be cast into the pure load buses with zero load demand. Synchronous generators (SGs) or inverters connect to generator buses, essentially determining the power system dynamics. Remember that load can also connect to a generator bus besides a pure load bus. In this section, we mainly introduce dynamic models of SGs and inverters briefly. Loads are treated as constants. These dynamic models may be simplified based on different considerations in the following chapters, with necessary explanations when needed.

2.8.1 Synchronous Generator Model

The model of a synchronous generator bus is illustrated in Figure 2.11. Here, we introduce the standard third-order generator model [17–19], which is depicted by (2.88a)–(2.88c). Equation (2.88d) is the simplified model of the governor and turbine, while Equation (2.88e) is the dynamics of excitation voltage:

$$\dot{\delta}_i = \omega_i \tag{2.88a}$$

$$\dot{\omega}_i = \frac{1}{M_i}(P_i^g - D_i\omega_i - P_{ei}) \tag{2.88b}$$

$$\dot{E}'_{qi} = -\frac{E_{qi}}{T'_{d0i}} + \frac{E_{fi}}{T'_{d0i}} \tag{2.88c}$$

$$\dot{P}_i^g = -\frac{P_i^g}{T_i} + u_i^g \tag{2.88d}$$

$$\dot{E}_{fi} = h(E_{fi}, E_{qi}) \tag{2.88e}$$

where M_i is the moment of inertia, D_i is the damping constant, T'_{d0i} is the d-axis transient time constant, δ_i is the power angle of generator i, ω_i is the generator's frequency deviation to the steady-state value ω_0, P_i^g is the mechanical power input, P_{Di} is the active load demand, Q_{Di} is the reactive load demand, P_{ei} is the active power injected to the network, E'_{qi} is the q-axis transient internal voltage, E_{qi} is the q-axis internal voltage, and E_{fi} is the excitation voltage.

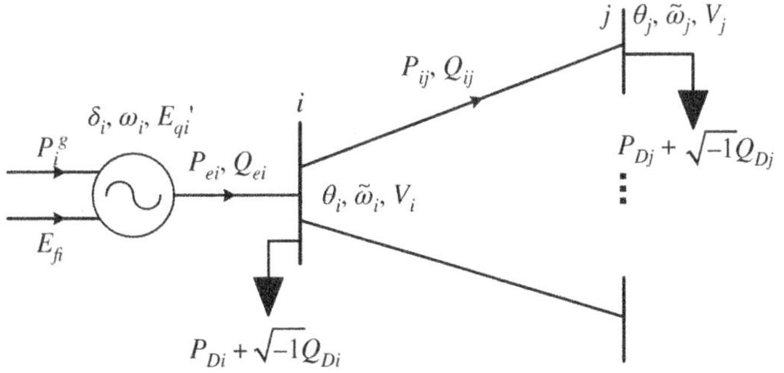

Figure 2.11 Summary of notations at a generator bus.

The relation between the q-axis internal voltage E_{qi} and the terminal voltage is given by

$$E_{qi} = \frac{x_{di}}{x'_{di}} E'_{qi} - \frac{x_{di} - x'_{di}}{x'_{di}} U_i \cos(\delta_i - \theta_i) \tag{2.89}$$

where x_{di} is the d-axis synchronous reactance and x'_{di} is the d-axis transient reactance.

The active and reactive power injections to the network are depicted by

$$P_{ei} = \frac{E'_{qi} U_i}{x'_{di}} \sin(\delta_i - \theta_i) \tag{2.90a}$$

$$Q_{ei} = \frac{U_i^2}{x'_{di}} - \frac{E'_{qi} U_i}{x'_{di}} \cos(\delta_i - \theta_i) \tag{2.90b}$$

As illustrated in Figure 2.11, we also have

$$P_{ei} = P_{Di} + \sum_{j \in N_i} P_{ij} \tag{2.91a}$$

$$Q_{ei} = Q_{Di} + \sum_{j \in N_i} Q_{ij} \tag{2.91b}$$

Then, the synchronous generator model can be rewritten as

$$\dot{\delta}_i = \omega_i \tag{2.92a}$$

$$\dot{\omega}_i = \frac{1}{M_i} \left(P_i^g - D_i \omega_i - p_i - \sum_{j \in N_i} P_{ij} \right) \tag{2.92b}$$

$$\dot{E}'_{qi} = -\frac{x_{di}}{x'_{di} T'_{d0i}} E'_{qi} - \frac{x_{di} - x'_{di}}{x'_{di} T'_{d0i}} U_i \cos(\delta_i - \theta_i) + \frac{E_{fi}}{T'_{d0i}} \tag{2.92c}$$

$$\dot{P}_i^g = -\frac{P_i^g}{T_i} + u_i^g \tag{2.92d}$$

$$\dot{E}_{fi} = h(E_{fi}, E_{qi}) \tag{2.92e}$$

The P_{ij} in (2.92b) could have the nonlinear form in (2.76a) or the linear form in (2.85). Both forms will be used and introduced in detail in the following chapters of this book.

2.8.2 Inverter Model

In this subsection, we introduce the dynamic model of DGs integrated to the power network by voltage source inverters, the main block diagram of which is shown in Figure 2.12 [20].

In Figure 2.12, the power processing section is composed of a three-leg inverter, an output LC filter, and the coupling inductor. The control of an inverter consists

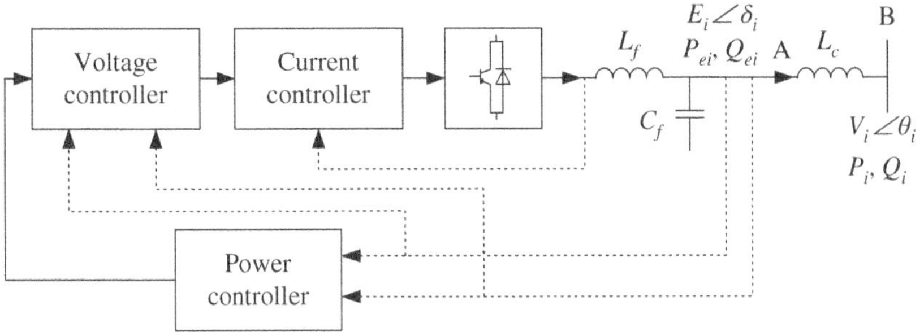

Figure 2.12 Block diagram of DG inverters.

of three loops, e.g. voltage control loop, current control loop, and power control loop [20]. The dynamics of *LC* filter, voltage control loop, and current control loop are much faster than the power control loop. Thus, we neglect their influences and only consider the dynamics of power controller [21]. In Figure 2.12, point A is the internal point of the inverter, and B is the network bus. Denote active and reactive power in point A as P_{ei}, Q_{ei} and in point B as P_i, Q_i. The voltage at point A is $E_i\angle\delta_i$, and the one in point B is $V_i\angle\theta_i$.

The power controller comprises three levels: primary control, secondary control, and tertiary control. Droop control is the most basic primary control, also the basis of secondary control and tertiary control.

Based on the droop control,[2] the dynamics of the inverter-based DGs is

$$\dot{\delta}_i = \omega_i \tag{2.93a}$$

$$\tau_i\dot{\omega}_i = -(\omega_i - \omega_i^d) - k_{P_i}\left(P_{ei} - P_i^d\right) + u_{pi} \tag{2.93b}$$

$$\tau_i\dot{E}_i = -(E_i - E_i^d) - k_{Q_i}\left(Q_{ei} - Q_i^d\right) + u_{qi} \tag{2.93c}$$

where ω_i is the frequency deviation to the steady-state frequency, τ_i, k_{P_i}, k_{Q_i} are positive constants, ω_i^d, E_i^d are the desired frequency and voltage amplitude, and P_i^d, Q_i^d are the desired active and reactive power, respectively. u_{pi}, u_{qi} are control inputs. If $u_{pi} = u_{qi} = 0$, (2.93) is exactly the droop control considering power measurement dynamics [21, 23]. Hence, (2.93) is a general form of state equations for inverter-based DGs. Recalling (2.76), we have

$$P_{ei} = \frac{E_iV_i\sin\left(\delta_i - \theta_i\right)}{X_{L_c}}$$

$$Q_{ei} = \frac{E_i^2 - E_iV_i\cos\left(\delta_i - \theta_i\right)}{X_{L_c}} \tag{2.94}$$

2 There are many control strategies for inverter based DGs. The droop control is the most widely used one. Other control methods can refer to [22].

where X_{L_c} is the reactance of L_c. Then, the final dynamic model is given by

$$\dot{\delta}_i = \omega_i \tag{2.95a}$$

$$\tau_i \dot{\omega}_i = -\omega_i - k_{P_i} \frac{E_i V_i \sin(\delta_i - \theta_i)}{X_{L_c}} + C_{P_i} + u_{pi} \tag{2.95b}$$

$$\tau_i \dot{E}_i = -E_i - k_{Q_i} \frac{E_i^2 - E_i V_i \cos(\delta_i - \theta_i)}{X_{L_c}} + C_{Q_i} + u_{qi} \tag{2.95c}$$

where $C_{P_i} = \omega_i^d + k_{P_i} P_i^d$, $C_{Q_i} = E_i^d + k_{Q_i} Q_i^d$ are constants.

The dynamic models of inverters heavily rely on their control strategies. Here, we only introduce the most commonly used one that employs the simplest droop control. This model will be used in Chapter 7, in which we will explain how to integrate the secondary and tertiary control into this model and merge optimization and control in a distributed framework.

Bibliography

1 R. A. Horn and C. R. Johnson, *Matrix Analysis*. Cambridge University Press, 2012.

2 Z. Li, Z. Wu, Z. Li, and Z. Ding, "Distributed optimal coordination for heterogeneous linear multiagent systems with event-triggered mechanisms," *IEEE Transactions on Automatic Control*, vol. 65, no. 4, pp. 1763–1770, 2020.

3 B. Brogliato, A. Daniilidis, C. Lemarechal, and V. Acary, "On the equivalence between complementarity systems, projected systems and differential inclusions," *Systems & Control Letters*, vol. 55, no. 1, pp. 45–51, 2006.

4 A. Ruszczynski, *Nonlinear Optimization*. Princeton University Press, 2006.

5 S. H. Low, "Analytical methods for network congestion control," *Synthesis Lectures on Communication Networks*, vol. 10, no. 1, pp. 1–213, 2017.

6 S. Boyd and L. Vandenberghe, *Convex Optimization*. Cambridge University Press, 2004.

7 H. Bauschke and P. L. Combettes, *Convex Analysis and Monotone Operator Theory in Hilbert Spaces*. Springer, 2017.

8 F. Facchinei and J.-S. Pang, *Finite-Dimensional Variational Inequalities and Complementarity Problems*. New York: Springer-Verlag, 2003.

9 P. Dupuis and H. J. Kushner, "Asymptotic behavior of constrained stochastic approximations via the theory of large deviations," *Probability Theory and Related Fields*, vol. 75, no. 2, pp. 223–244, 1987.

10 A. Nagurney and D. Zhang, *Projected Dynamical Systems and Variational Inequalities with Applications*. Springer Science & Business Media, 2012.

11 H. K. Khalil, *Nonlinear Systems*. Upper Saddle River: Prentice hall, 1996.

12 A. J. Van der Schaft, *L2-Gain and Passivity Techniques in Nonlinear Control*. Springer, 2017.

13 B. Zhang and Z. Yan, *Advanced Electric Power Network Analysis*. Cengage Learning Asia, 2011.

14 M. E. Baran and F. F. Wu, "Optimal capacitor placement on radial distribution systems," *IEEE Transactions on Power Delivery*, vol. 4, no. 1, pp. 725–734, 1989.

15 M. Baran and F. F. Wu, "Optimal sizing of capacitors placed on a radial distribution system," *IEEE Transactions on Power Delivery*, vol. 4, no. 1, pp. 735–743, 1989.

16 D. K. Molzahn, F. Dörfler, H. Sandberg, S. H. Low, S. Chakrabarti, R. Baldick, and J. Lavaei, "A survey of distributed optimization and control algorithms for electric power systems," *IEEE Transactions on Smart Grid*, vol. 8, no. 6, pp. 2941–2962, 2017.

17 T. Stegink, C. De Persis, and A. van der Schaft, "A unifying energy-based approach to stability of power grids with market dynamics," *IEEE Transactions Automatic Control*, vol. 62, no. 6, pp. 2612–2622, 2017.

18 P. Kundur, *Power System Stability and Control*. New York, USA: McGraw-Hill, 1994.

19 L. Yan-Hong, L. Chun-Wen, and W. Yu-Zhen, "Decentralized excitation control of multi-machine multi-load power systems using Hamiltonian function method," *Acta Automatica Sinica*, vol. 35, no. 7, pp. 919–925, 2009.

20 N. Pogaku, M. Prodanovic, and T. C. Green, "Modeling, analysis and testing of autonomous operation of an inverter-based microgrid," *IEEE Transactions on Power Electronics*, vol. 22, no. 2, pp. 613–625, 2007.

21 F. Guo, C. Wen, J. Mao, and Y. D. Song,"Distributed secondary voltage and frequency restoration control of droop-controlled inverter-based microgrids," *IEEE Transactions on Industrial Electronics*, vol. 62, no. 7, pp. 4355–4364, 2015.

22 C. De Persis and N. Monshizadeh, "Bregman storage functions for microgrid control," *IEEE Transactions on Automatic Control*, vol. 63, no. 1, pp. 53–68, 2017.

23 J. Schiffer, R. Ortega, A. Astolfi, J. Raisch, and T. Sezi, "Conditions for stability of droop-controlled inverter-based microgrids," *Automatica*, vol. 50, no. 10, pp. 2457–2469, 2014.

3

Bridging Control and Optimization in Distributed Optimal Frequency Control

Generally, there are two perspectives in merging control and optimization: "forward engineering" and "reverse engineering." The former suggests starting with a specified optimization problem and deriving the controller and dynamics of the physical system as an iterative algorithm to solve this problem. The latter, in contrast, suggests reshaping the objective function of the optimization problem subject to a given appropriate controller design such that the objective function is the utility (disutility) that the controller tries to maximize (minimize). No matter which perspective is concerned, the fundamental intuition is to figure out the underlying connection between optimization and control coupled through physical systems' dynamics. In this regard, this chapter presents a systematic method to design an optimization-guided feedback controller by taking the primary frequency control in power systems as an example, where both perspectives are involved.

In this chapter, a continuous fast-acting distributed load controller is designed by formulating an optimal load control (OLC) problem where the objective is to minimize the aggregate cost of tracking an operating point subject to power balance over the network. It is revealed with proofs that the swing dynamics and the branch power flows, coupled with frequency-based load control, serve as a distributed primal–dual algorithm to solve the OLC problem, bridging these two branches considered separately in tradition. Furthermore, we establish the globally asymptotic stability of a multi-machine network under such type of load-side primary frequency control. These results imply that the local frequency deviations on each bus convey exactly the right information about the global power imbalance for the loads to make individual decisions that turn out to be globally optimal.

Merging Optimization and Control in Power Systems: Physical and Cyber Restrictions in Distributed Frequency Control and Beyond, First Edition. Feng Liu, Zhaojian Wang, Changhong Zhao, and Peng Yang.

3.1 Background

3.1.1 Motivation

Frequency control maintains the frequency of a power system tightly around its nominal value when demand or supply fluctuates. Recalling Chapter 1, the frequency control is traditionally implemented on the generation side and consists of three mechanisms that work at different timescales in concert [1–3]. The primary frequency control operates at a timescale up to low tens of seconds and uses a governor to adjust, around a set point, the mechanical power input to a generator based on the local frequency deviation. It is called the droop control and is completely decentralized. The primary control can rebalance power and stabilize the frequency but does not in itself restore the nominal frequency. The secondary frequency control operates at a timescale up to a minute or so and adjusts the set points of governors in a control area in a centralized fashion to drive the frequency back to its nominal value and the inter-area power flows to their scheduled values. The tertiary frequency control (also called economic dispatch) operates at a timescale of several minutes or up and schedules the output levels of generators that are online and the inter-area power flows. This chapter focuses on load participation in the primary frequency control.

The needs and technologies for ubiquitous continuous fast-acting distributed load participation in frequency control at different timescales have started to mature in the last decade or so. The idea however dates back to the late 1970s. Schweppe *et al.* advocated in a 1980 paper [4] its deployment to "assist or even replace turbine-governed systems and spinning reserve." They also proposed to use spot prices to incentivize the users to adapt their consumption to the true cost of generation at the time of consumption. In contrast to direct load control, this approach allows the loads to choose their consumption pattern based on their need and the spot price, attaining with the generation a homeostatic equilibrium "to the benefit of both the utilities and their customers." Remarkably it was emphasized back then that such frequency adaptive loads would "allow the system to accept more readily a stochastically fluctuating energy source, such as wind or solar generation" [4]. This point is echoed recently in, e.g. [5–11], which argue for "grid-friendly" appliances, such as refrigerators, water or space heaters, ventilation systems, and air conditioners, as well as plug-in electric vehicles to help manage energy imbalance. For further references, see [10]. Simulations in all these studies have consistently shown significant improvement in performance and reduction in the need for spinning reserves. The benefit of this approach can thus be substantial as the total capacity of grid-friendly appliances in the United

States is estimated in [6] to be about 18% of the peak demand, comparable with the required operating reserve, currently at 13% of the peak demand. The feasibility of this approach is confirmed by experiments reported in [8] that measured the correlation between the frequency at a 230 kV transmission substation and the frequencies at the 120 V wall outlets at various places in a city in Montana. They show that local frequency measurements are adequate for loads to participate in primary frequency control and in the damping of electromechanical oscillations due to inter-area modes of large interconnected systems.

Indeed a small-scale demonstration project has been conducted by the Pacific Northwest National Lab during early 2006 to March 2007 where 200 residential appliances participated in primary frequency control by automatically reducing their consumption (e.g. the heating element of a clothes dryer was turned off while the tumble continued) when the frequency of the household dropped below a threshold (59.95 Hz) [12]. Field trials are also carried out in other countries around the globe, e.g. the UK Market Transformation Program [13]. Even though loads do not yet provide second-by-second or minute-by-minute continuous regulation service in any major electricity markets, the survey in [14] finds that they already provide 50% of the 2400 MW contingency reserve in Electric Reliability Council of Texas (ERCOT) and 30% of dispatched reserve energy (in between continuous reserve and economic dispatch) in the UK market. Long Island Power Authority (LIPA) developed LIPA Edge that provides 24.9 MW of demand reduction and 75 MW of spinning reserve by 23400 loads for peak power management [15].

While there are many simulation studies and field trials of frequency-based load control as discussed above, there is not much analytic study that relates the behavior of the loads and the equilibrium and dynamic behavior of a multi-machine power network. Indeed this has been recognized, e.g. in [5, 12, 13], as a major unanswered question that must be resolved before ubiquitous continuous fast-acting distributed load participation in frequency regulation will become widespread. Even though classical models for power system dynamics [1–3] that focus on the generator control can be adapted to include load adaptation, they do not consider the cost, or disutility, to the load in participating in frequency control, an important aspect of such an approach [4, 10–12].

In this chapter, we present a systematic method to design ubiquitous continuous fast-acting distributed load control and establish the globally asymptotic stability of a multi-machine network under this type of primary frequency control. Our approach allows the loads to choose their consumption pattern based on their need and the global power imbalance on the network, attaining with the generation what [4] calls a homeostatic equilibrium "to the benefit of both the utilities and their customers."

3.1.2 Summary

Specifically we consider a simple network model described by linearized swing dynamics on generator buses, power flow dynamics on the branches, and a measure of disutility to users when they participate in primary frequency control. At steady state, the frequencies on different buses are synchronized to a common value, and the mechanic power is balanced with the electric power on each bus. Suppose a small change in power injection occurs on an arbitrary subset of the buses, causing the bus frequencies to deviate from their nominal value. We assume the change is small and the DC power flow model is reasonably accurate. Instead of adjusting the generators as in the traditional approach, how should we adjust the controllable loads in the network to rebalance power in a way that minimizes the aggregate disutility of these loads? We formulate this question as an OLC problem, which informally takes the form

$$\begin{aligned} \min_{d} \quad & c(d) \\ \text{s.t.} \quad & \text{power rebalance} \end{aligned}$$

where d is the demand vector and c measures the cost of loads in participating in control. Even though neither frequency nor branch power flows appear in OLC, we will show that frequency deviations emerge as a measure of the cost of power imbalance and branch flow deviations as a measure of frequency asynchronism. More strikingly the swing dynamics together with local frequency-based load control serve as a distributed primal–dual algorithm to solve the dual of OLC. Then, it also solves OLC due to the strong duality. Moreover, the primal–dual algorithm is globally asymptotically stable, steering the network to the unique global optimal of OLC.

These results have four important implications:

- The local frequency deviation on each bus conveys exactly the right information about the global power imbalance for the loads themselves to make local decisions that turn out to be globally optimal. This allows a completely decentralized solution without explicit communication to or among the loads.
- The globally asymptotic stability of the primal–dual algorithm of OLC suggests that ubiquitous continuous decentralized load participation in primary frequency control is stable.
- We present a "forward engineering" perspective where we start with the basic goal of load control and derive the frequency-based controller and the swing dynamics as a distributed primal–dual algorithm to solve the dual of OLC. In this perspective, the controller design mainly boils down to specifying an appropriate optimization problem (OLC).
- The opposite perspective of "reverse engineering" is useful as well where, given an appropriate frequency-based controller design, the network dynamics will

converge to a unique equilibrium that *inevitably* solves OLC with an objective function that depends on the controller design. In this sense, any memoryless frequency adaptation implies a certain disutility function of the load that the control implicitly minimizes. For instance, the linear controller in [5, 8] implies a quadratic disutility function and hence a quadratic objective in OLC.

Our results confirm that frequency adaptive loads can rebalance power and resynchronize frequency, just as the droop control of the generators currently does.

3.1.3 Organization

The remainder of this chapter is organized as follows. Section 3.2 describes a dynamic model of power networks. Section 3.3 formulates OLC as a systematic method to design load-side primary frequency control and explains how the frequency-based load control and the system dynamics serve as a distributed primal–dual algorithm to solve OLC. Section 3.4 proves that the network equilibrium is globally asymptotically stable. Section 3.5 reports simulations of the IEEE 68-bus test system that uses a much more detailed and realistic model than our analytic model. The simulation results not only confirm the convergence of the primal-dual algorithm but also demonstrate significantly better transient performance. Section 3.6 concludes the chapter with notes.

3.2 Power System Model

Let $\mathbb{R}$ denote the set of real numbers and $\mathbb{N}$ denote the set of non-zero natural numbers. For a set $\mathcal{N}$, let $|\mathcal{N}|$ denote its cardinality. A variable without a subscript usually denotes a vector with appropriate components, e.g. $\omega = (\omega_j, j \in \mathcal{N}) \in \mathbb{R}^{|\mathcal{N}|}$. For $a, b \in \mathbb{R}$, $a \leq b$, the expression $[\cdot]_a^b$ denotes $\max\{\min\{\cdot, b\}, a\}$. For a matrix A, let A^{T} denote its transpose. For a signal $\omega(t)$ of time, let $\dot{\omega}$ denote its time derivative $\frac{\mathrm{d}\omega}{\mathrm{d}t}$.

The power transmission network is described by a graph $(\mathcal{N}, \mathcal{E})$, where $\mathcal{N} = \{1, \dots, |\mathcal{N}|\}$ is the set of buses and $\mathcal{E} \subseteq \mathcal{N} \times \mathcal{N}$ is the set of transmission lines connecting the buses. We make the following assumptions: [1]

- The lines $(i, j) \in \mathcal{E}$ are lossless and characterized by their reactances x_{ij}.
- The voltage magnitudes $|V_j|$ of buses $j \in \mathcal{N}$ are constants.
- Reactive power injections on the buses and reactive power flows on the lines are ignored.

1 These assumptions are similar to the standard DC approximation in Chapter 2 except that we do not assume the nominal phase angle difference is small across each link.

Assume that $(\mathcal{N}, \mathcal{E})$ is directed, with an arbitrary orientation, so that if $(i,j) \in \mathcal{E}$ then $(j,i) \notin \mathcal{E}$. Use (i,j) and $i \to j$ interchangeably to denote a link in $\mathcal{E}$, and use "$i : i \to j$" and "$k : j \to k$" to denote the set of buses i that are predecessors of bus j and the set of buses k that are successors of bus j, respectively. Without loss of generality, $(\mathcal{N}, \mathcal{E})$ is assumed to be connected.

The network has two types of buses: generator buses and load buses. A generator bus not only has loads but also an AC generator that converts mechanic power into electric power through a rotating prime mover. A load bus has only loads but no generator. We assume that the system is three-phase balanced. For a bus $j \in \mathcal{N}$, its phase voltage at time t is $\sqrt{2}|V_j| \cos(\omega^0 t + \theta_j^0 + \Delta\theta_j(t))$ where ω^0 is the nominal frequency, θ_j^0 is the nominal phase angle, and $\Delta\theta_j(t)$ is the time-varying phase angle deviation. The frequency on bus j is defined as $\omega_j := \omega^0 + \Delta\dot{\theta}_j$, and we call $\Delta\omega_j := \Delta\dot{\theta}_j$ the frequency deviation on bus j. Assume that $\Delta\omega_j$ is small for all the buses $j \in \mathcal{N}$ and the differences $\Delta\theta_i - \Delta\theta_j$ between phase angle deviations are small across all the links $(i,j) \in \mathcal{E}$. We adopt a standard dynamic model, e.g. in [2, Sec. 11.4].

3.2.1 Generator Buses

We assume coherency between the internal and terminal (bus) voltage phase angles of the generator. Then the dynamics on a generator bus j is modeled by the swing equation

$$M_j \Delta\dot{\omega}_j + D_j' \Delta\omega_j = P_j^{m\prime} - P_{\text{loss},j}^0 - P_j^e$$

where $M_j > 0$ is the inertia constant of the generator and $D_j'\Delta\omega_j$ with $D_j' > 0$ represents the (first-order approximation of) deviation in generator power loss due to friction [2] from its nominal value $P_{\text{loss},j}^0 := \frac{D_j'\omega^0}{2}$. Here $P_j^{m\prime}$ is the mechanic power injection to the generator, and P_j^e is the electric power export of the generator, which equals the sum of loads on bus j and the net branch power flow from bus j to other buses.

In general, load power may depend on both the bus voltage magnitude (which is assumed fixed) and frequency. We distinguish between three types of loads, *frequency sensitive*, *frequency insensitive but controllable*, and *uncontrollable loads*. Assume the power consumptions of frequency-sensitive (e.g., motor-type) loads increase linearly with frequency deviation and model the aggregate power consumption of these loads by $\hat{d}_j^0 + D_j''\Delta\omega_j$ with $D_j'' > 0$, where $\hat{d}_j^0$ is its nominal value. The frequency-insensitive loads can be actively controlled, and our goal is to design and analyze these control laws. Use d_j to denote the aggregate power of the controllable (but frequency-insensitive) loads on bus j. Finally let P_j^l denote the aggregate power consumption of uncontrollable (constant power) loads on bus j that are neither of the above two types of loads; P_j^l may change over time but is prespecified. Then the electric power P_j^e is the sum of frequency-sensitive loads, controllable loads, uncontrollable loads, and the net branch power flow from bus j to other buses:

$$P_j^e := \hat{d}_j^0 + D_j''\Delta\omega_j + d_j + P_j^l + \sum_{k:j\to k} P_{jk} - \sum_{i:i\to j} P_{ij}$$

where P_{jk} is the branch power flow from bus j to bus k.

Hence the dynamics on a generator bus j is

$$M_j\Delta\dot{\omega}_j = -\left(D_j\Delta\omega_j + d_j - P_j^m + P_j^{\text{out}} - P_j^{\text{in}}\right)$$

where

$$\begin{aligned}
D_j &:= D_j' + D_j'',\\
P_j^m &:= P_j^{m'} - P_{\text{loss},j}^0 - \hat{d}_j^0 - P_j^l,\\
P_j^{\text{out}} &:= \sum_{k:j\to k} P_{jk},\\
P_j^{\text{in}} &:= \sum_{i:i\to j} P_{ij}.
\end{aligned}$$

Note that P_j^l is integrated with $P_j^{m'}$ into a single term P_j^m, so that any change in power injection, whether on the generation side or the load side, is considered a change in P_j^m. Let $d_j^0, P_j^{m,0}, P_{ij}^0$ denote the nominal (operating) point at which

$$d_j^0 - P_j^{m,0} + P_j^{\text{out},0} - P_j^{\text{in},0} = 0.$$

Let

$$\begin{aligned}
&d_j(t) = d_j^0 + \Delta d_j(t), P_j^m(t) = P_j^{m,0} + \Delta P_j^m(t),\\
&P_{ij}(t) = P_{ij}^0 + \Delta P_{ij}(t).
\end{aligned}$$

Then the deviations satisfy

$$M_j\Delta\dot{\omega}_j = -\left(D_j\Delta\omega_j + \Delta d_j - \Delta P_j^m + \Delta P_j^{\text{out}} - \Delta P_j^{\text{in}}\right). \tag{3.1}$$

Figure 3.1 is a schematic of the generator bus model (3.1).

3.2.2 Load Buses

A load bus j that has no generator is modeled by the following algebraic equation that represents power balance at the bus[2]:

$$0 = D_j\Delta\omega_j + \Delta d_j - \Delta P_j^m + \Delta P_j^{\text{out}} - \Delta P_j^{\text{in}} \tag{3.2}$$

where $-\Delta P_j^m$ represents the change in the aggregate uncontrollable load.

2 There may be load buses with large inertia that can be modeled by swing dynamics (3.1) as proposed in [16]. We will treat them as generator buses mathematically.

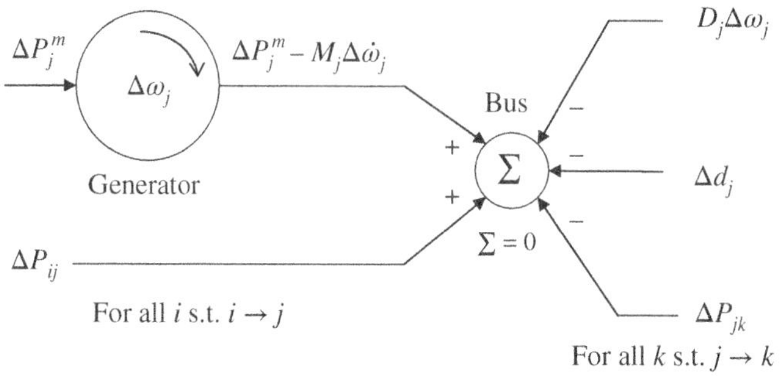

Figure 3.1 Schematic of a generator bus j, where $\Delta\omega_j$ is the frequency deviation, ΔP_j^m is the change in mechanic power minus aggregate uncontrollable load, $D_j\Delta\omega_j$ characterizes the effect of generator friction and frequency-sensitive loads, Δd_j is the change in aggregate controllable load, ΔP_{ij} is the deviation in branch power injected from another bus i to bus j, and ΔP_{jk} is the deviation in branch power delivered from bus j to another bus k.

3.2.3 Branch Flows

To derive the dynamics of branch flows, we assume that the system is always under the three-phase balanced condition, the frequency deviations $\Delta\omega_j$ are small, and the differences $\Delta\theta_i - \Delta\theta_j$ between phase angle deviations are small across all the links $(i,j) \in \mathcal{E}$. Specifically, $\Delta\omega_j$ is negligible compared with ω^0, and an approximation of a quantity to the first order of $\Delta\theta_i - \Delta\theta_j$ is reasonable.

Without loss of generality suppose that buses i and j are wye-connected and each of the three lines has the same inductance L and zero resistance. Let the phase a voltages at buses i and j at time t be

$$v_i^a(t) = \sqrt{2}\,|V_i|\cos\left(\omega^0 t + \theta_i^0 + \Delta\theta_i(t)\right)$$

and

$$v_j^a(t) = \sqrt{2}\left|V_j\right|\cos\left(\omega^0 t + \theta_j^0 + \Delta\theta_j(t)\right),$$

respectively, and assume the voltage magnitudes are fixed. Denote the phase a current from i to j at time t by $i_{ij}^a(t)$.

For $t \le 0$, suppose $\Delta\theta_j(t) = 0$ for all the buses j. Hence the system is at a steady state with $i_{ij}^a(t) = \sqrt{2}\,|I_0|\cos\left(\omega^0 t + \theta_c^0\right)$. From phasor calculations we have

$$|I_0| = \frac{|V_0|}{x_{ij}}$$

$$\theta_c^0 = \tan^{-1}\left[\frac{\left|V_j\right|\cos\theta_j^0 - |V_i|\cos\theta_i^0}{|V_i|\sin\theta_i^0 - \left|V_j\right|\sin\theta_j^0}\right] \tag{3.3}$$

where $x_{ij} := \omega^0 L$ and $|V_0| := \sqrt{|V_i|^2 + |V_j|^2 - 2|V_i||V_j|\cos\left(\theta_i^0 - \theta_j^0\right)}$. Then, we have

$$i_{ij}^a(0) = \sqrt{2}\,|I_0|\cos\theta_c^0$$
$$= \frac{\sqrt{2}\left(|V_i|\sin\theta_i^0 - |V_j|\sin\theta_j^0\right)}{x_{ij}} \tag{3.4}$$

From the property of the inductance, for $t \geq 0$, we have

$$L\frac{\mathrm{d}i_{ij}^a}{\mathrm{d}t} = v_i^a - v_j^a \tag{3.5}$$

whose solution is

$$\begin{aligned} i_{ij}^a(t) &= i_{ij}^a(0) + \frac{1}{L}\int_0^t \left(v_i^a(\tau) - v_j^a(\tau)\right)\mathrm{d}\tau \\ &\approx i_{ij}^a(0) + \frac{\sqrt{2}}{\omega^0 L}\left[|V_i|\sin\left(\omega^0 t + \theta_i^0 + \Delta\theta_i(t)\right) - |V_i|\sin\theta_i^0\right] \\ &\quad - \frac{\sqrt{2}}{\omega^0 L}\left[|V_j|\sin\left(\omega^0 t + \theta_j^0 + \Delta\theta_j(t)\right) - |V_j|\sin\theta_j^0\right] \\ &= \frac{\sqrt{2}}{x_{ij}}\left[|V_i|\sin\left(\omega^0 t + \theta_i^0 + \Delta\theta_i(t)\right) - |V_j|\sin\left(\omega^0 t + \theta_j^0 + \Delta\theta_j(t)\right)\right] \end{aligned} \tag{3.6}$$

where the approximate equality is due to the assumption that $\Delta\dot{\theta}_j = \Delta\omega_j$ are negligible compared with ω^0 and the last equality is due to (3.4).

From (3.6) the instantaneous real power injection from i to j at phase a is

$$\begin{aligned} p_{ij}^a &= v_i^a i_{ij}^a \\ &= 2\frac{|V_i|^2}{x_{ij}}\sin\left(\omega^0 t + \theta_i^0 + \Delta\theta_i\right)\cos\left(\omega^0 t + \theta_i^0 + \Delta\theta_i\right) \\ &\quad - 2\frac{|V_i||V_j|}{x_{ij}}\sin\left(\omega^0 t + \theta_j^0 + \Delta\theta_j\right)\cos\left(\omega^0 t + \theta_i^0 + \Delta\theta_i\right) \\ &= \frac{|V_i|^2}{x_{ij}}\sin\left(2\omega^0 t + 2\theta_i^0 + 2\Delta\theta_i\right) \\ &\quad + \frac{|V_i||V_j|}{x_{ij}}\sin\left(\theta_i^0 - \theta_j^0 + \Delta\theta_i - \Delta\theta_j\right) \\ &\quad - \frac{|V_i||V_j|}{x_{ij}}\sin\left(2\omega^0 t + \theta_i^0 + \theta_j^0 + \Delta\theta_i + \Delta\theta_j\right) \end{aligned} \tag{3.7}$$

Since we assumed the system is under the three-phase balanced condition, replacing θ_i^0 and θ_j^0 in (45) with $\theta_i^0 - \frac{2}{3}\pi$ and $\theta_j^0 - \frac{2}{3}\pi$, we get p_{ij}^b. Furthermore, replacing θ_i^0 and θ_j^0 in (45) with $\theta_i^0 + \frac{2}{3}\pi$ and $\theta_j^0 + \frac{2}{3}\pi$, we get p_{ij}^c. Hence the three-phase instantaneous real power flow is (to the first order of $\Delta\theta_i - \Delta\theta_j$)

$$\begin{aligned} P_{ij} &= p_{ij}^a + p_{ij}^b + p_{ij}^c \\ &= 3\frac{|V_i|\,|V_j|}{x_{ij}} \sin\left(\theta_i^0 - \theta_j^0 + \Delta\theta_i - \Delta\theta_j\right) \\ &\approx P_{ij}^0 + \Delta P_{ij} \end{aligned} \tag{3.8}$$

where

$$P_{ij}^0 = 3\frac{|V_i|\,|V_j|}{x_{ij}} \sin\left(\theta_i^0 - \theta_j^0\right) \tag{3.9}$$

is the nominal branch power flow and

$$\Delta P_{ij} = 3\frac{|V_i|\,|V_j|}{x_{ij}} \cos\left(\theta_i^0 - \theta_j^0\right)\left(\Delta\theta_i - \Delta\theta_j\right) \tag{3.10}$$

is the deviation in branch power flow. By $\Delta\omega_i = \Delta\dot{\theta}_i$, $\Delta\omega_j = \Delta\dot{\theta}_j$, we get the branch flow dynamics

$$\Delta\dot{P}_{ij} = B_{ij}\left(\Delta\omega_i - \Delta\omega_j\right) \tag{3.11}$$

with $B_{ij} := 3\frac{|V_i||V_j|}{x_{ij}}\cos(\theta_i^0 - \theta_j^0)$. The same model is studied in the literature [1, 2] based on quasi-steady-state assumptions. Here, we derive this model by solving the differential equation that characterizes the dynamics of three-phase instantaneous power flow on reactive lines, without explicitly using quasi-steady-state assumptions. Note that (3.11) omits the specification of the initial deviations $\Delta P(0)$ in branch flows. In practice $\Delta P(0)$ cannot be an arbitrary vector, but must satisfy

$$\Delta P_{ij}(0) = B_{ij}\left(\Delta\theta_i(0) - \Delta\theta_j(0)\right) \tag{3.12}$$

for some vector $\Delta\theta(0)$. In Remark 3.5 we discuss the implication of this omission on the convergence analysis.

3.2.4 Dynamic Network Model

We denote the set of generator buses by $\mathcal{G}$, the set of load buses by $\mathcal{L}$, and use $|\mathcal{G}|$ and $|\mathcal{L}|$ to denote the number of generator buses and load buses, respectively. Without loss of generality, label the generator buses so that $\mathcal{G} = \{1, \dots, |\mathcal{G}|\}$ and the load buses so that $\mathcal{L} = \{|\mathcal{G}| + 1, \dots, |\mathcal{N}|\}$. In summary, the dynamic model of the transmission network is specified by (3.1)–(3.11). To simplify notation, we drop the Δ from the variables denoting deviations and write (3.1)–(3.11) as

$$\dot{\omega}_j = -\frac{1}{M_j}(D_j\omega_j + d_j - P_j^m + P_j^{\text{out}} - P_j^{\text{in}}), \qquad \forall j \in \mathcal{G} \tag{3.13}$$

$$0 = D_j\omega_j + d_j - P_j^m + P_j^{\text{out}} - P_j^{\text{in}}, \qquad \forall j \in \mathcal{L} \tag{3.14}$$

$$\dot{P}_{ij} = B_{ij}\left(\omega_i - \omega_j\right), \qquad \forall (i,j) \in \mathcal{E} \tag{3.15}$$

Hence for the rest of this chapter, all variables represent *deviations* from their nominal values. We will refer to the term $D_j\omega_j$ as the deviation in the (aggregate) frequency-sensitive load even though it also includes the deviation in generator power loss due to friction. We will refer to P_j^m as a disturbance whether it is in generation or load.

An equilibrium point of the dynamic system (3.13)–(3.15) is a state (ω, P) where $\dot{\omega}_j = 0$ for $j \in \mathcal{G}$ and $\dot{P}_{ij} = 0$ for $(i,j) \in \mathcal{E}$, i.e. where all frequency deviations and branch power deviations are constant over time.

Remark 3.1 The model (3.13)–(3.15) captures the power system behavior at the timescale of seconds. In this chapter, we only consider a step change in generation or load (constant P^m), which implies that the model does not include the action of turbine–governor that changes the mechanic power injection in response to frequency deviation to rebalance power. Nor does it include any secondary frequency control mechanism such as automatic generation control that operates at a slower timescale to restore the nominal frequency. This model therefore explores the feasibility of fast timescale load control as a supplement to the turbine–governor mechanism to resynchronize frequency and rebalance power.

Here, we use a much more realistic simulation model developed in [17, 18] to validate our simple analytic model. We summarize the key conclusions from those simulations as follows:

1) In a power network with long transmission lines, the internal and terminal voltage phase angles of a generator swing coherently, i.e. the rotating speed of the generator is almost the same as the frequency on the generator bus even during transient.
2) Different buses, particularly those that are in different coherent groups [17] and far apart in electrical distance [19], may have different local frequencies for a duration similar to the time for them to converge to a new equilibrium, as opposed to resynchronizing almost instantaneously to a common system frequency that then converges to the equilibrium.
3) The simulation model and our analytic model exhibit similar transient behaviors and steady-state values for bus frequencies and branch power flows.

3.3 Design and Stability of Primary Frequency Control

Suppose a constant disturbance $P^m = (P_j^m, j \in \mathcal{N})$ is injected to the set $\mathcal{N}$ of buses. How should we adjust the controllable loads d_j in (3.13)–(3.15) to rebalance power in a way that minimizes the aggregate disutility of these loads? In general we can design state feedback controllers of the form $d_j(t) := d_j(\omega(t), P(t))$, prove the feedback system is globally asymptotically stable, and evaluate the aggregate disutility to the loads at the equilibrium point. Here we take an alternative approach by directly formulating our goal as an OLC problem and derive the feedback controller as a *distributed* algorithm to solve OLC.

We now formulate OLC and present our main results. These results are proved in Section 3.4.

3.3.1 Optimal Load Control

The objective function of OLC consists of two costs. First, suppose the (aggregate) controllable load on bus j incurs a cost (disutility) $\tilde{c}_j(d_j)$ when it is changed by d_j. Second, the frequency deviation ω_j causes the (aggregate) frequency-sensitive load on bus j to change by $\hat{d}_j := D_j\omega_j$. For reasons that will become clear later, we assume that this results in a cost to the frequency-sensitive load proportional to the squared frequency deviation weighted by its relative damping constant:

$$\frac{\kappa D_j}{\sum_{i\in\mathcal{N}} D_i}\omega_j^2 := \frac{\kappa}{D_j\left(\sum_{i\in\mathcal{N}} D_i\right)}\hat{d}_j^2$$

where $\kappa > 0$ is a constant. Hence the total cost is

$$\sum_{j\in\mathcal{N}}\left(\tilde{c}_j(d_j) + \frac{\kappa}{D_j\left(\sum_{i\in\mathcal{N}} D_i\right)}\hat{d}_j^2\right).$$

To simplify notation, we scale the total cost by $\frac{1}{2\kappa}\sum_{i\in\mathcal{N}} D_i$ without loss of generality and define $c_j(d_j) := \tilde{c}_j(d_j)\frac{1}{2\kappa}\sum_{i\in\mathcal{N}} D_i$. Then OLC minimizes the total cost over d and $\hat{d}$ while rebalancing generation and load across the network:

$$\textbf{OLC:}\quad \min_{\underline{d}\le d\le \overline{d},\hat{d}} \sum_{j\in\mathcal{N}}\left(c_j(d_j) + \frac{1}{2D_j}\hat{d}_j^2\right) \tag{3.16}$$

$$\text{s.t.}\quad \sum_{j\in\mathcal{N}}(d_j + \hat{d}_j) = \sum_{j\in\mathcal{N}} P_j^m \tag{3.17}$$

where $-\infty < \underline{d}_j \le \overline{d}_j < \infty$.

Remark 3.2 Note that (3.17) does not require the balance of generation and load on each individual bus, but only balance across the entire network. This constraint is less restrictive and offers more opportunity to minimize costs. Additional constraints can be imposed if it is desirable that certain buses, e.g. in the same control area, rebalance their own supply and demand, e.g. for economic or regulatory reasons.

We make the following assumption.

Assumption 3.1 OLC is feasible. The cost functions c_j are strictly convex and twice continuously differentiable on $\left[\underline{d}_j, \overline{d}_j\right]$.

The choice of cost functions is based on physical characteristics of loads and user comfort levels. Examples of cost functions can be found for air conditioners in [20] and plug-in electric vehicles in [21]. See, for example, [22–24] for other cost functions that satisfy Assumption 3.1.

3.3.2 Main Results

The objective function of the dual problem of OLC is

$$\sum_{j\in\mathcal{N}} \Phi_j(\nu) := \sum_{j\in\mathcal{N}} \min_{\underline{d}_j \le d_j \le \overline{d}_j, \hat{d}_j} \left(c_j(d_j) - \nu d_j + \frac{1}{2D_j}\hat{d}_j^2 - \nu \hat{d}_j + \nu P_j^m \right)$$

where the minimization can be solved explicitly as

$$\Phi_j(\nu) := c_j(d_j(\nu)) - \nu d_j(\nu) - \frac{1}{2}D_j\nu^2 + \nu P_j^m \tag{3.18}$$

with

$$d_j(\nu) := \left[c_j'^{-1}(\nu)\right]_{\underline{d}_j}^{\overline{d}_j}. \tag{3.19}$$

This objective function has a scalar variable ν and is not separable across buses $j \in \mathcal{N}$. Its direct solution hence requires coordination across buses. We propose the following distributed version of the dual problem over the vector $\nu := \left(\nu_j, j \in \mathcal{N}\right)$, where each bus j optimizes over its own variable ν_j that are constrained to be equal at optimality:

DOLC:

$$\max_{\nu} \quad \Phi(\nu) := \sum_{j\in\mathcal{N}} \Phi_j(\nu_j)$$

$$\text{s.t.} \quad \nu_i = \nu_j, \qquad \forall (i,j) \in \mathcal{E}.$$

Instead of solving OLC directly, they suggest solving DOLC and recovering the unique optimal point $(d^*, \hat{d}^*)$ of OLC from the unique dual optimal ν^*. We have the following results.

Lemma 3.1 *The objective function Φ of DOLC is strictly concave over $\mathbb{R}^{|\mathcal{N}|}$.*

Proof: From (3.19) either $c_j'(d_j(\nu)) = \nu$ or $d_j'(\nu) = 0$, and hence in (3.18), we have

$$\frac{\mathrm{d}}{\mathrm{d}\nu}\left(c_j(d_j(\nu)) - \nu d_j(\nu)\right) = c_j'(d_j(\nu))d_j'(\nu) - d_j(\nu) - \nu d_j'(\nu) = -d_j(\nu)$$

and therefore

$$\frac{\partial \Phi}{\partial \nu_j}(\nu) = \Phi_j'(\nu_j) = -d_j(\nu_j) - D_j\nu_j + P_j^m.$$

Hence the Hessian of Φ is diagonal. Moreover, since $d_j(\nu_j)$ defined in (3.19) is non-decreasing in ν_j, we have

$$\begin{aligned}\frac{\partial^2 \Phi}{\partial \nu_j^2}(\nu) &= \Phi_j''(\nu_j)\\ &= -d_j'(\nu_j) - D_j\\ &< 0\end{aligned} \tag{3.20}$$

and therefore Φ is strictly concave over $\mathbb{R}^{|\mathcal{N}|}$. ■

Lemma 3.2

1) *DOLC has a unique optimal point ν^* with $\nu_i^* = \nu_j^* = \nu^*$ for all $i, j \in \mathcal{N}$.* [3]
2) *OLC has a unique optimal point $(d^*, \hat{d}^*)$ where $d_j^* = d_j(\nu^*)$ and $\hat{d}_j^* = D_j\nu^*$ for all $j \in \mathcal{N}$.*

Proof: Let g denote the objective function of OLC with the domain $\mathcal{D} := [\underline{d}_1, \overline{d}_1] \times \cdots \times [\underline{d}_{|\mathcal{N}|}, \overline{d}_{|\mathcal{N}|}] \times \mathbb{R}^{|\mathcal{N}|}$. Since c_j is continuous on $[\underline{d}_j, \overline{d}_j]$, $\sum_j c_j(d_j)$ is lower bounded, i.e. $\sum_j c_j(d_j) > \underline{C}$ for some $\underline{C} > -\infty$. Let $(d', \hat{d}')$ be a feasible point of OLC (which exists by Assumption 3.1). Define the set

$$\mathcal{D}' := \left\{(d, \hat{d}) \in \mathcal{D} | \hat{d}_j^2 \le 2D_j\left(g(d', \hat{d}') - \underline{C}\right),\ \forall j \in \mathcal{N}\right\}$$

Note that for any $(d, \hat{d}) \in \mathcal{D}\mathcal{D}'$, there is some $i \in \mathcal{N}$ such that $\hat{d}_i^2 > 2D_i \left(g(d', \hat{d}') - \underline{C}\right)$, and thus

$$g(d, \hat{d}) > \underline{C} + \frac{\hat{d}_i^2}{2D_i} > g(d', \hat{d}').$$

3 For simplicity, we abuse the notation and use ν^* to denote both the vector $(\nu_j^*, j \in \mathcal{N})$ and the common value of its components. Its meaning should be clear from the context.

Hence any optimal point of OLC must lie in $\mathcal{D}'$. By Assumption 3.1 the objective function g of OLC is continuous and strictly convex over the compact convex set $\mathcal{D}'$ and thus has a minimum $g^* > -\infty$ attained at a unique point $(d^*, \hat{d}^*) \in \mathcal{D}'$.

Let $(d', \hat{d}') \in \mathcal{D}$ be a feasible point of OLC and then $d_j = \left(\underline{d}_j + \overline{d}_j\right)/2, \hat{d}_j = \hat{d}_j' - d_j + d_j'$ specify a feasible point $(d, \hat{d}) \in$ **relint** $\mathcal{D}$, where **relint** denotes the relative interior [25]. Moreover the only constraint of OLC is affine. Hence there is zero duality gap between OLC and its dual, and a dual optimal v^* is attained since $g^* > -\infty$ [[25], Section 5.2.3]. By (3.20), we have

$$\sum_{j\in\mathcal{N}} \Phi_j''(v) = -\sum_{j\in\mathcal{N}} \left(d_j'(v) + D_j\right) < 0$$

Namely, the objective function of the dual of OLC is strictly concave over $\mathbb{R}$, which implies the uniqueness of v^*. Then the optimal point $(d^*, \hat{d}^*)$ of OLC satisfies $d_j^* = d_j(v^*)$ given by (3.19) and $\hat{d}_j^* = D_j v^*$ for $j \in \mathcal{N}$. ■

To derive a distributed solution for DOLC, consider its Lagrangian:

$$L(v, \pi) := \sum_{j\in\mathcal{N}} \Phi_j(v_j) - \sum_{(i,j)\in\mathcal{E}} \pi_{ij}(v_i - v_j) \tag{3.21}$$

where $v \in \mathbb{R}^{|\mathcal{N}|}$ is the (vector) variable for DOLC and $\pi \in \mathbb{R}^{|\mathcal{E}|}$ is the associated dual variable for the dual of DOLC. Hence π_{ij}, for all $(i,j) \in \mathcal{E}$, measure the cost of not synchronizing the variables v_i and v_j across buses i and j. Using (3.18)–(3.21), a partial primal–dual algorithm for DOLC takes the form

$$\begin{aligned}\dot{v}_j &= \gamma_j \frac{\partial L}{\partial v_j}(v, \pi)\\ &= -\gamma_j\left(d_j(v_j) + D_j v_j - P_j^m + \pi_j^{\text{out}} - \pi_j^{\text{in}}\right), \quad \forall j \in \mathcal{G}\end{aligned} \tag{3.22}$$

$$\begin{aligned}0 &= \frac{\partial L}{\partial v_j}(v, \pi)\\ &= -\left(d_j(v_j) + D_j v_j - P_j^m + \pi_j^{\text{out}} - \pi_j^{\text{in}}\right), \quad \forall j \in \mathcal{L}\end{aligned} \tag{3.23}$$

$$\begin{aligned}\dot{\pi}_{ij} &= -\xi_{ij} \frac{\partial L}{\partial \pi_{ij}}(v, \pi)\\ &= \xi_{ij}(v_i - v_j), \qquad \forall (i,j) \in \mathcal{E}\end{aligned} \tag{3.24}$$

where $\gamma_j > 0$, $\xi_{ij} > 0$ are stepsizes and

$$\begin{aligned}\pi_j^{\text{out}} &:= \sum_{k:j\to k} \pi_{jk}\\ \pi_j^{\text{in}} &:= \sum_{i:i\to j} \pi_{ij}.\end{aligned}$$

We interpret (3.22)–(3.24) as an algorithm iterating on the primal variables ν and dual variables π over time $t \geq 0$. Set the stepsizes to be

$$\gamma_j = M_j^{-1}, \qquad \xi_{ij} = B_{ij}.$$

Then (3.22)–(3.24) become identical to (3.13)–(3.15) if we identify ν with ω and π with P and use $d_j(\omega_j)$ defined by (3.19) for d_j in (3.13), (3.14). This means that the frequency deviations ω and the branch flows P are, respectively, the primal and dual variables of DOLC, and the network dynamics together with frequency-based load control execute a primal–dual algorithm for DOLC.

Remark 3.3 Note the consistency of units between the following pairs of quantities: (i) γ_j and M_j^{-1}, (ii) ξ_{ij} and B_{ij}, (iii) ν and ω, and 4) π and P. Indeed, since the unit of D_j is [watt · s] from (3.13), the cost (3.16) is in [watt · s^{-1}]. From (3.18) and (3.21), ν and π are, respectively, in [s^{-1}] (or equivalently [rad · s^{-1}]) and [watt]. From (3.22), γ_j is in [watt^{-1} · s^{-2}] that is the same as the unit of M_j^{-1} from (3.13). From (3.24), ξ_{ij} is in [watt] that is the same as the unit of B_{ij} from (3.15).

For convenience, we collect here the system dynamics and load control equations:

$$\dot{\omega}_j = -\frac{1}{M_j}\left(d_j + \hat{d}_j - P_j^m + P_j^{\text{out}} - P_j^{\text{in}}\right), \quad \forall j \in \mathcal{G} \tag{3.25}$$

$$0 = d_j + \hat{d}_j - P_j^m + P_j^{\text{out}} - P_j^{\text{in}}, \quad \forall j \in \mathcal{L} \tag{3.26}$$

$$\dot{P}_{ij} = B_{ij}\left(\omega_i - \omega_j\right), \quad \forall (i,j) \in \mathcal{E} \tag{3.27}$$

$$\hat{d}_j = D_j \omega_j, \quad \forall j \in \mathcal{N} \tag{3.28}$$

$$d_j = \left[c_j'^{-1}\left(\omega_j\right)\right]_{\underline{d}_j}^{\overline{d}_j}, \quad \forall j \in \mathcal{N} \tag{3.29}$$

The dynamics (3.25)–(3.28) are automatically carried out by the system while the active control (3.29) needs to be implemented at each controllable load. Let $(d(t), \hat{d}(t), \omega(t), P(t))$ denote a trajectory of (deviations of) controllable loads, frequency-sensitive loads, Frequencies, and branch flows, generated by the dynamics (3.25)–(3.29) of the load-controlled system.

Theorem 3.1 *Starting with any* $(d(0), \hat{d}(0), \omega(0), P(0))$, *the solution trajectory* $(d(t), \hat{d}(t), \omega(t), P(t))$ *generated by (3.25)–(3.29) converges to a limit* $(d^*, \hat{d}^*, \omega^*, P^*)$ *as* $t \to \infty$ *such that:*

1) $(d^*, \hat{d}^*)$ *is the unique vector of optimal load control for OLC.*
2) ω^* *is the unique vector of optimal frequency deviations for DOLC.*
3) P^* *is a vector of optimal branch flows for the dual of DOLC.*

We will prove Theorem 3.1 and its related results in Section 3.4.

3.3.3 Implications

Our main results have several important implications:

1) *Ubiquitous continuous load-side primary frequency control.* Like the generator droop, frequency-adaptive loads can rebalance power and resynchronize frequencies after a disturbance. Theorem 3.1 implies that a multi-machine network under such control is globally asymptotically stable. The load-side control is often faster because of the larger time constants associated with valves and prime movers on the generator side. Furthermore OLC explicitly optimizes the aggregate disutility using the cost functions of heterogeneous loads.
2) *Complete decentralization.* The local frequency deviations $\omega_j(t)$ on each bus convey exactly the right information about global power imbalance for the loads to make local decisions that turn out to be globally optimal. This allows a completely decentralized solution without explicit communication among the buses.
3) *Equilibrium frequency.* The frequency deviations $\omega_j(t)$ on all the buses are synchronized to ω^* at optimality even though they can be different during transient. However ω^* at optimality is in general nonzero, implying that the new common frequency may be different from the common frequency before the disturbance. Mechanisms such as isochronous generators [1] or automatic generation control are needed to drive the new system frequency to its nominal value, usually through integral action on the frequency deviations. This will be introduced in the following chapters.
4) *Frequency and branch flows.* In the context of OLC, the frequency deviations $\omega_j(t)$ emerge as the Lagrange multipliers of OLC that measure the cost of power imbalance, whereas the branch flow deviations $P_{ij}(t)$ emerge as the Lagrange multipliers of DOLC that measure the cost of frequency asynchronism.
5) *Uniqueness of solution.* Lemma 3.2 implies that the optimal frequency deviation ω^* is unique and hence the OLC $(d^*, \hat{d}^*)$ is unique. As shown below, the vector P^* of optimal branch flows is unique if and only if the network is a tree. Nonetheless Theorem 3.1 says that, even for a mesh network, any trajectory of branch flows indeed converges to a limit point. See Remark 3.5 for further discussion.

3.4 Convergence Analysis

This section is devoted to the proof of Theorem 3.1 and other properties as given by Theorems 3.2 and 3.3. Before going into the details, we first sketch out the key steps in establishing Theorem 3.1, the convergence of the trajectories generated by (3.25)–(3.29).

1) Theorem 3.2: The set of optimal points (ω^*, P^*) of DOLC and its dual and the set of equilibrium points of (3.25)–(3.29) are nonempty and the same. Denote both of them by Z^*.
2) Theorem 3.3: If $(\mathcal{N}, \mathcal{E})$ is a tree network, Z^* is a singleton with a unique equilibrium point (ω^*, P^*); otherwise (if $(\mathcal{N}, \mathcal{E})$ is a mesh network) Z^* has an uncountably infinite number (a subspace) of equilibria with the same ω^* but different P^*. If we set a reference bus, P^* will be unique with respect to the reference bus.
3) Theorem 3.1: We use a Lyapunov argument to prove that every trajectory $(\omega(t), P(t))$ generated by (3.25)–(3.29) approaches a nonempty, compact subset Z^+ of Z^* as $t \to \infty$. Hence, if $(\mathcal{N}, \mathcal{E})$ is a tree network, Theorem 3.3 implies that any trajectory $(\omega(t), P(t))$ converges to the unique optimal point (ω^*, P^*). If $(\mathcal{N}, \mathcal{E})$ is a mesh network, we show with a more careful argument that $(\omega(t), P(t))$ still converges to a point in Z^+, as opposed to oscillating around Z^+. Theorem 1 then follows from Lemma 3.2.

We now elaborate on these ideas.

Given ω, the optimal loads $(d, \hat{d})$ are uniquely determined by (3.28), (3.29). Hence we focus on the variables (ω, P). Decompose $\omega^{\mathrm{T}} := \begin{bmatrix} \omega_{\mathcal{G}}^{\mathrm{T}} & \omega_{\mathcal{L}}^{\mathrm{T}} \end{bmatrix}$ into frequency deviations on generator buses and load buses. Let C be the $|\mathcal{N}| \times |\mathcal{E}|$ incidence matrix. We decompose C into a $|\mathcal{G}| \times |\mathcal{E}|$ submatrix $C_{\mathcal{G}}$ corresponding to generator buses and an $|\mathcal{L}| \times |\mathcal{E}|$ submatrix $C_{\mathcal{L}}$ corresponding to load buses, i.e. $C = \begin{bmatrix} C_{\mathcal{G}} \\ C_{\mathcal{L}} \end{bmatrix}$. Let

$$\begin{aligned}
\Phi_{\mathcal{G}}(\omega_{\mathcal{G}}) &:= \sum_{j \in \mathcal{G}} \Phi_j(\omega_j), \\
L_{\mathcal{G}}(\omega_{\mathcal{G}}, P) &:= \Phi_{\mathcal{G}}(\omega_{\mathcal{G}}) - \omega_{\mathcal{G}}^{\mathrm{T}} C_{\mathcal{G}} P \\
\Phi_{\mathcal{L}}(\omega_{\mathcal{L}}) &:= \sum_{j \in \mathcal{L}} \Phi_j(\omega_j) \\
L_{\mathcal{L}}(\omega_{\mathcal{L}}, P) &:= \Phi_{\mathcal{L}}(\omega_{\mathcal{L}}) - \omega_{\mathcal{L}}^{\mathrm{T}} C_{\mathcal{L}} P.
\end{aligned}$$

Identifying ν with ω and π with P, we rewrite the Lagrangian for DOLC defined in (3.21) in terms of $\omega_{\mathcal{G}}$ and $\omega_{\mathcal{L}}$, as

$$L(\omega, P) = \Phi(\omega) - \omega^{\mathrm{T}} C P = L_{\mathcal{G}}(\omega_{\mathcal{G}}, P) + L_{\mathcal{L}}(\omega_{\mathcal{L}}, P). \tag{3.30}$$

Then (3.25)–(3.29) (equivalently, (3.22)–(3.24)) can be rewritten in the vector form as

$$\begin{aligned}
\dot{\omega}_{\mathcal{G}} &= \Gamma_{\mathcal{G}} \left[\frac{\partial L_{\mathcal{G}}}{\partial \omega_{\mathcal{G}}} (\omega_{\mathcal{G}}, P) \right]^{\mathrm{T}} \\
&= \Gamma_{\mathcal{G}} \left(\left[\frac{\partial \Phi_{\mathcal{G}}}{\partial \omega_{\mathcal{G}}} (\omega_{\mathcal{G}}) \right]^{\mathrm{T}} - C_{\mathcal{G}} P \right) \qquad (3.31a)
\end{aligned}$$

$$0 = \frac{\partial L_{\mathcal{L}}}{\partial \omega_{\mathcal{L}}} (\omega_{\mathcal{L}}, P)$$

$$= \left[\frac{\partial \Phi_{\mathcal{L}}}{\partial \omega_{\mathcal{L}}} (\omega_{\mathcal{L}}) \right]^{\mathrm{T}} - C_{\mathcal{L}} P \tag{3.31b}$$

$$\dot{P} = -\Xi \left[\frac{\partial L}{\partial P} (\omega, P) \right]^{\mathrm{T}}$$

$$= \Xi \, C^{\mathrm{T}} \omega \tag{3.31c}$$

where $\Gamma_{\mathcal{G}} := \text{diag}(\gamma_j,\ j \in \mathcal{G})$ and $\Xi := \text{diag}(\xi_{ij},\ (i,j) \in \mathcal{E})$. The differential algebraic equations (3.31a)–(3.31c) describe the dynamics of the power network.

A pair (ω^*, P^*) is called a *saddle point* of L if

$$L(\omega, P^*) \leq L(\omega^*, P^*) \leq L(\omega^*, P), \qquad \forall (\omega, P). \tag{3.32}$$

By [25, Sec. 5.4.2], (ω^*, P^*) is primal–dual optimal for DOLC and its dual if and only if it is a saddle point of $L(\omega, P)$. The following theorem establishes the equivalence between the primal–dual optimal points and the equilibrium points of (3.31a)–(3.31c).

Theorem 3.2 *A point (ω^*, P^*) is primal–dual optimal for DOLC and its dual if and only if it is an equilibrium point of (3.31a)–(3.31c). Moreover, at least one primal–dual optimal point (ω^*, P^*) exists, and ω^* is unique among all possible points (ω^*, P^*) that are primal–dual optimal.*

Proof: Recall that we identified ν with ω and π with P. In DOLC, the objective function Φ is (strictly) concave over $\mathbb{R}^{|\mathcal{N}|}$ (by Lemma 3.1), its constraints are linear, and a finite optimal ω^* is attained (by Lemma 3.2). These facts imply that there is no duality gap between DOLC and its dual, and there exists a dual optimal point P^* [[25], Sec. 5.2.3]. Moreover, (ω^*, P^*) is optimal for DOLC and its dual if and only if the following Karush–Kuhn–Tucker (KKT) conditions [[25], Sec. 5.5.3] are satisfied:

$$\text{Stationarity:} \quad \frac{\partial \Phi}{\partial \omega}(\omega^*) = (CP^*)^{\mathrm{T}} \tag{3.33}$$

$$\text{Primal feasibility:} \quad \omega_i^* = \omega_j^*, \quad \forall (i,j) \in \mathcal{E} \tag{3.34}$$

On the other hand $(\omega^*, P^*) = (\omega_{\mathcal{G}}^*, \omega_{\mathcal{L}}^*, P^*)$ is an equilibrium point of (3.31a)–(3.31c) if and only if (3.33), (3.34) are satisfied. Hence (ω^*, P^*) is primal–dual optimal if and only if it is an equilibrium point of (3.31a)–(3.31c). The uniqueness of ω^* is given by Lemma 3.2.

From Lemma 3.2, we denote the unique optimal point of DOLC by $\omega^* \mathbf{1}_{\mathcal{N}} = \begin{bmatrix} \omega^* \mathbf{1}_{\mathcal{G}} \\ \omega^* \mathbf{1}_{\mathcal{L}} \end{bmatrix}$, where $\mathbf{1}_{\mathcal{N}} \in \mathbb{R}^{|\mathcal{N}|}$, $\mathbf{1}_{\mathcal{G}} \in \mathbb{R}^{|\mathcal{G}|}$, and $\mathbf{1}_{\mathcal{L}} \in \mathbb{R}^{|\mathcal{L}|}$ have all their elements equal to 1. From (3.33) and (3.34), define the nonempty set of equilibrium points of (3.31a)–(3.31c) (or equivalently, primal-dual optimal points of DOLC and its dual) as

$$Z^* := \left\{ (\omega, P) | \omega = \omega^* \mathbf{1}_{\mathcal{N}},\ CP = \left[\frac{\partial \Phi}{\partial \omega} \left(\omega^* \mathbf{1}_{\mathcal{N}} \right) \right]^{\mathrm{T}} \right\}. \tag{3.35}$$

Let $(\omega^* \mathbf{1}_{\mathcal{N}}, P^*) = (\omega^* \mathbf{1}_{\mathcal{G}}, \omega^* \mathbf{1}_{\mathcal{L}}, P^*) \in Z^*$ be *any* equilibrium point of (3.31a)–(3.31c). We consider a candidate Lyapunov function:

$$\begin{aligned} U(\omega, P) &= \frac{1}{2} \left(\omega_{\mathcal{G}} - \omega^* \mathbf{1}_{\mathcal{G}} \right)^{\mathrm{T}} \Gamma_{\mathcal{G}}^{-1} \left(\omega_{\mathcal{G}} - \omega^* \mathbf{1}_{\mathcal{G}} \right) \\ &\quad + \frac{1}{2} (P - P^*)^{\mathrm{T}} \Xi^{-1} (P - P^*) \end{aligned} \tag{3.36}$$

Obviously $U(\omega, P) \geq 0$ for all (ω, P) with equality if and only if $\omega_{\mathcal{G}} = \omega^* \mathbf{1}_{\mathcal{G}}$ and $P = P^*$. We will show below that $\dot{U}(\omega, P) \leq 0$ for all (ω, P), where $\dot{U}$ denotes the derivative of U over time along the trajectory $(\omega(t), P(t))$.

Even though U depends explicitly only on $\omega_{\mathcal{G}}$ and P, $\dot{U}$ depends on $\omega_{\mathcal{L}}$ as well through (3.31c). However, it will prove convenient to express $\dot{U}$ as a function of only $\omega_{\mathcal{G}}$ and P. To this end, write (3.31b) as $F(\omega_{\mathcal{L}}, P) = 0$. Then

$$\frac{\partial F}{\partial \omega_{\mathcal{L}}} (\omega_{\mathcal{L}}, P) = \frac{\partial^2 \Phi_{\mathcal{L}}}{\partial \omega_{\mathcal{L}}^2} (\omega_{\mathcal{L}})$$

is nonsingular for all $(\omega_{\mathcal{L}}, P)$ from the proof of Lemma 3.1. By the inverse function theorem [26], $\omega_{\mathcal{L}}$ can be written as a continuously differentiable function of P, denoted by $\omega_{\mathcal{L}}(P)$, with

$$\frac{\partial \omega_{\mathcal{L}}}{\partial P} (P) = \left(\frac{\partial^2 \Phi_{\mathcal{L}}}{\partial \omega_{\mathcal{L}}^2} \left(\omega_{\mathcal{L}}(P) \right) \right)^{-1} C_{\mathcal{L}}. \tag{3.37}$$

Then we rewrite $L(\omega, P)$ as a function of $(\omega_{\mathcal{G}}, P)$ as

$$\begin{aligned} L(\omega, P) &= L_{\mathcal{G}}(\omega_{\mathcal{G}}, P) + L_{\mathcal{L}} \left(\omega_{\mathcal{L}}(P), P \right) \\ &=: \tilde{L} \left(\omega_{\mathcal{G}}, P \right). \end{aligned} \tag{3.38}$$

We have the following lemma, regarding the properties of $\tilde{L}$.

Lemma 3.3 *$\tilde{L}$ is strictly concave in $\omega_{\mathcal{G}}$ and convex in P.*

Proof: From (3.20), the Hessian

$$\frac{\partial^2 \tilde{L}}{\partial \omega_{\mathcal{G}}^2} (\omega_{\mathcal{G}}, P) = \frac{\partial^2 \Phi_{\mathcal{G}}}{\partial \omega_{\mathcal{G}}^2} (\omega_{\mathcal{G}})$$

is diagonal and negative definite for all $\omega_{\mathcal{G}} \in \mathbb{R}^{|\mathcal{G}|}$. Therefore $\tilde{L}$ is strictly concave in $\omega_{\mathcal{G}}$. Moreover from (3.38) and the fact that $\frac{\partial L_{\mathcal{L}}}{\partial \omega_{\mathcal{L}}}\left(\omega_{\mathcal{L}}(P), P\right) = 0$, we have

$$\frac{\partial \tilde{L}}{\partial P}(\omega_{\mathcal{G}}, P) = -\omega_{\mathcal{G}}^{\mathrm{T}} C_{\mathcal{G}} - \omega_{\mathcal{L}}^{\mathrm{T}}(P) C_{\mathcal{L}}. \tag{3.39}$$

Therefore we have (using (3.37))

$$\begin{aligned} \frac{\partial^2 \tilde{L}}{\partial P^2}(\omega_{\mathcal{G}}, P) &= -C_{\mathcal{L}}^{\mathrm{T}} \frac{\partial \omega_{\mathcal{L}}}{\partial P}(P) \\ &= -C_{\mathcal{L}}^{\mathrm{T}} \left(\frac{\partial^2 \Phi_{\mathcal{L}}}{\partial \omega_{\mathcal{L}}^2} \left(\omega_{\mathcal{L}}(P)\right) \right)^{-1} C_{\mathcal{L}}. \end{aligned}$$

From (3.20), $\frac{\partial^2 \Phi_{\mathcal{L}}}{\partial \omega_{\mathcal{L}}^2}$ is diagonal and negative definite. Hence $\frac{\partial^2 \tilde{L}}{\partial P^2}(\omega_{\mathcal{G}}, P)$ is positive semidefinite, and $\tilde{L}$ is convex in P ($\tilde{L}$ may not be strictly convex in P because $C_{\mathcal{L}}$ is not necessarily of full rank). ■

Rewrite (3.31a)–(3.31c) as

$$\dot{\omega}_{\mathcal{G}} = \Gamma_{\mathcal{G}} \left[\frac{\partial \tilde{L}}{\partial \omega_{\mathcal{G}}} \left(\omega_{\mathcal{G}}, P\right) \right]^{\mathrm{T}} \tag{3.40}$$

$$\dot{P} = -\Xi \left[\frac{\partial \tilde{L}}{\partial P} \left(\omega_{\mathcal{G}}, P\right) \right]^{\mathrm{T}} \tag{3.41}$$

Then the derivative of U along any trajectory $(\omega(t), P(t))$ generated by (3.31a)–(3.31c) is

$$\begin{aligned} \dot{U}(\omega, P) &= \left(\omega_{\mathcal{G}} - \omega^* \mathbf{1}_{\mathcal{G}}\right)^{\mathrm{T}} \Gamma_{\mathcal{G}}^{-1} \dot{\omega}_{\mathcal{G}} + (P - P^*)^{\mathrm{T}} \Xi^{-1} \dot{P} \\ &= \frac{\partial \tilde{L}}{\partial \omega_{\mathcal{G}}}(\omega_{\mathcal{G}}, P) \left(\omega_{\mathcal{G}} - \omega^* \mathbf{1}_{\mathcal{G}}\right) - \frac{\partial \tilde{L}}{\partial P}(\omega_{\mathcal{G}}, P)(P - P^*) && (3.42) \\ &\leq \tilde{L}\left(\omega_{\mathcal{G}}, P\right) - \tilde{L}\left(\omega^* \mathbf{1}_{\mathcal{G}}, P\right) + \tilde{L}(\omega_{\mathcal{G}}, P^*) - \tilde{L}\left(\omega_{\mathcal{G}}, P\right) && (3.43) \\ &= L\left(\omega_{\mathcal{G}}, \omega^* \mathbf{1}_{\mathcal{L}}, P^*\right) - \tilde{L}\left(\omega^* \mathbf{1}_{\mathcal{G}}, P\right) && (3.44) \\ &\leq L\left(\omega^* \mathbf{1}_{\mathcal{N}}, P\right) - \tilde{L}\left(\omega^* \mathbf{1}_{\mathcal{G}}, P\right) && (3.45) \\ &= L_{\mathcal{G}}\left(\omega^* \mathbf{1}_{\mathcal{G}}, P\right) + L_{\mathcal{L}}\left(\omega^* \mathbf{1}_{\mathcal{L}}, P\right) \\ &\quad - \left[L_{\mathcal{G}}\left(\omega^* \mathbf{1}_{\mathcal{G}}, P\right) + L_{\mathcal{L}}\left(\omega_{\mathcal{L}}(P), P\right)\right] \\ &\leq 0 && (3.46) \end{aligned}$$

where (3.42) results from (3.40) and (3.41), the inequality in (3.43) results from Lemma 3.3, the equality in (3.44) holds since $\omega_{\mathcal{L}}(P^*) = \omega^* \mathbf{1}_{\mathcal{L}}$ by (3.33), the inequality in (3.45) holds since $L\left(\omega_{\mathcal{G}}, \omega^* \mathbf{1}_{\mathcal{L}}, P^*\right) \leq L\left(\omega^* \mathbf{1}_{\mathcal{N}}, P^*\right) \leq L\left(\omega^* \mathbf{1}_{\mathcal{N}}, P\right)$ from the saddle point condition (3.32), and the inequality in (3.46) holds since $\omega_{\mathcal{L}}(P)$ is the maximizer of $L_{\mathcal{L}}(\cdot, P)$ by the concavity of $L_{\mathcal{L}}$ in $\omega_{\mathcal{L}}$.

The next lemma characterizes the set in which the value of U does not change over time.

Lemma 3.4 $\dot{U}(\omega, P) = 0$ *if and only if either (3.47) or (3.48) holds:*

$$\begin{cases} \omega_{\mathcal{G}} = \omega^* \mathbf{1}_{\mathcal{G}} \\ C_{\mathcal{L}} P = \left[\dfrac{\partial \Phi_{\mathcal{L}}}{\partial \omega_{\mathcal{L}}} \left(\omega^* \mathbf{1}_{\mathcal{L}} \right) \right]^T \end{cases} \tag{3.47}$$

$$\begin{cases} \omega_{\mathcal{G}} = \omega^* \mathbf{1}_{\mathcal{G}} \\ \omega_{\mathcal{L}}(P) = \omega^* \mathbf{1}_{\mathcal{L}} \end{cases} \tag{3.48}$$

Proof: The equivalence of (3.48) and (3.47) follows directly from the definition of $\omega_{\mathcal{L}}(P)$. To prove that (3.48) is necessary and sufficient for $\dot{U}(\omega, P) = 0$, we first claim that the discussion preceding the lemma implies that $(\omega, P) = \left(\omega_{\mathcal{G}}, \omega_{\mathcal{L}}, P \right)$ satisfies $\dot{U}(\omega, P) = 0$ if and only if

$$\begin{cases} \omega_{\mathcal{G}} = \omega^* \mathbf{1}_{\mathcal{G}} \\ \dfrac{\partial \tilde{L}}{\partial P} (\omega_{\mathcal{G}}, P) \left(P - P^* \right) = 0 \end{cases} \tag{3.49}$$

Indeed if (3.49) holds, then the expression in (3.42) evaluates to zero. Conversely, if $\dot{U}(\omega, P) = 0$, then the inequality in (3.43) must hold with equality, which is possible only if $\omega_{\mathcal{G}} = \omega^* \mathbf{1}_{\mathcal{G}}$ since $\tilde{L}$ is *strictly* concave in $\omega_{\mathcal{G}}$. Then we must have $\frac{\partial \tilde{L}}{\partial P}(\omega_{\mathcal{G}}, P) \left(P - P^* \right) = 0$ since the expression in (3.42) needs to be zero. Hence we only need to establish the equivalence of (3.49) and (3.48). Indeed, with $\omega_{\mathcal{G}} = \omega^* \mathbf{1}_{\mathcal{G}}$, the other part of (3.49) becomes

$$\begin{aligned} & \frac{\partial \tilde{L}}{\partial P} (\omega^* \mathbf{1}_{\mathcal{G}}, P) \left(P - P^* \right) \\ & \quad = - \left[\omega^* \mathbf{1}_{\mathcal{G}}^{\mathrm{T}} \ \ \omega_{\mathcal{L}}^{\mathrm{T}}(P) \right] C \left(P - P^* \right) && (3.50) \\ & \quad = - \left[0 \ \ \omega_{\mathcal{L}}^{\mathrm{T}}(P) - \omega^* \mathbf{1}_{\mathcal{L}}^{\mathrm{T}} \right] C \left(P - P^* \right) && (3.51) \\ & \quad = - \left(\omega_{\mathcal{L}}(P) - \omega^* \mathbf{1}_{\mathcal{L}} \right)^{\mathrm{T}} \left[\frac{\partial \Phi_{\mathcal{L}}}{\partial \omega_{\mathcal{L}}} \left(\omega_{\mathcal{L}}(P) \right) - \frac{\partial \Phi_{\mathcal{L}}}{\partial \omega_{\mathcal{L}}} (\omega^* \mathbf{1}_{\mathcal{L}}) \right]^{\mathrm{T}} && (3.52) \end{aligned}$$

where (3.50) results from (3.39), the equality in (3.51) holds since $\mathbf{1}_{\mathcal{N}}^{\mathrm{T}} C = 0$, and (3.52) results from (3.31b), (3.33). Note that $\Phi_{\mathcal{L}}$ is separable over ω_j for $j \in \mathcal{L}$ and $\Phi_j'(\omega_j) = -d_j(\omega_j) - D_j \omega_j + P_j^m$. Writing $D_{\mathcal{L}} := \mathrm{diag}(D_j, j \in \mathcal{L})$, we have

$$\begin{aligned} & \frac{\partial \tilde{L}}{\partial P} (\omega^* \mathbf{1}_{\mathcal{G}}, P) \left(P - P^* \right) \\ & \quad = \left(\omega_{\mathcal{L}}(P) - \omega^* \mathbf{1}_{\mathcal{L}} \right)^{\mathrm{T}} D_{\mathcal{L}} \left(\omega_{\mathcal{L}}(P) - \omega^* \mathbf{1}_{\mathcal{L}} \right) \\ & \qquad + \sum_{j \in \mathcal{L}} \left(\omega_j(P) - \omega^* \right) \left(d_j \left(\omega_j(P) \right) - d_j(\omega^*) \right) && (3.53) \end{aligned}$$

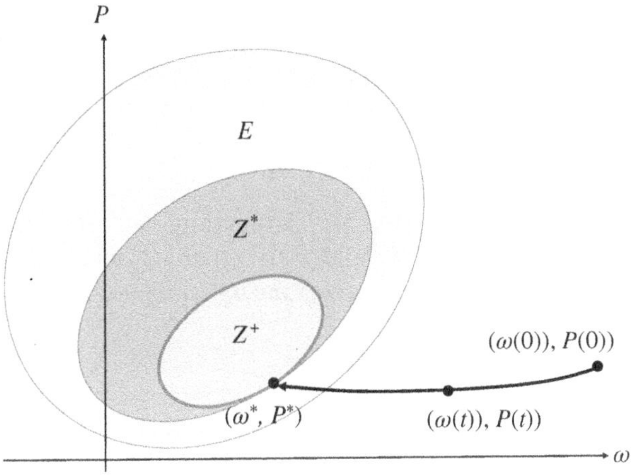

Figure 3.2 E is the set on which $\dot{U} = 0$, Z^* is the set of equilibrium points of (3.31a)–(3.31c), and Z^+ is a compact subset of Z^* to which all solutions $(\omega(t), P(t))$ approach as $t \to \infty$. Indeed every solution $(\omega(t), P(t))$ converges to a point $(\omega^*, P^*) \in Z^+$ that is dependent on the initial state.

Since $d_j(\omega_j)$ defined in (3.19) is nondecreasing in ω_j, each term in the summation above is nonnegative for all P. Hence (3.53) evaluates to zero if and only if $\omega_{\mathcal{L}}(P) = \omega^* \mathbf{1}_{\mathcal{L}}$, establishing the equivalence between (3.49) and (3.48). This completes the proof. ∎

Lemma 3.4 motivates the definition of the set

$$E := \{(\omega, P) \mid \dot{U}(\omega, P) = 0\}$$
$$= \left\{ (\omega, P) \mid \omega = \omega^* \mathbf{1}_{\mathcal{N}},\ C_{\mathcal{L}} P = \left[\frac{\partial \Phi_{\mathcal{L}}}{\partial \omega_{\mathcal{L}}} (\omega^* \mathbf{1}_{\mathcal{L}}) \right]^{\mathrm{T}} \right\} \tag{3.54}$$

in which $\dot{U} = 0$ along any trajectory $(\omega(t), P(t))$. The definition of Z^* in (3.35) implies that $Z^* \subseteq E$, as shown in Figure 3.2.

As shown in the figure E may contain points that are not in Z^*. Nonetheless every accumulation point (limit point of any convergent sequence sampled from the trajectory) of a trajectory $(\omega(t), P(t))$ of (3.31a)–(3.31c) lies in Z^*, as the next lemma shows.

Lemma 3.5 *Every solution $(\omega(t), P(t))$ of (3.31a)–(3.31c) approaches a nonempty, compact subset (denoted Z^+) of Z^* as $t \to \infty$.*

Proof: The proof of LaSalle's invariance principle [27, Theorem 3.4] shows that $(\omega(t), P(t))$ approaches its positive limit set Z^+ that is nonempty, compact, invariant, and a subset of E, as $t \to \infty$. It is then sufficient to show that $Z^+ \subseteq Z^*$, i.e. for any point $(\omega, P) = (\omega_\mathcal{G}, \omega_\mathcal{L}, P) \in Z^+$, to show that $(\omega, P) \in Z^*$. By (3.35), (3.54) and the fact that $(\omega, P) \in E$, we only need to show that

$$C_\mathcal{G} P = \left[\frac{\partial \Phi_\mathcal{G}}{\partial \omega_\mathcal{G}}(\omega_\mathcal{G})\right]^\mathrm{T}. \tag{3.55}$$

Since Z^+ is invariant with respect to (3.31a)–(3.31c), a trajectory $(\omega(t), P(t))$ that starts in Z^+ must stay in Z^+ and hence stay in E. By (3.54), $\omega_\mathcal{G}(t) = \omega^* \mathbf{1}_\mathcal{G}$ for all $t \geq 0$, and therefore $\dot{\omega}_\mathcal{G}(t) = 0$ for all $t \geq 0$. Hence by (3.31a) any trajectory $(\omega(t), P(t))$ in Z^+ must satisfy

$$C_\mathcal{G} P(t) = \left[\frac{\partial \Phi_\mathcal{G}}{\partial \omega_\mathcal{G}}(\omega_\mathcal{G}(t))\right]^\mathrm{T}, \quad \forall t \geq 0$$

which implies (3.55). ∎

The sets $Z^+ \subseteq Z^* \subseteq E$ are illustrated in Figure 3.2. Lemma 3.5 only guarantees that $(\omega(t), P(t))$ approaches Z^+ as $t \to \infty$ while we now show that $(\omega(t), P(t))$ indeed converges to a point in Z^+. The convergence is immediate in the special case when Z^* is a singleton but needs a more careful argument when Z^* has multiple points. The next theorem reveals the relation between the number of points in Z^* and the network topology.

Theorem 3.3

1) *If $(\mathcal{N}, \mathcal{E})$ is a tree, then Z^* is a singleton.*
2) *If $(\mathcal{N}, \mathcal{E})$ is a mesh (i.e. contains a cycle if regarded as an undirected graph), then Z^* has uncountably many points with the same ω^* but different P^*.*

Proof: From (3.35), the projection of Z^* on the space of ω is always a singleton $\omega^* \mathbf{1}_\mathcal{N}$, and hence we only consider the projection of Z^* on the space of P, which is

$$Z_P^* := \{P \mid CP = h^*\}$$

where $h^* := \left[\frac{\partial \Phi}{\partial \omega}(\omega^* \mathbf{1}_\mathcal{N})\right]^\mathrm{T}$. By Theorem 3.2, Z_P^* is nonempty, i.e. there is $P^* \in Z_P^*$ such that $CP^* = h^*$ and hence $\mathbf{1}_\mathcal{N}^\mathrm{T} h^* = \mathbf{1}_\mathcal{N}^\mathrm{T} CP^* = 0$. Therefore we have

$$Z_P^* := \{P \mid \tilde{C}P = \tilde{h}^*\} \tag{3.56}$$

where $\tilde{C}$ is the $(|\mathcal{N}| - 1) \times |\mathcal{E}|$ reduced incidence matrix obtained from C by removing any one of its rows and $\tilde{h}^*$ is obtained from h^* by removing the corresponding row. Note that $\tilde{C}$ has a full row rank of $|\mathcal{N}| - 1$ [28]. If $(\mathcal{N}, \mathcal{E})$ is a tree, then $|\mathcal{E}| = |\mathcal{N}| - 1$, so $\tilde{C}$ is square and invertible and Z_P^* is a singleton. If $(\mathcal{N}, \mathcal{E})$

is a (connected) mesh, then $|\mathcal{E}| > |\mathcal{N}| - 1$, so $\tilde{C}$ has a nontrivial null space and there are uncountably many points in Z_P^*. ■

We can now complete the proof of Theorem 3.1.

Proof: [Proof of Theorem 3.1]
For the case in which $(\mathcal{N}, \mathcal{E})$ is a tree, Lemma 3.5 and Theorem 3.3(1) guarantees that every trajectory$(\omega(t), P(t))$ converges to the unique primal–dual optimal point (ω^*, P^*) of DOLC and its dual, which, by Lemma 3.2, immediately implies Theorem 3.1.

For the case in which $(\mathcal{N}, \mathcal{E})$ is a mesh, since $\dot{U} \leq 0$ along any trajectory $(\omega(t), P(t))$, then $U(\omega(t), P(t)) \leq U(\omega(0), P(0))$ and hence $(\omega(t), P(t))$ stays in a compact set for $t \geq 0$. Therefore, there exists a convergent subsequence $\{(\omega(t_k), P(t_k)),\ k \in \mathbb{N}\}$, where $0 \leq t_1 < t_2 < \ldots$ and $t_k \to \infty$ as $k \to \infty$, such that

$$\lim_{k\to\infty} \omega(t_k) = \omega^\infty, \text{ and } \lim_{k\to\infty} P(t_k) = P^\infty$$

for some $(\omega^\infty, P^\infty)$. Lemma 3.5 implies that $(\omega^\infty, P^\infty) \in Z^*$, and hence $\omega^\infty = \omega^* \mathbf{1}_{\mathcal{N}}$ by (3.35). Recall that the Lyapunov function U in (3.36) can be defined in terms of any equilibrium point $(\omega^* \mathbf{1}_{\mathcal{N}}, P^*) \in Z^*$. In particular, select $(\omega^* \mathbf{1}_{\mathcal{N}}, P^*) = (\omega^* \mathbf{1}_{\mathcal{N}}, P^\infty)$, i.e.

$$\begin{aligned} U(\omega, P) := &\frac{1}{2}\left(\omega_{\mathcal{G}} - \omega^* \mathbf{1}_{\mathcal{G}}\right)^{\mathrm{T}} \Gamma_{\mathcal{G}}^{-1} \left(\omega_{\mathcal{G}} - \omega^* \mathbf{1}_{\mathcal{G}}\right) \\ &+ \frac{1}{2}(P - P^\infty)^{\mathrm{T}} \Xi^{-1} (P - P^\infty) \end{aligned}$$

Since $U \geq 0$ and $\dot{U} \leq 0$ along any trajectory $(\omega(t), P(t))$, $U(\omega(t), P(t))$ must converge as $t \to \infty$. Indeed it converges to 0 due to the continuity of U in both ω and P:

$$\lim_{t\to\infty} U(\omega(t), P(t)) = \lim_{k\to\infty} U\left(\omega(t_k), P(t_k)\right) = U(\omega^\infty, P^\infty) = 0$$

which implies that $(\omega_{\mathcal{G}}(t), P(t))$ converges to $\left(\omega^* \mathbf{1}_{\mathcal{G}}, P^\infty\right)$ and hence $(\omega(t), P(t))$ converges to $\left(\omega^* \mathbf{1}_{\mathcal{N}}, P^\infty\right)$, a primal–dual optimal point for DOLC and its dual. Theorem 3.1 then follows from Lemma 3.2.

Remark 3.4 The standard technique of using a Lyapunov function that is quadratic in both the primal and the dual variables was first proposed by Arrow *et al.* [29] and has been revisited recently, e.g. in [30, 31]. We apply a variation of this technique to our problem with the following features. First, because of the algebraic equation (3.31b) in the system, our Lyapunov function is not a function of all the primal variables, but only the part $\omega_{\mathcal{G}}$ corresponding to generator buses. Second, in the case of a mesh network when there is a subspace of equilibrium points, we show that the system trajectory still converges to one of the equilibrium points instead of oscillating around the equilibrium set.

Remark 3.5 Theorems 3.1–3.3 are based on our analytic model (3.25)–(3.29) that omits an important specification on the initial conditions $P(0)$ of the branch flows. As mentioned earlier, in practice, the initial branch flows must satisfy (3.12) for some $\theta(0)$ (with Δ dropped). With this requirement the branch flow model (3.11)–(3.12) implies $P(t) \in \text{Col}(BC^{\text{T}})$ for all t, where Col denotes the column space, B is the diagonal matrix with entries B_{ij}, and C is the incidence matrix. Indeed $P(t) \in \text{Col}(B\tilde{C}^{\text{T}})$ since $C^{\text{T}}\mathbf{1}_{\mathcal{N}} = 0$ and $\tilde{C}^{\text{T}}$ with one column from C^{T} removed has a full column rank. A simple derivation from (3.56) shows that $Z_P^* \cap \text{Col}(B\tilde{C}^{\text{T}}) = \left\{ B\tilde{C}^{\text{T}}(\tilde{C}B\tilde{C}^{\text{T}})^{-1}\tilde{h}^* \right\}$ is a singleton, where $\tilde{C}B\tilde{C}^{\text{T}}$ is invertible [28]. Moreover by (3.56) and Lemma 3.5, we have $P(t) \to B\tilde{C}^{\text{T}}(\tilde{C}B\tilde{C}^{\text{T}})^{-1}\tilde{h}^*$ as $t \to \infty$. In other words, though for a mesh network the dynamics (3.25)–(3.29) have a subspace of equilibrium points, all the practical trajectories, whose initial points $(\omega(0), P(0))$ satisfy (3.12) for some arbitrary $\theta(0)$, converge to a unique equilibrium point.

3.5 Case Studies

3.5.1 Test System

In this section we illustrate the performance of OLC through the simulation of the IEEE 68-bus New England/New York interconnection test system [17]. The single line diagram of the 68-bus system is given in Figure 3.3. We run the simulation on Power System Toolbox [18]. Unlike our analytic model, the simulation model is much more detailed and realistic, including two-axis subtransient reactance generator model, IEEE-type DC1 exciter model, classical power system stabilizer model, AC (nonlinear) power flows, and nonzero line resistances. The detail of the simulation model including parameter values can be found in the data files of the toolbox. It is shown in [32] that our analytic model is a good approximation of the simulation model.

In the test system there are 35 load buses serving different types of loads, including constant active current loads, constant impedance loads, and induction motor loads, with a total real power of 18.23 GW. In addition, we add three loads to buses 1, 7, and 27, each making a step increase of real power by 1 pu (based on 100 MVA), as the P^m in previous analysis. We also select 30 load buses to perform OLC. In the simulation we use the same bounds $\left[\underline{d}, \overline{d}\right]$ with $\underline{d} = -\overline{d}$ for each of the 30 controllable loads and call the value of $30 \times \overline{d}$ the *total size of controllable loads*. We present

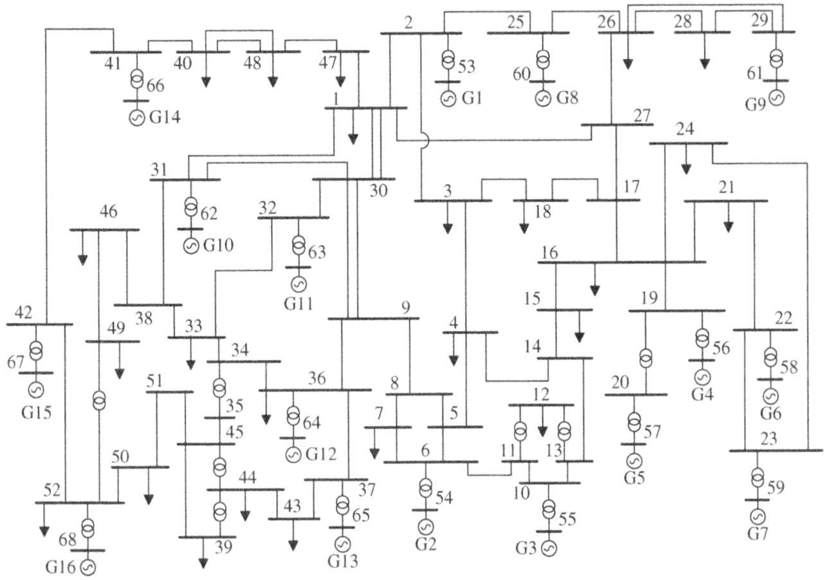

Figure 3.3 Single line diagram of the IEEE 68-bus test system.

simulation results below with different sizes of controllable loads. The disutility function of controllable load d_j is $c_j(d_j) = d_j^2/(2\alpha)$, with identical $\alpha = 100$ pu for all the loads. The loads are controlled every 250 ms, which is a relatively conservative estimate of the rate of load control in an existing testbed [33].

3.5.2 Simulation Results

We look at the impact of OLC on both the steady state and the transient response of the system, in terms of both frequency and voltage. We present the results with a widely used generation-side stabilizing mechanism known as power system stabilizer (PSS) either enabled or disabled. Figure 3.4a,b respectively, show the frequency and voltage on bus 66, under four cases: (i) no PSS, no OLC; (ii) with PSS, no OLC; (iii) no PSS, with OLC; and (iv) with PSS and OLC. In both (iii) and (iv) cases, the total size of controllable loads is 1.5 pu. We observe in Figure 3.4a that whether PSS is used or not, adding OLC always improves the transient response of frequency, in the sense that both the overshoot and the settling time (the time after which the difference between the actual frequency and its new steady-state value never goes beyond 5% of the difference between its old and new steady-state values) are decreased. Using OLC also results in a smaller steady-state frequency error. Cases (ii) and (iii) suggest that using OLC solely without PSS produces a much better performance than using PSS solely without OLC. The impact of OLC

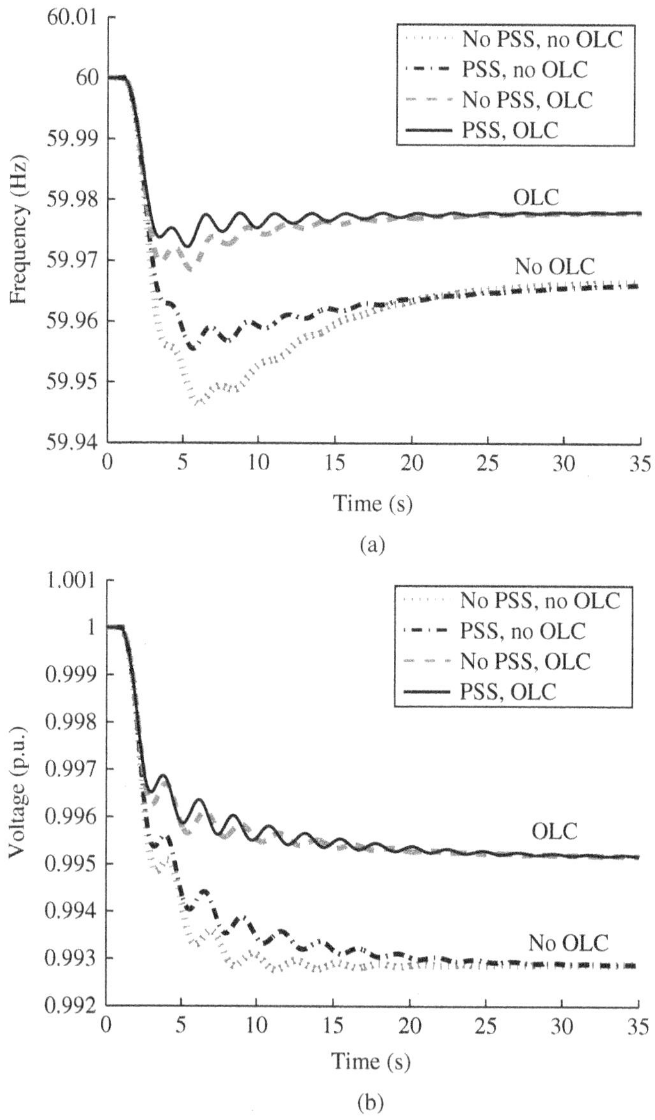

Figure 3.4 The (a) frequency and (b) voltage on bus 66 for the following cases: (i) no PSS, no OLC; (ii) with PSS, no OLC; (iii) no PSS, with OLC; and (iv) with PSS and OLC.

on voltage, with and without PSS, is qualitatively demonstrated in Figure 3.4b. Similar to its impact on frequency, OLC improves significantly both the transient and steady-state of voltage with or without PSS. For instance, the steady-state voltage is within 4.5% of the nominal value with OLC and 7% without OLC.

Figure 3.5 The (a) new steady-state frequency, (b) lowest frequency, and (c) settling time of frequency on bus 66, against the total size of controllable loads.

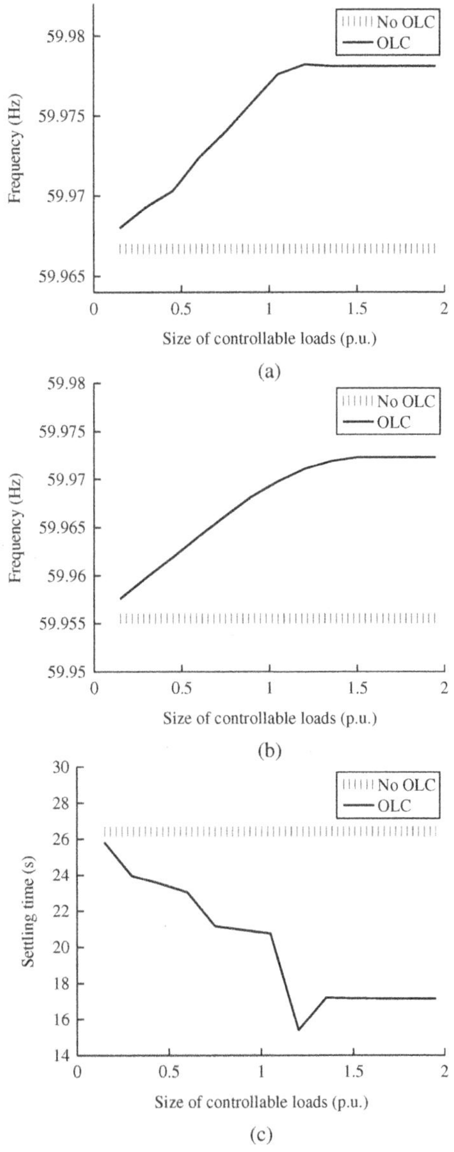

To better quantify the performance improvement due to OLC, we plot in Figure 3.5a–c the new steady-state frequency, the lowest frequency (which indicates overshoot), and the settling time of frequency on bus 66, against the total size of controllable loads. PSS is always enabled. We observe that using OLC always leads to

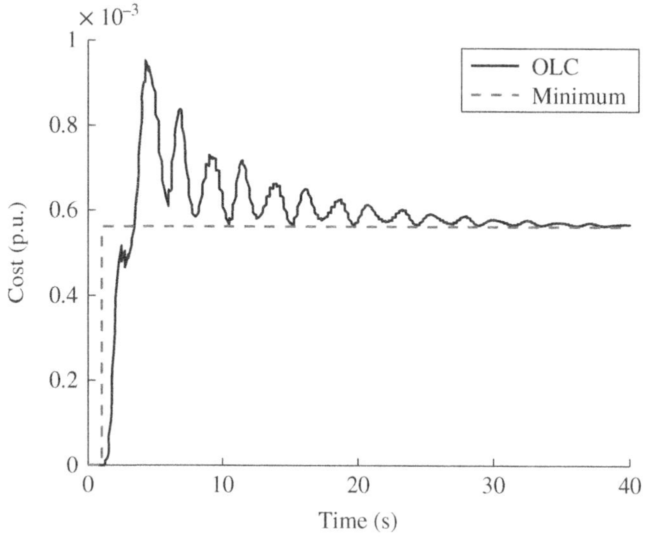

Figure 3.6 The cost trajectory of OLC (solid line) compared with the minimum cost (dashed line).

a higher new steady-state frequency (a smaller steady-state error), a higher lowest frequency (a smaller overshoot), and a shorter settling time, regardless of the total size of controllable loads. As the total size of controllable loads increases, the steady-state error and overshoot decrease almost linearly until a saturation of around 1.5 pu. There is a similar trend for the settling time, though the linear dependence is approximate. In summary OLC improves both the steady-state and transient performance of frequency, and in general deploying more controllable loads leads to bigger improvement.

To verify the theoretical result that OLC minimizes the aggregate cost of load control, Figure 3.6 shows the cost of OLC over time, obtained by evaluating the quantity defined in (3.16) using the trajectory of controllable and frequency-sensitive loads from the simulation. We see that the cost indeed converges to the minimum cost for the given change in P^m.

3.6 Conclusion and Notes

As mentioned in Chapter 1, the historical transition of our power system calls for a new diagram that allows merging control and optimization. This chapter exploits the unique structure of the OLC problem considering the frequency dynamics of power systems, revealing the intrinsic connection behind the optimization algorithm to solve the dual problem of OLC and the closed-loop frequency control system with branch power flows. This finding leads to a systematic methodology to design ubiquitous continuous fast-acting distributed load control for primary frequency regulation in power networks by formulating the OLC problem to minimize the aggregate control cost subject to power balance across the network. Furthermore, even though the system has multiple equilibrium points with nonunique branch power flows, we have proved that it, nonetheless, converges to a unique optimal point.

This methodology, involving both "forward engineering" and "reverse engineering" perspectives, virtually enables a framework for the design and convergence analysis of distributed optimal control in power systems. However, despite its mathematical elegance, many cyber and physical restrictions from practical applications have yet to be considered, which will be discussed in the following chapters.

Bibliography

1 A. J. Wood and B. F. Wollenberg, *Power Generation, Operation, and Control*, 2nd ed. New York, NY: John Wiley & Sons, Inc., 1996.

2 A. R. Bergen and V. Vittal, *Power Systems Analysis*, 2nd ed. Upper Saddle River, NJ: Prentice Hall, Inc., 2000.

3 J. Machowski, J. Bialek, and J. Bumby, *Power System Dynamics: Stability and Control*, 2nd ed. New York, NY: John Wiley & Sons, Inc., 2008.

4 F. C. Schweppe, R. D. Tabors, J. L. Kirtley, H. R. Outhred, F. H. Pickel, and A. J. Cox,"Homeostatic utility control," *IEEE Transactions on Power Apparatus and Systems*, vol.PAS-99, no. 3, pp. 1151–1163, 1980.

5 D. Trudnowski, M. Donnelly, and E. Lightner, "Power-system frequency and stability control using decentralized intelligent loads," in *Proceedings of IEEE Transmission and Distribution Conference and Exhibition*, Dallas, TX, USA, 2006, pp. 1453–1459.

6 N. Lu and D. Hammerstrom, "Design considerations for frequency responsive grid friendly appliances," in *Proceedings of IEEE Transmission and Distribution Conference and Exhibition*, Dallas, TX, USA, 2006, pp. 647–652.

7 J. Short, D. Infield, andF. Freris, "Stabilization of grid frequency through dynamic demand control," *IEEE Transactions on Power Systems*, vol. 22, no. 3, pp. 1284–1293, 2007.

8 M. Donnelly *et al.* "Frequency and stability control using decentralized intelligent loads: benefits and pitfalls," in *Proceedings of IEEE Power and Energy Society General Meeting*,Minneapolis, MN, USA, 2010, pp. 1–6.

9 A. Brooks *et al.*, "Demand dispatch," *IEEE Power & Energy Magazine*, vol. 8, no. 3, pp. 20–29, 2010.

10 D. S. Callaway and I. A. Hiskens, "Achieving controllability of electric loads," *Proceedings of the IEEE*, vol. 99, no. 1, pp. 184–199, 2011.

11 A. Molina-Garcia, F. Bouffard, and D. S. Kirschen, "Decentralized demand-side contribution to primary frequency control," *IEEE Transactions on Power Systems*, vol. 26, no. 1, pp. 411–419, 2011.

12 D. Hammerstrom *et al.*, "Pacific Northwest GridWise testbed demonstration projects, Part II: Grid Friendly Appliance project," Pacific Northwest National Laboratory, Technical Report PNNL-17079, 2007.

13 U. K. Market Transformation Programme, "Dynamic demand control of domestic appliances," U. K. Market Transformation Programme, Technical Report, 2008.

14 G. Heffner, C. Goldman, and M. Kintner-Meyer, "Loads providing ancillary services: review of international experience," Lawrence Berkeley National Laboratory, Berkeley, CA, USA, Technical Report, 2007.

15 B. J. Kirby, *Spinning Reserve from Responsive Loads.* United States Department of Energy, 2003.

16 M. D. Ilic, L. Xie, U. A. Khan, and J. M.Moura, "Modeling of future cyber–physical energy systems for distributed sensing and control," *IEEE Transactions on Systems, Man, and Cybernetics - Part A: Systems and Humans*, vol. 40, no. 4, pp. 825–838,2010.

17 G. Rogers, *Power System Oscillations.* Boston, MA: Kluwer Academic Publishers, 2000.

18 K. W. Cheung, J. Chow, and G. Rogers, Power System Toolbox, v 3.0. Rensselaer Polytechnic Institute and Cherry Tree Scientific Software, 2009.

19 M. D. Ilic and Q. Liu, "Toward sensing, communications and control architectures for frequency regulation in systems with highly variable resources," in *Control and Optimization Methods for Electric Smart Grids.* Springer, 2012, pp. 3–33.

20 B. Ramanathan and V. Vittal, "A framework for evaluation of advanced direct load control with minimum disruption," *IEEE Transactions on Power Systems*, vol. 23, no. 4, pp. 1681–1688, 2008.

21 Z. Ma, D. Callaway, and I. Hiskens, "Decentralized charging control for large populations of plug-in electric vehicles," in *Proceedings of IEEE Conference on Decision and Control (CDC)*, Atlanta, GA, USA, 2010, pp. 206–212.

22 A. Kiani andA. Annaswamy, "A hierarchical transactive control architecture for renewables integration in smart grids," in *Proceedings of IEEE Conference on Decision and Control (CDC)*, Maui, Hi, USA, 2012, pp. 4985–4990.

23 M. Fahrioglu and F. L. Alvarado, "Designing incentive compatible contracts for effective demand management," *IEEE Transactions on Power Systems*, vol. 15, no. 4, pp. 1255–1260, 2000.

24 P. Samadi *et al.*, "Optimal real-time pricing algorithm based on utility maximization for smart grid," in *Proceedings of IEEE SmartGridComm*, Gaithersburg, MD, USA, 2010, pp. 415–420.

25 S. P. Boyd and L. Vandenberghe, *Convex Optimization*. Cambridge University Press, 2004.

26 W. Rudin, *Principles of Mathematical Analysis*, 3rd ed. New York, NY: McGraw-Hill, 1976.

27 H. K. Khalil, *Nonlinear Systems*, 2nd ed. Upper Saddle River, NJ: Prentice Hall, Inc., 2002.

28 V. Mieghem, *Graph Spectra for Complex Networks*. Cambridge University Press, 2011.

29 K. J. Arrow *et al.*, *Studies in Linear and Non-Linear Programming*. Stanford University Press, 1958.

30 D. Feijer and F. Paganini, "Stability of primal–dual gradient dynamics and applications to network optimization," *Automatica*, vol. 46, no. 12, pp. 1974–1981, 2010.

31 A. Rantzer, "Dynamic dual decomposition for distributed control," in *Proceedings of American Control Conference (ACC)*, St. Louis, MO, USA, 2009, pp. 884–888.

32 C. Zhao *et al.*, "Power system dynamics as primal-dual algorithm for optimal load control," *arXiv:1305.0585v1*, 2013.

33 P. J. Douglass *et al.*, "Smart demand for frequency regulation: Experimental results," IEEE Transactions on Smart Grid, vol. 4, no. 3, pp. 1713–1720, 2013.

4

Physical Restrictions: Input Saturation in Secondary Frequency Control

The previous chapter establishes the connection between control and optimization of power systems with a particular interest in distributed primary frequency control. However, a practical control system inevitably suffers from various physical restrictions due to operational constraints that should be satisfied in steady-state, transient, or both. So how to cope with such constraints appears to be critical for applications in practice. This chapter considers the decentralized optimal secondary frequency control in multi-area power systems with regulation capacity limits of individual control areas. Unlike the previous chapter, both generators and controllable loads correspond to recover the nominal frequency while minimizing regulation cost. Specifically, this chapter considers the per-node balance case, where the tie-line flows are restored after disturbance, and each control area realizes power balance by itself. To this end, we derive a fully decentralized strategy capable of adapting to unknown load/generation disturbance without communication even between control areas. The major challenge is that the regulation capacity constraints enforce input saturation in the control system, which can undermine the optimality and stability of the closed-loop systems. To address this problem, we follow and extend the idea presented in Chapter 3 to the constrained secondary frequency problem. It is shown that the resulted decentralized optimal control strategy can still carry out (approximate) primal–dual updates for solving the associated centralized frequency optimization problem. However, the enforcement of input saturation significantly complicates the control design and the convergence proof due to the induced nonsmooth dynamics. This chapter shows that combining the projection technique with LaSalle's invariance principle can lead to a constructive proof of convergence.

Merging Optimization and Control in Power Systems: Physical and Cyber Restrictions in Distributed Frequency Control and Beyond, First Edition. Feng Liu, Zhaojian Wang, Changhong Zhao, and Peng Yang.

4.1 Background

In a modern large-scale power system, multiple regional grids are usually interconnected for improving operational reliability and economic efficiency [1, 2]. In each control area, power generation and controllable load can be utilized to eliminate power imbalance and maintain frequency stability in real time. Generally, frequency control is a paid service, and hence control areas always try to minimize their control cost. As different control areas may belong to different utilities and global information may not be accessible due to privacy and operational considerations, a distributed strategy is desirable. Roughly speaking, there are two possible modes of operation. In the first mode, each node area balances its own supply and demand after a disturbance. Then the power flow on each tie line should be regulated to its scheduled value, i.e. the deviation of power flows on the tie lines is eliminated in equilibrium. In the second mode, all nodes cooperate to rebalance power over the entire network after a disturbance. The power flows on the tie lines may deviate from their scheduled values but must satisfy line limits in equilibrium. We refer to the first case as per-node (power) balance and the second network (power) balance. Here we focus on the first case, while the second case will be addressed in Chapter 5. Specifically, this chapter shows how to design a decentralized optimal frequency controller for restoring frequency and tie-line power while satisfying regulation capacity constraints in transient.

Different distributed strategies have been developed in the literature for frequency control. They can roughly be divided into two categories in terms of different types of regulation resources: the automatic generation control (AGC), e.g. [3–9], and the load-side frequency control, e.g. [10–16]. The former focuses on generation regulation. For example, in [4] a flatness-based control combining trajectory generation and trajectory tracking is proposed for AGC in multi-area power systems. In [5, 7], the closed-loop system composed of power system dynamics and controller dynamics is formulated as a port-Hamiltonian system, and its stability is proved. In [6], generators are driven by AGC to restore frequencies. Correspondence between the (partial) primal–dual gradient algorithm for solving the associated optimization problem and the frequency control dynamics of the physical system is established. The resulting decomposition enables the system design of a fully distributed optimal frequency control.

For the load-side frequency control, load frequency dynamics are formulated similarly to the generator model in [10], leading to a distributed frequency control for both generation and controllable loads. A distributed adaptive control is presented in [11] to guarantee acceptable frequency deviation from the nominal value. In [12, 13], an optimal load control (OLC) problem is formulated, and a ubiquitous primary load-side control is derived as a partial primal–dual gradient algorithm for solving the OLC problem. It is decentralized but does not restore the nominal frequency. This design approach is extended in [14] to secondary control that restores nominal frequency and scheduled inter-area flows and enforces line limits in equilibrium. It is further extended to more general models in [15, 16], where a passivity condition guaranteeing stability is proposed for each local bus, and the conservativeness is significantly reduced.

In terms of methodology, there are mainly three types of distributed frequency control: the droop-based approach, e.g. [17, 18]; the consensus-based approach, e.g. [19–21]; and the primal–dual decomposition-based approach, e.g. [6, 12–14]. In the primal–dual decomposition approach, control goals such as rebalancing power after a disturbance and restoring nominal frequency and scheduled inter-area flows are formalized as a global optimization problem. The feedback control laws are designed so that the equilibrium of the closed-loop system solves the optimization problem and achieves the control objectives in equilibrium. Moreover, the closed-loop system is designed to be an asymptotically stable primal–dual algorithm for solving an associated optimization problem. This approach is mathematically similar to that taken in [22] where the real-time control is through nodal prices.

In all the primal–dual algorithms proposed in the literature, even though input constraints are usually enforced, constraints on states, such as power injections on buses, are enforced only in steady-state. In practice, however, a control area always maintains regulation capacity bounds that constrain power generations and controllable loads within available ranges, not only at equilibrium but also in transient. In this chapter, we design input-saturation controllers that maintain these capacity constraints during transient as well. We show that these controllers still carry out primal–dual updates of the associated optimization problem.

Specifically, this chapter investigates the approach to rebalancing power after a disturbance with the focus on the per-node power balance case, which means the power disturbance in each control area to be balanced by generations and controllable loads within that area. To this end, we construct a fully decentralized control to recover nominal frequencies and tie-line power flows. The regulation capacity constraints are also enforced in transient. We show that the controller and the physical dynamics serve as primal–dual updates with saturation for solving the optimization problem. Then we prove the optimality of our control by exploiting the equivalence between the equilibrium of the closed-loop frequency control system and the optimal solution of the optimization problem. We also show that the optimal solution to the primal–dual problem and equilibrium point of the closed-loop systems are both unique. Furthermore, we prove the stability of the closed-loop system by combining the projection technique with LaSalle's invariance principle, coping with the impact of nonsmooth dynamics created by the imposed transient constraints.

The remainder of this chapter is organized as follows. In Section 4.2, we introduce our network model. Section 4.3 formulates the optimal frequency control problem, presents our controller and its relationship with the primal–dual update, and proves the optimality, uniqueness, and stability of the closed-loop equilibrium point. We confirm the performance of controllers via simulations on a detailed power system model in Section 4.6. Section 4.7 concludes this chapter with notes.

4.2 Power System Model

A large-scale power network usually comprises multiple control areas, each with its own generators and loads. These control areas are interconnected with each other through tie lines. For simplicity, we treat each control area as a node with an aggregate power generation, an aggregate controllable load, and an aggregate uncontrollable load.[1] Then the power network is modeled by a graph $\mathcal{G} := (\mathcal{N}, \mathcal{E})$ where $\mathcal{N} = \{0, 1, 2, \dots n\}$ is the set of nodes (control areas) and $\mathcal{E} \subseteq \mathcal{N} \times \mathcal{N}$ is the set of edges (tie lines). If a pair of nodes i and j are connected by a tie line directly, we denote the tie line by $(i, j) \in \mathcal{E}$. Let $m := |\mathcal{E}|$ denote the number of tie lines. We treat $\mathcal{G}$ as directed with an arbitrary orientation, and we use $(i, j) \in \mathcal{E}$ or $i \to j$ interchangeably to denote a directed edge from i to j. It should be clear from the context which is the case. Without loss of generality, throughout this chapter we assume the graph is connected, and node 0 is the reference node.

We adopt a second-order linearized model to describe the frequency dynamics of each node and two first-order inertia equations to describe the dynamics of power generation regulation and load regulation at each node. We assume the tie lines are lossless and adopt the DC power flow model. Then for each node $j \in \mathcal{N}$,

$$\dot{\theta}_j = \omega_j(t) \tag{4.1a}$$

$$\begin{aligned} M_j \dot{\omega}_j = {} & P_j^g(t) - P_j^l(t) - p_j - D_j \omega_j(t) \\ & + \sum_{i: i \to j} B_{ij}(\theta_i(t) - \theta_j(t)) - \sum_{k: j \to k} B_{jk}(\theta_j(t) - \theta_k(t)) \end{aligned} \tag{4.1b}$$

$$T_j^g \dot{P}_j^g = -P_j^g(t) + u_j^g(t) - \frac{\omega_j(t)}{R_j} \tag{4.1c}$$

$$T_j^l \dot{P}_j^l = -P_j^l(t) + u_j^l(t) \tag{4.1d}$$

1 In our study, each of the nodes can be regarded as a control area, including controllable generation and load. All controllable generations in the same control area are aggregated into one equivalent generator, while all controllable loads are aggregated into one controllable load.

Let $x := (\theta, \omega, P^g, P^l)$ denote the state of the network and $u := (u^g, u^l)$ denote the control.[2]

Our goal is to design feedback control laws for the generation command $u^g(x(t))$ and load control $u^l(x(t))$. The operational constraints are

$$\underline{P}_j^g \leq P_j^g(t) \leq \overline{P}_j^g, \quad j \in \mathcal{N} \tag{4.2a}$$

$$\underline{P}_j^l \leq P_j^l(t) \leq \overline{P}_j^l, \quad j \in \mathcal{N} \tag{4.2b}$$

Differing from the literature, here (4.2a) and (4.2b) are *hard limits* on the regulation capacities of generation and controllable load at each node, which should not be violated at any time, even during transient states. Hence we will design controllers so that these constraints are satisfied not only at equilibrium, but also during transient.

We assume that the system operates in a steady state initially, i.e. the generation and the load are balanced, and the frequency is at its nominal value. All variables represent *deviations* from their nominal or scheduled values so that, e.g. $\omega_j(t) = 0$ means the frequency is at its nominal value.

As the generation P_j^g and load P_j^l in each area can increase or decrease, and a line flow P_{ij} can in either direction, we make the following assumption.

Assumption 4.1 The upper and lower bounds of P_j^g, P_j^l satisfy $\underline{P}_j^g < 0 < \overline{P}_j^g$ and $\underline{P}_j^l < 0 < \overline{P}_j^l$ for $\forall j \in \mathcal{N}$.

Assumption 4.2 The angle of reference bus is $\theta_0(t) := 0$ for all $t \geq 0$.

The assumption $\theta_0 \equiv 0$ amounts to using $\theta_0(t)$ as reference angles. It is made merely for notational convenience: as we will see, the equilibrium point is unique with this assumption (or unique up to reference angles without this assumption).

4.3 Control Design for Per-Node Power Balance

In this chapter, we are particularly interested in the per-node power balance case, which is modeled by the requirement:

$$P_j^g = P_j^l + p_j, \quad j \in \mathcal{N} \tag{4.3}$$

---

2 Given a collection of x_i for i in a certain set A, x denotes the column vector $x := (x_i, i \in A)$ of a proper dimension with x_i as its components.

4.3.1 Control Goals

The control goals are formalized as an optimization problem:

$$\text{PBO: } \min \quad \frac{1}{2}\sum_{j\in\mathcal{N}} \alpha_j \left(P_j^g\right)^2 + \frac{1}{2}\sum_{j\in\mathcal{N}} \beta_j \left(P_j^l\right)^2 + \frac{1}{2}\sum_{j\in\mathcal{N}} D_j \omega_j^2 \tag{4.4a}$$

$$\text{over} \quad x := (\theta, \omega, P^g, P^l) \text{ and } u := (u^g, u^l)$$

$$\text{s. t.} \quad (4.2), (4.3)$$

$$P_j^g = P_j^l + p_j + U_j(\theta, \omega_j), \quad j \in \mathcal{N} \tag{4.4b}$$

$$P_j^g = u_j^g, \quad j \in \mathcal{N} \tag{4.4c}$$

$$P_j^l = u_j^l, \quad j \in \mathcal{N} \tag{4.4d}$$

where

$$U_j(\theta, \omega_j) := D_j\omega_j - \sum_{i:i\to j} B_{ij}\theta_{ij} + \sum_{k:j\to k} B_{jk}\theta_{jk}$$

Here we have abused notation and use $\tilde{\theta}_{ij} := \theta_i - \theta_j$. In vector form, we have

$$U(\theta, \omega) := D\omega + CBC^{\mathrm{T}}\theta \tag{4.5}$$

where

$$D := \operatorname{diag}(D_i, i \in \mathcal{N}),$$
$$B := \operatorname{diag}(B_{ij}, (i,j) \in \mathcal{E}),$$

and C is the $(n+1) \times m$ incidence matrix.

We comment on the optimization problem (4.4).

Remark 4.1 ***(Control Goals)*** The following remarks are helpful for a better understanding of the goals of control:

1) Since the variables are deviations from their nominal values, the parameters (α_j, β_j) in the objective function (4.4a) are not electricity costs. Minimizing the objective aims to track generation and consumption that have been scheduled at a slower timescale, e.g. to optimize economic efficiency or user utility. The parameters (α_j, β_j, D_j) weigh the relative costs of deviating from scheduled generation and load and the nominal frequency. In the next subsection, we show that, for every optimal solution, the corresponding frequency deviation must be zero, provided a feasible solution exists.
2) For the definition of (4.4), the regulation capacity limits (4.2) apply only at optimality. As we will see below, our controller, however, enforces (4.2) even during transient.

3) The per-node balance requirement (4.3) and the constraint (4.4b) imply $U(\theta, \omega) = 0$ at any feasible x. This will drive the power flow on *every* tie line to its scheduled value, i.e. $P^*_{ij} = 0$ in equilibrium (see Theorem 4.2), even though this is not included in (4.4) as a constraint.
4) The constraints (4.4c) and (4.4d) require that, at optimality, the power injection P^g_j and controllable load P^l_j are equal to their control commands u^g_j and u^l_j, respectively.

In the rest of this chapter, we make one of the following assumptions (recall that $(\underline{P}^g, \underline{P}^l) < 0 < (\overline{P}^g, \overline{P}^l)$ under Assumption 4.1).

Assumption 4.3 The PBO problem (4.4) is feasible, i.e.

$$\underline{P}^g_j - \overline{P}^l_j \leq p_j \leq \overline{P}^g_j - \underline{P}^l_j, \quad \forall j \in \mathcal{N}$$

Moreover (4.4) has a finite optimal solution.

Note that the feasibility of (4.4) is equivalent to the inequalities in Assumption 4.2 because the per-node balance constraint (4.3) requires $p = P^g - P^l$ in equilibrium. In what follows below, we sometimes strengthen the inequalities in Assumption 4.2 to strict inequalities. Strict inequalities mean that each area has a certain power margin. If there is no margin, the system may have no feasible solution after a small load disturbance. For example, if $\overline{P}^g_j - \underline{P}^l_j = p_j$ for any area j, then any feasible solution must have $P^g_j = \overline{P}^g_j$ and $P^l_j = \underline{P}^l_j$, i.e. there is no more regulation capacity in area j so that, if the load p_j further increases, then the frequency will drop and cannot be restored.

4.3.2 Decentralized Optimal Controller

Our control laws for u^g and u^l are as follows: for each node $j \in \mathcal{N}$,

$$\dot{\lambda}_j = \gamma^\lambda_j \left(P^g_j(t) - P^l_j(t) - p_j \right) \tag{4.6a}$$

$$u^g_j(t) = \left[P^g_j(t) - \gamma^g_j \left(\alpha_j P^g_j(t) + \omega_j(t) + \lambda_j(t) \right) \right]^{\overline{P}^g_j}_{\underline{P}^g_j} + \frac{\omega_j(t)}{R_j} \tag{4.6b}$$

$$u^l_j(t) = \left[P^l_j(t) - \gamma^l_j \left(\beta_j P^l_j(t) - \omega_j(t) - \lambda_j(t) \right) \right]^{\overline{P}^l_j}_{\underline{P}^l_j} \tag{4.6c}$$

For any $x_i, a_i, b_i \in \mathbb{R}$ with $a_i \leq b_i$, we denote

$$[x_i]^{b_i}_{a_i} := \min\{b_i, \max\{a_i, x_i\}\}.$$

For vectors x, a, b, $[x]^b_a$ is defined accordingly component wise.

It is interesting that the controller (4.6) has a simple proportional–integral (PI) structure with saturation. Moreover, it is *completely decentralized* where each node j updates its internal state $\lambda_j(t)$ in (4.6a) based only on the generation $P_j^g(t)$, the controllable load $P_j^l(t)$, and the uncontrolled load p_j that are all local at j (within a control area). The control inputs $u_j^g(t)$ and $u_j^l(t)$ in (4.6b) and (4.6c) are then static functions of the local state $(P_j^g(t), P_j^l(t), \omega_j(t))$ and the internal state $\lambda_j(t)$. Therefore, no communication is required even between neighboring nodes.

We often write u_j^g and u_j^l as functions of $(P_j^g, P_j^l, \omega_j, \lambda_j)$:

$$u_j^g(t) := u_j^g\left(P_j^g(t), \omega_j(t), \lambda_j(t)\right) \tag{4.7a}$$

$$u_j^l(t) := u_j^l\left(P_j^l(t), \omega_j(t), \lambda_j(t)\right) \tag{4.7b}$$

for $j \in \mathcal{N}$, where these functions are given by the right-hand side of (4.6b) and (4.6c), respectively. We now comment on measurements required to implement the controller (4.6).

Remark 4.2 ***(Implementation)*** The variable $\lambda_j(t)$ in (4.6a) is a cyber quantity that is computed at each node j based on $(P_j^g(t), P_j^l(t), p_j)$ locally at j (within a control area). These quantities can in principle be measured at j. We would however like to avoid measuring the uncontrolled load change p_j for ease of implementation. To this end, let

$$\Delta P_j(t) := P_j^g(t) - P_j^l(t) - p_j \quad j \in \mathcal{N}$$

denote the surplus generation at node j. We then have from (4.1b) and (4.5) that

$$\Delta P_j(t) = M_j\dot{\omega}_j + U_j(\theta, \omega_j(t))$$

Since $\dot{\lambda}_j = \gamma_j^\lambda \Delta P_j(t)$, (4.6a) becomes

$$\dot{\lambda}_j = \gamma_j^\lambda M_j\dot{\omega}_j + \gamma_j^\lambda D_j\omega_j(t) - \gamma_j^\lambda \left(\sum_{i:i\to j} P_{ij}(t) - \sum_{k:j\to k} P_{jk}(t) \right)$$

where $P_{ij}(t) := B_{ij}(\theta_i(t) - \theta_j(t))$ are the tie-line flows from nodes i to j according to the DC power flow model. Hence, to update the internal state $\lambda_j(t)$, we only need to measure the local frequency deviation $\omega_j(t)$, its derivative $\dot{\omega}_j(t)$, and the tie-line flows $P_{ij}(t)$ incident on node j, and not the uncontrolled load p_j in area j. An important advantage is that the controller naturally adapts to unknown load changes p_j. This feature will be illustrated in the section of case studies.

The control inputs $u_j^g(t)$ and $u_j^l(t)$ in (4.7) can then be implemented using measurements of the local generation $P_j^g(t)$, controlled load $P_j^l(t)$, frequency deviation $\omega_j(t)$, and tie-line powers $P_{ij}(t), P_{jk}(t)$.

Since p_j is difficult to measure, especially with large penetration of renewable generations, one cannot solve the problem (4.4) locally to obtain controllers. The proposed controller naturally adapts to unknown load changes.

4.3.3 Design Rationale

The controller design (4.6) is motivated by an approximate primal–dual algorithm for (4.4). We first review the form of a standard primal–dual algorithm and then explain that the closed-loop dynamics (4.1) and (4.6) carry out an approximate version for (4.4) in real time over the closed-loop system.

4.3.3.1 Primal–Dual Algorithms

Consider a general constrained convex optimization:

$$\min_{x \in X} \; f(x) \qquad s.t. \quad g(x) = 0$$

where $f : \mathbb{R}^n \to \mathbb{R}$, $g : \mathbb{R}^n \to \mathbb{R}^k$, and $X \subseteq \mathbb{R}^n$ is closed and convex. Let $\rho \in \mathbb{R}^k$ be the Lagrange multiplier associated with the equality constraint $g(x) = 0$. Define the Lagrangian $L(x; \rho) := f(x) + \rho^{\mathrm{T}} g(x)$. A standard primal–dual algorithm takes the form

$$x(t+1) := \mathcal{P}_X\left(x(t) - \Gamma^x \, \nabla_x L(x(t); \rho(t))\right) \tag{4.8a}$$

$$\rho(t+1) := \rho(t) + \Gamma^\rho \, \nabla_\rho L(x(t); \rho(t)) \tag{4.8b}$$

where $\mathcal{P}_X(a)$ projects $a \in \mathbb{R}^n$ to the closest point in X under the Euclidean norm and the gain matrices Γ^x, Γ^ρ are (strictly) positive definite. Hence the iterate $(x(t), \rho(t))$ stays in the set $X \times \mathbb{R}^k$ for all t and, under appropriate assumptions, converges to a primal–dual optimal point.

In contrast a standard dual algorithm takes the form

$$\rho(t+1) := \rho(t) + \Gamma^\rho \, \nabla_\rho L(x(t); \rho(t)) \tag{4.9a}$$

$$x(t) := \min_{x \in X} \; L(x; \rho(t)) \tag{4.9b}$$

As we will see below, almost all primal variables in $x(t)$ are updated according to (4.8a) except $\omega(t)$ that is updated according to (4.9a).

4.3.3.2 Design of Controller (4.6)

Let λ and μ be the Lagrange multipliers associated with constraints (4.3) and (4.4b), respectively, and let $\rho := (\lambda, \mu)$. Define the Lagrangian of (4.4) as

$$L_1(x; \rho) = \frac{1}{2} \sum_{j \in \mathcal{N}} \alpha_j \left(P_j^g\right)^2 + \frac{1}{2} \sum_{j \in \mathcal{N}} \beta_j \left(P_j^l\right)^2$$

$$+\frac{1}{2}\sum_{j\in\mathcal{N}} D_j\omega_j^2 + \sum_{j\in\mathcal{N}} \lambda_j \left(P_j^g - P_j^l - p_j\right)$$

$$+\sum_{j\in\mathcal{N}} \mu_j \left(P_j^g - P_j^l - p_j - D_j\omega_j + \sum_{i:i\to j} B_{ij}\theta_{ij} - \sum_{k:j\to k} B_{jk}\theta_{jk}\right) \tag{4.10}$$

The Lagrangian is defined to be only a function of (x, ρ) and independent of $u := (u^g, u^l)$ as we treat u as a function of (x, ρ) defined by the right-hand side of (4.6b) and (4.6c). The set X in (4.8a) is defined by the constraints (4.2):

$$X := \left\{(P^g, P^l) \, : \, (\underline{P}^g, \underline{P}^l) \leq (P^g, P^l) \leq (\overline{P}^g, \overline{P}^l)\right\} \tag{4.11}$$

We now explain how the closed-loop system (4.1) and (4.6) implements an approximate primal–dual algorithm for solving (4.4) in real time. We first show that the control (4.6a) and the swing dynamic (4.1b) implement the dual update (4.8b) on dual variables $\rho = (\lambda(t), \mu(t))$. We then show that (4.1a), (4.1c), and (4.1d) implement a mix of the primal updates (4.8a) and (4.9b) on the primal variables $x = (\theta(t), \omega(t), P^g(t), P^l(t))$.

First the variable λ is the Lagrange multiplier vector for the per-node power balance constraint (4.3). The control law (4.6a) implements part of the dual update (4.8b) in continuous time:

$$\dot{\lambda} = \Gamma^{\lambda} \; \nabla_{\lambda} L_1(x(t), \rho(t)) \tag{4.12a}$$

where $\Gamma^{\lambda} := \mathrm{diag}(\gamma_j^{\lambda}, j \in \mathcal{N})$.

The variable μ is the Lagrange multiplier vector for the constraint (4.4b). It can be identified with the frequency deviation ω as the KKT condition [23]

$$\frac{\partial L_1}{\partial \omega_j}(x^*, \rho^*) = D_j(\omega_j^* - \mu_j^*) = 0$$

implies $\mu_j^* = \omega_j^*$ at optimality since $D_j > 0$. Moreover we can identify $\mu(t) \equiv \omega(t)$ during transient if we update the cyber quantity $\mu(t)$ according to

$$\begin{aligned}\dot{\mu} &= M^{-1} \left(P^g(t) - P^l(t) - p_j(t) - U(\theta(t), \omega(t))\right)\\ &= M^{-1} \; \nabla_{\mu} L_1(x(t), \rho(t))\end{aligned} \tag{4.12b}$$

where $M := \mathrm{diag}(M_j, j \in \mathcal{N})$. Then μ and ω have the same dynamics (compared with (4.1b)) and hence $\mu(t) \equiv \omega(t)$ as long as $\mu(0) = \omega(0)$. Therefore the swing dynamic (4.1b) is equivalent to (4.12b) and carries out the dual update (4.8b) on μ when we take $\mu(t) \equiv \omega(t)$.

Second, to see how (4.1a), (4.1c), and (4.1d) implement the primal updates, note that the last term in the definition (4.10) of the Lagrangian L_1 is

$$\sum_{j\in\mathcal{N}} \mu_j \left(\sum_{i:i\to j} B_{ij}\theta_{ij} - \sum_{k:j\to k} B_{jk}\theta_{jk}\right)$$

$$= -\sum_{(i,j)\in\mathcal{E}} B_{ij}\left(\mu_i - \mu_j\right)\left(\theta_i - \theta_j\right)$$
$$= -\mu^{\mathrm{T}} C B C^{\mathrm{T}} \theta$$

We fix $\theta_0 := 0$ to be the reference angle. Then there is a bijection between θ and $\tilde{\theta}$ that is in the column space of C^{T}, given by $\tilde{\theta} = C^{\mathrm{T}}\theta$. Hence we can work with either variable. For stability proof we use $\tilde{\theta}$. In vector form,

$$L_1 = \frac{1}{2}\left((P^g)^{\mathrm{T}} A^g P^g + (P^l)^{\mathrm{T}} A^l P^l + \omega^{\mathrm{T}} D\omega\right)$$
$$+ \lambda^{\mathrm{T}}\left(P^g - P^l - p\right) + \mu^{\mathrm{T}}\left(P^g - P^l - p - D\omega - CB\tilde{\theta}\right)$$

where $A^g := \mathrm{diag}(\alpha_j, j \in \mathcal{N})$, $A^l := \mathrm{diag}(\beta_j, j \in \mathcal{N})$, $B := \mathrm{diag}(B_{ij}, (i,j) \in \mathcal{E})$, and

$$\nabla_{\tilde{\theta}} L_1 = -BC^{\mathrm{T}}\mu = -BC^{\mathrm{T}}\omega$$

Since $\dot{\tilde{\theta}} = C^{\mathrm{T}}\dot{\theta} = C^{\mathrm{T}}\omega$, we have

$$\dot{\tilde{\theta}} = -B^{-1}\nabla_{\tilde{\theta}} L_1 \tag{4.13a}$$

i.e. (4.1a) implements the primal update (4.8a) on $\tilde{\theta}$.

Identification of $\omega(t)$ with $\mu(t)$ means that, given the dual variable $\rho(t)$, we update $\omega(t)$ as in the dual algorithm (4.9b):

$$\omega(t) = \mu(t) = \arg\min_{\omega} \nabla_{\omega} L_1(x, \rho(t)) \tag{4.13b}$$

instead of (4.8a). Moreover we have

$$\nabla_{P^g} L_1(x(t), \rho(t)) = A^g P^g(t) + \omega(t) + \lambda(t)$$

Therefore the control law (4.6b) is equivalent to

$$u^g(t) = \left[P^g(t) - \Gamma^g \nabla_{P^g} L_1(x(t), \rho(t))\right]_{\underline{P}^g}^{\overline{P}^g} + R^{-1}\omega(t)$$

where $\Gamma^g := \mathrm{diag}(\gamma_j^g, j \in \mathcal{N})$ and $R := \mathrm{diag}(R_j, j \in \mathcal{N})$. Then the generation dynamic (4.1c) becomes

$$T^g \dot{P}^g = \left[P^g(t) - \Gamma^g\, \nabla_{P^g} L_1(x(t), \rho(t))\right]_{\underline{P}^g}^{\overline{P}^g} - P^g(t) \tag{4.13c}$$

where $T^g := \mathrm{diag}(T_j^g, j \in \mathcal{N})$. Similarly the control law (4.6c) is equivalent to

$$u^l(t) = \left[P^l(t) - \Gamma^l \nabla_{P^l} L_1(x(t), \rho(t))\right]_{\underline{P}^l}^{\overline{P}^l}$$

where $\Gamma^l := \mathrm{diag}(\gamma_j^l, j \in \mathcal{N})$. The controllable load dynamic (4.1d) is equivalent to

$$T^l \dot{P}^l = \left[P^l(t) - \Gamma^l \nabla_{P^l} L_1(x(t), \rho(t))\right]_{\underline{P}^l}^{\overline{P}^l} - P^l(t) \tag{4.13d}$$

where $T^l := \mathrm{diag}(T_j^l, j \in \mathcal{N})$.

Writing $P := (P^g, P^l)$, $T^{gl} := \text{diag}(T^g, T^l)$, and $\Gamma^{gl} := \text{diag}(\Gamma^g, \Gamma^l)$, the dynamics (4.13c) and (4.13d) become

$$T^{gl}\dot{P} = \mathcal{P}_X \left(P(t) - \Gamma^{gl}\nabla_P L_1(x(t), \rho(t))\right) - P(t) \tag{4.13e}$$

where X is defined in (4.11). Informally (4.13e) can be interpreted as a continuous-time version of the primal update (4.8a) since the right-hand side can be interpreted as $P(t+1) - P(t)$ in the discrete-time version (4.8a). While it is clear from (4.8a) that $P(t)$ in the discrete-time formulation stays in X for all t, it may not be obvious that $P(t)$ in the continuous-time formulation (4.13e) stays in X for all t. This is proved formally in Lemma 4.3.

In summary the closed-loop system (4.1) and (4.6) carries out an approximate primal–dual algorithm (4.8) in continuous time. The dual updates (4.12a) and (4.12b) on $(\lambda(t), \mu(t))$ are implemented by (4.6a) and (4.1b), respectively. The primal updates (4.13a) and (4.13e) on $(\theta(t), P^g(t), P^l(t))$ are implemented by (4.1a), (4.1c), and (4.1d), respectively. We refer to this as an *approximate* primal–dual algorithm because the identification of $\omega(t) \equiv \mu(t)$ implements the update (4.9b) on $\omega(t)$ instead of (4.8a).

4.4 Optimality and Uniqueness of Equilibrium

In this subsection, we address the optimality of the equilibrium point of the closed-loop system (4.1) and (4.6). Given an $(x, \rho) := (\theta, \omega, P^g, P^l, \lambda, \mu)$, recall that the control input $u(x, \rho)$ is given by (4.7).

Definition 4.1 A point $(x^*, \rho^*) := (\theta^*, \omega^*, P^{g*}, P^{l*}, \lambda^*, \mu^*)$ is an *equilibrium* of the closed-loop system (4.1) and (4.6) if the following conditions are met:

1) The right-hand side of (4.1) vanishes at x^* and $u(x^*, \rho^*)$.
2) The right-hand side of (4.6a) vanishes at (x^*, ρ^*).

Definition 4.2 A point (x^*, ρ^*) is *primal–dual optimal* if $(x^*, u(x^*, \rho^*))$ is optimal for (4.4) and ρ^* is optimal for its dual problem.

Section 4.3.3 says that the closed-loop system (4.1) and (4.6) carries out an (approximate) primal–dual algorithm in real time to solve (4.4). In this subsection we prove that a point (x^*, ρ^*) is an equilibrium of the closed-loop system if and only if it is primal–dual optimal. Moreover the equilibrium is unique. In the next subsection we prove that the closed-loop system converges to an equilibrium point starting from any initial point that satisfies constraint (4.2).

Theorem 4.1 *Suppose Assumption 4.3 holds. A point (x^*, ρ^*) is primal–dual optimal if and only if (x^*, ρ^*) is an equilibrium of the closed-loop system* (4.1) *and* (4.6) *that satisfies* (4.2) *and $\mu^* = 0$.*

Theorem 4.1 reveals the equivalence between the equilibrium of the closed-loop system and the primal–dual optimal solution. It also implies that, in equilibrium, the per-node power balance (4.3) is achieved and constraints (4.2) are satisfied. The next theorem shows that the equilibrium point is almost unique and has a simple and intuitive structure.

Theorem 4.2 *Suppose Assumptions 4.1–4.3 hold. Let (x^*, ρ^*) be primal–dual optimal:*

1) *x^* and μ^* are unique, with θ^* being unique up to an (equilibrium) reference angle θ_0^*.*
2) *λ^* is also unique if strict inequalities hold in Assumption 4.2. In that case, λ_j^* equals the (negative of the) marginal generation/load regulation cost at node j, i.e. $\alpha_j P_j^{g*} = -\beta_j P_j^{l*} = -\lambda_j^*$.*
3) *Nominal frequencies are restored, i.e. $\omega_j^* = 0$ for all $j \in \mathcal{N}$; moreover $\theta_j^* = \theta_0^*$ for all $j \in \mathcal{N}$.*
4) *The power flow $P_{ij}^* := B_{ij}(\theta_i^* - \theta_j^*) = 0$ on every line $(i, j) \in \mathcal{E}$.*

The rest of this subsection is devoted to the proof of Theorems 4.1 and 4.2. We start with the following two lemmas.

Lemma 4.1 *Suppose (x^*, u^*) is optimal for* (4.4)*. Then $\omega^* = 0$ and $\theta^* = \theta_0^* \mathbf{1}$ where $\mathbf{1}$ is the vector with all entries being 1.*

Proof: Suppose for the sake of contradiction that $\omega^* \neq 0$. Construct from x^* another point $\hat{x}$ by setting $\hat{\theta} = 0$, $\hat{\omega} = 0$ and keeping the other components of x^* unchanged. Since x^* satisfies both (4.3) and (4.4b), we must have $U(\theta^*, \omega^*) = 0$. This also holds for $\hat{x}$, i.e. $U(\hat{\theta}, \hat{\omega}) = 0$, and hence $(\hat{x}, u^*)$ remains feasible since other components of $\hat{x}$ are the same as those of x^*. Moreover $(\hat{x}, u^*)$ has a strictly lower cost than (x^*, u^*), contradicting the optimality of (x^*, u^*). Hence any optimal (x^*, u^*) must have $\omega^* = 0$.

We claim that $\omega^* = 0$ implies that $\theta^* = \theta_0^* \mathbf{1}$. For any feasible point x, (4.3) and (4.4b) imply that

$$U(\theta, \omega) = D\omega + CBC^{\mathrm{T}}\theta = 0$$

Hence we have $CBC^{\mathrm{T}}\theta^* = 0$ at an optimal x^*. Since CBC^{T} is an $(n+1) \times (n+1)$ matrix with rank n, its null space has dimension 1. The vector $\mathbf{1}$ is in its null space because $C^{\mathrm{T}}\mathbf{1} = 0$. Hence $\theta^* = \theta_0^* \mathbf{1}$. ■

Lemma 4.1 implies that the frequencies are restored at the optimality. Noticing that $\theta^* = \theta_0^* \mathbf{1}$ means all angles are equal, such an optimal solution also implies that all the tie-line power flows are restored to their original values. At the optimal solution, we have the following results.

Lemma 4.2 *Suppose* (x^*, ρ^*) *is primal–dual optimal. Then*

$$u_j^{g*} = P_j^{g*} = \left[P_j^{g*} - \gamma_j^g \left(\alpha_j P_j^{g*} + \omega_j^* + \lambda_j^*\right)\right]_{\underline{P}_j^g}^{\overline{P}_j^g}$$

$$u_j^{l*} = P_j^{l*} = \left[P_j^{l*} - \gamma_j^l \left(\beta_j P_j^{l*} - \omega_j^* - \lambda_j^*\right)\right]_{\underline{P}_j^l}^{\overline{P}_j^l}$$

for any $\gamma_j^g > 0$ *and* $\gamma_j^l > 0$.

Proof: Since (4.4) is convex with linear constraints, strong duality holds. Hence (x^*, ρ^*) is a primal–dual optimal if and only if it satisfies the KKT condition: $(x^*, u(x^*, \rho^*))$ is primal feasible and

$$x^* = \arg\min_x \left\{L_1(x; \rho^*) | \ (x, u(x, \rho^*)) \text{ satisfies (4.2), (4.4c), (4.4d)}\right\} \tag{4.14}$$

From the definition (4.10) of L_1, x^* satisfies (4.14) if and only if $(x^*, u(x^*, \rho^*))$ satisfies (4.2), (4.4c), and (4.4d) and the first-order stationarity condition, i.e. for all $j \in \mathcal{N}$, we have

$$\alpha_j P_j^{g*} + \mu_j^* + \lambda_j^* \begin{cases} \geq 0 & \text{if } P_j^{g*} = \underline{P}_j^g \\ = 0 & \text{if } \underline{P}_j^g < P_j^{g*} < \overline{P}_j^g \\ \leq 0 & \text{if } P_j^{g*} = \overline{P}_j^g \end{cases} \tag{4.15a}$$

$$\beta_j P_j^{l*} - \mu_j^* - \lambda_j^* \begin{cases} \geq 0 & \text{if } P_j^{l*} = \underline{P}_j^l \\ = 0 & \text{if } \underline{P}_j^l < P_j^{l*} < \overline{P}_j^l \\ \leq 0 & \text{if } P_j^{l*} = \overline{P}_j^l \end{cases} \tag{4.15b}$$

$$D_j(\omega_j^* - \mu_j^*) = 0 \tag{4.15c}$$

$$\sum_{i:i\to j} B_{ij}(\mu_i^* - \mu_j^*) = \sum_{k:j\to k} B_{jk}(\mu_j^* - \mu_k^*) \tag{4.15d}$$

From Lemma 4.1 we have $\omega^* = 0$, and hence (4.15) reduces to $\mu^* = \omega^* = 0$ since $D_j > 0$ and

$$\alpha_j P_j^{g*} + \omega_j^* + \lambda_j^* \begin{cases} \geq 0 & \text{if } P_j^{g*} = \underline{P}_j^g \\ = 0 & \text{if } \underline{P}_j^g < P_j^{g*} < \overline{P}_j^g \\ \leq 0 & \text{if } P_j^{g*} = \overline{P}_j^g \end{cases} \tag{4.16a}$$

$$\beta_j P_j^{l*} - \omega_j^* - \lambda_j^* \begin{cases} \geq 0 & \text{if } P_j^{l*} = \underline{P}_j^l \\ = 0 & \text{if } \underline{P}_j^l < P_j^{l*} < \overline{P}_j^l \\ \leq 0 & \text{if } P_j^{l*} = \overline{P}_j^l \end{cases} \tag{4.16b}$$

It can be checked that (4.16) is equivalent to

$$P_j^{g*} = \left[P_j^{g*} - \gamma_j^g \left(\alpha_j P_j^{g*} + \omega_j^* + \lambda_j^*\right)\right]_{\underline{P}_j^g}^{\overline{P}_j^g}$$

$$P_j^{l*} = \left[P_j^{l*} - \gamma_j^l \left(\beta_j P_j^{l*} - \omega_j^* - \lambda_j^*\right)\right]_{\underline{P}_j^l}^{\overline{P}_j^l}$$

for any $\gamma_j^g > 0$ and $\gamma_j^l > 0$. The lemma then follows from (4.4c) and (4.4d). ∎

Now we can complete the proofs of Theorems 4.1 and 4.2.

Proof: [Proof of Theorems 4.1]

Sufficiency. Suppose a point (x^*, ρ^*) is primal–dual optimal. Then x^* satisfies the operational constraints (4.2). Moreover the right-hand side of (4.1) vanishes because of the following reasons:

- $\dot{\theta} = 0$ since $\omega^* = 0$ from Lemma 4.1.
- $\dot{\omega} = 0$ because of (4.4b).
- $\dot{P}^g = \dot{P}^l = 0$ since $\omega^* = 0$ and x^* satisfies (4.4c) and (4.4d).

The right-hand side of (4.6a) vanishes because x^* satisfies per-node power balance (4.3). By Lemma 4.2 (x^*, ρ^*) satisfies (4.6b) and (4.6c). Hence (x^*, ρ^*) is an equilibrium of the closed-loop system (4.1) and (4.6) that satisfies the operational constraints (4.2). Moreover $\mu^* = \omega^* = 0$ by (4.15c) since $D_j > 0$ for all $j \in \mathcal{N}$.

Necessity. Suppose now (x^*, ρ^*) is an equilibrium of the closed-loop system (4.1) and (4.6) and satisfies (4.2) with $\mu^* = 0$. Since the optimization problem (4.4) is convex with linear constraints, (x^*, ρ^*) is a primal–dual optimal if and only if $(x^*, u(x^*, \rho^*))$ is primal feasible and satisfies (4.14) (note that $\nabla_\rho L_1(x^*, \rho^*) = 0$ since $\dot{\mu} = \dot{\lambda} = 0$).

To show that $(x^*, u(x^*, \rho^*))$ is primal feasible, note that since $(x^*, u(x^*, \rho^*))$ is an equilibrium of (4.1), it satisfies $\omega^* = 0$, (4.4b), (4.4c), and (4.4d), in addition to (4.2). Since (x^*, ρ^*) is a closed-loop equilibrium, we have $\dot{\lambda}^* \equiv 0$ in (4.6a), implying (4.3). Hence x^* is primal feasible.

To show that (x^*, ρ^*) satisfies (4.14), note that (4.4c), (4.4d) and (4.6b), (4.6c) imply that

$$P_j^{g*} = \left[P_j^{g*} - \gamma_j^g \left(\alpha_j P_j^{g*} + \omega_j^* + \lambda_j^*\right)\right]_{\underline{P}_j^g}^{\overline{P}_j^g}$$

$$P_j^{l*} = \left[P_j^{l*} - \gamma_j^l \left(\beta_j P_j^{l*} - \omega_j^* - \lambda_j^*\right)\right]_{\underline{P}_j^l}^{\overline{P}_j^l}$$

for any $\gamma_j^g > 0$ and $\gamma_j^l > 0$. This is equivalent to (4.16). Since $\mu^* = \omega^* = 0$, (4.16) is equivalent to (4.15) that is equivalent to (4.14). This proves that (x^*, ρ^*) is primal–dual optimal and completes the proof of Theorem 4.1. ■

Proof: [Proof of Theorems 4.2]

Let $(x^*, \rho^*) = (\theta^*, \omega^*, P^{g*}, P^{l*}, \lambda^*, \mu^*)$ be primal–dual optimal. Lemma 4.1 implies that $\omega^* = 0$ and θ^* are unique up to the reference angle θ_0^*. It directly follows the assertions 3) and 4) of Theorem 4.2.

It is easy to see the optimal values $\left(P^{g*}, P^{l*}\right)$ are unique since the objective function is strictly convex in $\left(P^g, P^l\right)$. As a consequence, x^* is unique (up to θ_0^*), which proves the assertion 1).

As for the assertion 2) of the theorem, from (4.15c) in the proof of Lemma 4.2, we have $\mu^* = \omega^* = 0$, implying the uniqueness of μ^*. From (4.15a) and (4.15b), λ_j^* is unique if either $\underline{P}_j^g < P_j^{g*} < \overline{P}_j^g$ or $\underline{P}_j^l < P_j^{l*} < \overline{P}_j^l$. We now prove that this is indeed the case by showing that the other four cases cannot hold:

(i) $P_j^{g*} = \underline{P}_j^g$ and $P_j^{l*} = \underline{P}_j^l$.
(ii) $P_j^{g*} = \underline{P}_j^g$ and $P_j^{l*} = \overline{P}_j^l$.
(iii) $P_j^{g*} = \overline{P}_j^g$ and $P_j^{l*} = \underline{P}_j^l$.
(iv) $P_j^{g*} = \overline{P}_j^g$ and $P_j^{l*} = \overline{P}_j^l$.

Since $P_j^{g*} - P_j^{l*} = p_j$ for per-node power balance, (ii) and (iii) cannot hold since the inequalities in Assumption 4.2 are strict. Suppose (i) holds. Then there exists an $\epsilon_j > 0$ such that $\hat{P}_j^g = \underline{P}_j^g + \epsilon_j < 0$ and $\hat{P}_j^l = \underline{P}_j^l + \epsilon_j < 0$, together with other components of x^*, remain a feasible primal solution. However this new feasible solution attains a strictly smaller objective value, contradicting the optimality of x^*. Thus (i) cannot hold. Similarly (iv) cannot hold. This proves that λ^* is unique.

Finally, if $\underline{P}_j^g < P_j^{g*} < \overline{P}_j^g$, then λ_j^* is uniquely determined by $\lambda_j^* = -\alpha_j P_j^{g*}$ according to (4.15a). If $\underline{P}_j^l < P_j^{l*} < \overline{P}_j^l$, then λ_j^* is uniquely determined by $\lambda_j^* = \beta P_j^{l*}$ according to (4.15b). The proof of Theorem 4.2 completes. ■

4.5 Stability Analysis

To justify the asymptotic stability, we begin with the following assumption.

Assumption 4.4 The initial state of the closed-loop system (4.1) and (4.6) is finite, and $(P_j^g(0), P_j^l(0))$ satisfy constraint (4.2).

Motivated by (4.13e), we will write the closed-loop system (4.1) and (4.6) in a similar form that will turn out to be critical for our stability analysis. To do this we first prove the boundedness property of $(P^g(t), P^l(t))$, as stated in Lemma 4.3.

Lemma 4.3 *Suppose Assumptions 4.1 and 4.4 hold. Then constraint* (4.2) *is satisfied for all* $t \geq 0$*, i.e.* $(P^g(t), P^l(t)) \in X$ *for all* $t \geq 0$ *where* X *is defined in* (4.11).

We prove Lemma 4.3 using the first-order inertia dynamics of (4.6b) and (4.6c). The detailed proof is as follows.

Proof: Set

$$\hat{u}_j^g(t) = \left[P_j^g(t) - \gamma_j^g\left(\alpha_j P_j^g(t) + \omega_j(t) + \lambda_j(t)\right)\right]_{\underline{P}_j^g}^{\overline{P}_j^g}$$

Then (4.1c) can be rewritten as

$$T_j^g \dot{P}_j^g(t) + P_j^g(t) = \hat{u}_j^g(t) \tag{4.17}$$

Apply the Laplace transform to (4.17) to obtain

$$\mathcal{L}\left(P_j^g\right)(s) = \mathcal{L}\left(\hat{u}_j^g\right)\frac{s}{T_j^g s + 1}$$

In the time domain, $P_j^g(t)$ is then given by convolution:

$$P_j^g(t) = \frac{1}{T_j^g}\int_{0^-}^{+\infty} \hat{u}_j^g(t-\tau)e^{-\tau/T_j^g}d\tau = \int_0^{\frac{t}{T_j^g}} \hat{u}_j^g\left(t - T_j^g\tau\right)e^{-\tau}d\tau$$

Since $e^{-\tau} > 0$ we can replace $\hat{u}_j^g$ in the integrand by its lower and upper bounds $\underline{P}_j^g$ and $\overline{P}_j^g$, respectively, to conclude

$$\int_0^{\frac{t}{T_j^g}} \underline{P}_j^g \cdot e^{-\tau}d\tau \leq P_j^g(t) \leq \int_0^{\frac{t}{T_j^g}} \overline{P}_j^g \cdot e^{-\tau}d\tau$$

Hence

$$\underline{P}_j^g\left(1 - e^{-t/T_j^g}\right) \leq P_j^g(t) \leq \overline{P}_j^g\left(1 - e^{-t/T_j^g}\right)$$

and $P_j^g \leq P_j^g(t) \leq \overline{P}_j^g$ for all $t \geq 0$ under Assumptions 4.1 and 4.4. That $\underline{P}_j^l \leq P_j^l(t) \leq \overline{P}_j^l$ can be proved similarly. ∎

We set the control gains for $(\hat{u}^g, \hat{u}^l)$ in (4.6) as $\gamma_j^g = (T_j^g)^{-1}, \gamma_j^l = (T_j^l)^{-1}$ Identifying $\mu(t) \equiv \omega(t)$, the closed-loop system (4.1) and (4.6) is (in vector form)

$$\dot{\theta}(t) = C^{\mathrm{T}}\omega(t) \tag{4.18a}$$

$$\dot{\omega}(t) = M^{-1}\left(P^g(t) - P^l(t) - p(t) - D\omega(t) - CB\tilde{\theta}(t)\right)$$

$$\dot{P}^g(t) = (T^g)^{-1}\left(-P^g(t) + \hat{u}^g(t)\right) \tag{4.18b}$$

$$\dot{P}^l(t) = (T^l)^{-1}\left(-P^l(t) + \hat{u}^l(t)\right) \tag{4.18c}$$

$$\dot{\lambda}(t) = \Gamma^\lambda\left(P^g(t) - P^l(t) - p\right) \tag{4.18d}$$

Here

$$\hat{u}^g(t) = \left[P^g(t) - (T^g)^{-1}\left(A^g P^g(t) + \omega(t) + \lambda(t)\right)\right]_{\underline{P}^g}^{\overline{P}^g}$$

$$\hat{u}^l(t) = \left[P^l(t) - (T^l)^{-1}\left(A^l P^l(t) - \omega(t) - \lambda(t)\right)\right]_{\underline{P}^l}^{\overline{P}^l}$$

Denote $w := (\tilde{\theta}, \omega, P^g, P^l, \lambda)$, and define

$$F(w) := \begin{bmatrix} -B^{1/2}C^{\mathsf{T}}\omega \\ -M^{-1/2}\left(P^g - P^l - p - D\omega - CB\tilde{\theta}\right) \\ (T^g)^{-1}\left(A^g P^g + \omega + \lambda\right) \\ (T^l)^{-1}\left(A^l P^l - \omega - \lambda\right) \\ -(\Gamma^\lambda)^{1/2}\left(P^g - P^l - p\right) \end{bmatrix} \tag{4.19}$$

We further define

$$S := \mathbb{R}^{m+n+1} \times X \times \mathbb{R}^n$$

where the closed convex set X is defined in (4.11). For any w define the projection of $w - F(w)$ onto S as

$$\begin{aligned} H(w) &= \mathcal{P}_S(w - F(w)) \\ &= \arg\min_{y \in S} \| y - (w - F(w))\|_2 \end{aligned}$$

where $\| \cdot \|_2$ is the Euclidean norm. Then the closed-loop system (4.18) can be written as

$$\dot{w}(t) = \Gamma_1\left(H(w(t)) - w(t)\right) \tag{4.20}$$

where the positive definite gain matrix is

$$\Gamma_1 := \operatorname{diag}\left(B^{-1/2}, M^{-1/2}, (T^g)^{-1}, (T^l)^{-1}, (\Gamma^\lambda)^{1/2}\right)$$

Note that the projection operation H has an effect only on $(\dot{P}^g, \dot{P}^l)$. Lemma 4.3 implies that $w(t) \in S$ for all t, justifying the equivalence of (4.18) and (4.20).

A point $w^* \in S$ is an *equilibrium* of the closed-loop system (4.20) if and only if it is a fixed point of the projection:

$$H(w^*) = w^*$$

Let

$$E_1 := \{w \mid H(w(t)) - w(t) = 0\}$$

be the set of equilibrium points. Then we have the following theorem.

Theorem 4.3 *Suppose Assumptions 4.1–4.4 hold. Starting from any initial point $w(0)$, $w(t)$ remains in a bounded set for all t and $w(t) \to w^*$ as $t \to \infty$ for some equilibrium $w^* \in E_1$ that is optimal for problem (4.4). If strict inequalities hold in Assumption 4.3, then the equilibrium point w^* of the closed-loop system* (4.20) *is unique.*

Theorem 4.3 justifies the asymptotic stability of the closed-loop system. Before giving the detailed proof, we first provide the readers with a short description for a better understanding. Unlike the conventional quadratic Lyapunov function used in [6, 12, 14, 24–26] for the analysis of primal–dual gradient dynamics, we use the following special Lyapunov function:

$$\begin{aligned} V_1(w) = &-(H(w) - w)^{\mathrm{T}} F(w) - \frac{1}{2} ||H(w) - w||_2^2 \\ &+ \frac{1}{2} k(w - w^*)^{\mathrm{T}} \Gamma_1^{-2} (w - w^*) \end{aligned} \tag{4.21}$$

where w^* is an equilibrium point (to be determined later) and k is small enough (as indicated in the proof later on, it is required that the diagonal matrix $\Gamma_1 - k\Gamma_1^{-1} > 0$, i.e. is strictly positive definite). The first part of V_1 is motivated by the observation in [27] that $H(w) - w$ with a stepsize computed from an exact line search defines an iterative descent algorithm for minimizing the following function over S:

$$\hat{V}_1(w) = -(H(w) - w)^{\mathrm{T}} F(w) - \frac{1}{2} ||H(w) - w||_2^2$$

It has been proved in [[27], Theorem 3.1] that $\hat{V}_1(w) \geq 0$ on S and $\hat{V}_1(w) = 0$ holds only at any equilibrium $w^* = H(w^*)$. The use of $\tilde{\theta}$ instead of θ in (4.18) and the definitions of F and Γ_1 in (4.19), and (4.20) are carefully chosen to prove that $\dot{V}_1(w(t)) \leq 0$ along any solution trajectory. The second part

$$\frac{1}{2} k(w - w^*)^{\mathrm{T}} \Gamma_1^{-2} (w - w^*)$$

of V_1 is motivated by the conventional quadratic Lyapunov function. While the first part $\hat{V}_1$ is critical for proving $\dot{V}_1 \leq 0$, implying that any trajectory $w(t)$ of the closed-loop system converges to a set of equilibrium points by LaSalle's invariance principle, the quadratic term $(w - w^*)^{\mathrm{T}} \Gamma_1^{-2} (w - w^*)$ in V_1 is used to prove that $w(t)$ actually converges to a limit point, using the technique due to [6, 12].

To prove Theorem 4.3, we start with the following lemmas.

Lemma 4.4 *Suppose Assumptions 4.1–4.4 hold. Given any $w(0) \in S$, we have the following assertions:*

1) $\dot{V}_1(w(t)) \leq 0, \forall t > 0.$
2) *The trajectory $w(t)$ is bounded, i.e. there exists $\overline{w}$ such that $\| w(t) \| \leq \overline{w}$ for all $t \geq 0$.*

Proof: We omit t in the proof for simplicity. According to [[27], Theorem 3.2], since $F(w)$ is continuously differentiable, $V_1(w)$ defined by (4.21) is also continuously differentiable. Moreover its gradient is given by

$$\nabla_w V_1(w) = F(w) - (\nabla_w F(w) - I)(H(w) - w) + k\Gamma_1^{-2}(w - w^*)$$

Then the derivative of $V_1(w)$ along the solution trajectory is

$$\begin{aligned}
\dot{V}_1(w) &= \nabla_w^{\mathrm{T}} V_1(w) \cdot \dot{w} \\
&= \nabla_w^{\mathrm{T}} V_1(w) \cdot \Gamma_1(H(w) - w) \\
&= \left(F(w) - (\nabla_w F(w) - I)(H(w) - w)\right)^{\mathrm{T}} \Gamma_1 (H(w) - w) \\
&\quad + k(w - w^*) \cdot \Gamma_1^{-1}(H(w) - w) \\
&= -(H(w) - w)^{\mathrm{T}} \nabla_w F(w) \Gamma_1 (H(w) - w) \\
&\quad - (H(w) - (w - F(w)))^{\mathrm{T}} \Gamma_1 (w - H(w)) \\
&\quad + k(w - w^*)^{\mathrm{T}} \cdot \Gamma_1^{-1}(H(w) - F(w) + F(w) - w) \\
&= -(H(w) - w)^{\mathrm{T}} \nabla_w F(w) \Gamma_1 (H(w) - w) \\
&\quad - (H(w) - (w - F(w)))^{\mathrm{T}} \Gamma_1 (w - H(w)) \\
&\quad - k(w - w^*)^{\mathrm{T}} \cdot \Gamma_1^{-1} F(w) \\
&\quad + k(w - H(w) + H(w) - w^*)^{\mathrm{T}} \Gamma_1^{-1}(H(w) - (w - F(w))) \\
&= k(H(w) - w^*)^{\mathrm{T}} \cdot \Gamma_1^{-1}(H(w) - (w - F(w))) && \text{(4.22a)} \\
&\quad - (H(w) - (w - F(w)))^{\mathrm{T}} (\Gamma_1 - k\Gamma_1^{-1})(w - H(w)) && \text{(4.22b)} \\
&\quad - (H(w) - w)^{\mathrm{T}} \nabla_w F(w) \cdot \Gamma_1 (H(w) - w) && \text{(4.22c)} \\
&\quad - k(w - w^*)^{\mathrm{T}} \cdot \Gamma_1^{-1} F(w) && \text{(4.22d)}
\end{aligned}$$

where

$$\Gamma_1 := \mathrm{diag}\left(B^{-1/2}, M^{-1/2}, (T^g)^{-1}, (T^l)^{-1}, (\Gamma_1^{\lambda})^{1/2}\right)$$

is diagonal and positive definite. We now prove that all terms on the right-hand side are nonpositive and hence $\dot{V}_1(w) \leq 0$:

$$\nabla_w F(w) :=$$

$$\underbrace{\begin{bmatrix} B^{-1/2} & 0 & 0 & 0 & 0 \\ 0 & M^{-1/2} & 0 & 0 & 0 \\ 0 & 0 & (T^g)^{-1} & 0 & 0 \\ 0 & 0 & 0 & (T^l)^{-1} & 0 \\ 0 & 0 & 0 & 0 & (\Gamma^\lambda)^{1/2} \end{bmatrix}}_{\Gamma_1} \cdot \begin{bmatrix} 0 & -BC^{\mathrm{T}} & 0 & 0 & 0 \\ CB & D & -I & I & 0 \\ 0 & I & A^g & 0 & I \\ 0 & -I & 0 & A^l & -I \\ 0 & 0 & -I & I & 0 \end{bmatrix} \tag{4.23}$$

For the term in (4.22a), denote the projection of any w onto S under the norm defined by a (symmetric) positive definite matrix Γ as

$$\mathcal{P}_{S,\Gamma}(w) := \arg\min_{y \in S} \ (y - w)^{\mathrm{T}}\Gamma(y - w)$$

By the projection theorem, a vector $\hat{w}_\Gamma$ is equal to the projection $\mathcal{P}_{S,\Gamma}(w)$ if and only if

$$\left(\hat{w}_\Gamma - w\right)^{\mathrm{T}}\Gamma\left(y - \hat{w}_\Gamma\right) \geq 0, \quad y \in S \tag{4.24}$$

Note that $S =: \prod_i S_i$ is a direct Cartesian product of intervals S_i and $\Gamma = \mathrm{diag}(\Gamma_{ii})$ is diagonal. Hence the projection under the Γ-norm coincides with the projection under the Euclidean norm

$$\begin{aligned} \mathcal{P}_{S,\Gamma}(w) &= \arg\min_{y:y_i \in S_i} \sum_i \Gamma_{ii}(y_i - w_i)^2 \\ &= \arg\min_{y:y_i \in S_i} \sum_i (y_i - w_i)^2 \\ &= \mathcal{P}_S(w) \end{aligned}$$

Substituting into (4.24), we have, for any diagonal positive definite Γ,

$$\left(\mathcal{P}_S(w) - w\right)^{\mathrm{T}}\Gamma\left(y - \mathcal{P}_S(w)\right) \geq 0, \quad y \in S \tag{4.25}$$

for any w. The projection

$$H(w) := \mathcal{P}_S(w - F(w))$$

of $w - F(w)$ therefore satisfies (for $\Gamma := \Gamma_1^{-1}$)

$$(H(w) - (w - F(w))^{\mathrm{T}}\Gamma_1^{-1}\,(w^* - H(w)) \geq 0 \tag{4.26}$$

since $w^* \in S$, which proves that the right-hand side of (4.22a) is nonpositive.

To show that (4.22b) is nonpositive, we use a similar argument. Since $\Gamma_0 := \Gamma_1 - k\Gamma_1^{-1} > 0$ we can define the projection $\mathcal{P}_{S,\Gamma}(w)$ with respect to this Γ_0. As explained above, $\mathcal{P}_{S,\Gamma}(w) = \mathcal{P}_S(w)$, and hence as before, we have

$$(H(w) - (w - F(w))^{\mathrm{T}}\Gamma_0\,(w - H(w)) \geq 0$$

since the solution trajectory $w(t) \in S$ for all $t \geq 0$ by Lemma 4.3. This proves that the term in (4.22b) is nonpositive.

Next, we will prove that (4.22d) is nonpositive. Along any solution trajectory, we always have $\mu(t) \equiv \omega(t)$. Substituting into the Lagrangian $L_1(x, \rho)$ in (4.10), we obtain the function

$$\begin{aligned}\hat{L}_1(\tilde{\theta}, P^g, P^l, \lambda, \omega) &:= L_1(\theta, \omega, P^g, P^l, \lambda, \omega)\\ &= \frac{1}{2}\left((P^g)^T A^g P^g + (P^l)^T A^l P^l - \omega^T D\omega\right)\\ &\quad + \lambda^T\left(P^g - P^l - p\right) + \omega^T\left(P^g - P^l - p - CB\tilde{\theta}\right)\end{aligned}$$

Write $w_1 := (\tilde{\theta}, P^g, P^l)$, $w_2 := (\lambda, \omega)$. Then $\hat{L}_1(w_1, w_2)$ is convex in w_1 and concave in w_2. It can be verified that[3]

$$\begin{aligned}\Gamma_1^{-1}F(w) &= \begin{bmatrix} \nabla_{\tilde{\theta}}\hat{L}_1 \\ \nabla_{P^g}\hat{L}_1 \\ \nabla_{P^l}\hat{L}_1 \\ -\nabla_{\lambda}\hat{L}_1 \\ -\nabla_{\omega}\hat{L}_1 \end{bmatrix}(w_1, w_2)\\ &= \begin{bmatrix} \nabla_{w_1}\hat{L}_1 \\ -\nabla_{w_2}\hat{L}_1 \end{bmatrix}(w_1, w_2)\end{aligned}$$

Hence, we have

$$\begin{aligned}&-k(w - w^*)^T \cdot \Gamma_1^{-1}F(w)\\ &= -k(w_1 - w_1^*)^T \nabla_{w_1}\hat{L}_1(w_1, w_2) + k(w_2 - w_2^*)^T \nabla_{w_2}\hat{L}_1(w_1, w_2)\\ &\le k\left(\hat{L}_1(w_1^*, w_2) - \hat{L}_1(w_1, w_2) + \hat{L}_1(w_1, w_2) - \hat{L}_1(w_1, w_2^*)\right)\\ &= k\left(\hat{L}_1(w_1^*, w_2) - \hat{L}_1(w_1^*, w_2^*) + \hat{L}_1(w_1^*, w_2^*) - \hat{L}_1(w_1, w_2^*)\right)\\ &\le 0 \end{aligned} \tag{4.27}$$

where the first inequality follows because $\hat{L}_1$ is convex in w_1 and concave in w_2 and the second inequality follows because (w_1^*, w_2^*) is a saddle point. Therefore (4.22d) is nonpositive.

Finally to prove that (4.22c) is nonpositive, note that

$$(H(w) - w)^T \nabla_w F(w) \Gamma_1 (H(w) - w) = \dot{w}^T\left(\Gamma_1^{-1}\nabla_w F(w)\right)\dot{w}$$

where $\dot{w} := (\dot{\tilde{\theta}}, \dot{\omega}, \dot{P}^g, \dot{P}^l, \dot{\lambda})$. From (4.19), $\nabla_w F(w)$ is given by (4.23).

Hence

$$\dot{w}^T\left(\Gamma_1^{-1}\nabla_w F(w)\right)\dot{w} = \dot{\omega}^T D\dot{\omega} + \dot{P}^{gT} A^g \dot{P}^g + \dot{P}^{lT} A^l \dot{P}^l \ge 0$$

and hence (4.22c) is nonpositive.

3 For notational simplicity, we have rearranged the order of the variables in w to $w := (\tilde{\theta}, P^g, P^l, \lambda, \omega)$ and components of F to match the order of (w_1, w_2).

This also implies that

$$\dot{V}_1(w(t)) \leq -\left(\dot{\omega}^{\mathrm{T}} D\dot{\omega} + \dot{P}^{g\mathrm{T}} A^g \dot{P}^g + \dot{P}^{l\mathrm{T}} A^l \dot{P}^l\right) \leq 0 \tag{4.28}$$

for all $t \geq 0$. This proves the first assertion of the lemma.

To prove that the trajectory $w(t)$ is bounded, note that [[27], Theorem 3.1] has proved that

$$\hat{V}_1(w) := -(H(w) - w)^{\mathrm{T}} F(w) - \frac{1}{2}||H(w) - w||_2^2$$

satisfies $\hat{V}_1(w) \geq 0$ over S. Hence, there must be

$$\frac{1}{2}k(w(t) - w^*)^{\mathrm{T}}\Gamma_1^{-2}(w(t) - w^*) \leq V_1(w(t)) \leq V_1(w(0))$$

indicating the trajectory $w(t)$ is bounded, which completes the proof. ∎

Lemma 4.5 *Suppose Assumptions 4.1–4.4 hold. Given any initial state $w(0) \in S$, we have the following assertions:*

1) *The trajectory $w(t)$ converges to the largest invariant set W_1^* contained in $W_1 = \{w \in S|\ \dot{P}^g = \dot{P}^l = \dot{\omega} = 0\}$.*
2) *Every point $w^* \in W_1^*$ is an equilibrium point of (4.20).*

Proof: **Assertion 1).** Fix any initial state $w(0)$ of interest, and consider the trajectory $(w(t), t \geq 0)$ of the closed-loop system (4.20). Lemma 4.4 implies a compact set $\Omega_0 := \Omega(w(0)) \subset S$ such that $w(t) \in \Omega_0$ for $t \geq 0$ and $\dot{V}_1(w) \leq 0$ in Ω_0. Let

$$W_1 := \{w \in \Omega_0|\ \dot{P}^g = \dot{P}^l = \dot{\omega} = 0\}.$$

Then (4.28) implies that $w \in W_1$ if and only if $\dot{V}_1(w) = 0$. According to LaSalle's invariance principle [[28], Theorem 4.4] the solution trajectory $(w(t), t \geq 0)$ converges to the largest invariant set contained in W_1, proving the first assertion.

Assertion 2). Fix any $w(0) \in W_1^*$. We claim that $w(0)$ must be an equilibrium point of (4.20). Since W_1^* is invariant, we have

$$\dot{P}^g(t) = \dot{P}^l(t) = \dot{\omega}(t) = 0, \quad t \geq 0 \tag{4.29}$$

It suffices to prove that $\dot{w}(t) = 0$ for $t \geq 0$, i.e. $\dot{\theta} = 0$ and $\dot{\lambda} = 0$ for $t \geq 0$.

Since $P^g(t), P^l(t), \omega(t)$ are bounded (Lemma 4.4), (4.29) implies that

$$(P^g(t), P^l(t), \omega(t)) \equiv (P^{g\infty}, P^{l\infty}, \omega^\infty)$$

for some finite constants $(P^{g\infty}, P^{l\infty}, \omega^\infty)$. Hence

$$\dot{\theta}(t) = C^{\mathrm{T}}\omega^\infty = \text{constant}$$

implying that $\tilde{\theta}(t)$ grows linearly in t, contradicting that $\tilde{\theta}(t)$ is bounded unless $\dot{\tilde{\theta}} = 0$ for $t \geq 0$. Similarly

$$\dot{\lambda}(t) = \Gamma^{\lambda}\left(P^{g\infty} - P^{l\infty} - p\right) = \text{constant}$$

Hence the boundedness of $\lambda(t)$ implies that $\dot{\lambda}(t) = 0$ for $t \geq 0$. This proves that any $w(0) \in W_1^*$ is an equilibrium point. ■

Note that the equilibrium point w^* of the closed-loop system (4.20) is unique (Theorem 4.2) if all inequalities in Assumption 4.3 are strictly satisfied. Furthermore, Lemma 4.5 implies that $w(t)$ converges to w^* [[28], Corollary 4.1, p. 128] as $t \to \infty$. When there are multiple equilibrium points, however, Lemma 4.5 is not adequate to conclude asymptotic stability. In this situation, we use instead a more direct argument to prove Theorem 4.3 by invoking [6, 12].

Proof: [Proof of Theorem 4.3]

Fix any initial state $w(0)$, and consider the trajectory $(w(t), t \geq 0)$ of the closed-loop system (4.20). As mentioned in the proof of Lemma 4.5, $w(t)$ stays entirely within a compact set Ω_0. Hence there exists an infinite sequence of time instants t_k such that $w(t_k) \to \hat{w}^*$ as $k \to \infty$, for some $\hat{w}^*$ in W_1^*. Lemma 4.5 guarantees that $\hat{w}^*$ is an equilibrium point of the closed-loop system (4.20) and hence $H(\hat{w}^*) = \hat{w}^*$. Use this specific equilibrium point $\hat{w}^*$ in the definition of V_1 in (4.21) to construct the Lyapunov function

$$V_1(w) = -(H(w) - w)^T F(w) - \frac{1}{2}||H(w) - w||_2^2 + \frac{1}{2}k(w - \hat{w}^*)^T \Gamma_1^{-2}(w - \hat{w}^*)$$

Since $\dot{V}_1 \leq 0$, $V_1(w(t))$ converges. Moreover it follows from the continuity of V_1 that

$$\lim_{t\to\infty} V_1(w(t)) = \lim_{k\to\infty} V_1(w(t_k)) = V_1(\hat{w}^*) = 0.$$

The quadratic term $(w - \hat{w}^*)^T \Gamma_1^{-2}(w - \hat{w}^*)$ in V_1 then implies that $w(t) \to \hat{w}^*$ as $t \to \infty$. ■

4.6 Case Studies

4.6.1 Test System

To test the optimal frequency controller, we modify Kundur's four-machine, two-area system [29, 30] by expanding it to a four-area system. Each area has one (aggregate) generator (Gen1–Gen4), one (aggregate) controllable load (L1c–L4c),

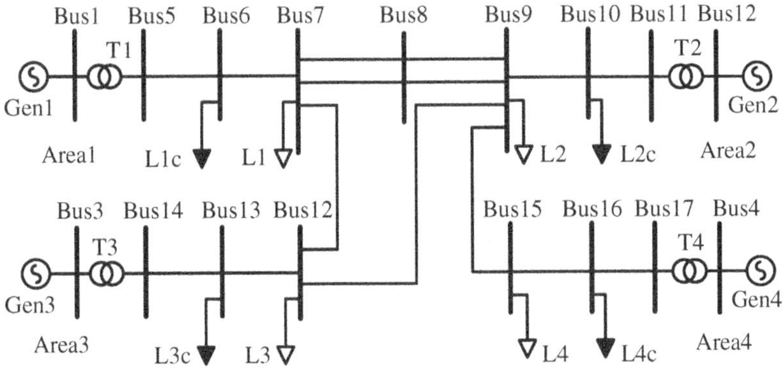

Figure 4.1 Four-area power system.

Table 4.1 System parameters.

Area j	D_j	R_j	α_j	β_j	T_j^g	T_j^l
1	0.04	0.04	2	2.5	4	4
2	0.045	0.06	2.5	4	6	5
3	0.05	0.05	1.5	2.5	5	4
4	0.055	0.045	3	3	5.5	5

and one (aggregate) uncontrollable load (L1–L4), as shown in Figure 4.1. The parameters of generators and controllable loads are given in Table 4.1. For others one can refer to [30]. The total uncontrollable load in each area is identically 480 MW. At time $t = 20$ s, we add step changes on the uncontrollable loads in four areas to test the performance of our controllers.

All the simulations are implemented in PSCAD[4] [31]. The laptop is with 8 GB memory and 2.39 GHz CPU. The detailed electromagnetic transient model of three-phase synchronous machines is adopted to simulate generators with both governors and exciters. The uncontrollable loads L1–L4 are modeled by the fixed load in PSCAD, while controllable load L1c–L4c are formulated by the self-defined controlled current source. The closed-loop system diagram is shown in Figure 4.2. We only need measure local frequency, generation, controllable load, and tie-line power flows to compute control demands. There are no need of uncontrollable load and communication from other areas. Note that in the

4 PSCAD is recognized as a professional power system electromagnetic transient simulator, which has been widely used in power system simulation.

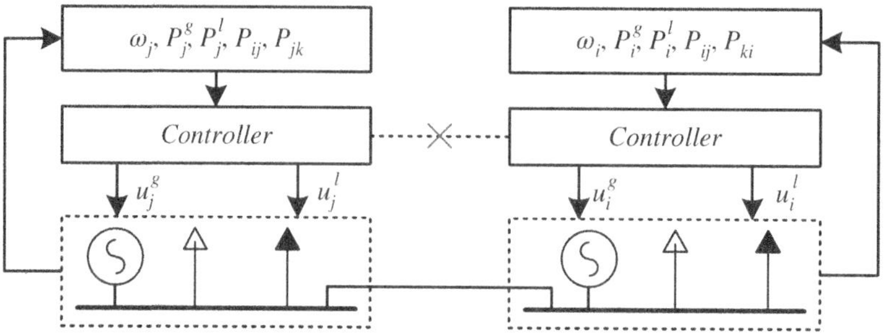

Figure 4.2 Closed-loop system diagram.

simulation, all variables are added by their initial steady-state values to explicitly show the actual values.

4.6.2 Simulation Results

In the simulation, the generations in each area are [625.9, 562.7, 701.7, 509.6] MW, and the controllable loads are all 120 MW. The load changes are given in Table 4.2, which are unknown to the controllers. Here we use the method mentioned in Remark 4.2 to estimate the load change in the dynamic simulations. In the transient process, the estimation may vary around the real value, which will converge to it in the steady state. We also show the operational constraints on generations and controllable loads in individual control areas in Table 4.2, which are shown in the third and fourth columns. These bounds are hard even in the transient process.

4.6.2.1 Stability and Optimality

The dynamics of local frequencies and tie-line power flows are illustrated in Figures 4.3 and 4.4, respectively. The frequency is restored in all the four control areas, while power deviations of tie lines between these areas are also recovered. The recovery of frequency implies that the system works in a new steady state and

Table 4.2 Capacity limits and load disturbance.

Area j	**Load changes**	$[\underline{P}_j^g, \overline{P}_j^g]$ **(MW)**	$[\underline{P}_j^l, \overline{P}_j^l]$ **(MW)**
1	90 MW	[600, 700]	[75, 120]
2	90 MW	[550, 680]	[80, 120]
3	90 MW	[650, 800]	[80, 120]
4	120 MW	[500, 600]	[55, 120]

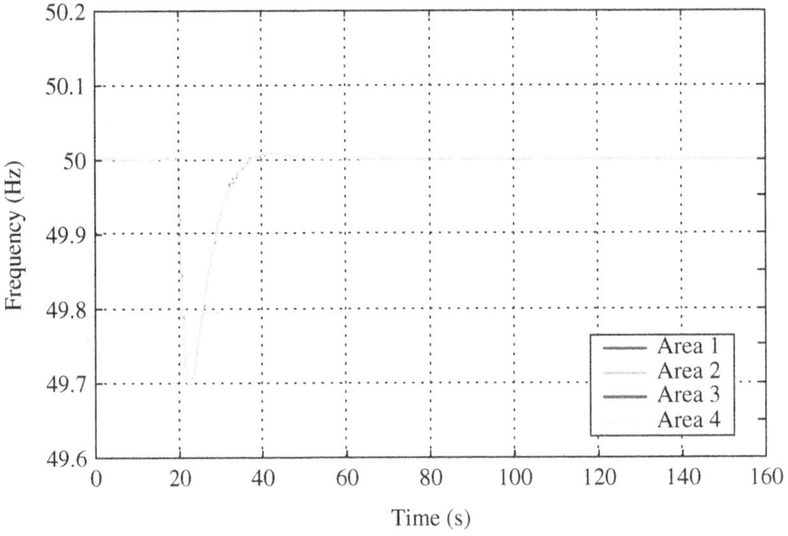

Figure 4.3 Frequency dynamics in network balance case.

Table 4.3 Equilibrium points.

	Area 1	**Area 2**	**Area 3**	**Area 4**
P_j^{g*} (MW)	676	618	758	570
P_j^{l*} (MW)	80	85.3	86.2	60

the power rebalance is achieved. The nadir of the frequency is about 49.7 Hz, and the recovery time is 30 s. The tie-line power further implies that the rebalance is realized in each individual area.

The generations and controllable loads are different from those before disturbance, indicating that the system is stabilized at a new equilibrium point. The resulting equilibrium point is given in Table 4.3, which is identical to the optimal solution of (4.4) computed by centralized optimization using Yalmip. The simulation results confirm our theoretic analyses, verifying that our controller can autonomously guarantee the frequency stability while achieving optimal operating point with the recovery of tie-line power in a completely decentralized manner.

4.6.2.2 Dynamic Performance

In this subsection, we analyze the impacts of regulation capacity constraints on the dynamic performance. To this end, we compare the dynamic responses of

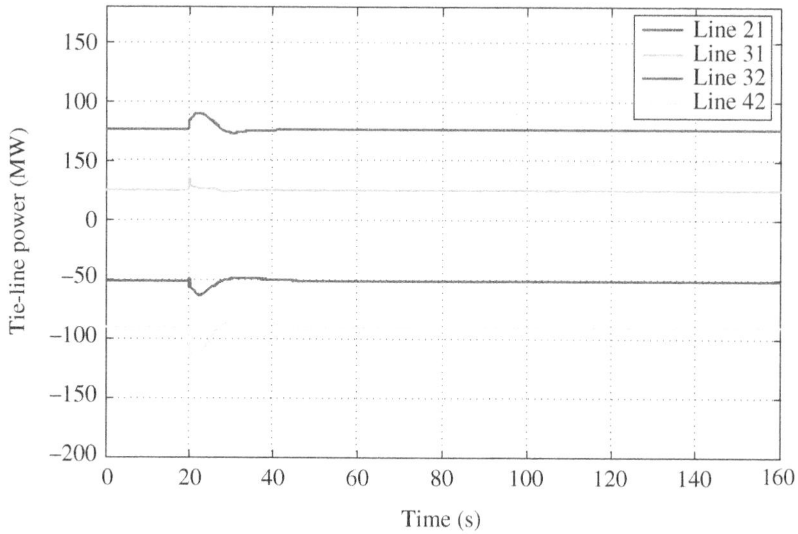

Figure 4.4 Tie-line power dynamics in network balance case.

the frequency controllers with and without input saturations. The trajectories of mechanical powers of turbines and controllable loads are shown in Figures 4.5 and 4.6, respectively. In this case, the system frequency and tie-line flows are restored, and the same optimal equilibrium point is achieved. With the saturated controller, the mechanical power of turbines and controllable loads are strictly within the limits in transient. On the contrary, the controller without saturation results in considerable violation of the capacity constraints during transient, which is practically infeasible.

4.6.2.3 Comparison with AGC

AGC is often utilized in the conventional secondary frequency control. To compare performance of our controller, we give the frequency dynamics of proposed controller (left) and AGC (right) in Figure 4.7.

The results show that the frequency nadir under the proposed controller is a bit lower than that under AGC. Generally, AGC does not cause frequency overshoot while the proposed controller may cause a small overshoot. In addition, the convergence time is shorter under our controller than that under AGC. Overall the frequency dynamics in both methods are smooth without large oscillation, and our controller has a similar frequency performance to AGC.

4.6.2.4 Digital Implementation

Practically, the proposed continuous-time controller needs to be digitally implemented. The update rates could influence the system's performance. To compute

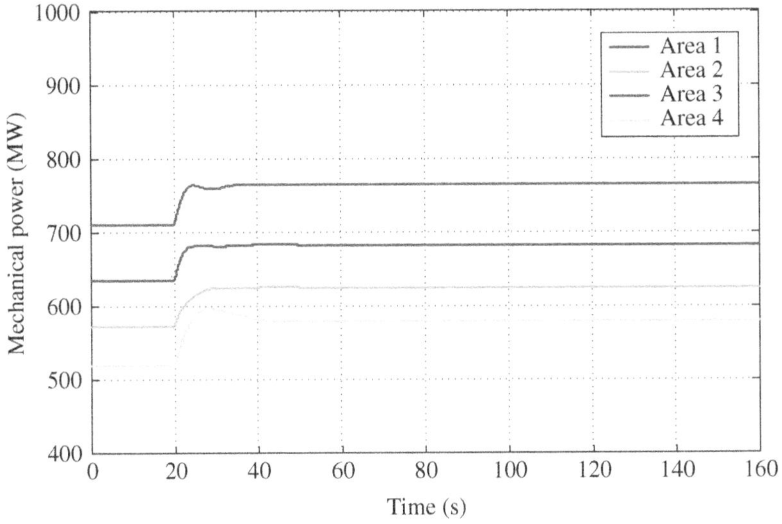

(a) With constraints

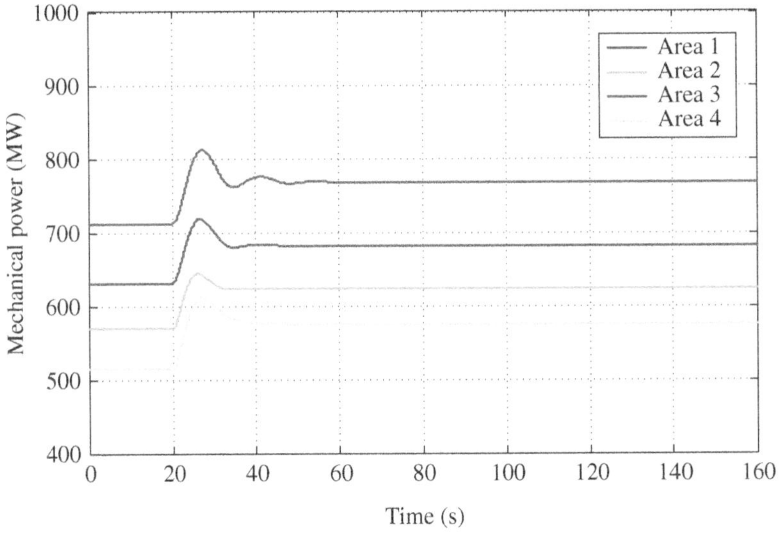

(b) Without constraints

Figure 4.5 Mechanical outputs with/without capacity constraints.

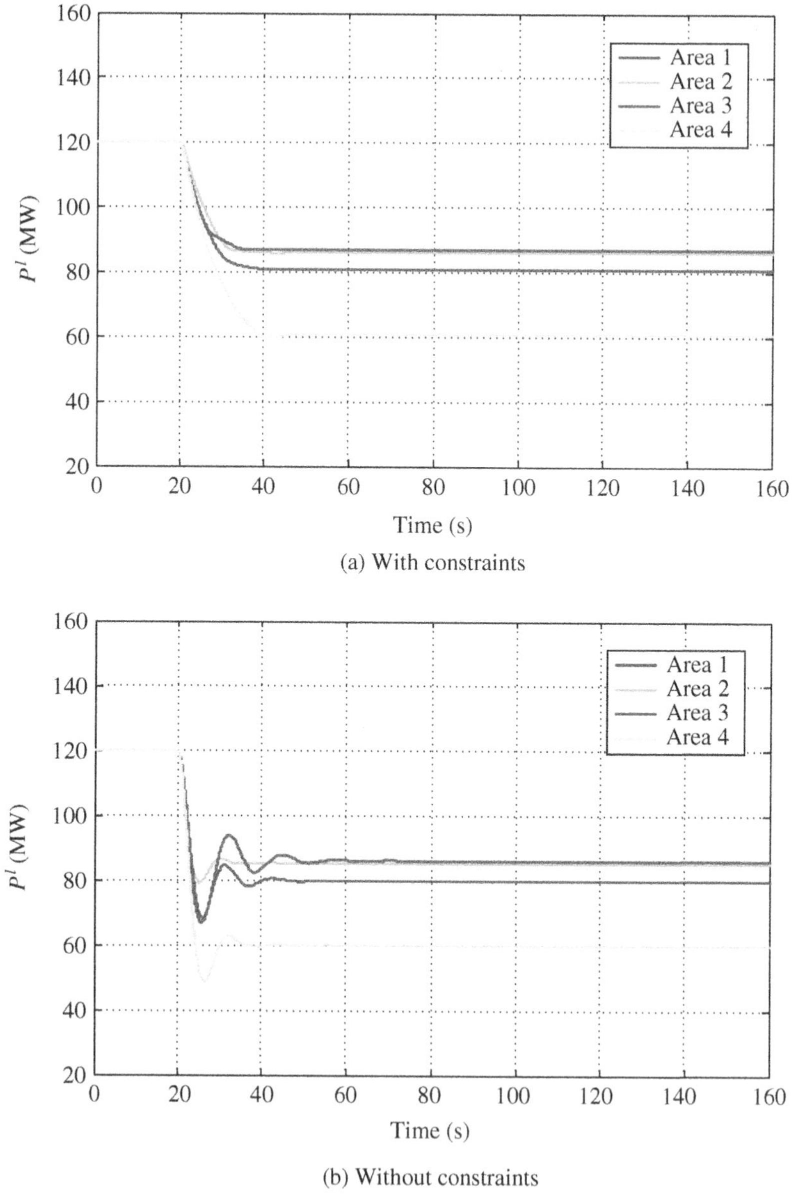

Figure 4.6 Controllable loads with/without capacity constraints.

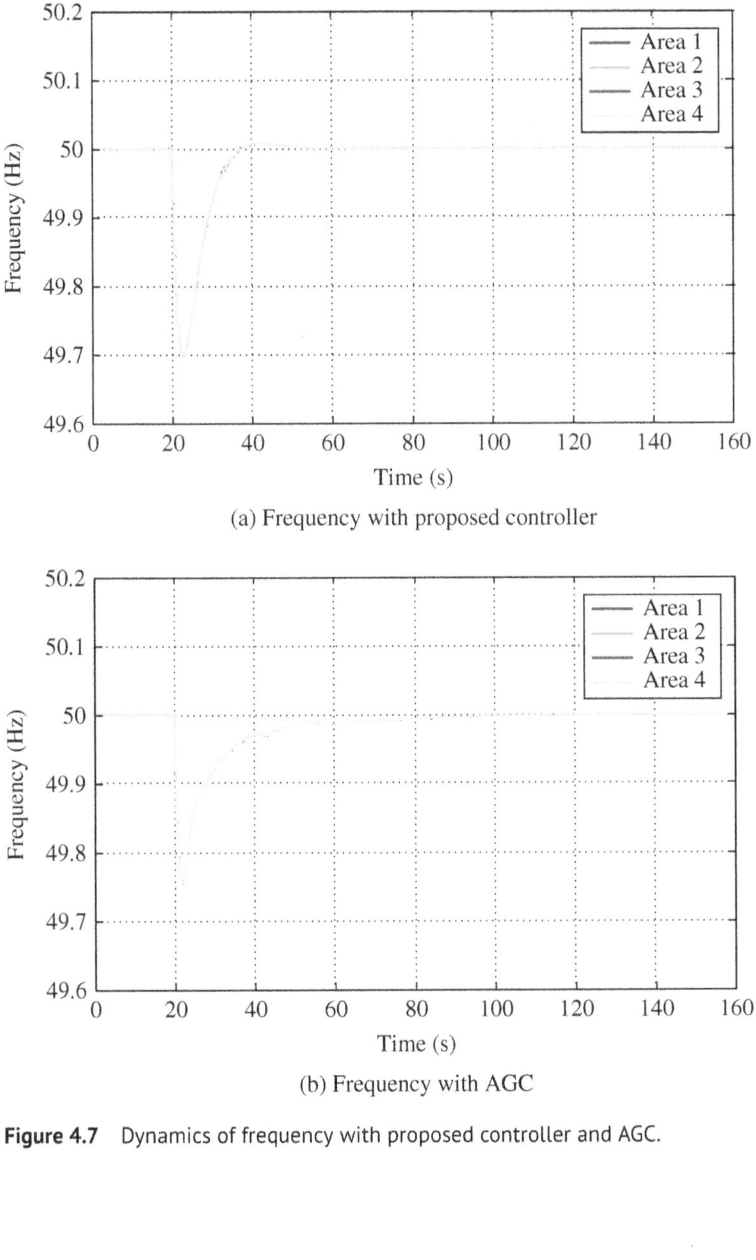

Figure 4.7 Dynamics of frequency with proposed controller and AGC.

λ_j, u_j^g, u_j^l in (4.6), phasor measurement units (PMUs) are needed to get $P_j^g, P_j^l, p_j, \omega_j$. By "IEEE Standard for Synchrophasor Measurements for Power Systems," the PMU can return 10–30 numerical values per second, i.e. (10–30 Hz). Thus, we first simulate with two different update frequencies, i.e. 10 and 30 Hz, where u_j^g, u_j^l are sent to the generator and controllable load with these frequencies. Then, we further investigate the impact of lower update frequency with 1 Hz. Figures 4.8 and 4.9 are dynamics of mechanical power and controllable load.

From Figures 4.8 and 4.9, the dynamics with update rates as 10 and 30 Hz are almost identical, where the system are stable smoothly. When we continue to reduce the update rate to 1 Hz, there is small oscillation in the transient although the system is still stable. This implies that the controller is robust to different digital implementation frequencies.

4.7 Conclusion and Notes

This chapter has devised a fully decentralized optimal secondary frequency control with aggregate generators and controllable loads, mainly thanks to the intrinsic connection between frequency control and its associated optimization problem uncovered in Chapter 3.[5] The proposed controller can autonomously restore the nominal frequencies and tie-line flows after unknown load disturbances while minimizing the regulation cost. The capacity constraints on the generations and the controllable loads can always be satisfied, even in transient. We have revealed that the closed-loop system carries out an (approximate) primal–dual algorithm to solve the associated optimal dispatch problem, guaranteeing the optimality of closed-loop equilibria. We have proved the asymptotically stability of the closed-loop system as well. Simulations on the modified Kundur's power system verify the effectiveness of our designs.

Technically, the regulation capacity limits result in nonsmooth dynamics due to input saturation, which is challenging to distributed optimal control design and stability analysis. In this context, this chapter has demonstrated that, from the forward and reverse engineering perspectives, the primal–dual updates to solve the associated optimization problem intrinsically provide a guideline to devise the controller. It also has suggested proving the convergence/stability by combining the projection technique with the argument of LaSalle's invariance principle. The way of constructing the Lyapunov function presented in Section 4.5 appears to be promising to distributed optimal controller design in the presence of complicated constraints. In the next chapter, this approach will further extend to cope

5 Reference [6] provides a more detailed investigation on this connection.

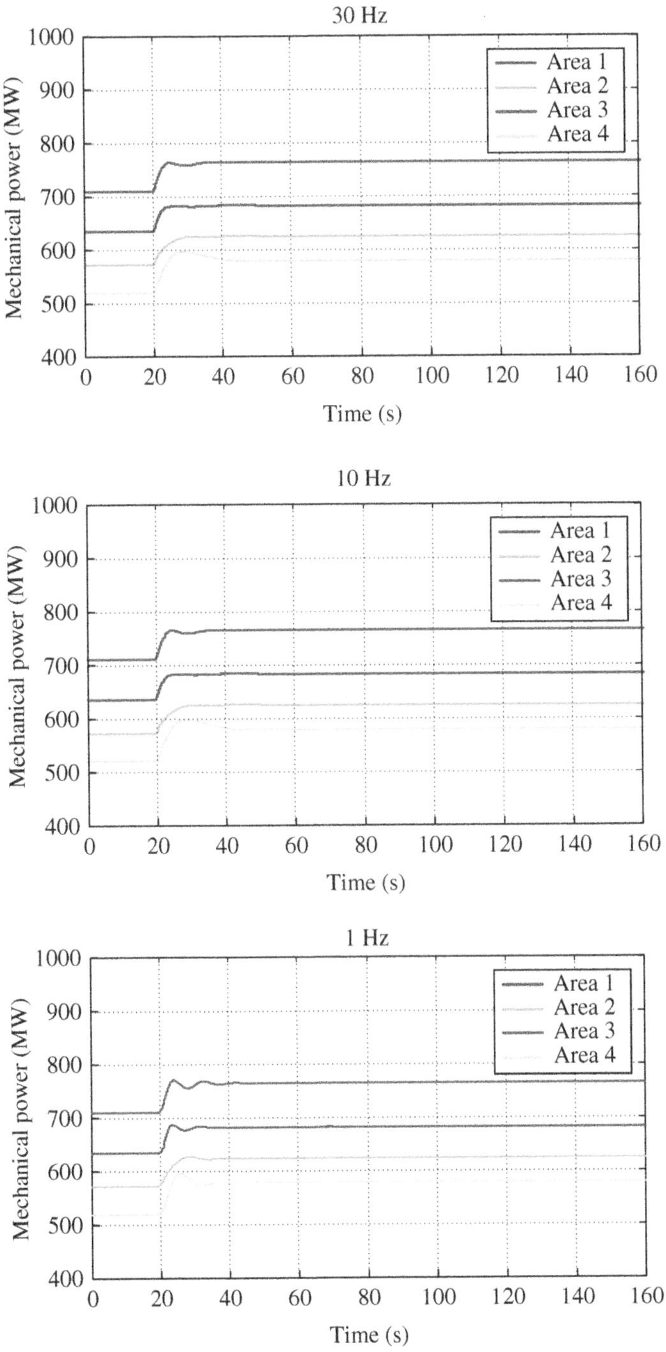

Figure 4.8 Dynamics of mechanical power with different update rates.

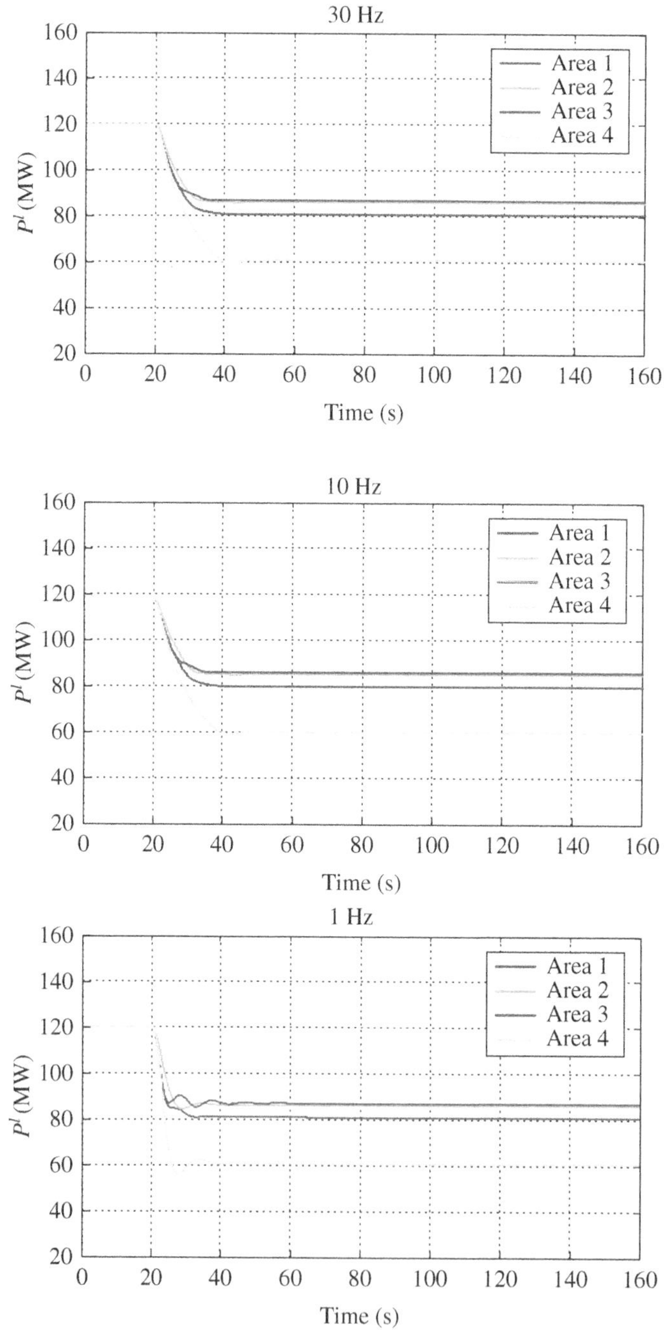

Figure 4.9 Dynamics of controllable load with different update rates.

with a more challenging secondary frequency control problem, where tie-line flow limits are coupled with input saturation constraints.

Bibliography

1 L. Min and A. Abur, "Total transfer capability computation for multi-area power systems," *IEEE Transactions on Power Systems*, vol. 21, no. 3, pp. 1141–1147, 2006.

2 A. Ahmadi-Khatir, M. Bozorg, and R. Cherkaoui, "Probabilistic spinning reserve provision model in multi-control zone power system,"*IEEE Transactions on Power Systems*, vol. 28, no. 3, pp. 2819–2829, 2013.

3 I. Ibraheem, P. Kumar, and D. P. Kothari, "Recent philosophies of automatic generation control strategies in power systems,"*IEEE Transactions on Power Systems*, vol. 20, no. 1, pp. 346–357, 2005.

4 M. H. Variani and K. Tomsovic, "Distributed automatic generation control using flatness-based approach for high penetration of wind generation," *IEEE Transactions on Power Systems*, vol. 28, no. 3, pp. 3002–3009, 2013.

5 T. Stegink, C. De Persis, and A. van der Schaft, "A unifying energy-based approach to stability of power grids with market dynamics," *IEEE Transactions Automatic Control*, vol. 62, no. 6, pp. 2612–2622, 2017.

6 N. Li, C. Zhao, and L. Chen, "Connecting automatic generation control and economic dispatch from an optimization view," *IEEE Transactions on Control of Network Systems*, vol. 3, no. 3, pp. 254–264, 2016.

7 T. Stegink, C. D. Persis, and A. van der Schaft, "A port-Hamiltonian approach to optimal frequency regulation in power grids," in *Proceedings of 54th IEEE Conference on Decision and Control (CDC)*, (Osaka, Japan), pp. 3224–3229, IEEE, Dec 2015.

8 Z. Wang, F. Liu, L. Chen, and S. Mei, "Distributed economic automatic generation control: a game theoretic perspective," in *Power Energy Society General Meeting, 2015 IEEE*, (Denver, CO), pp. 1–5, July 2015.

9 Z. Wang, F. Liu, S. H. Low, C. Zhao, and S. Mei, "Decentralized optimal frequency control of interconnected power systems with transient constraints," in *2016 IEEE 55th Conference on Decision and Control (CDC)*, pp. 664–671, Dec 2016.

10 M. D. Ilic, L. Xie, U. A. Khan, and J. M. F. Moura, "Modeling of future cyber-physical energy systems for distributed sensing and control," *IEEE Transactions on Systems, Man, and Cybernetics - Part A: Systems and Humans*, vol. 40, no. 4, pp. 825–838, 2010.

11 M. Zribi, M. Al-Rashed, and M. Alrifai, "Adaptive decentralized load frequency control of multi-area power systems," *International Journal of Electrical Power & Energy Systems*, vol. 27, no. 8, pp. 575–583, 2013.

12 C. Zhao, U. Topcu, N. Li, and S. H. Low.,"Design and stability of load-side primary frequency control in power systems," *IEEE Transactions on Automatic Control*, vol. 59, no. 5, pp. 1177–1189, 2014.

13 C. Zhao and S. H. Low, "Decentralized primary frequency control in power networks," in *Proceedings of the 53rd IEEE Conference on Decision and Control (CDC)*, pp. 2467–2473, December 2014.

14 E. Mallada, C. Zhao, and S. Low, "Optimal load-side control for frequency regulation in smart grids," *IEEE Transactions on Automatic Control*, vol. 62, no. 12, pp. 6294–6309, 2017.

15 A. Kasis, E. Devane, C. Spanias, and I. Lestas, "Primary frequency regulation with load-side participation–part I: stability and optimality," *IEEE Transactions on Power Systems*, vol. 32, no. 5, pp. 3505–3518, 2017.

16 E. Devane, A. Kasis, M. Antoniou, and I. Lestas, "Primary frequency regulation with load-side participation– Part II: beyond passivity approaches," *IEEE Transactions on Power Systems*, vol. 32, no. 5, pp. 3519–3528, 2017.

17 A. Maknouninejad, Z. Qu, F. L. Lewis, and A. Davoudi, "Optimal, nonlinear, and distributed designs of droop controls for DC microgrids," *IEEE Transactions on Smart Grid*, vol. 5, no. 5, pp. 2508–2516, 2014.

18 V. Nasirian, A. Davoudi, F. L. Lewis, and J. M. Guerrero, "Distributed adaptive droop control for DC distribution systems," *IEEE Transactions on Energy Conversion*, vol. 29, no. 4, pp. 944–956, 2014.

19 R. Olfati-Saber, J. A. Fax, and R. M. Murray, "Consensus and cooperation in networked multi-agent systems," *Proceedings of the IEEE*, vol. 95, no. 1, pp. 215–233, 2007.

20 G. Binetti, A. Davoudi, F. L. Lewis, D. Naso, and B. Turchiano, "Distributed consensus-based economic dispatch with transmission losses," *IEEE Transactions on Power Systems*, vol. 29, no. 4, pp. 1711–1720, 2014.

21 H. Xin, Z. Lu, Z. Qu, D. Gan, and D. Qi, "Cooperative control strategy for multiple photovoltaic generators in distribution networks," *IET Control Theory and Applications*, vol. 5, no. 14, pp. 1617–1629, 2011.

22 A. Jokić, M. Lazar, and P. P. van den Bosch, "Real-time control of power systems using nodal prices," *International Journal of Electrical Power & Energy Systems*, vol. 31, no. 9, pp. 522–530, 2009.

23 D. P. Bertsekas., *Nonlinear Programming*, 2nd ed., Athena Scientific, 2008.

24 K. J. Arrow, L. Hurwicz, H. Uzawa, and H. B. Chenery, *Studies in Linear and Non-Linear Programming*. Redwood City, CA, USA: Stanford University Press, 1958.

25 D. Feijer and F. Paganini, "Stability of primal-dual gradient dynamics and applications to network optimization,"*Automatica*, vol. 46, no. 12, pp. 1974–1981, 2010.

26 A. Rantzer, "Dynamic dual decomposition for distributed control," in *Proceedings of American Control Conference (ACC)*, (St. Louis, MO, USA), pp. 884–888, IEEE, 2009.

27 M. Fukushima, "Equivalent differentiable optimization problems and descent methods for asymmetric variational inequality problems," *Mathematical Programming*, vol. 53, pp. 99–110, 1992.

28 H. K. Khalil, *Nonlinear Systems*, Upper Saddle River, NJ: Prentice hall, 1996.

29 J. Fang, W. Yao, Z. Chen, J. Wen, and S. Cheng, "Design of anti-windup compensator for energy storage-based damping controller to enhance power system stability," *IEEE Transactions on Power Systems*, vol. 29, no. 3, pp. 1175–1185, 2014.

30 P. Kundur, *Power System Stability and Control*, New York, USA: McGraw-hill, 1994.

31 "Home of pscad." https://hvdc.ca/pscad/, December 2017.

5

Physical Restrictions: Line Flow Limits in Secondary Frequency Control

In many networked control systems, the agents are coupled by the physical network flows, such as a power system governed by Kirchhoff's law, constituting a so-called cyber-physical networked control system. In this circumstance, the flow limits should also be satisfied in the distributed controller design. In Chapter 4, we have investigated the decentralized optimal secondary frequency control of a multi-area power system, where the line flow delivered from one area to another is fixed, i.e. the per-node balance case. If the tie-line flows, however, are allowed to be adjustable within certain given ranges, the power mismatch will be rebalanced by all generations and controllable loads across the overall power network, which is referred to as the network balance case. In this situation, additional operational constraints should be imposed since all the adjustable tie-line flows must satisfy preset limits for ensuring the security of operation. As an extension of Chapter 4, this chapter shows how to design distributed optimal secondary frequency control considering both generation capacity constraints and tie-line flow limits. With the proposed controller, the tie-line flows can remain within the acceptable range at equilibrium, while the regulation capacity constraints are satisfied both at equilibrium and during transient. Moreover, it can adapt to unknown load disturbances. It is revealed that the closed-loop system with the proposed controller carries out (approximate) primal–dual updates with input saturation for solving an associated optimization problem. Nevertheless, the coupling of line limits and input saturation creates discontinuity in the closed-loop dynamics, which is more challenging than the problem discussed in the previous chapter. To cope with the induced discontinuous dynamics, we deploy the arguments of LaSalle's invariance principle with a nonpathological Lyapunov function to address the asymptotic stability.

Merging Optimization and Control in Power Systems: Physical and Cyber Restrictions in Distributed Frequency Control and Beyond, First Edition. Feng Liu, Zhaojian Wang, Changhong Zhao, and Peng Yang.

5.1 Background

In multi-area secondary frequency control, the control areas are desired to cooperate for better frequency recovery or regulation cost reduction. In this case, power mismatch will be balanced by all generations and controllable loads among all control areas in cooperation. Similar situations also appear in one control area with multiple generators and controllable loads that would cooperate to eliminate the power mismatch in the area. It is referred to as the *network power balance* case. Compared with the per-node power balance case discussed in Chapter 4, the new challenge is that the tie-line power flows are changeable, and hence congestion may occur. As tie-line flows are generally determined by all generations and loads, it imposes additional global constraints. Thereby local information turns out to be insufficient, and communication becomes necessary. Unlike input saturation, the tie-line flow constraints are not hard limits, which only need to be satisfied at equilibrium. In this chapter, both the tie-line flow limits and input saturation are considered.

In the literature on frequency control, tie-line power constraints are considered in [1–6]. In [1], tie-line power constraints are included in the load-side secondary frequency control. A virtual variable is used to estimate the tie-line power, whose value is identical to the tie-line power at equilibrium. In [2, 3], an optimal economic dispatch problem including tie-line power constraints is formulated; then the solution dynamics derived from a primal–dual algorithm is shaped as a port-Hamiltonian form. The power system dynamics also have a port-Hamiltonian form, which are interconnected with the solution dynamics to constitute a closed-loop Hamiltonian system. Then, the optimality and stability are proved. In [5], a unified method is proposed for primary and secondary frequency controls, where congestion management is implemented in the secondary control. In [6], a real-time control framework is proposed for tree power networks, where transmission capacities are considered.

Similar to the per-node balance case, hard limits, such as capacity constraints of power injections on buses, are enforced only in steady-state in the literature, which may fail if such constraints are violated in transient. Here we construct a fully distributed control to recover nominal frequency while eliminating congestion. Differing from the literature, it enforces regulation capacity constraints not only at equilibrium but also during transient. We show that the controllers, together with the physical dynamics, serve as primal–dual updates with saturation for solving the optimization problem. The optimal solution to the optimization problem and the equilibrium point of the closed-loop system are identical.

The enforcement of capacity constraints during transient and tie-line flow limits in steady-state simultaneously makes the stability proof very challenging. Specifically, the Lyapunov function is not continuous anymore, as in the per-node case in

Chapter 4. In this situation, the conventional LaSalle's invariance principle does not apply. To overcome the difficulty, we construct a nonpathological Lyapunov function to mitigate the impacts of nonsmooth dynamics and adopt the weak invariance principle to prove the asymptotic stability of the closed-loop system.

The rest of this chapter is organized as follows. In Section 5.2, we describe our model. Section 5.3 formulates the optimal frequency control problem in the network balance case, presents the distributed frequency controller, and proves the optimality, uniqueness, and stability of the closed-loop equilibrium. Simulation results are given in Section 5.6. Section 5.7 gives the conclusion with notes.

5.2 Power System Model

The power network is modeled by a directed graph $\mathcal{G} := (\mathcal{N}, \mathcal{E})$ where $\mathcal{N} = \{0, 1, 2, \ldots n\}$ is the set of nodes (control areas) and $\mathcal{E} \subseteq \mathcal{N} \times \mathcal{N}$ is the set of edges (tie lines). If a pair of nodes i and j are connected by a tie line directly, we denote the tie line by $(i, j) \in \mathcal{E}$. Let $m := |\mathcal{E}|$ denote the number of tie lines. Use $(i, j) \in \mathcal{E}$ or $i \to j$ interchangeably to denote a directed edge from i to j. Assume the graph is connected and node 0 is the reference node.

The power system dynamics for each node $j \in \mathcal{N}$ is

$$\dot{\theta}_j = \omega_j(t) \tag{5.1a}$$

$$\begin{aligned} M_j\dot{\omega}_j = {} & P_j^g(t) - P_j^l(t) - p_j - D_j\omega_j(t) \\ & + \sum_{i:i\to j} B_{ij}(\theta_i(t) - \theta_j(t)) \\ & - \sum_{k:j\to k} B_{jk}(\theta_j(t) - \theta_k(t)) \end{aligned} \tag{5.1b}$$

$$T_j^g \dot{P}_j^g = -P_j^g(t) + u_j^g(t) - \frac{\omega_j(t)}{R_j} \tag{5.1c}$$

$$T_j^l \dot{P}_j^l = -P_j^l(t) + u_j^l(t) \tag{5.1d}$$

Let $x := (\theta, \omega, P^g, P^l)$ denote the state of the network and $u := (u^g, u^l)$ denote the control.[1]

The capacity constraints are as follows:

$$\underline{P}_j^g \le P_j^g(t) \le \overline{P}_j^g, \quad j \in \mathcal{N} \tag{5.2a}$$

$$\underline{P}_j^l \le P_j^l(t) \le \overline{P}_j^l, \quad j \in \mathcal{N} \tag{5.2b}$$

---

1 Given a collection of x_i for i in a certain set A, x denotes the column vector $x := (x_i, i \in A)$ of a proper dimension with x_i as its components.

Here (5.2a) and (5.2b) are hard limits on the regulation capacities of generation and controllable load at each node, which should not be violated at any time, even during transient.

The system operates in a steady state initially, i.e. the generation and the load are balanced and the frequency is at its nominal value. All variables represent deviations from their nominal or scheduled values so that, for example, $\omega_j(t) = 0$ means the frequency is at its nominal value.

In this chapter, all nodes cooperate to rebalance power over the entire network after a disturbance. The power flows P_{ij} on the tie lines may deviate from their scheduled values, and we require that they satisfy line limits, i.e.

$$\underline{P}_{ij} \leq P_{ij} \leq \overline{P}_{ij} \qquad \forall (i,j) \in \mathcal{E} \tag{5.3}$$

for some upper and lower bounds $\underline{P}_{ij}, \overline{P}_{ij}$.

In DC approximation the power flow on line (i,j) is given by

$$P_{ij} = B_{ij}(\theta_i - \theta_j).$$

Hence line flow constraints in the per-node balance case are $\theta_i = \theta_j$ for all $(i,j) \in \mathcal{E}$ and in the network balance case are

$$\underline{\theta}_{ij} \leq \theta_i - \theta_j \leq \overline{\theta}_{ij} \qquad \forall (i,j) \in \mathcal{E} \tag{5.4}$$

where $\underline{\theta}_{ij} = \underline{P}_{ij}/B_{ij}, \overline{\theta}_{ij} = \overline{P}_{ij}/B_{ij}$.

As the generation P_j^g and controllable load P_j^l in each area can increase or decrease, and a line flow P_{ij} can in either direction, we make the following assumption.

Assumption 5.1 Throughout this chapter, the following conditions hold:

- $\underline{P}_j^g < 0 < \overline{P}_j^g$ and $\underline{P}_j^l < 0 < \overline{P}_j^l$ for $\forall j \in \mathcal{N}$.
- $\underline{\theta}_{ij} \leq 0 \leq \overline{\theta}_{ij}$ for $(i,j) \in \mathcal{E}$.
- $\theta_0(t) := 0$ and $\phi_0(t) := 0$ for all $t \geq 0$.

The assumption $\theta_0 \equiv \phi_0 \equiv 0$ amounts to using $(\theta_0(t), \phi_0(t))$ as reference angles. It is made merely for notational convenience: as we will see, the equilibrium point is unique with this assumption (or unique up to reference angles without this assumption).

5.3 Control Design for Network Power Balance

In the per-node balance case, individual control areas rebalance power within their own areas after disturbances. However, in many circumstances, it may be more efficient for all control areas to eliminate power imbalance of the overall system in a coordinated manner. This can be modeled as the condition:

$$\sum_{j\in\mathcal{N}} P_j^g = \sum_{j\in\mathcal{N}} \left(P_j^l + p_j\right) \tag{5.5}$$

In this case the tie-line flows may not be restored to their pre-disturbance values. To ensure that they are within operational limits, the constraints (5.4) are imposed.

Even though the philosophy of the controller design and the proofs are similar to the per-node case, the details are much more complicated. Our presentation will however be brief where there is no confusion.

5.3.1 Control Goals

In the network power balance case, the control goals are formalized as the following optimization problem (network balance optimization [NBO]):

$$\text{NBO:}\ \min\ \sum_{j\in\mathcal{N}} \frac{\alpha_j}{2}\left(P_j^g\right)^2 + \sum_{j\in\mathcal{N}} \frac{\beta_j}{2}\left(P_j^l\right)^2 + \sum_{j\in\mathcal{N}} \frac{D_j}{2}\omega_j^2 + \sum_{j\in\mathcal{N}} \frac{z_j^2}{2} \tag{5.6a}$$

$$\text{over}\quad x := (\theta, \phi, \omega, P^g, P^l)\ \text{and}\ u := (u^g, u^l)$$

$$\text{s. t.}\quad (5.2),$$

$$P_j^g = P_j^l + p_j + U_j(\theta, \omega) \quad j \in \mathcal{N} \tag{5.6b}$$

$$P_j^g = P_j^l + p_j + \hat{U}_j(\phi) \quad j \in \mathcal{N} \tag{5.6c}$$

$$\underline{\theta}_{ij} \le \phi_i - \phi_j \le \overline{\theta}_{ij}, \quad (i,j) \in \mathcal{E} \tag{5.6d}$$

$$P_j^g = u_j^g, \quad j \in \mathcal{N} \tag{5.6e}$$

$$P_j^l = u_j^l, \quad j \in \mathcal{N} \tag{5.6f}$$

where z_j is a shorthand defined for convenience as

$$z_j := P_j^g - P_j^l - p_j - \hat{U}_j(\phi) \tag{5.6g}$$

$U(\theta, \omega) := D\omega + CBC^{\mathrm{T}}\theta$, and $\hat{U}(\phi) := CBC^{\mathrm{T}}\phi$.

As in the per-node case, we define the variables $\tilde{\theta}_{ij} := \theta_i - \theta_j$, or in vector form, $\tilde{\theta} := C^{\mathrm{T}}\theta$. As we fix $\theta_0 := 0$ to be the reference angle under Assumption 5.1, $\tilde{\theta} = C^{\mathrm{T}}\theta$ defines a bijection between θ and $\tilde{\theta}$. Similarly we define $\tilde{\phi}_{ij} := \phi_i - \phi_j$ or $\tilde{\phi} := C^{\mathrm{T}}\phi$, and $\phi_0 := 0$ so there is a bijection between ϕ and $\tilde{\phi}$. Note that both $\tilde{\theta}$ and $\tilde{\phi}$ are restricted to the column space of C^{T}. We will use (θ, ϕ) and $(\tilde{\theta}, \tilde{\phi})$ interchangeably. For instance, we will abuse notation and write $\hat{U}(\phi) := CBC^{\mathrm{T}}\phi$ or $\hat{U}(\tilde{\phi}) := CB\tilde{\phi}$.

We now summarize some of the interesting properties of NBO (5.6) that will be proved formally in the next two subsections. We first compare the NBO (5.6) with per-node balance optimization (PBO) presented in Chapter 4 for the per-node balance case.

Remark 5.1 ***(Comparison of NBO and PBO)*** Intuitively the network balance condition (5.5) is a relaxation of the per-node balance condition (4.3) in Chapter 4, and hence we expect that the optimal cost of NBO gives a lower bound of that of PBO. This is indeed the case, as we now argue. Constraint (5.6c) implies that any feasible point of (5.6) has $z_j = 0$ and hence these two optimization problems have the same objective function. Their variables and constraints are different in that PBO directly enforces the per-node balance condition while NBO (5.6) has the additional variable ϕ and constraints (5.6c) and (5.6d). Any optimal point $(\theta^*, \omega^*, P^{g*}, P^{l^*})$ for PBO however defines a feasible point $(\theta^*, \phi, \omega^*, P^{g*}, P^{l*})$ for NBO (5.6) with the same cost where $\phi = \theta^*$. The point $(\theta^*, \phi, \omega^*, P^{g*}, P^{l*})$ satisfies (5.6c) and (5.6d) because $(\theta^*, \omega^*, P^{g*}, P^{l^*})$ satisfies (5.6b), $\omega^* = 0$ and $\theta_i^* = \theta_j^*$ by Theorem 4.2 in Chapter 4, and $\underline{\theta}_{ij} \leq 0 \leq \overline{\theta}_{ij}$ by Assumption 4.1. Furthermore, reference [7] reveals that the objective function augmented with z_j^2 can improve the convergence even if a feasible point of (5.6) has $z_j = 0$.

Even though neither the network balance condition (5.5) nor the line flow limits (5.4) are explicitly enforced in (5.6), they are satisfied at optimality (Theorem 5.2). Indeed, the virtual phase angles ϕ and the conditions (5.6b)–(5.6d) are carefully designed to enforce these conditions and to restore the nominal frequency $\omega^* = 0$ at optimality. This technique is previously used in [1].

Remark 5.2 ***(Virtual Phase Angles ϕ)*** We have the following helpful remarks on the virtual phase angles ϕ:

1) Under mild conditions, $\omega^* = 0$ at optimality for both PBO and NBO. For NBO, this is a consequence of the constraint (5.6c) on ϕ; see Lemma 5.1. Summing (5.6c) over all $j \in \mathcal{N}$ also implies the network balance condition (5.5) since $\mathbf{1}^T \hat{U}(\phi) = \mathbf{1}^T CBC^T \phi = 0$.
2) In PBO, $\theta_i^* = \theta_j^*$ at optimality (i.e. tie-line flows are restored $P_{ij}^* = 0$) and $U(\theta^*, \omega^*) = 0$. This does not necessarily hold in NBO. However ϕ is regarded as virtual phase angles because, at optimality, ϕ^* differs from the real phase angles θ^* only by a constant, $\phi^* - \theta^* = \mathbf{1}(\phi_0 - \theta_0)$ (see Lemma 5.1). Hence $C^T\phi^* - C^T\theta^* = C^T \cdot \mathbf{1}(\phi_0 - \theta_0) = 0$, implying $\tilde{\phi}_{ij}^* = \tilde{\theta}_{ij}^*$. Then the constraints (5.6d) are exactly the flow constraints (5.4). In other words, we impose the flow constraints on $\tilde{\theta}$ indirectly by enforcing such constraints on the virtual angle $\tilde{\phi}$.

5.3.2 Distributed Optimal Controller

The control laws are given by

$$\dot{\lambda}_j = \gamma_j^{\lambda}\left(P_j^g(t) - P_j^l(t) - p_j - \hat{U}_j(\tilde{\phi}(t))\right), \quad j \in \mathcal{N} \tag{5.7a}$$

$$\dot{\eta}_{ij}^{+} = \gamma_{ij}^{\eta}[\tilde{\phi}_{ij}(t) - \overline{\theta}_{ij}]^{+}_{\eta_{ij}^{+}}, \qquad \forall (i,j) \in \mathcal{E} \tag{5.7b}$$

$$\dot{\eta}_{ij}^{-} = \gamma_{ij}^{\eta}[\underline{\theta}_{ij} - \tilde{\phi}_{ij}(t)]^{+}_{\eta_{ij}^{-}}, \qquad \forall (i,j) \in \mathcal{E} \tag{5.7c}$$

$$\dot{\tilde{\phi}}_{ij} = \gamma_{ij}^{\phi}\left(B_{ij}[\lambda_i(t) - \lambda_j(t) + z_i(t) - z_j(t)] + \eta_{ij}^{-}(t) - \eta_{ij}^{+}(t)\right), \qquad \forall (i,j) \in \mathcal{E} \tag{5.7d}$$

$$u_j^g(t) = \left[P_j^g(t) - \gamma_j^g\left(\alpha_j P_j^g(t) + \omega_j(t) + z_j(t) + \lambda_j(t)\right)\right]_{\underline{P}_j^g}^{\overline{P}_j^g} + \frac{\omega_j(t)}{R_j}, \qquad j \in \mathcal{N} \tag{5.7e}$$

$$u_j^l(t) = \left[P_j^l(t) - \gamma_j^l\left(\beta_j P_j^l(t) - \omega_j(t) - z_j(t) - \lambda_j(t)\right)\right]_{\underline{P}_j^l}^{\overline{P}_j^l}, \quad j \in \mathcal{N} \tag{5.7f}$$

For any $x_i, a_i \in \mathbb{R}$, the operator $[x_i]^{+}_{a_i}$ is defined by

$$[x_i]^{+}_{a_i} := \begin{cases} x_i & \text{if } a_i > 0 \text{ or } x_i > 0; \\ 0, & \text{otherwise} \end{cases} \tag{5.8}$$

For a vector case, $[x]^{+}_{a}$ is defined accordingly component wise [8].

Here we assume that each node j updates a set of internal states $(\lambda_j(t), \eta_{ij}^{+}(t), \eta_{ji}^{-}(t), \tilde{\phi}_{ij}(t))$ according to (5.7a)–(5.7d).[2] In contrast to the *completely decentralized* control derived in the per-node balance case, here the control is *distributed* where each node j updates $(\lambda_j(t), \eta_{ij}^{+}(t), \eta_{ij}^{-}(t))$ using only local measurements or computation but requires the information $(\lambda_j(t), z_j(t))$ from its neighbors i to update $\tilde{\phi}_{ij}(t)$. Note that $z_j(t)$ is not a variable but a shorthand for (function) $P_j^g(t) - P_j^l(t) - p_j - \hat{U}_j(\tilde{\phi}(t))$. The control inputs $(u_j^g(t), u_j^l(t))$ in (5.7e) (5.7f) are functions of the network state $(P_j^g(t), P_j^l(t), \omega_j(t))$ and the internal state $(\lambda_j(t), \eta_{ij}^{+}(t), \eta_{ij}^{-}(t), \tilde{\phi}_{ij}(t))$. We write u_j^g and u_j^l as functions of $(P_j^g, P_j^l, \omega_j, \lambda_j)$: for $j \in \mathcal{N}$,

$$u_j^g(t) := u_j^g\left(P_j^g(t), \omega_j(t), \lambda_j(t), z_j(t)\right) \tag{5.9a}$$

$$u_j^l(t) := u_j^l\left(P_j^l(t), \omega_j(t), \lambda_j(t), z_j(t)\right) \tag{5.9b}$$

where the functions are defined by the right-hand sides of (5.7e) and (5.7f).

Now we comment on the implementation of the control (5.7).

2 For each (directed) link $(i,j) \in \mathcal{E}$, we assume that only node i maintains the variables $(\eta_{ij}^{+}(t), \eta_{ji}^{-}(t), \tilde{\phi}_{ij}(t))$. In practice, node j will probably maintain symmetric variables to reduce communication burden or for other reasons outside our mathematical model here.

Remark 5.3 ***(Implementation)*** Here, we give an introduction about the practical implementation:

- As discussed above, communication is needed only between neighboring nodes (areas) to update the variables $\tilde{\phi}_{ij}(t)$.
- Similar to the per-node power balance case, we can avoid measuring the load change p_j by using (5.1b) and the definition of $z_j(t)$ to replace (5.7a) with

$$z_j(t) = M_j\dot{\omega}_j + D_j\omega_j(t) - \sum_{i:i\to j} P_{ij}(t) + \sum_{k:j\to k} P_{jk}(t) - \hat{U}_j(\tilde{\phi}(t)) \tag{5.10a}$$

$$\dot{\lambda}_j = \gamma_j^{\lambda}\, z_j(t) \tag{5.10b}$$

The controller (5.7) can also achieve per-node balance by setting $\underline{\theta}_{ij} = \overline{\theta}_{ij} = 0$; it is still distributed but needs more computation and communication compared with the controller in Chapter 4.

5.3.3 Design Rationale

The controller design (5.7) is also motivated by the (partial) primal–dual gradient algorithm for (5.6), as for the per-node power balance case.

5.3.3.1 Primal–Dual Gradient Algorithms

The optimization problem in the network balance case differs from that in the per-node balance case in the inequalities (5.6d) on $\tilde{\phi}$. Consider a general constrained convex optimization with equality and inequality constraints:

$$\begin{aligned} \min_{x\in X} \quad & f(x) \\ \text{s.t.} \quad & g(x) = 0 \\ & h(x) \le 0 \end{aligned} \tag{5.11}$$

where $f : \mathbb{R}^n \to \mathbb{R}$, $g : \mathbb{R}^n \to \mathbb{R}^{k_1}$, $h : \mathbb{R}^n \to \mathbb{R}^{k_2}$, and $X \subseteq \mathbb{R}^n$ are closed and convex. Here an inequality constraint $h(x) \le 0$ is imposed explicitly. Let $\rho_1 \in \mathbb{R}^{k_1}$ be the Lagrange multiplier associated with the equality constraint $g(x) = 0$, $\rho_2 \in \mathbb{R}^{k_2}$ that associated with the inequality constraint $h(x) \le 0$ and $\rho := (\rho_1, \rho_2)$. Define the Lagrangian as

$$L(x; \rho) := f(x) + \rho_1^{\mathrm{T}} g(x) + \rho_2^{\mathrm{T}} h(x).$$

A standard primal–dual algorithm takes the form

$$x(t+1) := \mathcal{P}_X\left(x(t) - \Gamma^x\, \nabla_x L(x(t); \rho(t))\right) \tag{5.12a}$$

$$\rho_1(t+1) := \rho_1(t) + \Gamma^{\rho_1}\, \nabla_{\rho_1} L(x(t); \rho(t)) \tag{5.12b}$$

$$\rho_2(t+1) := \left(\rho_2(t) + \Gamma^{\rho_2}\, \nabla_{\rho_2} L(x(t); \rho(t))\right)^+ \tag{5.12c}$$

where Γ^x, Γ^{ρ_1}, and Γ^{ρ_2} are strictly positive diagonal gain matrices. Here, if a is a scalar, then $(a)^+ := \max\{a, 0\}$, and if a is a vector, then $(a)^+$ is defined accordingly component wise. For a dual algorithm, (5.12a) is replaced by

$$x(t) := \min_{x \in X} L(x; \rho(t)) \tag{5.12d}$$

As for the per-node balance case, all variables in $x(t)$ are updated according to (5.12a) except $\omega(t)$ that is updated according to (5.12d), as we see below.

The set X in (5.12a) is defined by the constraints (5.2):

$$X := \left\{ (P^g, P^l) : (\underline{P}^g, \underline{P}^l) \le (P^g, P^l) \le (\overline{P}^g, \overline{P}^l) \right\} \tag{5.13}$$

5.3.3.2 Controller Design

Let $\rho_1 := (\lambda, \mu)$ be the Lagrange multipliers associated with constraints (5.6c) and (5.6b), respectively, $\rho_2 := (\eta^+, \eta^-)$ the multipliers associated with constraints (5.6d), and $\rho := (\rho_1, \rho_2)$. Define the Lagrangian of (5.6) by (5.14). Note that it is only a function of (x, ρ) and independent of $u := (u^g, u^l)$ as we treat u as a function of (x, ρ) defined by the right-hand sides of (5.7e) and (5.7f).

$$\begin{aligned} L_2(x;\rho) = {} & \frac{1}{2}\left(\sum_{j\in\mathcal{N}} \alpha_j \left(P_j^g\right)^2 + \sum_{j\in\mathcal{N}} \beta_j \left(P_j^l\right)^2 + \sum_{j\in\mathcal{N}} D_j \omega_j^2 + \sum_j z_j^2 \right) \\ & + \sum_{j\in\mathcal{N}} \mu_j \left(P_j^g - P_j^l - p_j - D_j\omega_j + \sum_{i:i\to j} B_{ij}\theta_{ij} - \sum_{k:j\to k} B_{jk}\theta_{jk} \right) \\ & + \sum_{j\in\mathcal{N}} \lambda_j \left(P_j^g - P_j^l - p_j + \sum_{i:i\to j} B_{ij}\phi_{ij} - \sum_{k:j\to k} B_{jk}\phi_{jk} \right) \\ & + \sum_{(i,j)\in\mathcal{E}} \eta_{ij}^- \left(\underline{\theta}_{ij} - \phi_{ij} \right) + \sum_{(i,j)\in\mathcal{E}} \eta_{ij}^+ \left(\phi_{ij} - \overline{\theta}_{ij} \right) \end{aligned} \tag{5.14}$$

The closed-loop dynamics (5.1) and (5.7) carry out an approximate primal–dual algorithm (5.12) for solving (5.6) in real time over the coupled physical power network and cyber computation. Since the reasoning is similar to the per-node balance case, we only provide a summary. Rewrite the Lagrangian L_2 in vector form

$$\begin{aligned} L_2(x;\rho) = {} & \frac{1}{2}\left((P^g)^{\mathrm{T}} A^g P^g + (P^l)^{\mathrm{T}} A^l P^l + \omega^{\mathrm{T}} D\omega + z^{\mathrm{T}} z \right) \\ & + \lambda^{\mathrm{T}} \left(P^g - P^l - p - CB\tilde{\phi} \right) \\ & + \mu^{\mathrm{T}} \left(P^g - P^l - p - D\omega - CB\tilde{\theta} \right) \\ & + (\eta^+)^{\mathrm{T}} \left(\tilde{\phi} - \overline{\theta} \right) + (\eta^-)^{\mathrm{T}} \left(\underline{\theta} - \tilde{\phi} \right) \end{aligned} \tag{5.15}$$

where $A^l := \mathrm{diag}(\beta_j, j \in \mathcal{N})$ and $B := \mathrm{diag}(B_{ij}, (i,j) \in \mathcal{E})$.

First, the control (5.7b) and (5.7c) can be interpreted as a continuous-time version of the dual update (5.12c) on the dual variable $\rho_2 := (\eta^+(t), \eta^-(t))$:

$$\dot{\eta}^+ = \Gamma^\eta \left[\nabla_{\eta^+} L_2(x(t); \rho(t))\right]^+_{\eta^+(t)} \tag{5.16a}$$

$$\dot{\eta}^- = \Gamma^\eta \left[\nabla_{\eta^-} L_2(x(t); \rho(t))\right]^+_{\eta^-} \tag{5.16b}$$

where $\Gamma^\eta := \mathrm{diag}(\gamma^\eta_{ij}, (i,j) \in \mathcal{E})$.

Second, the control (5.7a) carries out the dual update (5.12b) on $\lambda(t)$:

$$\dot{\lambda} = \Gamma^\lambda \nabla_\lambda L_2(x(t); \rho(t)) \tag{5.16c}$$

where $\Gamma^\lambda := \mathrm{diag}(\gamma^\lambda_j, j \in \mathcal{N})$. The swing dynamic (5.1b) carries out the dual update (5.12b) on $\mu(t)$ because, as in the per-node balance case, we can identify $\mu(t) \equiv \omega(t)$ so that

$$\dot{\mu} = \dot{\omega} = M^{-1} \nabla_\mu L_2(x(t); \rho(t)) \tag{5.16d}$$

where $M := \mathrm{diag}(M_j, j \in \mathcal{N})$.

Finally we show that (5.1a), (5.1c), (5.1d), and (5.7d) implement a mix of the primal updates (5.12a) and (5.12d) on the primal variables denoted by

$$x := (\tilde{\theta}(t); \tilde{\phi}(t); \omega(t); P^g(t); P^l(t)).$$

Setting $\omega(t) \equiv \mu(t)$ is equivalent to the primal update (5.12d) on $\omega(t)$, as in the per-node balance case. Moreover the control laws (5.7e) and (5.7f) are then equivalent to

$$T^g \dot{P}^g = \left[P^g(t) - \Gamma^g \nabla_{P^g} L_2(x(t); \rho(t))\right]^{\overline{P}^g}_{\underline{P}^g} - P^g(t) \tag{5.16e}$$

$$T^l \dot{P}^l = \left[P^l(t) - \Gamma^l \nabla_{P^l} L_2(x(t); \rho(t))\right]^{\overline{P}^l}_{\underline{P}^l} - P^l(t) \tag{5.16f}$$

i.e. the generator and controllable load at each node j carry out the primal update (5.12a). For $(\tilde{\theta}, \tilde{\phi})$, (5.1a) and (5.7d) are equivalent to the primal update (5.12a):

$$\dot{\tilde{\theta}} = -B^{-1} \nabla_{\tilde{\theta}} L_2(x(t); \rho(t)) \tag{5.16g}$$

$$\dot{\tilde{\phi}} = -\Gamma^\phi \nabla_{\tilde{\phi}} L_2(x(t); \rho(t)) \tag{5.16h}$$

where $\Gamma^\phi := \mathrm{diag}(\gamma^\phi_{ij}, (i,j) \in \mathcal{E})$.

5.4 Optimality of Equilibrium

In this subsection, we address the optimality of the equilibrium point of the closed-loop system (5.1) and (5.7). Denote

$$(x, \rho) := \left((\tilde{\theta}, \tilde{\phi}, \omega, P^g, P^l), (\lambda, \mu), (\eta^-, \eta^+)\right).$$

Then given a point (x, ρ), recall the control input $u(x, \rho_1, \rho_2)$ is given by (5.9).

Definition 5.1 A point $(x^*, \rho^*) := (\tilde{\theta}^*, \tilde{\phi}^*, \omega^*, P^{g*}, P^{l*}, \lambda^*, \eta^{+*}, \eta^{-*}, \mu^*)$ is an equilibrium point or an equilibrium of the closed-loop system (5.1) and (5.7) if the following are met:

1) The right-hand side of (5.1) vanishes at x^* and $u(x^*, \rho^*)$.
2) The right-hand side of (5.7a)–(5.7d) vanishes at (x^*, ρ^*).

Definition 5.2 A point (x^*, ρ^*) is *primal–dual optimal* if $(x^*, u(x^*, \rho^*))$ is optimal for (5.6) and ρ^* is optimal for its dual problem.

We make the following assumption:

Assumption 5.2 The problem (5.6) is feasible.

The following theorem characterizes the correspondence between the equilibrium of the closed-loop system (5.1) and (5.7) and the primal–dual optimal solution to (5.6).

Theorem 5.1 *Suppose Assumption 5.2 holds. A point (x^*, ρ^*) is primal–dual optimal if and only if (x^*, ρ^*) is an equilibrium of closed-loop system* (5.1) and (5.7) *satisfying* (5.2) *and $\mu^* = 0$.*

Next result says that, at equilibrium, the network balance condition (5.5) and line flow limits (5.4) are satisfied and the nominal frequency is restored. Moreover the equilibrium is unique.

Theorem 5.2 *Suppose Assumptions 5.1 and 5.2 hold. Let (x^*, ρ^*) be primal–dual optimal:*

1) *The equilibrium (x^*, μ^*) is unique, with (θ^*, ϕ^*) being unique up to reference angles (θ_0, ϕ_0).*
2) *The nominal frequency is restored, i.e. $\omega_j^* = 0$ for all $j \in \mathcal{N}$; moreover $\tilde{\phi}_{ij}^* = \tilde{\theta}_{ij}^*$ for all $(i,j) \in \mathcal{E}$.*
3) *The network balance condition (5.5) is satisfied by x^*.*
4) *The line limits (5.4) are satisfied by x^*, implying $\underline{P}_{ij} \leq P_{ij} \leq \overline{P}_{ij}$ on every tie line $(i,j) \in \mathcal{E}$.*

Theorem 5.2 shows that the equilibrium point has a simple yet intuitive structure. Moreover, it implies that the closed-loop system can autonomously eliminate congestions on tie lines. This feature has important implications. It means our distributed frequency control is capable of serving as a corrective redispatch without the coordination of dispatch centers if a congestion arises.

This can enlarge the feasible region for economic dispatch, since corrective redispatch has been naturally taken into account.

To prove Theorem 5.1 and 5.2, we start with the following two lemmas:

Lemma 5.1 *Suppose (x^*, u^*) is optimal for (5.6):*

1) $\omega^* = 0$, *i.e. the nominal frequency is restored.*
2) *The network balance condition (5.5) is satisfied by x^*.*
3) $\phi^* - \theta^* = (\phi_0^* - \theta_0^*)\,\mathbf{1}$.
4) $\underline{\theta}_{ij} \leq \theta_{ij}^* \leq \overline{\theta}_{ij}$, *i.e. the line limits (5.4) are satisfied.*

Proof: Suppose (x^*, u^*) is optimal but $\omega^* \neq 0$. Then (5.6c) implies

$$P^{g*} - P^{l*} - p = CBC^{\mathrm{T}}\phi^* \tag{5.17}$$

Consider $\hat{x} := (\hat{\theta}, \phi^*, \hat{\omega}, P^{g*}, P^{l*})$ with $\hat{\theta} := \phi^*$ and $\hat{\omega} := 0$. Then $\hat{x}$ satisfies (5.6b) (5.6c) due to (5.17). Hence $(\hat{x}, u^*)$ is feasible for (5.6) but has a strictly lower cost, contradicting the optimality of (x^*, u^*). Hence $\omega^* = 0$, which proves assertion 1).

Multiplying both sides of (5.17) by $\mathbf{1}^{\mathrm{T}}$ yields the network balance condition (5.5), proving assertion 2).

To prove assertion 3), setting $\omega^* = 0$ in (5.6b) and combining with (5.6c) yield

$$CBC^{\mathrm{T}}\theta^* = P^{g*} - P^{l*} - p = CBC^{\mathrm{T}}\phi^* \tag{5.18}$$

Since CBC^{T} is an $(n+1) \times (n+1)$ matrix with rank n, its null space has dimension 1 and is spanned by $\mathbf{1}$ because $C^{\mathrm{T}}\mathbf{1} = 0$. Hence $CBC^{\mathrm{T}}(\phi^* - \theta^*) = 0$ implies that $\phi^* - \theta^* = (\phi_0^* - \theta_0^*)\,\mathbf{1}$. To prove 4), note that $\tilde{\phi} = C^{\mathrm{T}}\phi$ and $\tilde{\theta} = C^{\mathrm{T}}\theta$, and hence we have

$$\tilde{\phi}^* - \tilde{\theta}^* = C^{\mathrm{T}}(\phi^* - \theta^*) = (\phi_0^* - \theta_0^*)C^{\mathrm{T}}\mathbf{1} = 0 \tag{5.19}$$

i.e. $\tilde{\phi}^* = \tilde{\theta}^*$. We conclude from (5.6d) that $\underline{\theta}_{ij} \leq \theta_{ij}^* \leq \overline{\theta}_{ij}$. This completes the proof. ■

Lemma 5.2 *Suppose (x^*, ρ^*) is primal–dual optimal for (5.6). Then*

$$\begin{aligned} u_j^{g*} &= P_j^{g*} = \left[P_j^{g*} - \gamma_j^g\left(\alpha_j P_j^{g*} + \omega_j^* + z_j^* + \lambda_j^*\right)\right]_{\underline{P}_j^g}^{\overline{P}_j^g} \\ u_j^{l*} &= P_j^{l*} = \left[P_j^{l*} - \gamma_j^l\left(\beta_j P_j^{l*} - \omega_j^* - z_j^* - \lambda_j^*\right)\right]_{\underline{P}_j^l}^{\overline{P}_j^l} \end{aligned} \tag{5.20}$$

for any $\gamma_j^g > 0$ and $\gamma_j^l > 0$.

Remark 5.4 Since (5.6) is (strictly) convex with linear constraints, strong duality holds. Hence (x^*, ρ^*) is primal–dual optimal if and only if it satisfies the KKT condition: $(x^*, u(x^*, \rho^*))$ is primal feasible and

$$x^* = \arg\min_x \{L_2(x; \rho^*) | (x, u(x, \rho^*)) \text{ satisfies } (5.2), (5.6e), (5.6f)\} \tag{5.21}$$

The proof is the same as the proof of Lemma 4.2, which is omitted here.

Lemma 5.2 shows that the saturation of control input does not impact the optimal solution to optimization problem (5.6).

With Lemmas 5.1, 5.2, and 5.3, we now can prove Theorems 5.1 and 5.2.

Proof:

1) **Sufficiency.** Suppose (x^*, ρ^*) is primal–dual optimal. Then x^* satisfies the operational constraints (5.2). Moreover the right-hand side of (5.1) vanishes because of the following:
   - $\dot{\theta} = 0$ since $\omega^* = 0$ from Lemma 5.1.
   - $\dot{\omega} = 0$ since constraint (5.6b) holds for x^*.
   - $\dot{P}^g = \dot{P}^l = 0$ since $\omega^* = 0$ and x^* satisfies (5.6e) and (5.6f).
   - $\dot{\lambda} = 0$ since (5.6c) holds for x^*.
   - $\dot{\eta}^+ = \dot{\eta}^- = 0$ since (5.6d) holds for x^*.
   - From (5.16h) we have

   $$\dot{\tilde{\phi}} = -\Gamma^{\tilde{\phi}}\ \nabla_{\tilde{\phi}} L_2(x^*; \rho^*) \tag{5.22}$$

   Since (x^*, ρ^*) is a saddle point, we have $\nabla_{\tilde{\phi}} L_2(x^*; \rho^*) = 0$, implying $\dot{\tilde{\phi}} = 0$.

   Therefore, (x^*, ρ^*) is an equilibrium of the closed-loop system (5.1) and (5.7) that satisfies the operational constraints (5.2). Moreover $\mu^* = \omega^* = 0$ since $\frac{\partial L_2}{\partial \omega_j}(x^*, \rho^*) = D_j(\omega_j^* - \mu_j^*) = 0$ and $D_j > 0$ for all $j \in \mathcal{N}$.

2) **Necessity.** Suppose now (x^*, ρ^*) is an equilibrium of the closed-loop system (5.1) and (5.7) that satisfies (5.2) and $\mu^* = 0$. Since (5.6) is convex with linear constraints, (x^*, ρ^*) is a primal–dual optimal if and only if $(x^*, u(x^*, \rho^*))$ is primal feasible and satisfies

   $$x^* = \arg\min_{x} \{L_2(x; \rho^*) | (x, u(x, \rho^*)) \text{ satisfies (5.2), (5.6e), and (5.6f)}\} \tag{5.23}$$

   This is because $\nabla_{\rho_1} L_2(x^*; \rho^*) = 0$ due to $\dot{\mu} = \dot{\lambda} = 0$, $\eta^{+*} \geq 0$, $\eta^{-*} \geq 0$, and the complementary slackness condition $\eta_{ij}^{+*}(\tilde{\phi}_{ij}^* - \overline{\theta}_{ij}) = 0$, $\eta_{ij}^{-*}(\underline{\theta}_{ij} - \tilde{\phi}_{ij}^*) = 0$ is satisfied since $\dot{\eta}^+ = \dot{\eta}^- = 0$.

   To show that $(x^*, u(x^*, \rho^*))$ is primal feasible, note that since $(x^*, u(x^*, \rho^*))$ is an equilibrium of (5.1), it satisfies $\omega^* = 0$ and hence (5.6e) (5.6f), in addition to (5.2). Moreover $\dot{\omega} = 0$ means $(x^*, u(x^*, \rho^*))$ satisfies (5.6b), $\dot{\lambda} = 0$ implies (5.6c), and $\dot{\eta}^+ = \dot{\eta}^- = 0$ implies (5.6d).

   To show that (x^*, ρ^*) satisfies (5.23), note that (5.6e), (5.6f), and (5.7e), (5.7f) imply that

   $$\begin{aligned} P_j^{g*} &= \left[P_j^{g*} - \gamma_j^g \left(\alpha_j P_j^{g*} + \omega_j^* + z_j^* + \lambda_j^*\right)\right]_{\underline{P}_j^g}^{\overline{P}_j^g} \\ P_j^{l*} &= \left[P_j^{l*} - \gamma_j^l \left(\beta_j P_j^{l*} - \omega_j^* - z_j^* - \lambda_j^*\right)\right]_{\underline{P}_j^l}^{\overline{P}_j^l} \end{aligned} \tag{5.24}$$

The rest of the proof follows the same line of argument as that in Theorem 4.1. This proves that (x^*, ρ^*) is primal–dual optimal and completes the proof of Theorem 5.1. ■

Next we prove Theorem 5.2.

Proof: [Proof Theorem 5.2]
Let $(x^*, \rho^*) = (\tilde{\theta}^*, \tilde{\phi}^*, \omega^*, P^{g*}, P^{l*}, \lambda^*, \eta^{+*}, \eta^{-*}, \mu^*)$ be primal–dual optimal. For the uniqueness of x^*, $(\omega^*, P^{g*}, P^{l*})$ are unique because the objective function in (5.6) is strictly convex in (ω, P^g, P^l). Hence $\mu^* = \omega^*$ is unique as well. Assumption 5.1 that $\phi_0^* = 0$ and (5.17) imply that ϕ^* is uniquely determined by the equilibrium $(\omega^*, P^{g*}, P^{l*})$. Since $\theta^* - \phi^* = (\theta_0^* - \phi_0^*)\mathbf{1}$, Assumption 5.1 is that $\theta_0^* = 0$ and then implies that θ^* is unique. This proves the uniqueness of (x^*, μ^*).
The remaining three parts of the theorem follow from Lemma 5.1. ■

5.5 Asymptotic Stability

In this subsection, we address the asymptotic stability of the closed-loop system (5.1) and (5.7), under an additional assumption.

Assumption 5.3 The initial state of the closed-loop system (5.1) (5.7) is finite, and $p_j^g(0), p_j^l(0)$ satisfy constraint (5.2).

Similar to that in the per-node balance case, the closed-loop system (5.1) and (5.7) satisfies the constraint (5.2) even during transient, which is introduced in the following lemma.

Lemma 5.3 *Suppose Assumptions 5.1 and 5.3 hold. Then constraint* (5.2) *is satisfied for all* $t > 0$, *i.e.* $(P^g(t), P^l(t)) \in X$ *for all* $t \geq 0$ *where* X *is defined in* (5.13).

The proof is similar to that of Theorem 4.3, which is omitted here.
Similar to the per-node balance case, we first rewrite the closed-loop system using states $\tilde{\theta}, \tilde{\phi}$ instead of θ, ϕ (they are equivalent under Assumption 5.1). Setting $\mu \equiv \omega$, the closed-loop system (5.1) and (5.7) is equivalent to (in vector form)

$$\dot{\tilde{\theta}}(t) = C^{\mathrm{T}}\omega(t) \tag{5.25a}$$

$$\dot{\omega}(t) = M^{-1}\left(P^g(t) - P^l(t) - p - D\omega(t) - CB\tilde{\theta}(t)\right) \tag{5.25b}$$

$$\dot{P}^g(t) = (T^g)^{-1}\left(-P^g(t) + \hat{u}^g(t)\right) \tag{5.25c}$$

$$\dot{P}^l(t) = (T^l)^{-1}\left(-P^l(t) + \hat{u}^l(t)\right) \tag{5.25d}$$

$$\dot{\eta}^+(t) = \Gamma^\eta[\tilde{\phi}(t) - \overline{\theta}]^+_{\eta^+} \tag{5.25e}$$

$$\dot{\eta}^-(t) = \Gamma^\eta[\underline{\theta} - \tilde{\phi}(t)]^+_{\eta^-} \tag{5.25f}$$

$$\dot{\lambda}(t) = \Gamma^\lambda \left(P^g(t) - P^l(t) - p - CB\tilde{\phi}(t)\right) \tag{5.25g}$$

$$\dot{\tilde{\phi}}(t) = \Gamma^{\tilde{\phi}} \left(BC^{\mathrm{T}}\lambda(t) + BC^{\mathrm{T}}z(t) + \eta^-(t) - \eta^+(t)\right) \tag{5.25h}$$

where

$$z(t) := P^g(t) - P^l(t) - p - CB\tilde{\phi}(t)$$

and

$$\hat{u}^g(t) = \left[P^g(t) - \Gamma^g\left(A^g P^g_j(t) + \omega(t) + z(t) + \lambda(t)\right)\right]^{\overline{P}^g}_{\underline{P}^g} \tag{5.25i}$$

$$\hat{u}^l_j(t) = \left[P^l(t) - \Gamma^l\left(A^l P^l_j(t) - \omega(t) - z(t) - \lambda(t)\right)\right]^{\overline{P}^l}_{\underline{P}^l} \tag{5.25j}$$

where $A^g := \mathrm{diag}\{\alpha_j, j \in \mathcal{N}\}$ and $A^l := \mathrm{diag}\{\beta_j, j \in \mathcal{N}\}$.

For simplicity, denote $w := (\tilde{\theta}, \omega, P^g, P^l, \lambda, \eta^+, \eta^-, \tilde{\phi})$. Note that the right-hand sides of (5.25e) and (5.25f) are discontinuous due to projection to the nonnegative quadrant for $(\eta^+(t), \eta^-(t))$. The system (5.25) is called a projected dynamical system, and we adopt the concept of Caratheodory solutions for such a system where a trajectory $(w(t), t \geq 0)$ is called a Caratheodory solution, or just a solution, to (5.25) if it is absolutely continuous in t and satisfies (5.25) almost everywhere. The result in [[9], Theorems 2 and 3] implies that, given any initial state, there exists a unique solution trajectory to the closed-loop system (5.25) as the unprojected system is Lipschitz and the nonnegative quadrant is closed and convex. See [[10], Theorem 3.1] for extension of this result to the Hilbert space.

With regard to system (5.25), we first define two sets σ^+ and σ^- according to reference [7], which are given by

$$\sigma^+ := \{(i,j) \in \mathcal{E} \mid \eta^+_{ij} = 0,\ \tilde{\phi}_{ij} - \overline{\theta}_{ij} < 0\} \tag{5.26}$$

$$\sigma^- := \{(i,j) \in \mathcal{E} \mid \eta^-_{ij} = 0,\ \underline{\theta}_{ij} - \tilde{\phi}_{ij} < 0\} \tag{5.27}$$

Then (5.7b) and (5.7c) are equivalent to

$$\dot{\eta}^+_{ij} = \begin{cases} \gamma^\eta_{ij}(\tilde{\phi}_{ij} - \overline{\theta}_{ij}), & \text{if } (i,j) \notin \sigma^+; \\ 0, & \text{if } (i,j) \in \sigma^+. \end{cases} \tag{5.28a}$$

$$\dot{\eta}^-_{ij} = \begin{cases} \gamma^\eta_{ij}(\underline{\theta}_{ij} - \tilde{\phi}_{ij}), & \text{if } (i,j) \notin \sigma^-; \\ 0, & \text{if } (i,j) \in \sigma^-. \end{cases} \tag{5.28b}$$

In a fixed σ^+, σ^-, define $F(w)$.

$$F(w) = \begin{bmatrix} -B^{1/2}C^{\mathrm{T}}\omega \\ -M^{-1/2}\left(P^g - P^l - p - D\omega - CB\tilde{\theta}\right) \\ (T^g)^{-1}\left(A^g P^g + \omega + z + \lambda\right) \\ (T^l)^{-1}\left(A^l P^l - \omega - z - \lambda\right) \\ -(\Gamma^{\eta})^{1/2}[\tilde{\phi} - \overline{\theta}]^+_{\eta^+} \\ -(\Gamma^{\eta})^{1/2}[\underline{\theta} - \tilde{\phi}]^+_{\eta^-} \\ -(\Gamma^{\lambda})^{1/2}\left(P^g - P^l - p - CB\tilde{\phi}\right) \\ -(\Gamma^{\tilde{\phi}})^{1/2}\left(BC^{\mathrm{T}}\lambda + BC^{\mathrm{T}}z + \eta^- - \eta^+\right) \end{bmatrix} \tag{5.29}$$

If σ^+ and σ^- do not change, $F(w)$ is continuously differentiable in w.

Similarly, we define $S := \mathbb{R}^{m+n+1} \times X \times \mathbb{R}^{2m+n+1+m}$, where the closed convex set X is defined in (5.13). Then for any w we define the projection of $w - F(w)$ onto S as

$$\begin{aligned} H(w) &= \mathcal{P}_S(w - F(w)) \\ &= \arg\min_{y \in S} \| y - (w - F(w))\|_2 \end{aligned}$$

Then the closed-loop system (5.25) is equivalent to

$$\dot{w}(t) = \Gamma_2(H(w(t)) - w(t)) \tag{5.30}$$

where the positive definite gain matrix is

$$\Gamma_2 = \mathrm{diag}\left(B^{-1/2}, M^{-1/2}, (T^g)^{-1}, (T^l)^{-1}, (\Gamma^{\lambda})^{1/2}, (\Gamma^{\eta})^{1/2}, (\Gamma^{\eta})^{1/2}, (\Gamma^{\tilde{\phi}})^{1/2}\right)$$

Note that the projection operation H has an effect only on $(P^g; P^l)$ and Lemma 5.3 indicates that $w(t) \in S$ for all $t > 0$, justifying the equivalence of (5.25) and (5.30).

A point $w^* \in S$ is an equilibrium of the closed-loop system (5.30) if and only if it is a fixed point of the projection $H(w^*) = w^*$. Let

$$E_2 := \{ \ w \in S \mid H(w(t)) - w(t) = 0 \ \}$$

be the set of equilibrium points. Then we have the following theorem.

Theorem 5.3 *Suppose Assumptions 5.1, 5.2, and 5.3 hold. Starting from any initial point $w(0)$, $w(t)$ remains in a bounded set for all t and $w(t) \to w^*$ as $t \to \infty$ for some equilibrium $w^* \in E_2$ that is optimal for problem (5.6).*

For any equilibrium point w^*, we define the following function taking the same form as the per-node case:

$$\begin{aligned} \tilde{V}_2(w) = &-(H(w) - w)^{\mathrm{T}}F(w) - \frac{1}{2}||H(w) - w||_2^2 \\ &+ \frac{1}{2}k(w - w^*)^{\mathrm{T}}\Gamma_2^{-2}(w - w^*) \end{aligned} \tag{5.31}$$

where k is small enough such that $\Gamma_2 - k\Gamma_2^{-1} > 0$ is strictly positive definite.

For any fixed σ^+ and σ^-, $\tilde{V}_2$ is continuously differentiable as $F(w)$ is also continuously differentiable in this situation. Similar to $V_1(w)$ used in Chapter 4, we know $\tilde{V}_2(w) \geq 0$ on S and $\tilde{V}_2(w) = 0$ holds only at any equilibrium $w^* = H(w^*)$ [11]. Moreover, $\tilde{V}_2$ is nonincreasing for fixed σ^+ and σ^-, which will be proved in Lemma 5.4.

It is worth to note that the index sets σ^+ and σ^- may change sometimes, resulting in discontinuity of $\tilde{V}_2(w)$. To circumvent such an issue, we slightly modify the definition of $V_2(w)$ at the discontinuous points as follows:

1) $V_2(w) := \tilde{V}_2(w)$, if $\tilde{V}_2(w)$ is continuous at w.
2) $V_2(w) := \lim\sup_{v \to w} \tilde{V}_2(v)$, if $\tilde{V}_2(w)$ is discontinuous at w.

Then $V_2(w)$ is upper semicontinuous in w, and $V_2(w) \geq 0$ on S and $V_2(w) = 0$ holds only at any equilibrium $w^* = H(w^*)$. As $V_2(w)$ is not differentiable for w at discontinuous points, we use the Clarke generalized gradient at these points [12, Page 27].

Note that $\tilde{V}_2$ is continuous almost everywhere except the switching points. Hence both $V_2(w)$ is *nonpathological* [13, 14]. With these definitions and notations above, we can prove Theorem 5.3.

Now, we will prove Theorem 5.3 by starting with a lemma.

Lemma 5.4 *Suppose Assumptions 5.1, 5.2, and 5.3 hold:*

1) *$\dot{V}_2(w(t)) \leq 0$ in a fixed σ^+ and σ^-.*
2) *The trajectory $w(t)$ is bounded, i.e. there exists $\overline{w}$ such that $\|w(t)\| \leq \overline{w}$ for all $t \geq 0$.*

Proof: Given fixed σ^+, σ^-, for all $(i,j) \notin \sigma^+$, $(i,j) \notin \sigma^-$, we have

$$\dot{V}_2(w) \leq k(H(w) - w^*)^{\mathrm{T}} \cdot \Gamma_2^{-1}(H(w) - (w - F(w))) \tag{5.32a}$$

$$-(H(w) - w)^{\mathrm{T}}\Gamma_2 \cdot Q \cdot \Gamma_2(H(w) - w) \tag{5.32b}$$

$$-(H(w) - (w - F(w)))^{\mathrm{T}}(\Gamma_2 - k\Gamma_2^{-1})(w - H(w)) \tag{5.32c}$$

$$-k(w - w^*)^{\mathrm{T}} \cdot \Gamma_2^{-1}F(w) \tag{5.32d}$$

where Q is a semidefinite positive matrix and $\Gamma_2 Q = \nabla_w F(w)$, which given in (5.33). Here, the subscript of I means its dimension, and $|A|$ means the cardinality of set A.

$$Q = \begin{bmatrix} 0 & -BC^{\mathrm{T}} & 0 & 0 & 0 & 0 & 0 & 0 \\ CB & D & -I_{|\mathcal{N}|} & I_{|\mathcal{N}|} & 0 & 0 & 0 & 0 \\ 0 & I_{|\mathcal{N}|} & A^g + I_{|\mathcal{N}|} & -I_{|\mathcal{N}|} & 0 & 0 & 0 & -CB \\ 0 & -I_{|\mathcal{N}|} & -I_{|\mathcal{N}|} & A^l + I_{|\mathcal{N}|} & 0 & 0 & 0 & CB \\ 0 & 0 & -I_{|\mathcal{N}|} & I_{|\mathcal{N}|} & 0 & 0 & 0 & CB \\ 0 & 0 & 0 & 0 & 0 & 0 & 0 & -I_{|\mathcal{E}|-|\sigma^+|} \\ 0 & 0 & 0 & 0 & 0 & 0 & 0 & I_{|\mathcal{E}|-|\sigma^-|} \\ 0 & 0 & -BC^{\mathrm{T}} & BC^{\mathrm{T}} & -BC^{\mathrm{T}} & I_{|\mathcal{E}|-|\sigma^+|} & -I_{|\mathcal{E}|-|\sigma^-|} & BC^{\mathrm{T}}CB \end{bmatrix} \tag{5.33}$$

Given fixed σ^+ and σ^-, $F(w)$ is continuous differentiable. In this case, (5.32a) and (5.32c) are nonpositive due to as discussed in the proof of Lemma 4.4 in Chapter 4. (5.32b) is also nonpositive as Q is semidefinite positive (see Eq. (5.33)).

Next, we prove that (5.32d) is nonpositive. Similar to the per-node case, substituting $\mu(t) \equiv \omega(t)$ into the Lagrangian $L_2(x;\rho)$ in (5.15), we obtain a function $\hat{L}_2(w)$ defined as follows:

$$
\begin{aligned}
\hat{L}_2(w) :=& L_2\left(\tilde{\theta}, \omega, P^g, P^l, \lambda, \mu, \eta^+, \eta^-, \tilde{\phi}\right)_{\mu=\omega} \\
=& \frac{1}{2}\left((P^g)^T A^g P^g + (P^l)^T A^l P^l - \omega^T D\omega + z^T z\right) \\
&+ \lambda^T\left(P^g - P^l - p - CB\tilde{\phi}\right) + \omega^T\left(P^g - P^l - p - CB\tilde{\theta}\right) \\
&+ (\eta^-)^T\left(\underline{\theta} - \tilde{\phi}\right) + (\eta^+)^T\left(\tilde{\phi} - \overline{\theta}\right)
\end{aligned}
$$

In addition, denote $w_1 = (\tilde{\theta}, P^g, P^l, \tilde{\phi})$ and $w_2 = (\lambda, \omega, \eta^+, \eta^-)$. Then $\hat{L}_2$ is convex in w_1 and concave in w_2.

In $F(w)$, $[\tilde{\phi} - \overline{\theta}]^+_{\eta^+}$ and $[\underline{\theta} - \tilde{\phi}]^+_{\eta^-}$ have unknown dimensions (up to σ^+ and σ^-). Fortunately, we have

$$
(\eta^+ - \eta^{+*})^T[\tilde{\phi} - \overline{\theta}]^+_{\eta^+} \le (\eta^+ - \eta^{+*})^T(\tilde{\phi} - \overline{\theta}) = (\eta^+ - \eta^{+*})^T \nabla_{\eta^+}\hat{L}_2
$$

where the inequality holds since $\eta^+_{ij} = 0 \le \eta^{+*}_{ij}$ and $\tilde{\phi}_{ij} - \overline{\theta}_{ij} < 0$ for $(i,j) \in \sigma^+$, i.e. $(\eta^+_{ij} - \eta^{+*}_{ij}) \cdot (\tilde{\phi}_{ij} - \overline{\theta}_{ij}) \ge 0$. Similarly,

$$
(\eta^- - \eta^{-*})^T[\underline{\theta} - \tilde{\phi}]^+_{\eta^-} \le (\eta^- - \eta^{-*})^T(\underline{\theta} - \tilde{\phi}) = (\eta^- - \eta^{-*})^T \nabla_{\eta^-}\hat{L}_2.
$$

Consequently, it can be verified that

$$
(w - w^*)^T \Gamma_2^{-1} F(w) \le (w - w^*)^T \nabla_w \hat{L}_2(w) = (w - w^*)^T \begin{bmatrix} \nabla_{w_1}\hat{L}_2 \\ -\nabla_{w_2}\hat{L}_2 \end{bmatrix} (w_1, w_2) \tag{5.34}
$$

where $\nabla_{w_1}\hat{L}_2 := \begin{bmatrix} \nabla_{\tilde{\theta}}\hat{L}_2 \\ \nabla_{P^g}\hat{L}_2 \\ \nabla_{P^l}\hat{L}_2 \\ \nabla_{\tilde{\phi}}\hat{L}_2 \end{bmatrix}$ and $\nabla_{w_2}\hat{L}_2 := \begin{bmatrix} \nabla_{\lambda}\hat{L}_2 \\ \nabla_{\omega}\hat{L}_2 \\ \nabla_{\eta^+}\hat{L}_2 \\ \nabla_{\eta^-}\hat{L}_2 \end{bmatrix}$.

Then we have

$$
\begin{aligned}
& -k(w - w^*)^T \Gamma_2^{-1} F(w) \le -k(w - w^*)^T \cdot \Gamma_2^{-1} F(w) \\
=& -k(w_1 - w_1^*)^T \nabla_{w_1}\hat{L}_2(w_1, w_2) + k(w_2 - w_2^*)^T \nabla_{w_2}\hat{L}_2(w_1, w_2) \\
\le& k\left(\hat{L}_2(w_1^*, w_2) - \hat{L}_2(w_1, w_2) + \hat{L}_2(w_1, w_2) - \hat{L}_2(w_1, w_2^*)\right) \\
=& k\left(\underbrace{\hat{L}_2(w_1^*, w_2) - \hat{L}_2(w_1^*, w_2^*)}_{\le 0} + \underbrace{\hat{L}_2(w_1^*, w_2^*) - \hat{L}_2(w_1, w_2^*)}_{\le 0}\right) \le 0
\end{aligned} \tag{5.35}
$$

where the first inequality holds because $\hat{L}_2$ is convex in w_1 and concave in w_2 and the second inequality follows because (w_1^*, w_2^*) is a saddle point. Therefore (5.32d) is nonpositive, proving the first assertion.

To prove the second assertion, we further investigate the situation that σ^+ or σ^- changes. We only consider the set σ^+ since it is the same to σ^-. We have the following observations:

- The set σ^+ is reduced, which only happens when $\tilde{\phi}_{ij} - \overline{\theta}_{ij}$ goes through zero, from negative to positive. Hence an extra term will be added to V_2. As this term is initially zero, there is no discontinuity of V_2 in this case.
- The set σ^+ is enlarged when η_{ij}^+ goes to zero from positive while $\tilde{\phi}_{ij} < \overline{\theta}_{ij}$. Here V_2 will lose a positive term $(\gamma_{ij}^{\eta})^2(\tilde{\phi}_{ij} - \overline{\theta}_{ij})^2/2$, causing discontinuity.

In the context, we conclude that V_2 is always nonincreasing along the trajectory even when σ^+ or σ^- changes and discontinuity occurs.

To prove that the trajectory $w(t)$ is bounded note that [[11], Theorem 3.1] proves that $\hat{V}_2(w) := -(H(w) - w)^{\mathrm{T}}F(w) - \frac{1}{2}||H(w) - w||_2^2$ satisfies $\hat{V}_2(w) \geq 0$ over S. Therefore, we have

$$\frac{1}{2}k(w(t) - w^*)^{\mathrm{T}}\Gamma_2^{-2}(w(t) - w^*) \leq V_2(w(t)) \leq V_2(w(0)) \tag{5.36}$$

indicating the trajectory $w(t)$ is bounded. ■

Lemma 5.5 *Suppose Assumptions 5.1, 5.2, and 5.3 hold:*

1) *The trajectory $w(t)$ converges to the largest weakly invariant subset W_2^* contained in $W_2 := \{w \in S | \dot{V}_2(w) = 0\}$.*
2) *Every point $w^* \in W_2^*$ is an equilibrium point of (5.30).*

Proof: Given an initial point $w(0)$, there is a compact set $\Omega_0 := \Omega(w(0)) \subset S$ such that $w(t) \in \Omega_0$ for $t \geq 0$ and $\dot{V}_2(w) \leq 0$ in Ω_0.

Invoking the proof of Lemma 5.4, V_2 is radially unbounded and positive definite except at equilibrium. As V_2 and $\dot{V}_2$ are nonpathological, we conclude that any trajectory $w(t)$ starting from Ω_0 converges to the largest weakly invariant subset W_2^* contained in $W_2 = \{\ w \in \Omega_0 \mid \dot{V}_2(w) = 0\ \}$ [13, Proposition 3], proving the first assertion.

For the second assertion, We fix $w(0) \in W_2^*$ and then prove that $w(0)$ must be an equilibrium point.

From (5.32b), direct computing yields

$$\begin{aligned}\dot{V}_2(w(t)) &\leq -\dot{\omega}^{\mathrm{T}}D\dot{\omega} - (\dot{P}^g)^{\mathrm{T}}A^g\dot{P}^g - (\dot{P}^l)^{\mathrm{T}}A^l\dot{P}^l \\ &\quad - (CB\dot{\tilde{\phi}} - \dot{P}^g + \dot{P}^l)^{\mathrm{T}}(CB\dot{\tilde{\phi}} - \dot{P}^g + \dot{P}^l) \leq 0\end{aligned} \tag{5.37}$$

Since A^g, A^l, and D are positive definite diagonal matrices, $\dot{V}_2(w) = 0$ holds only when $\dot{P}^g = \dot{P}^l = \dot{\omega} = 0$. Therefore, for any $w(0) \in W_2^*$, the trajectory $w(t)$ satisfies

$$\dot{P}^g(t) = \dot{P}^l(t) = \dot{\omega}(t) = 0, \qquad t \geq 0 \tag{5.38}$$

Hence $P^g(t)$, $P^l(t)$, and $\omega(t)$ are all constants due to the boundedness property guaranteed by Lemma 5.4.

On the other hand, for $\dot{V}_2(w) = 0$, both terms in (5.35) have to be zero, implying that

$$\hat{L}_2(w_1(t), w_2^*) = \hat{L}_2(w_1^*, w_2^*)$$

must hold in W_2. Differentiating with respect to t gives

$$\left(\frac{\partial}{\partial w_1}\hat{L}_2(w_1(t), w_2^*)\right)^{\mathrm{T}} \cdot \dot{w}_1(t) = 0 = -\dot{\tilde{\phi}}^{\mathrm{T}}(\Gamma^{\phi})^{-1}\dot{\tilde{\phi}} \tag{5.39}$$

The second equality holds due to Eqs. (5.38) and (5.25h). Then we can conclude $\dot{\tilde{\phi}} = 0$ immediately, implying $\tilde{\phi}$ is also constant in W^* due to its boundedness.

Invoking the close-loop dynamics (5.25), $\dot{\tilde{\theta}}(t)$, $\dot{\eta}^+(t)$, $\dot{\eta}^-(t)$, and $\dot{\lambda}(t)$ must be constants in W_2^* as $P^g(t)$, $P^l(t)$, $\omega(t)$, and $\tilde{\phi}(t)$ are all constants. Then we conclude that $\dot{\tilde{\theta}}(t) = \dot{\eta}^+(t) = \dot{\eta}^-(t) = \dot{\lambda}(t) = 0$ holds for all $t \geq 0$ due to the boundedness property of $w(t)$ (Lemma 5.4). This implies that any $w(0) \in W_2^*$ must be an equilibrium point, completing the proof. ■

Now, we can complete the proof of Theorem 5.3 by using Lemma 5.5.

Proof: [Proof of Theorem 5.3.]
Fix any initial state $w(0)$ and consider the trajectory $(w(t), t \geq 0)$ of the z (5.25). As mentioned in the proof of Lemma 5.5, $w(t)$ stays entirely in a compact set Ω_0. Hence there exists an infinite sequence of time instants t_k such that $w(t_k) \to \hat{w}^*$ as $t_k \to \infty$, for some $\hat{w}^* \in W_2^*$. Lemma 5.5 guarantees that $\hat{w}^*$ is an equilibrium point of the closed-loop system (5.25), and hence $\hat{w}^* = H(\hat{w}^*)$. Thus, using this specific equilibrium point $\hat{w}^*$ in the definition of V_2, we have

$$V_2^* = \lim_{t\to\infty} V_2(w(t)) = \lim_{t_k\to\infty} V_2(w(t_k)) = \lim_{w(t_k)\to\hat{w}^*} V_2\left(w(t_k)\right) = V_2(\hat{w}^*) = 0$$

Here, the first equality uses the fact that $V_2(t)$ is nonincreasing in t; the second equality uses the fact that t_k is the infinite sequence of t; the third equality uses the fact that $w(t)$ is absolutely continuous in t; the fourth equality is due to the upper semicontinuity of $V_2(w)$; and the last equality holds as $\hat{w}^*$ is an equilibrium point of V_2.

The quadratic term $(w - \hat{w}^*)^{\mathrm{T}}\Gamma_2^{-2}(w - \hat{w}^*)$ in V_2 then implies that $w(t) \to \hat{w}^*$ as $t \to \infty$, which completes the proof.

5.6 Case Studies

5.6.1 Test System

A four-area system based on Kundur's four-machine, two-area system [15, 16] is used to test our optimal frequency controller. There are one (aggregate) generator (Gen1–Gen4), one controllable (aggregate) load (L1c–L4c), and one uncontrollable (aggregate) load (L1–L4) in each area, which is shown in Figure 5.1. The parameters of generators and controllable loads are given in Table 5.1. The total uncontrollable load in each area are identically 480 MW. At time $t = 20$ s, we add step changes on the uncontrollable loads in four areas to test the performance of our controllers.

All the simulations are implemented in power systems computer aided design (PSCAD) with 8 GB memory and 2.39 GHz CPU. We use the detailed electromagnetic transient model of three-phase synchronous machines to simulate generators with both governors and exciters. The uncontrollable loads L1–L4 are modeled by the fixed load in PSCAD, while controllable loads L1c–L4c are formulated by the self-defined controlled current source. The closed-loop system diagram is shown

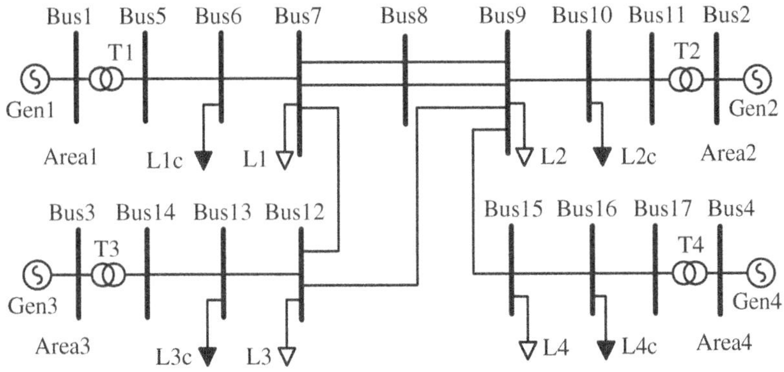

Figure 5.1 Four-area power system.

Table 5.1 System parameters.

Area j	D_j	R_j	α_j	β_j	T_j^g	T_j^l
1	0.04	0.04	2	2.5	4	4
2	0.045	0.06	2.5	4	6	5
3	0.05	0.05	1.5	2.5	5	4
4	0.055	0.045	3	3	5.5	5

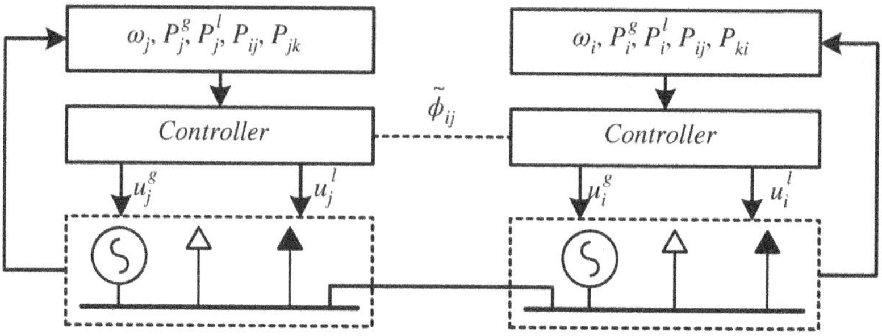

Figure 5.2 Closed-loop system diagram.

in Figure 5.2. We need measure local frequency, generation, controllable load, and tie-line power flows to compute control demands. Only $\tilde{\phi}_{ij}$ are exchanged between neighbors. All variables are added by their initial steady-state values to explicitly show the actual values.

5.6.2 Simulation Results

In this case, the generations in each area are initiated as 560.9, 548.7, 581.2, and 540.6 MW, and the controllable loads are 70.8, 89.6, 71.3, and 79.4 MW. The load changes are identical to those in Table 5.2, which are also unknown to the controllers. We use method in Remark 5.3 to estimate the load changes. Operational constraints on generations, controllable loads, and tie lines are shown in Table 5.2.

5.6.2.1 Stability and Optimality

The dynamics of local frequencies and tie-line power flows are illustrated in Figures 5.3 and 5.4, respectively. The frequencies are well restored in all four control areas while the tie-line powers are remained within their acceptable ranges. The generations and controllable loads are different from that before disturbance, indicating that the system is stabilized at a new steady state. The resulting equilibrium point is given in Table 5.3, which is identical to the optimal

Table 5.2 Capacity limits in network case.

	Area 1	Area 2	Area 3	Area 4
$[\underline{P}_j^g, \overline{P}_j^g]$ (MW)	[550, 650]	[530, 650]	[550, 700]	[530, 670]
$[\underline{P}_j^l, \overline{P}_j^l]$ (MW)	[20, 80]	[60, 100]	[20, 80]	[35, 80]
Tie line	(2,1)	(3,1)	(3,2)	(4,2)
$[\underline{P}_{ij}, \overline{P}_{ij}]$ (MW)	[−65, 65]	[−65, 65]	[−65, 65]	[−65, 65]

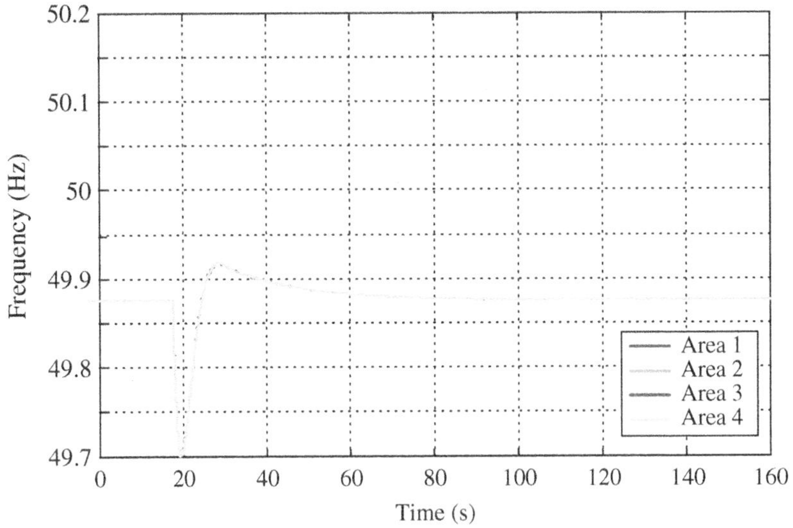

Figure 5.3 Frequency dynamics in network balance case.

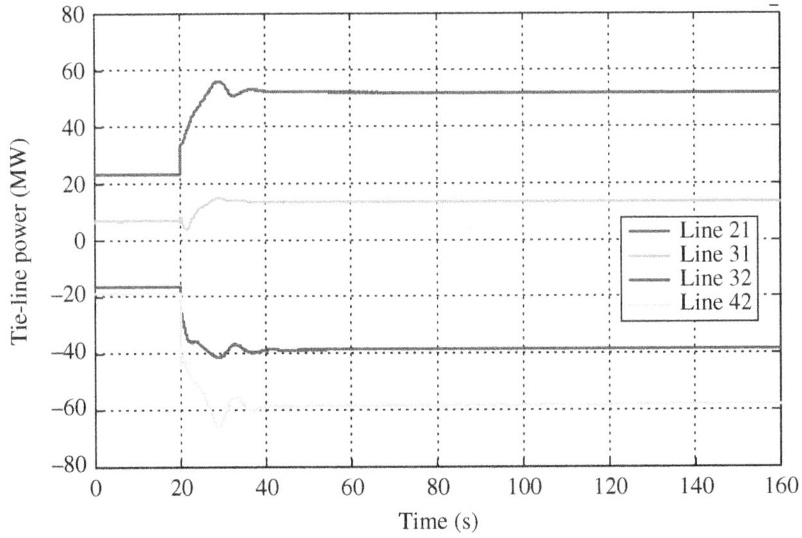

Figure 5.4 Tie-line power dynamics in network balance case.

Table 5.3 Equilibrium points.

	Area 1	Area 2	Area 3	Area 4
P_j^{g*} (MW)	620	596	660	580
P_j^{l*} (MW)	23.6	59.8	23.6	39.7
Tie line	(2,1)	(3,1)	(3,2)	(4,2)
P_{ij}^* (MW)	−39.94	13.35	53.27	−59.6

solution to (5.6) computed by centralized optimization using Yalmip. These simulation results confirm that our controller can autonomously guarantee the frequency stability while achieving optimal operating point in the overall system.

5.6.2.2 Dynamic Performance

In this subsection, we analyze the impacts of operational (capacity and line power) constraints on the dynamic property of the proposed distributed optimal frequency controller. Similarly, we compare the responses of frequency controllers with and without considering input saturations. The trajectories of mechanical power of turbines and controllable loads are shown in Figures 5.5 and 5.6, respectively. In this case, the system frequency is restored, and the same optimal equilibrium point is achieved.

5.6.2.3 Comparison with AGC

AGC is often utilized in the conventional secondary frequency control. To compare performance of our controller, we give the frequency dynamics of proposed controller and AGC in Figure 5.7.

The results show that frequency nadirs under the proposed controller similar to that under AGC. The AGC does not cause frequency overshoot, while the proposed controller causes a small overshoot (about 0.04 Hz). In addition, the convergence times are also similar. To achieve the optimal regulation with capacity constraints and tie-line congestions, our controller may cause a small frequency overshoot, but it still has a pretty smooth transient performance.

5.6.2.4 Congestion Analysis

In this scenario, we reduce tie-line power constraints to $\overline{P}_{ij} = -\underline{P}_{ij} = 50$ MW, which causes congestions in tie lines (2,3) and (2,4). The steady states under the distributed control are listed in Table 5.4. Note that the power balance equation

$$P_j^{g*} - p_j + P_j^{l*} - \sum_{k:j\to k} P_{jk}^* + \sum_{i:i\to j} P_{ij}^* = 0$$

holds in each area.

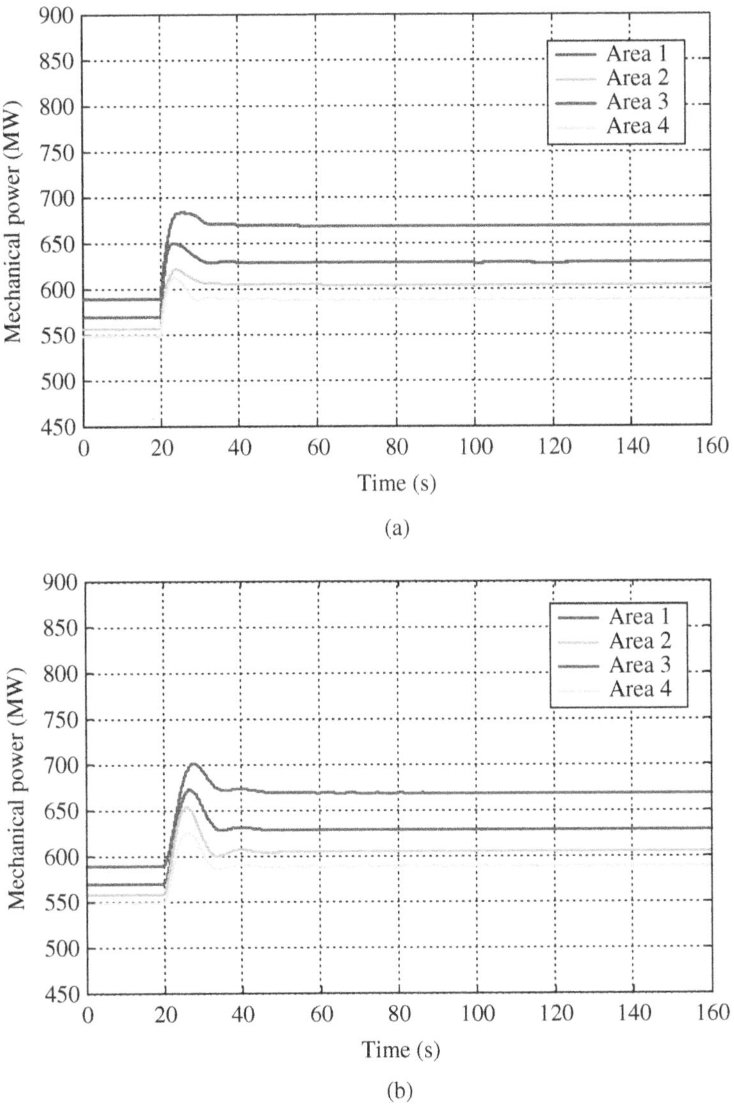

Figure 5.5 Mechanical outputs (a) with/(b) without capacity constraints.

The dynamics of tie-line power flows in the two different scenarios shown in Figure 5.8 indicate that the tie line (2,4) reaches the limit in steady state. However, by adopting the proposed fully distributed optimal frequency control, the congestion is eliminated, and all the tie line power flows remain within their limits. Thus, the proposed controller can achieve congestion management optimally in a distributed manner.

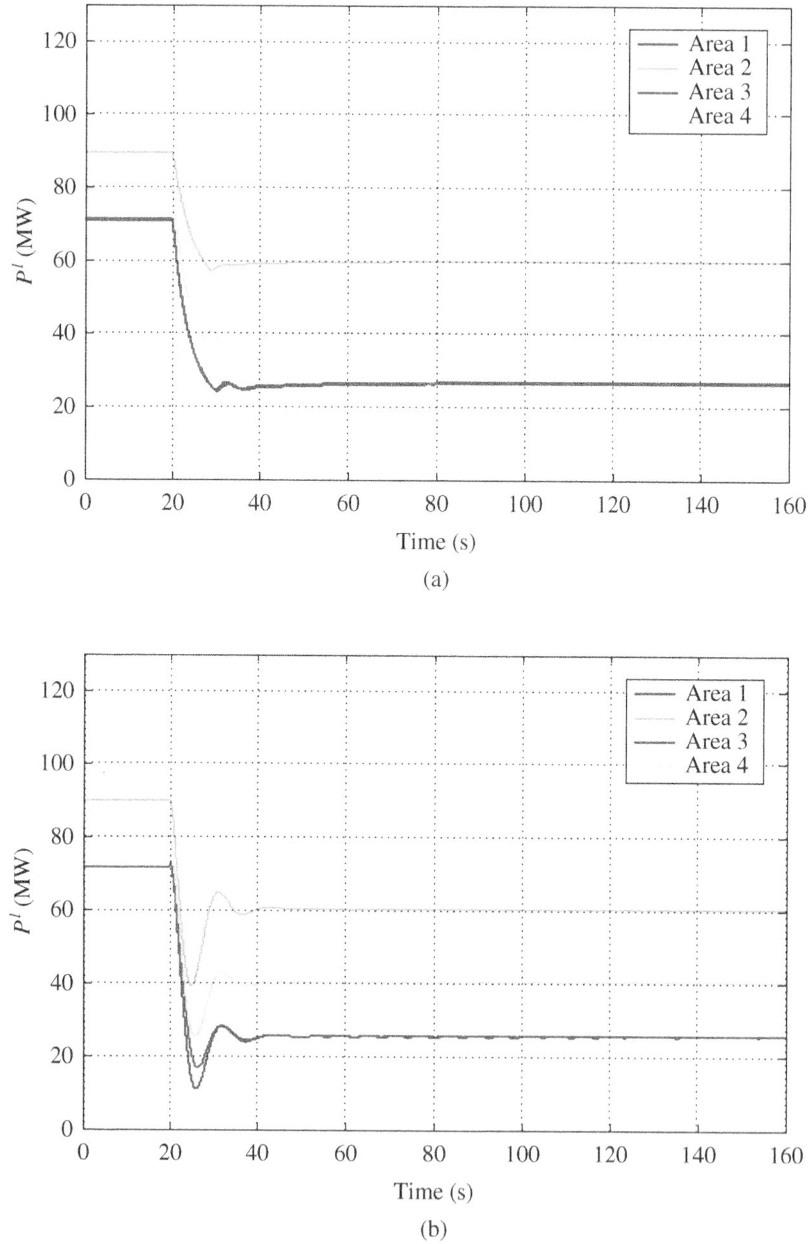

Figure 5.6 Controllable loads (a) with/(b) without capacity constraints.

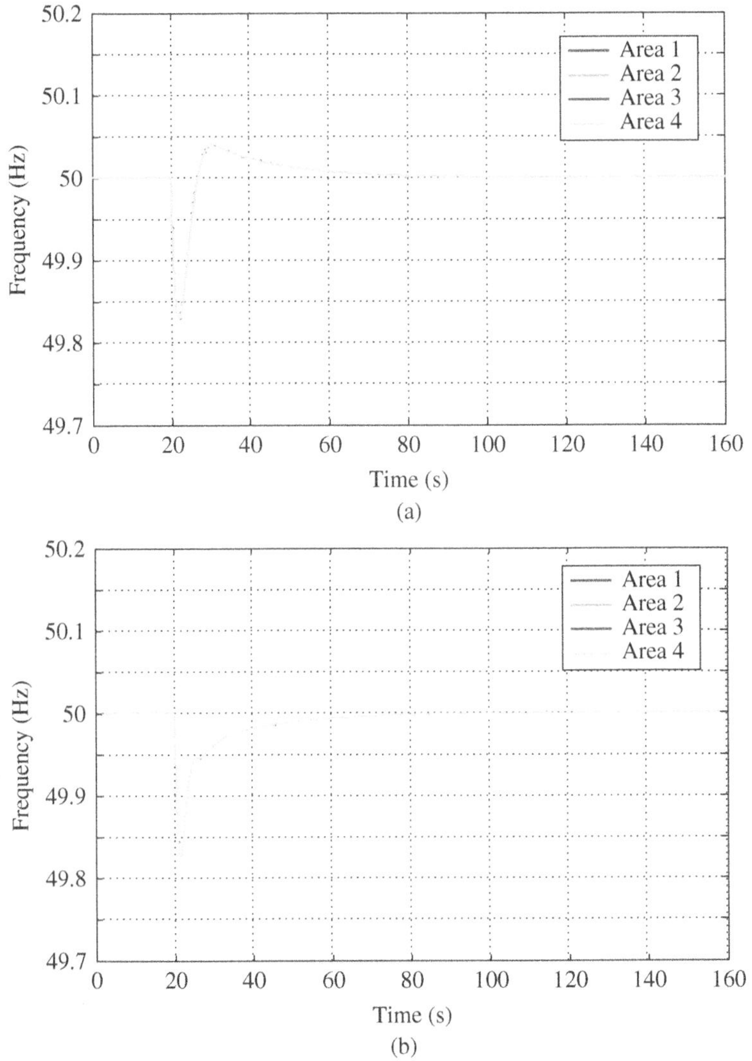

Figure 5.7 Frequency dynamics with proposed controller and AGC. (a) Frequency with proposed controller. (b) Frequency with AGC.

5.6.2.5 Time Delay Analysis

As shown in Figure 5.2, $\tilde{\phi}_{ij}$ is conveyed between areas i and j. In the real power system, there may be communication delays between areas, which vary from tens of milliseconds to hundreds of milliseconds [17]. In this chapter, time delays are set to be 100 and 500 ms. Dynamics of mechanical power outputs and controllable loads with different time delays are shown in Figures 5.9 and 5.10, respectively.

Table 5.4 Simulation results with congestion.

	Area 1	Area 2	Area 3	Area 4
P_j^{g*} (MW)	618	595	658	585
P_j^{l*} (MW)	25.1	60.7	25.1	34.9
Tie line	(2,1)	(3,1)	(3,2)	(4,2)
P_{ij}^* (MW)	−36.4	13.1	49.5	−49.9

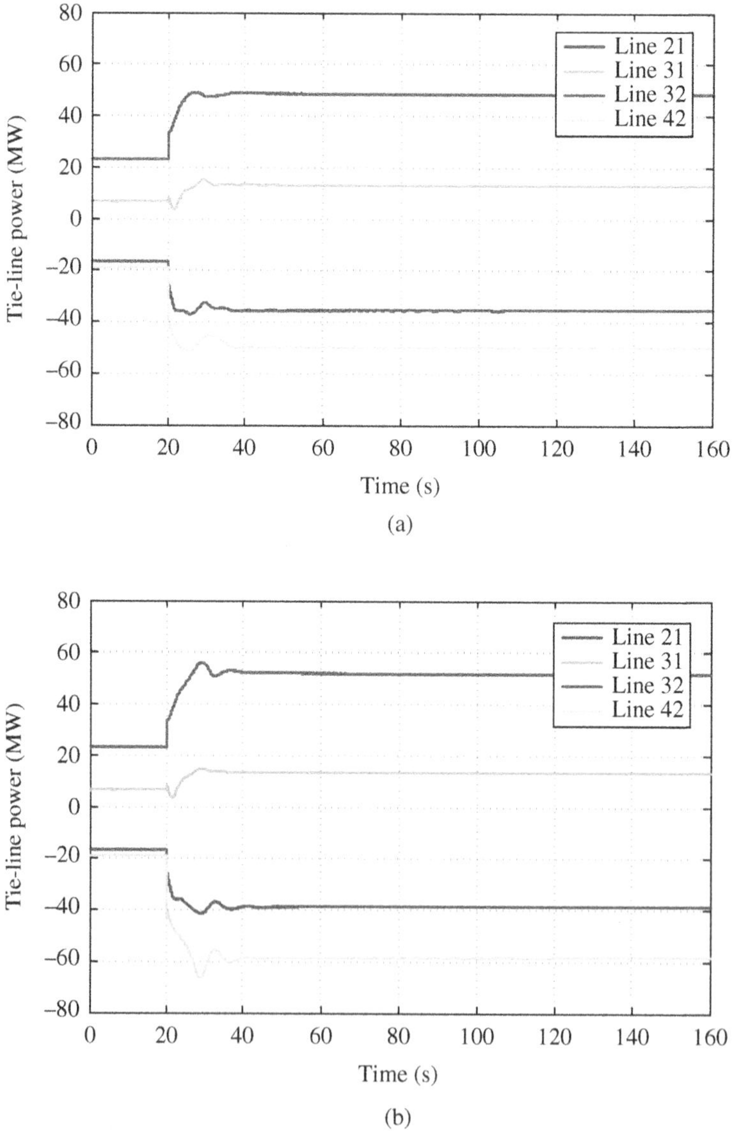

Figure 5.8 Tie-line power (a) with/(b) without capacity constraints.

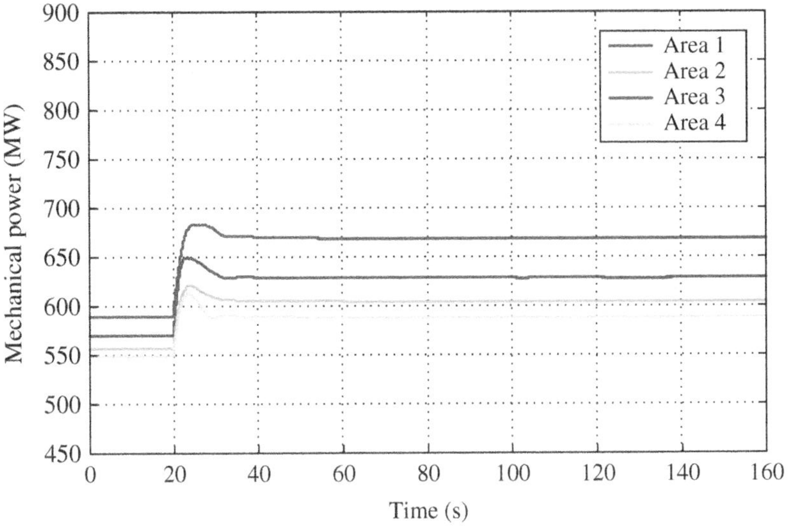

(a)

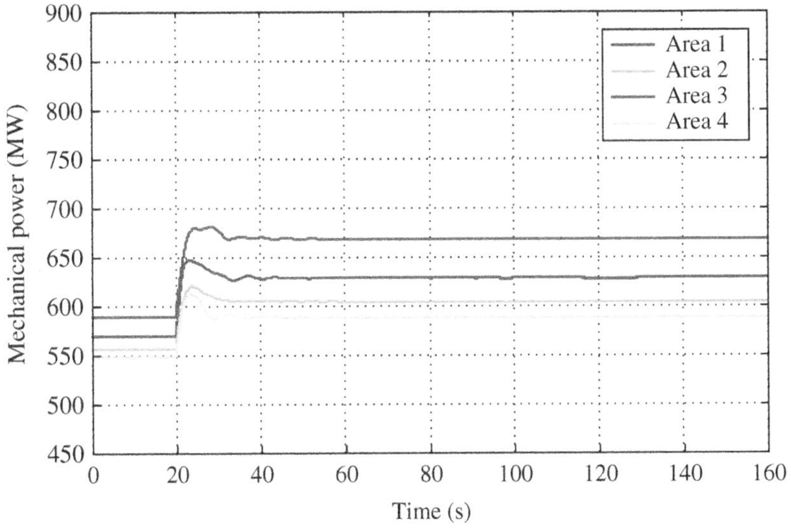

(b)

Figure 5.9 Mechanical outputs with different time delays. (a) With 100 ms delay. (b) With 500 ms delay.

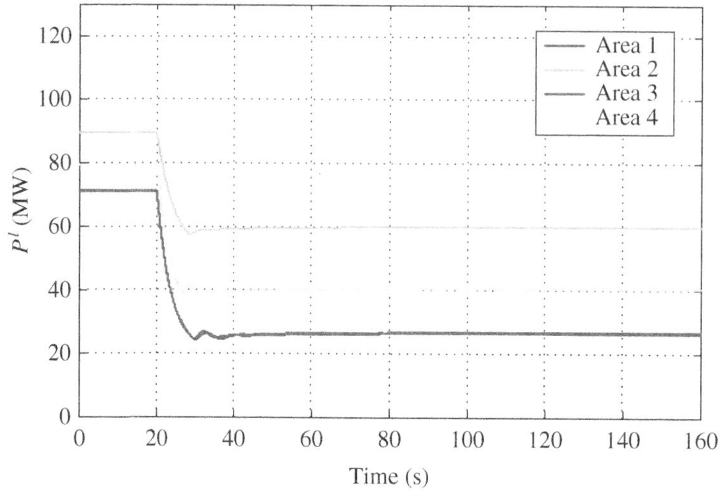

(a)

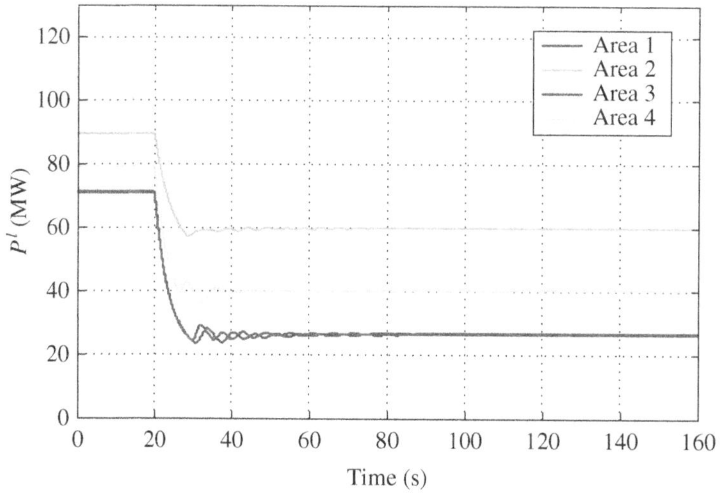

(b)

Figure 5.10 Controllable loads with different time delays. (a) With 100 ms delay. (b) With 500 ms delay.

It is shown in Figures 5.9 and 5.10 that dynamics of mechanical power outputs and controllable loads converge to the same value in both cases. However, the convergence time of 500 ms delay is a little longer than that of 100 ms delay. These results also show that our controller adapt to common time delays in the real power system.

5.7 Conclusion and Notes

In this chapter, we have devised a distributed optimal frequency control in the network balance case, which can autonomously restore the nominal frequencies after unknown load disturbances while minimizing the regulation cost. The capacity constraints on the generations and controllable loads can also be satisfied even during transient. In addition, congestions can be eliminated automatically, implying tie-line powers can be remained within given ranges. Only neighborhood communication is required in this case. Like the per-node case, here the closed-loop system again carries out an approximate primal–dual algorithm with saturation to solve the associated optimal problem. To cope with the discontinuity introduced due to enforcing different types of constraints, we have constructed a nonpathological Lyapunov function to prove the asymptotically stability of the closed-loop system. Simulations on a modified Kundur's power system validate the effectiveness of our controller.

This approach is also applicable to other problem involving frequency regulation, e.g. control of standalone microgrids or demand-side management. We highlight two crucial implications of our work: first, the suggested distributed frequency control is capable of serving as an automatically corrective redispatch without the coordination of dispatch center when certain congestion happens and second, the feasible region of economic decisions can be enlarged benefiting from the corrective redispatch. In this sense, this methodology may provide a systematic way to bridge the gap between the (secondary) frequency control in a fast timescale and the economic dispatch in a slow timescale, hence breaking the traditional hierarchy of the power system frequency control and economic dispatch. Nevertheless, many issues remain in both theoretic and practical aspects, such as the ways to improve the robustness to stochastic perturbations due to volatile renewable generations, to consider the effects of nonlinearity of power systems, and to theoretically guarantee the stability and optimality in the presence of time delays or other imperfect communications. We leave these issues to the following chapters.

Bibliography

1 E. Mallada, C. Zhao, and S. Low, "Optimal load-side control for frequency regulation in smart grids," *IEEE Transactions on Automatic Control*, vol. 62, no. 12, pp. 6294–6309, 2017.

2 T. Stegink, C. De Persis, and A. van der Schaft, "A unifying energy-based approach to stability of power grids with market dynamics," *IEEE Transactions on Automatic Control*, vol. 62, no. 6, pp. 2612–2622, 2017.

3 T. Stegink, C. D. Persis, and A. van der Schaft, "A port-Hamiltonian approach to optimal frequency regulation in power grids," in *Proceedings of 54th IEEE Conference on Decision and Control (CDC)*, (Osaka, Japan), pp. 3224–3229, IEEE, Dec 2015.
4 P. Yi, Y. Hong, and F. Liu, "Distributed gradient algorithm for constrained optimization with application to load sharing in power systems," *System & Control Letters*, vol. 83, pp. 45–52, 2015.
5 C. Zhao, E. Mallada, S. Low, and J. Bialek,"A unified framework for frequency control and congestion management," in *Power Systems Computation Conference (PSCC), 2016*, (Genoa, Italy), pp. 1–7, IEEE, 2016.
6 X. Zhang and A. Papachristodoulou, "A real-time control framework for smart power networks: design methodology and stability," *Automatica*, vol. 58, pp. 43–50, 2015.
7 D. Feijer and F. Paganini., "Stability of primal-dual gradient dynamics and applications to network optimization," *Automatica*, vol. 46, no. 12, pp. 1974–1981, 2010.
8 A. Cherukuri, E. Mallada, and J. Cortés, "Asymptotic convergence of constrained primaldual dynamics," *System & Control Letters*, vol. 87, pp. 10–15, 2016.
9 P. Dupuis and A. Nagurney, "Dynamical systems and variational inequalities," *Annals of Operations Research*, vol. 44, no. 1, pp. 9–42, 1993.
10 M.-G. Cojocaru and L. B. Jonker, "Existence of solutions to projected differential equations in Hilbert spaces," *Proceedings of the American Mathematical Society*, vol. 132, no. 1, pp. 183–193, 2004.
11 M. Fukushima, "Equivalent differentiable optimization problems and descent methods for asymmetric variational inequality problems," *Mathematical Programming*, vol. 53, no. 1, pp. 99–110, 1992.
12 F. H. Clarke, *Optimization and Nonsmooth Analysis*, Philadelphia, PA, USA: Siam, 1990.
13 A. Bacciotti and F. Ceragioli,"Nonpathological Lyapunov functions and discontinuous Carathéodory systems," *Automatica*, vol. 42, no. 3, pp. 453–458, 2006.
14 A. Bacciotti and F. Ceragioli, "Stability and stabilization of discontinuous systems and nonsmooth Lyapunov functions," *ESAIM: Control, Optimisation and Calculus of Variations*, vol. 4, pp. 361–376, 1999.
15 J. Fang, W. Yao, Z. Chen, J. Wen, and S. Cheng, "Design of anti-windup compensator for energy storage-based damping controller to enhance power system stability," *IEEE Transactions on Power Systems*, vol. 29, no. 3, pp. 1175–1185, 2014.
16 P. Kundur, *Power System Stability and Control*, New York, USA: McGraw-hill, 1994.
17 X. Zhang, C. Lu, X. Xie, and Z. Y. Dong, "Stability analysis and controller design of a wide-area time-delay system based on the expectation model method," *IEEE Transactions on Smart Grid*, vol. 7, no. 1, pp. 520–529, 2016.

6

Physical Restrictions: Nonsmoothness of Objective Functions in Load-Frequency Control

In distributed optimization, the smoothness of objective functions plays a crucial role in the proofs of convergence and stability. Usually, the objective functions are assumed to be differential so that their gradients can be utilized to guide the searching directions and facilitate the convergence. Nevertheless, unfortunately, many practical problems are inherently associated with fully or partly nonsmooth objective functions, such as tiered pricing in power systems. In this circumstance, there does not exist a conventional gradient with respect to the objective function. This chapter, therefore, discusses how to deal with such an issue in distributed optimal control design, with a particular interest in the distributed frequency control of power systems with (aggregate) controllable load participating. Here, the cost function of frequency regulation is nonsmooth and hence is not differentiable. To address this issue, we first formulate the optimization problem that minimizes the nonsmooth cost of frequency regulation, where both capacity limits of controllable loads and tie-line flows are considered. Then, a distributed optimal controller is derived based on the (approximate) primal–dual gradient dynamics, where the Clark generalized gradient is utilized to handle the nonsmooth objective function. By doing so, we prove the optimality of the equilibrium of the closed-loop dynamics and theoretically justify its stability.

6.1 Background

In previous chapters, the idea of "forward and reverse engineering" has been leveraged to merge fast time-scale frequency control with slow time-scale optimal operation of power systems. Explicitly or implicitly, it has inspired a number of works in the literature [1–11]. In [5], an optimal load frequency control (OLC) problem is formulated, and a distributed controller is derived using controllable loads to realize primary frequency control. To eliminate the steady-state frequency deviation, the method is extended in [6, 10] to realize a secondary

Merging Optimization and Control in Power Systems: Physical and Cyber Restrictions in Distributed Frequency Control and Beyond, First Edition. Feng Liu, Zhaojian Wang, Changhong Zhao, and Peng Yang.

load frequency control. At the same time, the tie-line power limit is taken into account. The design approach is further generalized in [8], where the specific requirements in modeling are relaxed. It only requires that the bus dynamics satisfy a passivity condition to guarantee the asymptotic stability. In [7, 9], the operational constraints, including regulation capacity limits and tie-line power limits, are considered to guarantee that both the steady-state and transient capacity limit constraints are satisfied. In [11], the distributed load frequency control under time-varying and unknown power injection is investigated, which can recover the nominal frequency even under unknown disturbances.

In practice, the distributed load frequency control is, in most cases, a paid service, i.e. the system operator needs to pay for controlling deferrable load to support the system frequency. In the literature, the cost of the controllable load is usually assumed to be differentiable, or equivalently, the price of the controllable load is continuous. However, it is not true for many cases, e.g. the price may have step changes when controllable load values are in different intervals. In such a situation, the regulation is inherently nonsmooth, which makes conventional methods challenging to apply.

To deal with the nonsmooth objective function, many methods are proposed, where the projection-based continuous algorithm is widely adopted, especially for problems with domain set constraints [12–19]. These works can be roughly divided into two categories in terms of how the set is used: (i) projection onto the tangent set [13–15, 18] and (ii) projection onto the domain set [12, 16, 17, 19]. In the first category, the primal–dual dynamics are projected onto the tangent set of the domain, which is also called orthogonal set projection [18]. In the second type, the dynamics are projected onto the domain set itself, guaranteeing the variables stay within the domain. In most works except [14, 17], the objective function is required to be strictly convex [12, 15, 16, 19], or its (generalized) gradient is strictly monotone [13, 18]. In [14, 17], the objective function is only required to be convex, where, however, the decision variables of one agent should be the same as those of others.

This chapter designs a distributed controller for the optimal load frequency control in power systems, where the regulation cost can be nonsmooth. The most critical feature distinguishing it from others [12–19] is that it counts for the interplay between the solving algorithm and the power system dynamics and proves the stability of the closed-loop dynamics. In other words, we intend to drive the physical system to the optimal solution to the corresponding optimization problem by the controller design instead of devising an algorithm only. Another difference is that the objective function is not required to be strictly convex with respect to all decision variables. Particularly, some variables even do not appear in the objective function at all. Compared with [14, 17], each agent has its own decision variables that may be different from others. Thus, the model in our work is more general as we relax the domain constraint. In such a situation, we prove the stability of the closed-loop system and the optimality of the equilibrium.

The rest of this chapter is organized as follows. In Section 6.3, we introduce necessary preliminaries and system models. Section 6.4 formulates the optimal load frequency control problem and introduces the distributed controller. In Section 6.5, the convergence of the closed-loop system and optimality of the equilibrium point is proved. We confirm the performance of the controller via simulations on the IEEE 68-bus system in Section 6.6. Finally, Section 6.7 concludes the chapter with notes.

6.2 Notations and Preliminaries

In this chapter, we use $\mathbb{R}^n$ ($\mathbb{R}^n_+$) to denote the n-dimensional (nonnegative) Euclidean space. For a column vector $x \in \mathbb{R}^n$ (matrix $A \in \mathbb{R}^{m\times n}$), $x^{\mathrm{T}}(A^{\mathrm{T}})$ denotes its transpose. For vectors $x, y \in \mathbb{R}^n$, $x^{\mathrm{T}}y = \langle x, y\rangle$ denotes the inner product of x, y. $\|x\| = \sqrt{x^{\mathrm{T}}x}$ denotes the Euclidean norm of x. Use $\mathbf{1}$ to denote the vector with all 1 elements. For a matrix $A = [a_{ij}]$, a_{ij} is the entry in the ith row and jth column of A. Use $\prod_{i=1}^{n} \Omega_i$ to denote the Cartesian product of the sets $\Omega_i, i = 1, \dots, n$. Given a collection of y_i for $i \in Y$, y denotes the column vector, i.e. $y := (y_i, i \in Y)$, with a proper dimension, and y_i is the ith component.

Let $f(x) : \mathbb{R}^n \to \mathbb{R}$ be a locally Lipschitz continuous function and denote its Clarke generalized gradient by $\partial f(x)$ [Page 27] [20], which is defined as

Definition 6.1 The Clarke generalized gradient of $f(x)$ at x, denoted by $\partial f(x)$, is defined as the convex hull of the set of limits of the form $\lim \nabla f(x + h_i)$, where $h_i \to 0$ as $i \to \infty$.

This definition is crucial for optimization problems with nonsmooth objective functions since it enables the gradient calculations when $f(x)$ fails to be differentiable somewhere. It can be extended to the functions that are even not almost everywhere differentiable. However, such circumstances seldom happen in engineering problems.

For a continuous and strictly convex function $f(x) : \mathbb{R}^n \to \mathbb{R}$, we have

$$(g_x - g_y)^{\mathrm{T}}(x - y) > 0, \quad \forall x \neq y$$

where $g_x \in \partial f(x)$ and $g_y \in \partial f(y)$ are both Clarke generalized gradients.

Recalling the definition of projection in Chapter 2, the projection of x onto a closed convex set Ω is

$$\mathcal{P}_{\Omega}(x) := \arg\min_{y\in\Omega} \|x - y\| \tag{6.1}$$

Use Id to denote the identity operator, i.e. $\mathrm{Id}(x) = x, \forall x \in \mathbb{R}^n$. We further define the operator $N_\Omega(x)$ as

$$N_\Omega(x) := \{v | \langle v, y - x \rangle \leq 0, \forall y \in \Omega\}$$

$N_\Omega(x)$ is also the normal cone to Ω at x.

According to [Chapter 23.1] [21], we have

$$\mathcal{P}_\Omega(x) = (\mathrm{Id} + N_\Omega)^{-1}(x).$$

A fundamental property of the projection $\mathcal{P}_\Omega(x)$ is

$$(x - \mathcal{P}_\Omega(x))^{\mathrm{T}}(y - \mathcal{P}_\Omega(x)) \leq 0, \quad \forall x \in \mathbb{R}^n, \ y \in \Omega \tag{6.2}$$

Moreover, we also have [Theorem 1.5.5] [22]

$$(\mathcal{P}_\Omega(x) - \mathcal{P}_\Omega(y))^{\mathrm{T}}(x - y) \geq \|\mathcal{P}_\Omega(x) - \mathcal{P}_\Omega(y)\|^2 \tag{6.3}$$

Define

$$V(x) := \frac{1}{2}\left(\|x - \mathcal{P}_\Omega(y)\|^2 - \|x - \mathcal{P}_\Omega(x)\|^2\right)$$

Reference [Lemma 4] [23] has proved that $V(x)$ is differentiable and convex with respect to x. Moreover, there are

$$\begin{aligned} V(x) &= \frac{1}{2}\|\mathcal{P}_\Omega(x) - \mathcal{P}_\Omega(y)\|^2 \\ &\quad - (x - \mathcal{P}_\Omega(x))^{\mathrm{T}}(\mathcal{P}_\Omega(y) - \mathcal{P}_\Omega(x)) && (6.4) \\ &\geq \frac{1}{2}\|\mathcal{P}_\Omega(x) - \mathcal{P}_\Omega(y)\|^2 \geq 0 && (6.5) \\ \nabla V(x) &= \mathcal{P}_\Omega(x) - \mathcal{P}_\Omega(y) && (6.6) \end{aligned}$$

where the inequality is due to (6.2). From (6.4), $V(x) = 0$ holds only when $\mathcal{P}_\Omega(x) = \mathcal{P}_\Omega(y)$.

Recalling the definition of Lyapunov function, $V(x)$ can serve as a Lyapunov-like function, which is helpful in the analyses of stability and convergence.

6.3 Power System Model

A power network is composed of a set of buses interconnected with each other through tie lines. It can be modeled as a graph $\mathcal{G} := (\mathcal{N}, \mathcal{E})$, where $\mathcal{N} = \{0, 1, 2, \dots n\}$ is the set of buses and $\mathcal{E} \subseteq \mathcal{N} \times \mathcal{N}$ is the set of edges (transmission lines). Let $m = |\mathcal{E}|$ denote the number of lines. The buses are cast into two types: generator buses, denoted by $\mathcal{N}_g$, and load buses, denoted by $\mathcal{N}_l$. A generator bus has a generator (possibly with a particular aggregate load). A load bus has pure load with no generator. The graph $\mathcal{G}$ is treated as directed with an arbitrary orientation, and use $(i,j) \in \mathcal{E}$ or $i \to j$ interchangeably to denote a directed edge from i to j. Without loss of generality, we assume the graph is connected and bus 0 is set as the reference bus. C denotes the incidence matrix of the graph, and we have $\mathbf{1}^{\mathrm{T}}C = 0$.

We adopt a second-order linearized model to describe the frequency dynamics of each (aggregate) bus. We assume that the lines are lossless and adopt the DC power flow model [5, 7]. For each bus $j \in \mathcal{N}$, let $\theta_j(t)$ denote the power angle of bus j at time t and $\omega_j(t)$ the corresponding frequency. Let $P_j^l(t)$ denote the controllable load at bus j. P_j^m denotes the change in power injection, which could occur on the generation side or the load side, or both.

Define $\theta_{ij} := \theta_i - \theta_j$ as the angle difference between buses i and j, and its compact form is denoted by $\theta_e := (\theta_{ij}, (i,j) \in \mathcal{E})$. Then for each bus $j \in \mathcal{N}$, its dynamics are depicted by

$$\dot{\theta}_{ij} = \omega_i - \omega_j, \quad j \in \mathcal{N} \tag{6.7a}$$

$$\dot{\omega}_j = \frac{1}{M_j}\left(P_j^m - P_j^l - D_j\omega_j + \sum_{i:i\to j} B_{ij}\theta_{ij} - \sum_{k:j\to k} B_{jk}\theta_{jk}\right), \quad j \in \mathcal{N}_g \tag{6.7b}$$

$$0 = P_j^m - P_j^l - D_j\omega_j + \sum_{i:i\to j} B_{ij}\theta_{ij} - \sum_{k:j\to k} B_{jk}\theta_{jk}, \quad j \in \mathcal{N}_l \tag{6.7c}$$

where $M_j > 0$ are inertia constants, $D_j > 0$ is damping constants, and $B_{jk} > 0$ are line parameters that depend on the reactance of the line (j, k).

In this chapter, we are interested in the following scenario: the system operates in a steady state at first. Then a certain power imbalance occurs due to the variation of power injection P_j^m. Afterward, the controllable load changes its output accordingly to eliminate the power imbalance and finally restores the frequency to the nominal value.

6.4 Control Design

In this section, we first formulate the OLC problem with a nonsmooth objective function. Then, we devise a distributed optimal controller based on the primal–dual gradient method to drive the power system to the optimal working point determined by the OLC problem, where the Clark generalized gradient is utilized to address the nonsmoothness.

6.4.1 Optimal Load Frequency Control Problem

Define the sets Ω_j as

$$\Omega_j := \left\{ P_j^l \mid \underline{P}_j^l \le P_j^l \le \overline{P}_j^l \right\}, \quad \Omega = \prod_{j=1}^{n} \Omega_j$$

where $\underline{P}_j^l \le \overline{P}_j^l$ are constants, denoting the lower and upper bounds of P_j^l.

The OLC problem is formulated as

$$\textbf{OLC}: \quad \min_{P_j^l \in \Omega_j,\ \phi_j} \quad f(P^l) = \sum_{j \in \mathcal{N}} f_j(P_j^l) \tag{6.8a}$$

$$\text{s.t.} \quad 0 = P_j^l - P_j^m - \sum_{i:i \to j} B_{ij}(\phi_i - \phi_j) + \sum_{k:j \to k} B_{jk}(\phi_j - \phi_k), \quad j \in \mathcal{N} \tag{6.8b}$$

$$\underline{\theta}_{ij} \le \phi_i - \phi_j \le \overline{\theta}_{ij}, \quad (i,j) \in \mathcal{E} \tag{6.8c}$$

where $\underline{\theta}_{ij} \le \overline{\theta}_{ij}$ are also constants, denoting the lower and upper bounds of the angle differences between buses i and j. The first constraint (6.8b) is the local power balance at bus j. ϕ_j are introduced as virtual phase angles, which are equal to θ_j at the optimal solution. Use $\phi_{ij} = \phi_i - \phi_j$ to denote the virtual phase angle differences. In DC power flow, we have $P_{ij} = B_{ij}\theta_{ij}$, where P_{ij} is the power of line (i,j). Thus, (6.8c) are in fact the tie-line power limit constraints.

Throughout the rest of this chapter, we make the following fundamental assumptions.

Assumption 6.1 The objective functions $f_j(P_j^l)$ are strictly convex, but not necessarily smooth.

Assumption 6.2 The Slater's condition [Chapter 5.2.3] [24] of (6.8) holds.

Assumption 6.2 means that the OLC problem (6.8) is feasible, provided that the constraints are affine. Regarding these two assumptions, here we present some helpful remarks as below.

Remark 6.1 In the existing literature of frequency control, such as [5–11], the objective function is assumed to be smooth, which may be restrictive in many practical circumstances. As a generalization, the problem (6.8) relaxes this assumption and allows the cost function $f_j(P_j^l)$ to be nonsmooth and, hence, can admit a wider variety of practical problems. A typical example is that when the tiered pricing scheme is adopted in frequency control, the price discontinuously increases with respect to the amount of the usage of controllable load. Particularly, it may take the following form:

$$f(P_j^l) = \begin{cases} \left(P_j^l\right)^2 - b, & P_j^l \leq -a \\ \frac{1}{2}\left(P_j^l\right)^2, & -a < P_j^l \leq a \\ \left(P_j^l\right)^2 - b, & a < P_j^l \end{cases}$$

where $a, b > 0$ are preset parameters up to the pricing scheme. It also should be noted that the decision variable ϕ_j is absent in the objective function of (6.8). Different from the work in presented [12], here the objective function is not required to be strictly convex to all the decision variables. This adaptation makes the convergence proof challenging.

Remark 6.2 The problem (6.8) is also different from those in [14, 17] in the sense that each bus has its own decision variable (P_j^l, ϕ_j), whose value is different from that of other buses. However, there is only one global decision variable in [14, 17]. Although each agent can compute its decision variables in a distributed way, their values are the same thanks to the consensus method. In this sense, the formulation (6.8) is more general.

Remark 6.3 In the literature, the suggested controllers usually involve the projection of a gradient onto a convex set. However, if the objective function is nonsmooth, it becomes the projection of a subdifferential set onto a convex set. In this situation, the existence of the solution trajectories may even not be guaranteed [12, 25], which makes most OLC methods inapplicable.

Remark 6.4 Assumption 6.1 could be further relaxed to be merely convex rather than strictly convex. It can be understood by noting that one can use a nonlinear perturbation to strictly convexify it [26].

6.4.2 Distributed Controller Design

Following the idea of "foreword and reverse engineering", we modify the objective function by adding an augmented quadratic term to facilitate the controller design, which yields

$$\min_{P_j^l \in \Omega_j,\ \phi_j} \quad \tilde{f}(P^l, \phi) = \sum_{j \in \mathcal{N}} f_j(P_j^l) + \frac{1}{2} \sum_{j \in \mathcal{N}} z_j^2 \tag{6.9a}$$

$$\text{s.t.} \quad (6.8b), (6.8c) \tag{6.9b}$$

where

$$z_j = P_j^l - P_j^m - \sum_{i:i\to j} B_{ij}\phi_{ij} + \sum_{k:j\to k} B_{jk}\phi_{jk}.$$

Note that $z_j = 0$ holds for any feasible solution to (6.8). Thus, the problems (6.8) and (6.9) have the same optimal solutions.

The Lagrangian of (6.9) is

$$\begin{aligned} L(\underset{P^l\in\Omega}{P^l}, \phi, \mu, \varphi^-, \varphi^+) &= \sum_{j\in\mathcal{N}} f_j(P_j^l) + \frac{1}{2}\sum_{j\in\mathcal{N}} z_j^2 \\ &+ \sum_{j\in\mathcal{N}} \mu_j\left(P_j^l - P_j^m - \sum_{i:i\to j} B_{ij}\phi_{ij} + \sum_{k:j\to k} B_{jk}\phi_{jk}\right) \\ &+ \sum_{(i,j)\in\mathcal{E}} \varphi_{ij}^-\left(\underline{\theta}_{ij} - \phi_{ij}\right) + \sum_{(i,j)\in\mathcal{E}} \varphi_{ij}^+\left(\phi_{ij} - \overline{\theta}_{ij}\right) \end{aligned} \tag{6.10}$$

where $\mu_j, \varphi_{ij}^-, \varphi_{ij}^+$ are Lagrangian multipliers.

Then, we present the controller for each controllable load as follows:

$$\dot{d}_j \in \left\{ p \,\middle|\, \begin{array}{rcl} p &=& -d_j + P_j^l + \omega_j - g_j(P_j^l) - z_j - \mu_j \\ g_j(P_j^l) &\in& \partial f_j(P_j^l) \end{array} \right\} \tag{6.11a}$$

$$\dot{\mu}_j = P_j^l - P_j^m - \sum_{i:i\to j} B_{ij}\phi_{ij} + \sum_{k:j\to k} B_{jk}\phi_{jk} \tag{6.11b}$$

$$\begin{aligned} \dot{\phi}_j &= \sum_{i:i\to j} B_{ij}(\mu_i - \mu_j) - \sum_{k:j\to k} B_{jk}(\mu_j - \mu_k) \\ &- \sum_{(i,j)\in\mathscr{E}} \eta_{ij}^- + \sum_{(j,k)\in\mathscr{E}} \eta_{jk}^- + \sum_{(i,j)\in\mathscr{E}} \eta_{ij}^+ - \sum_{(j,k)\in\mathscr{E}} \eta_{jk}^+ \\ &+ \sum_{i:i\to j} B_{ij}(z_i - z_j) - \sum_{k:j\to k} B_{jk}(z_j - z_k) \end{aligned} \tag{6.11c}$$

$$\dot{\varphi}_{ij}^+ = -\varphi_{ij}^+ + \eta_{ij}^+ + \phi_{ij} - \overline{\theta}_{ij} \tag{6.11d}$$

$$\dot{\varphi}_{ij}^- = -\varphi_{ij}^- + \eta_{ij}^- + \underline{\theta}_{ij} - \phi_{ij} \tag{6.11e}$$

$$P_j^l = \mathcal{P}_{\Omega_j}(d_j) \tag{6.11f}$$

$$\eta_{ij}^+ = \mathcal{P}_{\mathbb{R}_+}\left(\varphi_{ij}^+\right) \tag{6.11g}$$

$$\eta_{ij}^- = \mathcal{P}_{\mathbb{R}_+}\left(\varphi_{ij}^-\right) \tag{6.11h}$$

where d_j is the computed value of load before projection, P_l^j is the projection of d_j on Ω_j, η_{ij}^+ is the projection of φ_{ij}^+, and η_{ij}^- is the projection of φ_{ij}^-. The equation

(6.11a) is the type of differential inclusion. For simplicity, without confusion, we abuse the term OLC to denote the controller (6.11) in the rest of this chapter.

The controller is mainly derived from the primal–dual gradient dynamics of the solving algorithm, where (6.11a) and (6.11c) are from the primal updates and (6.11b), (6.11d), and (6.11e) are from the dual updates. Moreover, to address the nonsmooth objective function and the regulation capacity constraint, the projected output feedback method is adopted, i.e. (6.11f), (6.11g), and (6.11h). To this end, ω_j in (6.11a) is added to connect the optimal controller to the physical power system. In this sense, the suggested controller is a combination of optimization-guided control and feedback-based optimization. As a consequence, we have the closed-loop dynamics (6.7), (6.11). However, this formulation, differing from the models presented in the previous chapters, does not serve as an exact primal–dual update but an approximate one. We will prove the optimality of the equilibrium point and the stability of the closed-loop dynamics in Section 6.5.

Remark 6.5 ***(Load demand estimate)*** In power systems, P_j^m may involves components of uncontrollable load, which is technically challenging to accurately measure. As an alternative, following the line of [6, 9, 10], the term $P_j^l - P_j^m$ in (6.11b) can be substituted equivalently in following ways. For $j \in \mathcal{N}_g$, we have

$$P_j^l - P_j^m = -M_j\dot{\omega}_j - D_j\omega_j + \sum_{i:i\to j} P_{ij} - \sum_{k:j\to k} P_{jk}$$

For $j \in \mathcal{N}_l$,

$$P_j^l - P_j^m = -D_j\omega_j + \sum_{i:i\to j} P_{ij} - \sum_{k:j\to k} P_{jk}$$

By doing so, the measurement of P_j^m is not required any longer. Thus, we only need to measure ω_j and P_{ij}, which are much easier to realize in practice. Moreover, in this way, the power loss can be treated as unknown load demand that can be automatically compensated for by the proposed controller.

Remark 6.6 ***(Control architecture)*** The overall architecture of the controller (6.11) is illustrated in Figure 6.1. For bus j, it measures the local frequency ω_j and the line flow P_{jk} from the physical system for computing the control demand. Then, the control command P_j^l is sent to regulate the controllable load. Here, a peer-to-peer communication is needed for agents to exchange μ_k, z_k, ϕ_k with their neighbors. This diagram also shows the difference between the proposed controller and the works in literature [12–19]. Particularly, here, we design the controller by jointly considering both the cyber (communication and algorithm) and the physical (power system) dynamics, instead of designing the optimization algorithm only.

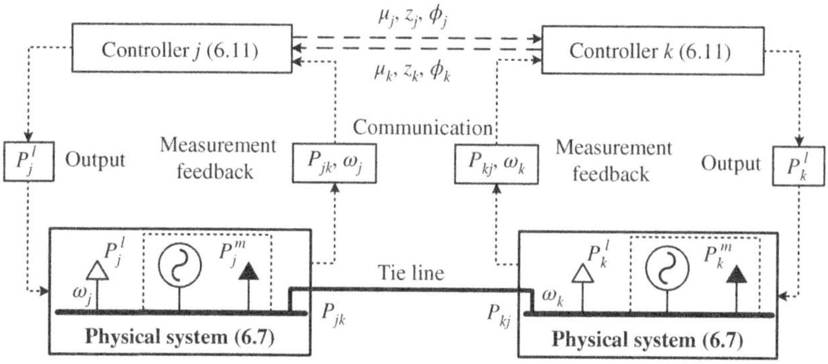

Figure 6.1 Distributed control architecture.

It should be noted that if ω_j is not included in (6.11a), the algorithm can also solve the problem (6.8). In this situation, it has no connection with the physical dynamics (6.7), which is not in keeping with our intention. The desired controller is expected to automatically drive the physical system to the optimal operation state, where (P^l, ϕ) approaches to the optimal solution to the OLC problem (6.8). In this regard, we design the distributed controller (6.11) for each controllable load, where the measurement of local frequency ω_j establishes the connection between the cyber and the physical systems. We will prove in Section 6.5.1 that, in the steady state, there must be $\omega_j = 0$ and (P^l, ϕ) is identical to the optimal solution to (6.8). Hence, ω_j vanishes at the optimum and does not affect the optimal solution in theory.

6.5 Optimality and Convergence

In this section, we prove the optimality of the equilibrium point and the convergence of the closed-loop dynamics.

6.5.1 Optimality

Denote

$$x := (\theta, \omega_g, d, \mu, \phi, \varphi^-, \varphi^+)$$

and

$$y := (x, P^l, \eta^+, \eta^-).$$

Let

$$x^* := (\theta^*, \omega_g^*, d^*, \mu^*, \phi^*, \varphi^{-*}, \varphi^{+*})$$

be an equilibrium of the closed-loop system (6.7) and (6.11). Then, there exists $g(P^{l*}) \in \partial f(P^{l*})$ such that

$$0 = C^{\mathrm{T}}\omega^* \tag{6.12a}$$

$$0 = P^m - P^{l*} - D\omega^* - CBC^{\mathrm{T}}\theta^* \tag{6.12b}$$

$$0 = -d^* + P^{l*} + \omega^* - g(P^{l*}) - \mu^* \tag{6.12c}$$

$$0 = P^{l*} - P^m + CBC^{\mathrm{T}}\phi^* \tag{6.12d}$$

$$0 = -CBC^{\mathrm{T}}\mu^* - C\eta^{-*} + C\eta^{+*} \tag{6.12e}$$

$$0 = -\varphi^{+*} + \eta^{+*} - C^{\mathrm{T}}\phi^* - \overline{\theta} \tag{6.12f}$$

$$0 = -\varphi^{-*} + \eta^{-*} + \underline{\theta} + C^{\mathrm{T}}\phi^* \tag{6.12g}$$

$$P^{l*} = \mathcal{P}_{\Omega}(d^*) \tag{6.12h}$$

$$\eta^{+*} = \mathcal{P}_{\mathbb{R}^m_+}\left(\varphi^{+*}\right) \tag{6.12i}$$

$$\eta^{-*} = \mathcal{P}_{\mathbb{R}^m_+}\left(\varphi^{-*}\right) \tag{6.12j}$$

The following theorem characterizes the properties of the equilibrium to the closed-loop dynamics.

Theorem 6.1 *Suppose Assumptions 6.1 and 6.2 hold for the closed-loop dynamics (6.7) and (6.11). Then we have the following assertions:*

1) *The nominal frequency is restored, i.e.* $\omega_j^* = 0$ *for all* $j \in \mathcal{N}$.
2) *If* x^* *is an equilibrium of (6.7) and (6.11), then* (P^{l*}, ϕ^*) *is an optimal solution to (6.8), and* $(\mu^*, \eta^{+*}, \eta^{-*})$ *is an optimal solution to its dual problem.*
3) $\phi_{ij}^* = \theta_{ij}^*$ *for all* $(i,j) \in \mathcal{E}$. *Moreover, the line flow limits are satisfied at* x^*, *implying* $\underline{\theta}_{ij} \leq \theta_{ij}^* \leq \overline{\theta}_{ij}$ *hold for every tie line* $(i,j) \in \mathcal{E}$.
4) *At the equilibrium,* $(\theta^*, \phi^*, \omega_g^*, P^{l*})$ *is unique, with* (θ^*, ϕ^*) *being unique up to (equilibrium) reference angles* (θ_0, ϕ_0).

Proof: **Assertion 1).** According to (6.12b) and (6.12d), we have $\mathbf{1}^{\mathrm{T}}D\omega^* = 0$. Then from the equation (6.12a), there must be $\omega^* = \omega_0 \cdot \mathbf{1}$ with a constant ω_0. By noting D is a diagonal positive definite matrix, it immediately follows that $\omega_0 = 0$, which is the assertion 1).

Assertion 2). We prove this assertion by checking the KKT condition. Combining (6.12c) and (6.12f)–(6.12j) yields

$$P^{l*} = \mathcal{P}_{\Omega}\left(P^{l*} - g(P^{l*}) - \mu^*\right) \tag{6.13a}$$

$$\eta^{+*} = \mathcal{P}_{\mathbb{R}^m_+}\left(\eta^{+*} - C^{\mathrm{T}}\phi^* - \overline{\theta}\right) \tag{6.13b}$$

$$\eta^{-*} = \mathcal{P}_{\mathbb{R}^m_+}\left(\eta^{-*} + \underline{\theta} + C^{\mathrm{T}}\phi^*\right) \tag{6.13c}$$

Or, equivalently,

$$-g(P^{l*}) - \mu^* \in N_{\Omega}(P^{l*}) \tag{6.14a}$$

$$-C^{\mathrm{T}}\phi^* - \overline{\theta} \in N_{\mathbb{R}^m_+}(\eta^{+*}) \tag{6.14b}$$

$$\underline{\theta} + C^{\mathrm{T}}\phi^* \in N_{\mathbb{R}^m_+}(\eta^{-*}) \tag{6.14c}$$

By the KKT condition in the form of normal cone [Theorem 3.34] [27], (6.12d), (6.12e), and (6.14) coincide with the KKT optimality condition of the problem (6.8). Hence, the assertion 2) is proved.

Assertion 3). From (6.12b) and (6.12d), we have $CBC^{\mathrm{T}}(\theta^* - \phi^*) = 0$, which holds for any incidence matrix C. It directly follows that $\theta^* - \phi^* = c_0 \cdot \mathbf{1}$ holds with a constant c_0, indicating $\theta^*_{ij} - \phi^*_{ij} = 0$. Moreover, by Assertion 2), we know $\underline{\theta}_{ij} \leq \phi^*_{ij} \leq \overline{\theta}_{ij}$, which implies that $\underline{\theta}_{ij} \leq \theta^*_{ij} \leq \overline{\theta}_{ij}$. Hence the assertion 3) is proved.

Assertion 4). Note that P^{l*} is unique because the objective function in (6.8a) is strictly convex in P^l and ω^* is unique due to $\omega^* = 0$. By (6.12d), we know ϕ^* is unique (in the sense of modulo in 2π). Since $\theta^* - \phi^* = c_0 \cdot \mathbf{1}$, it implies that θ^* is also unique. This guarantees the uniqueness of $(\theta^*, \phi^*, \omega_g^*, P^{l*})$, which completes the proof. ■

6.5.2 Convergence

Define the following function:

$$V(x) := V_1(x) + V_2(x) \tag{6.15a}$$

where

$$\begin{aligned} V_1(x) = {} & \frac{1}{2}\left\|P^l - P^{l*}\right\|^2 + \frac{1}{2}\|\mu - \mu^*\|^2 + \frac{1}{2}\|\phi - \phi^*\|^2 \\ & + \frac{1}{2}\|\eta^+ - \eta^{+*}\|^2 + \frac{1}{2}\|\eta^- - \eta^{-*}\|^2 \\ & + \frac{1}{2}(\theta - \theta^*)^{\mathrm{T}}B(\theta - \theta^*) \\ & + \frac{1}{2}\left(\omega_g - \omega_g^*\right)^{\mathrm{T}}M\left(\omega_g - \omega_g^*\right) \end{aligned} \tag{6.15b}$$

$$\begin{aligned} V_2(x) = {} & -(d - P^l)^{\mathrm{T}}(P^{l*} - P^l) \\ & - (\varphi^+ - \eta^+)^{\mathrm{T}}(\eta^{+*} - \eta^+) \\ & - (\varphi^- - \eta^-)^{\mathrm{T}}(\eta^{-*} - \eta^-) \end{aligned} \tag{6.15c}$$

Then, we have the following result about $V(x)$.

Lemma 6.1 *Suppose Assumptions 6.1 and 6.2 hold for the closed-loop dynamics (6.7), (6.11). Then the function $V(x)$ defined in (6.15) has the following properties:*

1) *$V(x) \geq 0$ holds for all x, and $V(x) = 0$ holds only at the equilibrium.*
2) *The time derivative of $V(x(t))$ satisfies $\dot{V}(x(t)) \leq 0$.*

Proof: **Assertion 1)**. By (6.2), we know that $V_2(x) \geq 0$. Invoking (6.15a), (6.15b), and (6.15c), we know $V(x) \geq 0$ and $V(x) = 0$ holds only at the equilibrium point.

Assertion 2). By (6.6), the gradient of V is

$$\nabla V = \begin{bmatrix} \nabla_d V \\ \nabla_\mu V \\ \nabla_\phi V \\ \nabla_{\eta^+} V \\ \nabla_{\eta^-} V \\ \nabla_\theta V \\ \nabla_{\omega_g} V \end{bmatrix} = \begin{bmatrix} P^l - P^{l*} \\ \mu - \mu^* \\ \phi - \phi^* \\ \eta^+ - \eta^{+*} \\ \eta^- - \eta^{-*} \\ B(\theta - \theta^*) \\ M\left(\omega_g - \omega_g^*\right) \end{bmatrix} \tag{6.16}$$

Then, there is $g(P^l) \in \partial f(P^l)$ such that the time derivative of V is

$$\begin{aligned} \dot{V} = {} & (P^l - P^{l*})^{\mathrm{T}}(-d + P^l + \omega - g(P^l) - z - \mu) \\ & + (\mu - \mu^*)^{\mathrm{T}}(P^l - P^m + CBC^{\mathrm{T}}\phi) + (\theta - \theta^*)^{\mathrm{T}} BC^{\mathrm{T}}\omega \\ & + (\phi - \phi^*)^{\mathrm{T}} \left(-CBC^{\mathrm{T}}\mu - C\eta^- + C\eta^+ - CBC^{\mathrm{T}}z\right) \\ & + \left(\eta^+ - \eta^{+*}\right)^{\mathrm{T}} \left(-\varphi^+ + \eta^+ - C^{\mathrm{T}}\phi - \overline{\theta}\right) \\ & + (\eta^- - \eta^{-*})^{\mathrm{T}} \left(-\varphi^- + \eta^- + \underline{\theta} + C^{\mathrm{T}}\phi\right) \\ & + (\omega - \omega^*)^{\mathrm{T}}(P^m - P^l - D\omega - CB\theta) \end{aligned} \tag{6.17}$$

where the last item is due to the fact that, for each $j \in \mathcal{N}_l$,

$$0 = \left(\omega_j - \omega_j^*\right)\left(P_j^m - P_j^l - D_j\omega_j + \sum_{i:i\to j} B_{ij}\theta_{ij} - \sum_{k:j\to k} B_{jk}\theta_{jk}\right) \tag{6.18}$$

Combing (6.17) and (6.12) and denoting $\tilde{x} = x - x^*$, we have

$$\begin{aligned} \dot{V} = {} & (\tilde{P}^l)^{\mathrm{T}}(-\tilde{d} + \tilde{P}^l + \tilde{\omega} - g(P^l) + g(P^{l*}) - \tilde{z} - \tilde{\mu}) \\ & + \tilde{\varphi}^{\mathrm{T}} \left(-CBC^{\mathrm{T}}\tilde{\mu} - C\tilde{\eta}^- + C\tilde{\eta}^+ - CBC^{\mathrm{T}}\tilde{z}\right) \\ & + \tilde{\mu}^{\mathrm{T}}(\tilde{P}^l + CBC^{\mathrm{T}}\tilde{\varphi}) + \left(\tilde{\eta}^+\right)^{\mathrm{T}} \left(-\tilde{\phi}^+ + \tilde{\eta}^+ - C^{\mathrm{T}}\tilde{\varphi}\right) \\ & + (\tilde{\eta}^-)^{\mathrm{T}} \left(-\tilde{\phi}^- + \tilde{\eta}^- + C^{\mathrm{T}}\tilde{\varphi}\right) + \tilde{\theta}^{\mathrm{T}} BC^{\mathrm{T}}\tilde{\omega} \\ & + \tilde{\omega}^{\mathrm{T}}\left(-\tilde{P}^l - D\tilde{\omega} - CB\tilde{\theta}\right) \end{aligned}$$

$$= -(P^l - P^{l*})^{\mathrm{T}}(d - d^*) + \left\|P^l - P^{l*}\right\|^2 \tag{6.19a}$$

$$- (\eta^+ - \eta^{+*})^{\mathrm{T}}(\phi^+ - \phi^{+*}) + \|\eta^+ - \eta^{+*}\|^2 \tag{6.19b}$$

$$- (\eta^- - \eta^{-*})^{\mathrm{T}}(\phi^- - \phi^{-*}) + \|\eta^- - \eta^{-*}\|^2 \tag{6.19c}$$

$$- (P^l - P^{l*})^{\mathrm{T}} \left(g(P^l) - g(P^{l*})\right) - \tilde{\omega}^{\mathrm{T}} D\tilde{\omega} \tag{6.19d}$$

$$- (\tilde{P}^l)^{\mathrm{T}}\tilde{z} - \tilde{\varphi}^{\mathrm{T}} CBC^{\mathrm{T}}\tilde{z} \tag{6.19e}$$

By the [Theorem 1.5.5] [22], we know

$$-(P^l - P^{l*})^{\mathrm{T}}(d - d^*) + \left\|P^l - P^{l*}\right\|^2 \leq 0 \tag{6.20}$$

Similarly, we have

$$-(\eta^+ - \eta^{+*})^{\mathrm{T}}(\varphi^+ - \varphi^{+*}) + \|\eta^+ - \eta^{+*}\|^2 \leq 0$$

and

$$-(\eta^- - \eta^{-*})^{\mathrm{T}}(\varphi^- - \varphi^{-*}) + \|\eta^- - \eta^{-*}\|^2 \leq 0.$$

The convexity of f implies that

$$-(P^l - P^{l*})^{\mathrm{T}} \left(g(P^l) - g(P^{l*})\right) \leq 0 \tag{6.21}$$

The positive definiteness of D yields

$$-\tilde{\omega}^{\mathrm{T}} D\tilde{\omega} \leq 0.$$

In addition, there is

$$-(\tilde{P}^l)^{\mathrm{T}}\tilde{z} - \tilde{\phi}^{\mathrm{T}} CBC^{\mathrm{T}}\tilde{z} = -\tilde{z}^{\mathrm{T}} \cdot \tilde{z} \leq 0 \tag{6.22}$$

Therefore, the terms (6.19a)–(6.19e) are all nonpositive, i.e. $\dot{V}(x(t)) \leq 0$, which completes the proof. ■

The following theorem justifies the asymptotic stability of the closed-loop system (6.7) and (6.11), which also confirms the convergence of the proposed algorithm.

Theorem 6.2 *Suppose Assumptions 6.1 and 6.2 hold. Then the trajectory of the closed-loop system* (6.7) *and* (6.11) *has following properties:*

1) *$(x(t), P^l(t), \eta^+(t), \eta^-(t))$ is bounded.*
2) *$(x(t), P^l(t), \eta^+(t), \eta^-(t))$ converges to an equilibrium of the closed-loop system* (6.7) *and* (6.11).
3) *The convergence of $x(t)$ is to a point, i.e. $x(t) \to x^*$ as $t \to \infty$ for some equilibrium point x^*.*

Proof: **Assertion 1)**. From Lemma 6.1, we know that $(\theta(t), \omega_g(t), \mu(t), \phi(t), P^l(t), \eta^+(t), \eta^-(t))$ is bounded. By (6.7c), $\omega_l(t)$ is also bounded. Since $\partial f(P^l)$ is compact, there exists a constant a_1 such that

$$\left\| P^l + \omega - g(P^l) - \mu \right\| < a_1 \tag{6.23}$$

Define the following function:

$$\tilde{V}_d(d) := \frac{1}{2}\|d\|^2 \tag{6.24}$$

The time derivative of $\tilde{V}_d(d)$ along the closed-loop system is

$$\begin{aligned}\dot{\tilde{V}}_d &= d^{\mathrm{T}}(-d + P^l + \omega - g(P^l) - \mu)\\ &= -\|d\|^2 + d^{\mathrm{T}}(P^l + \omega - g(P^l) - \mu)\\ &\leq -\|d\|^2 + a_1\,\|d\|\\ &= -2\tilde{V}_d + a_1\sqrt{2\tilde{V}_d}\end{aligned} \tag{6.25}$$

Thus, $\tilde{V}_d(d(t)), t \geq 0$ is bounded, so is $d(t), t \geq 0$. Similarly, we can also have that $\varphi^+(t), t \geq 0$ and $\varphi^-(t), t \geq 0$ are bounded.

Assertion 2). Invoking the invariance principle presented in [Theorem 2] [28], we know that the trajectory $x(t)$ converges to the largest (weakly) invariant subset W^* contained in

$$W := \{\ x \mid \dot{V}(x(t)) = 0\ \}.$$

From $\tilde{\omega}^{\mathrm{T}} D\tilde{\omega} = 0$, we know $\tilde{\omega} = 0$, i.e. $\omega(t) = \omega^*$. Due to the strict convexity of $f(P^l)$, the inequality

$$(P^l - P^{l*})^{\mathrm{T}}\left(g(P^l) - g(P^{l*})\right) > 0$$

holds for $P^l \neq P^{l*}$. Therefore, there must be $P^l(t) = P^{l*}$ in the set W. Moreover, from (6.22), we have

$$\tilde{P}^l(t) = CBC^{\mathrm{T}}\tilde{\phi}(t),$$

or

$$0 = \dot{\tilde{P}}^l(t) = CBC^{\mathrm{T}}\dot{\tilde{\phi}}(t),$$

which implies that $\dot{\tilde{\phi}}(t) = 0$. Then, we have $\dot{\mu}_j(t) = 0$ from (6.11b). Similarly, $\dot{d}_j(t) = 0$ by (6.11a). Up to now, we have known that $x(t)$ is constant except $\varphi^-(t), \varphi^+(t)$ for $t \to \infty$.

Note that the equality in (6.3) holds only when $\mathcal{P}_\Omega(x) = \mathcal{P}_\Omega(y)$ or $x = \mathcal{P}_\Omega(x)$ and $y = \mathcal{P}_\Omega(y)$. Thus, for $\eta^+(t),\ \eta^-(t), t \to \infty$, there are only four types of combinations:

$$\textbf{Type 1:}\quad \begin{cases}\eta^+(t) &= \eta^{+*}\\ \eta^-(t) &= \eta^{-*}\end{cases}$$

$$\textbf{Type 2:} \begin{cases} \eta^{+}(t) &= \eta^{+*} \\ \varphi^{-}(t) &= \mathcal{P}_{\mathbb{R}_{+}^{m}}(\varphi^{-}(t)) &= \eta^{-}(t) \\ \varphi^{-*} &= \mathcal{P}_{\mathbb{R}_{+}^{m}}(\varphi^{-*}) &= \eta^{-*} \end{cases}$$

$$\textbf{Type 3:} \begin{cases} \eta^{-} &= \eta^{-*} \\ \varphi^{+}(t) &= \mathcal{P}_{\mathbb{R}_{+}^{m}}\left(\varphi^{+}(t)\right) &= \eta^{+}(t) \\ \varphi^{+*} &= \mathcal{P}_{\mathbb{R}_{+}^{m}}\left(\varphi^{+*}\right) &= \eta^{+*} \end{cases}$$

$$\textbf{Type 4:} \begin{cases} \varphi^{+}(t) &= \mathcal{P}_{\mathbb{R}_{+}^{m}}\left(\varphi^{+}(t)\right) &= \eta^{+}(t) \\ \varphi^{+*} &= \mathcal{P}_{\mathbb{R}_{+}^{m}}\left(\varphi^{+*}\right) &= \eta^{+*} \\ \varphi^{-}(t) &= \mathcal{P}_{\mathbb{R}_{+}^{m}}(\varphi^{-}(t)) &= \eta^{-}(t) \\ \varphi^{-*} &= \mathcal{P}_{\mathbb{R}_{+}^{m}}(\varphi^{-*}) &= \eta^{-*} \end{cases}$$

For the first type, we know $\eta^{+}(t), \eta^{-}(t), t \to \infty$ are constants. It follows that $\varphi^{+}(t), \varphi^{-}(t), t \to \infty$ are also constants by (6.11d) and (6.11e). For the second type, we know $\eta^{+}(t), \varphi^{+}(t), t \to \infty$ are constants by (6.11d), and $\dot{\varphi}^{-}(t)$ is constant by (6.11e). Due to the boundedness of $\varphi^{-}(t)$, we have $\varphi^{-}(t), \eta^{-}(t), t \to \infty$ that are constants as well. The other two combinations can be analyzed similarly, which are omitted here for simplicity. To summarize, $\varphi^{+}(t), \varphi^{-}(t), \eta^{+}(t), \eta^{-}(t)$ all approach to some constants as $t \to \infty$.

Therefore, $(x(t), P^{l}(t), \eta^{+}(t), \eta^{-}(t))$ converges to an equilibrium of the closed-loop system.

Assertion 3). Next, we prove the convergence is to a point following the idea given in [3]. Fix any initial state $x(0)$, and consider the trajectory $(x(t), t \geq 0)$ of the closed-loop system. As $x(t)$ is bounded, there exists an infinite sequence of time instants t_k such that $x(t_k) \to \hat{x}^*$ as $t_k \to \infty$, for some $\hat{x}^* \in W$ by the Bolzano–Weierstrass theorem. Using this specific equilibrium point $\hat{x}^*$ in the definition of V, we have

$$\begin{aligned} V^* &= \lim_{t\to\infty} V(x(t)) = \lim_{t_k\to\infty} V(x(t_k)) \\ &= \lim_{x(t_k)\to\hat{x}^*} V_2\left(x(t_k)\right) = V_2(\hat{x}^*) \\ &= 0 \end{aligned}$$

Here, the first equality holds by noting the fact that $V(t)$ is nonincreasing in t while lower-bounded and hence must render a limit value V^*, the second equality holds by noting that t_k is the infinite subsequence of t, the third equality uses the fact that $x(t)$ is absolutely continuous in t, the fourth equality is due to the continuity of $V(x)$, and the last equality holds as $\hat{x}^*$ is an equilibrium point of V.

The quadratic part V_1 implies that

$$\begin{aligned} &\lim_{t\to\infty}(\theta(t), \omega_g(t), P^{l}(t), \mu(t), \phi(t), \eta^{-}(t), \eta^{+}(t)) \\ &\qquad = (\theta^*, \omega_g^*, P^{l*}, \mu^*, \phi^*, \eta^{-*}, \eta^{+*}) \end{aligned} \tag{6.26}$$

Moreover, from (6.12c), (6.12f), and (6.12g), we can get the corresponding $d^*, \varphi^{-*}, \varphi^{+*}$. This completes the proof. ∎

Remark 6.7 To summarize, there are three difficulties in the convergence proof compared with that in [12]. The first is that it is difficult to prove $\phi(t) = \phi^*$ directly, as there is no strictly convex term in the objective function associated with ϕ. The second is the convergence proof of Lagrangian multipliers $\varphi^-(t), \varphi^+(t)$, which is closely related to the convergence of ϕ. Because the uniqueness of the equilibrium cannot be guaranteed, it is nontrivial to prove that trajectory must converge to a point, which is the third difficulty as the convergence to a point is crucial for the power system stability.

To address the first challenge, we intentionally add a quadratic term $\frac{1}{2}\sum_{j\in\mathcal{N}} z_j^2$ into the objective function. It helps prove the convergence of ϕ when P^l converges. To address the second challenge, we use the property of the projection. To address the third challenge, we employ the Bolzano–Weierstrass theorem. The approach to the convergence proof can be used to other nonsmooth optimization problems with non-strictly convex objective functions.

6.6 Case Studies

6.6.1 Test System

In this section, the IEEE 68-bus New England/New York interconnection test system [5] is utilized to illustrate the performance of the proposed controller. The diagram of the 68-bus system is shown in Figure 6.2. We simulate with the Power System Toolbox [29] on Matlab. Although the theory and algorithm are built on a linear model, the simulations are processed on a much more detailed and realistic model. Specifically, the generator includes the two-axis subtransient reactance model, the IEEE type DC1 exciter model, and a classical power system stabilizer model. In addition, AC (nonlinear) power flows are utilized with nonzero line resistances. The upper bound of P_j^l is the load demand value at each bus. A detailed simulation model including parameter values can be found in the data files of PST.

In the simulations, the cost function of each controllable load is

$$f(P_j^l) = \begin{cases} \left(P_j^l\right)^2 - 0.02, & P_j^l \le -0.2 \\ \frac{1}{2}\left(P_j^l\right)^2, & -0.2 < P_j^l \le 0.2 \\ \left(P_j^l\right)^2 - 0.02, & 0.2 < P_j^l \end{cases} \tag{6.27}$$

It can be verified that $f(P_j^l)$ is continuous, strictly convex, but nonsmooth.

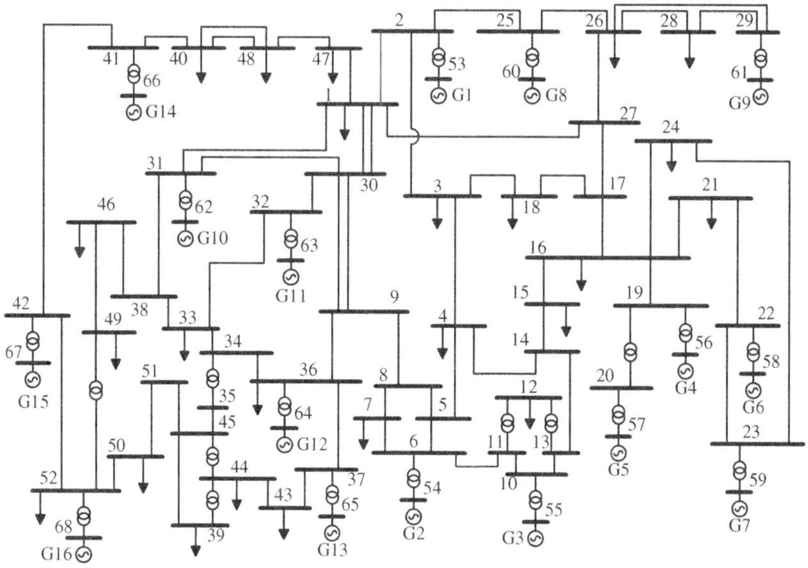

Figure 6.2 The IEEE 68-bus system for test.

6.6.2 Simulation Results

We consider the following scenario: at $t = 1$s, there is a step change of [3.5, 3.5, 3.5, 3.5, 3.5, 7] p.u. in the load demands at buses 4, 8, 20, 37, 42, and 52, respectively. However, neither the original load demands nor the changes are known to the controller. To estimate the load, the estimation method presented in Remark 6.5 is utilized.

At first, we do not set any limit to the tie-line power flow and analyze the dynamic performance of the closed-loop system under the proposed distributed controller, OLC. For comparison, a well-designed automatic generation control (AGC) is used for test in the same scenario as the benchmark. The setting of the AGC is the same as that in [10]. The frequency dynamics under the OLC and the AGC are given in Figure 6.3. It is shown that both the AGC and the OLC can recover the frequency to the nominal value. They also have similar frequency nadirs. However, compared with the AGC, the frequency under the proposed OLC has a faster convergence speed. This is one of the advantages benefiting from the fast response of load-side control.

The dynamics of μ are illustrated in Figure 6.4. If tie-line flow limits are not considered, then there will be $\eta^{-*} = \eta^{+*} = 0$. From (6.12e), we know that $CBC^{\mathrm{T}}\mu^* = 0$, i.e. μ of each bus will converge to the same value. This is verified in Figure 6.4 by taking buses 1–5 as examples, where the values of μ of buses 1–5 converge to the

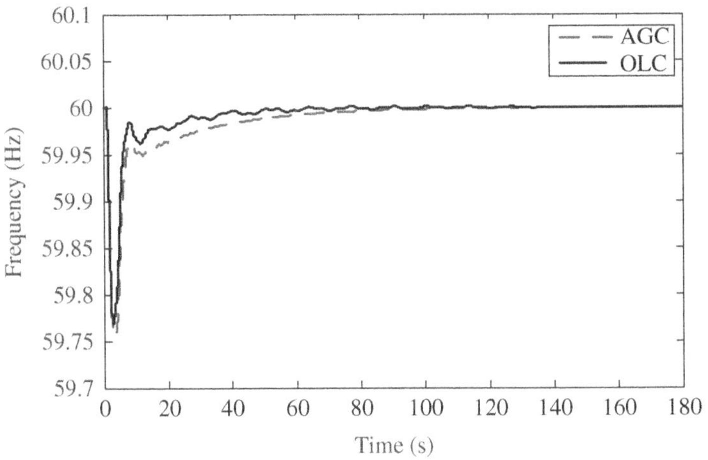

Figure 6.3 Comparison of the frequency dynamics under AGC and OLC.

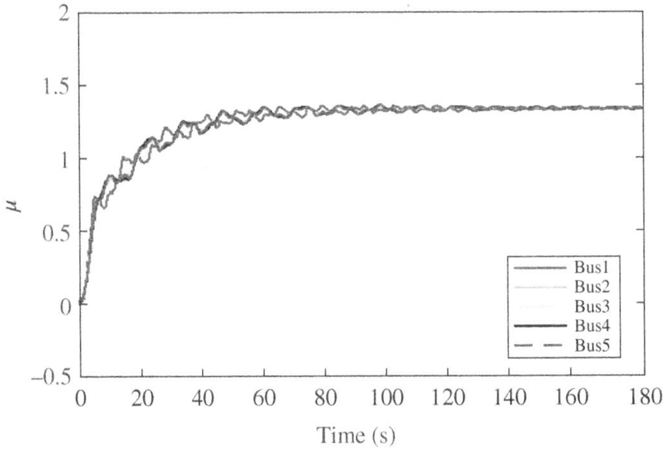

Figure 6.4 Dynamics of μ.

same value. Figure 6.5 illustrates the dynamics of controllable loads at these buses. They also converge to the same value since their cost functions are identical. As there is no controllable load on buses 2 and 5, their values are always zero. This result validates the correctness of the theoretic analyses.

We further take the tie-line flow limits into account. In this scenario, the active power limit of the line $(1, 2)$ is set as 0. Then, its tie-line flow dynamics with and without limit is given in Figure 6.6, where the black line is the result without the line limit and the dotted line is that with the limit. If there is no limit, the active

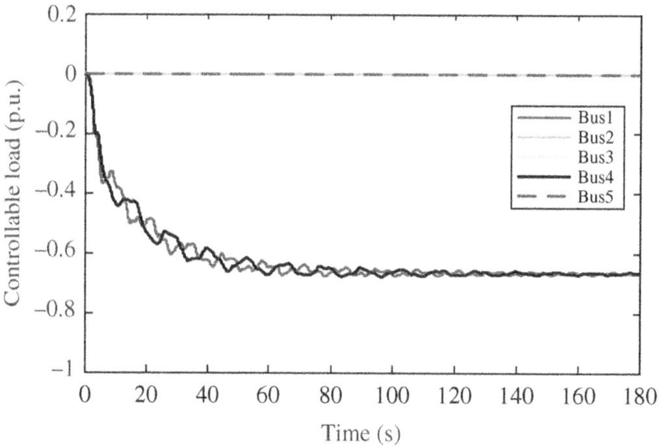

Figure 6.5 Dynamics of controllable loads.

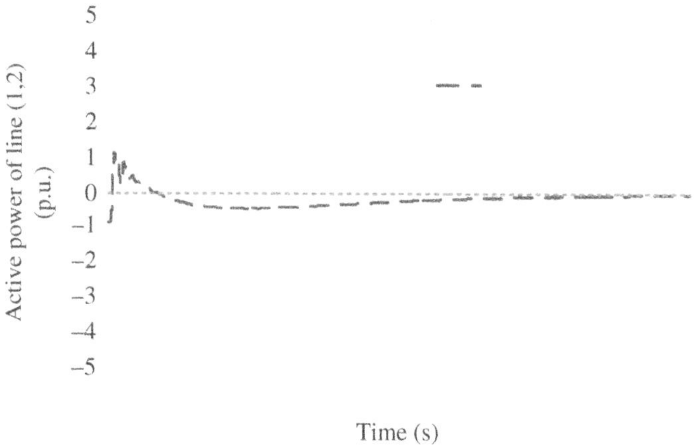

Figure 6.6 Active power dynamics of line 1 and controllable load.

power over line (1, 2) is −2.1 p.u. It decreases to zero if the limit is considered, which verifies the effectiveness of the line power control in relieving line flow congestions.

It is further observed that, if one of the tie-line limits is reached, the μ of those buses near this line will diverge due to the line congestion. This phenomenon is illustrated in Figure 6.7a, where μ_1, μ_3, μ_4 are all different. We also find that the μ of buses far from the congested line still converges to the same value, which is given in Figure 6.7b. It is shown that μ_{19}–μ_{23} converge to 0.83 identically.

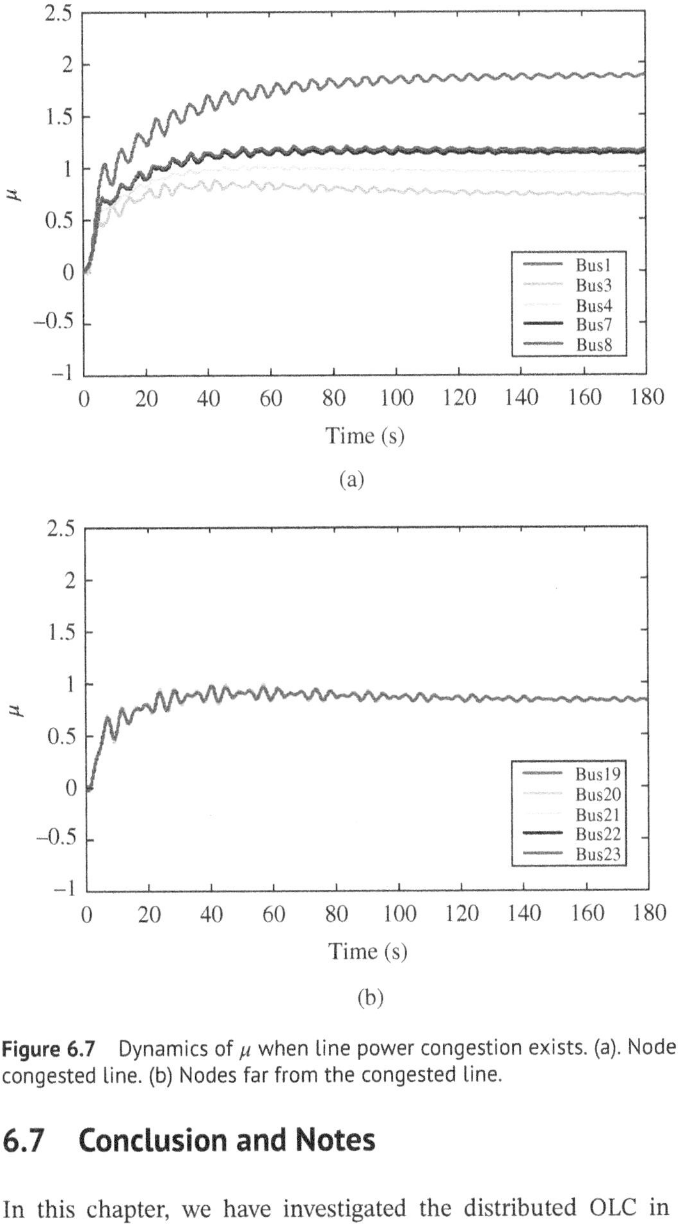

Figure 6.7 Dynamics of μ when line power congestion exists. (a). Nodes near the congested line. (b) Nodes far from the congested line.

6.7 Conclusion and Notes

In this chapter, we have investigated the distributed OLC in power systems when the regulation cost is nonsmooth. In our formulation, both the capacity limits of controllable loads and the tie-line flow limits have been considered. We have developed a distributed optimal controller based on (approximate) primal–dual algorithm, where the Clarke generalized gradient is utilized to address the nonsmoothness of the objective function. In addition, we have proved the optimality of the equilibrium of the closed-loop system and its asymptotic stability. Moreover, we have also proved that the convergence is to a point. Finally, we have conducted numerical experiments on the IEEE 68-bus system to demonstrate that the dynamic performance of the proposed controller outperforms the conventional AGC.

Although this chapter pays particular attention to load-frequency control, the proposed methodology can be extended and applied to many problems involving nonsmooth objective functions. Moreover, as mentioned in Remark 6.4, one can further relax Assumption 6.1 to a non-strict convexity condition by using the perturbation technique. Another interesting trick is augmenting the term $\frac{1}{2}\sum_{j\in\mathcal{N}} z_j^2$ into the objective function of the OLC problem. From a "forward engineering" perspective, it helps accelerate the convergence of the solution algorithm. On the other hand, it can also be understood from the perspective of "reverse engineering," which provides additional flexibility to controller design. This technique has been successfully applied in Chapter 5, showing its appealing merit in dealing with complicated constraints.

Bibliography

1 X. Zhang and A. Papachristodoulou, "A real-time control framework for smart power networks: design methodology and stability," *Automatica*, vol. 58, pp. 43–50, 2015.

2 T. Stegink, C. De Persis, and A. van der Schaft, "A unifying energy-based approach to stability of power grids with market dynamics," *IEEE Transactions Automatic Control*, vol. 62, no. 6, pp. 2612–2622, 2017.

3 N. Li, C. Zhao, and L. Chen, "Connecting automatic generation control and economic dispatch from an optimization view," *IEEE Transactions on Control of Network Systems*, vol. 3, no. 3, pp. 254–264, 2016.

4 D. Cai, E. Mallada, and A. Wierman, "Distributed optimization decomposition for joint economic dispatch and frequency regulation," *IEEE Transactions on Power Systems*, vol. 32, no. 6, pp. 4370–4385, 2017.

5 C. Zhao, U. Topcu, N. Li, and S. H. Low.,"Design and stability of load-side primary frequency control in power systems," *IEEE Transactions on Automatic Control*, vol. 59, no. 5, pp. 1177–1189, 2014.

6 E. Mallada, C. Zhao, and S. Low, "Optimal load-side control for frequency regulation in smart grids," *IEEE Transactions on Automatic Control*, vol. 62, no. 12, pp. 6294–6309, 2017.

7 Z. Wang, F. Liu, S. H. Low, C. Zhao, and S. Mei, "Distributed frequency control with operational constraints, Part I: Per-node power balance," *IEEE Transactions on Smart Grid*, vol. 10, no. 1, pp. 40–52, 2019.

8 A. Kasis, E. Devane, C. Spanias, and I. Lestas, "Primary frequency regulation with load-side participationpart I: stability and optimality," *IEEE Transactions on Power Systems*, vol. 32, no. 5, pp. 3505–3518, 2017.

9 Z. Wang, F. Liu, S. H. Low, C. Zhao, and S. Mei, "Distributed frequency control with operational constraints, Part II: network power balance," *IEEE Transactions on Smart Grid*, vol. 10, no. 1, pp. 53–64, 2019.

10 C. Zhao, E. Mallada, S. H. Low, and J. Bialek, "Distributed plug-and-play optimal generator and load control for power system frequency regulation," *International Journal of Electrical Power & Energy Systems*, vol. 101, pp. 1–12, 2018.

11 Z. Wang, F. Liu, S. H. Low, P. Yang, and S. Mei, "Distributed load-side control: coping with variation of renewable generations," *Automatica*, 10.1016/j.automatica.2019.108556 vol. 109, 108556, 2019.

12 X. Zeng, P. Yi, Y. Hong, and L. Xie, "Distributed continuous-time algorithms for nonsmooth extended monotropic optimization problems," *SIAM Journal on Control and Optimization*, vol. 56, no. 6, pp. 3973–3993, 2018.

13 S. K. Niederländer, F. Allgöwer, and J. Cortés, "Exponentially fast distributed coordination for nonsmooth convex optimization," in *2016 IEEE 55th Conference on Decision and Control (CDC)*, pp. 1036–1041, December 2016.

14 X. Zeng, Y. Peng, and Y. Hong, "Distributed continuous-time algorithm for constrained convex optimizations via nonsmooth analysis approach," *IEEE Transactions on Automatic Control*, vol. 62, no. 10, pp. 5227–5233, 2017.

15 S. Liang, X. Zeng, and Y. Hong, "Distributed nonsmooth optimization with coupled inequality constraints via modified lagrangian function," *IEEE Transactions on Automatic Control*, vol. 63, no. 6, pp. 1753–1759, 2018.

16 Z. Deng, X. Nian, and C. Hu, "Distributed algorithm design for nonsmooth resource allocation problems," *IEEE Transactions on Cybernetics*, vol. 50, no. 7, pp. 1753–1759, 2020.

17 X. Li, L. Xie, and Y. Hong, "Distributed continuous-time nonsmooth convex optimization with coupled inequality constraints," *IEEE Transactions on Control of Network Systems*, vol. 7, no. 1, pp. 74–84, 2020.

18 J. Cortés and S. K. Niederländer, "Distributed coordination for nonsmooth convex optimization via saddle-point dynamics," *Journal of Nonlinear Science*, vol. 29, no. 4, pp. 1247–1272, 2019.

19 Y. Wei, S. Ding, H. Fang, X. Zeng, and Q. Yang, "Distributed nonsmooth robust resource allocation with cardinality constrained uncertainty," *Chinese Control Conference (CCC)*, pp. 5758–5763, 2019.

20 F. H. Clarke, *Optimization and Nonsmooth Analysis*, Philadelphia, PA: Siam, 1990.

21 H. Bauschke and P. L. Combettes, *Convex Analysis and Monotone Operator Theory in Hilbert Spaces*. Springer, 2017.

22 F. Facchinei and J.-S. Pang, *Finite-Dimensional Variational Inequalities and Complementarity Problems*. New York: Springer-Verlag, 2003.

23 Q. Liu and J. Wang, "A one-layer projection neural network for nonsmooth optimization subject to linear equalities and bound constraints," *IEEE Transactions on Neural Networks & Learning Systems*, vol. 24, no. 5, pp. 812–824, 2013.

24 S. Boyd andL. Vandenberghe, *Convex Optimization*. Cambridge University Press, 2004.

25 H. Zhou, X. Zeng, and Y. Hong, "Adaptive exact penalty design for constrained distributed optimization," *IEEE Transactions on Automatic Control*, vol. 64, no. 11, pp. 4661–4667, 2019.

26 O. L. Mangasarian and R. Meyer, "Nonlinear perturbation of linear programs," *SIAM Journal on Control and Optimization*, vol. 17, no. 6, pp. 745–752, 1979.

27 A. P. Ruszczyński and A. Ruszczynski, *Nonlinear Optimization*, Princeton University Press, 2006.

28 J. Cortes, "Discontinuous dynamical systems," *IEEE Control Systems Magazine*, vol. 28, no. 3, pp. 36–73, 2008.

29 K. W. Cheung, J. Chow, and G. Rogers, "Power system toolbox ver. 3.0.," Rensselaer Polytechnic Institute and Cherry Tree Scientific Software, 2009.

7

Cyber Restrictions: Imperfect Communication in Power Control of Microgrids

The previous chapters focus on physical restrictions in distributed optimal control. However, since distributed control relies on communication systems, necessary information exchange is required. For simplifying the analysis and design, researchers usually assume that the communication is always perfect in many theoretical works, which means there are no time delays and no packet drops, and all agents make decisions simultaneously at the same pace. It leads to the so-called synchronous algorithms. Nevertheless, unfortunately, perfect communication does not exist in practice, making this hypothesis questionable. For facilitating the implementation and application of distributed optimal control, the impact of imperfect communication must be taken into account in the design, which calls for asynchronous algorithms. This chapter takes microgrids (MGs) as an example to demonstrate how to deal with asynchrony in distributed optimal control. Specifically, this chapter addresses the asynchronous distributed power control problem in hybrid AC/DC MGs, considering different kinds of asynchrony, such as non-identical sampling rates and random time delays. To this end, we first formulate the economic dispatch problem of MGs and devise a synchronous algorithm. Then, we analyze the impact of asynchrony and propose an asynchronous iteration algorithm based on the synchronous version. By introducing a virtual global clock with randomization, different types of asynchrony are fitted into a unified framework, where the asynchronous algorithm is converted into a fixed-point iteration problem with a nonexpansive operator, leading to a convergence proof. We further provide an upper bound estimation of the time delay of the communication that guarantees the convergence. Moreover, we present a real-time implementation of the proposed asynchronous distributed algorithm in both AC and DC MGs as well. Note that the inner dynamics of inverter-based MGs are very fast, which can be omitted in power control. Therefore, this chapter mostly involves feedback-based optimization, where the frequency measurements of AC MGs and the bus voltage of DC MGs are used as feedback to facilitate fast and adaptive distributed power control.

Merging Optimization and Control in Power Systems: Physical and Cyber Restrictions in Distributed Frequency Control and Beyond, First Edition. Feng Liu, Zhaojian Wang, Changhong Zhao, and Peng Yang.

7.1 Background

Multi-microgrid systems, or MGs, are clusters of distributed generators (DGs), energy storage systems, and loads, which are generally categorized into three types: AC MGs, DC MGs, and hybrid AC/DC MGs [1, 2]. A hybrid AC/DC MGs has the advantage of reducing processes of multiple inverse conversions in the involved individual AC or DC grids [3]. Recently, the distributed power control for MGs has attracted more and more attention due to its fast response, privacy preservation, and immunization of single-point-of-failure [4–10]. In the implementation of distributed control, communication plays a very crucial role. However, the communication channel is never perfect in practice, noting time delay, packet drops, congestion, and even failures [11–13]. In addition, non-identical sampling/computation rates of different MGs also exist [14]. As a consequence, the raised asynchrony issues can cause a detrimental impact on the control response speed, closed-loop system stability, and operation optimality [15–17]. This fact motivates to provide new understandings and design methods to facilitate the implementation of distributed control in real-world MGs. This chapter tries to simplify, and to some extent, unify the design of asynchronous distributed control that is resilient to different kinds of asynchrony.

To address the asynchrony in distributed power control, some methods have been proposed in [18–25]. In [18, 19], the energy trading game is investigated using Bayesian game theory, where the communication package loss is considered. In [21, 22], primal-dual gradient and consensus-based methods are used to solve the optimal power flow problem considering asynchronous iterations. In [20], a distributed algorithm is proposed to solve the optimal distributed energy resource (DER) coordination problem over lossy communication networks with packet-dropping communication links, where diminishing step sizes are required to guarantee the convergence. Reference [23] proposes a consensus-based economic dispatch algorithm under constant time delays. An explicit form of the upper bound for the gain parameter is given to guarantee the convergence. In reference [24], the authors devise a consensus-based method to achieve the proportional load sharing in DC MGs under a constant time delay. Zholbaryssov *et al.* [25] suggests a subgradient-based distributed algorithm to consider the network loss and communication delay in optimal generation dispatch.

Despite these very inspiring works, which have addressed many communication issues, some critical problems are still under-explored. In many works, constant time delays are necessary [23, 24]. The convergence proof is not provided in [18, 19], while [20, 25] can only guarantee the convergence with diminishing step sizes. In [21, 22], a stricter requirement on the step size is required, which needs to perform an extra line search at every iteration. In addition, the load demand in these works is usually assumed to be known and not varying. However, the load is difficult to measure and is always time-varying, particularly when demand response and electric vehicles are present. Moreover, the fast varying environment calls for real-time implementations of asynchronous distributed algorithms, which is also challenging.

In this chapter, we develop an asynchronous distributed algorithm to solve the economic dispatch problem of hybrid AC/DC MGs. The proposed algorithm can admit different kinds of asynchrony, such as non-identical sampling rates and random communication delays, which are very common in the practical operation of MGs. Thus, this work is different from the literature [23, 24], which only addresses constant time delays. After introducing a virtual global clock for analysis purposes, we prove the convergence of the distributed algorithm by converting it into a fixed-point iteration problem with a non-expansive operator. It greatly simplifies the convergence proof of asynchronous distributed algorithms. Moreover, in this way, constant stepsizes can be used under a convergence guarantee, which is much easier to implement. This is different from the literature [20–22, 25]. In [20, 25], diminishing step sizes are required, while an extra line search needs to be performed at every iteration in [21, 22] to determine the step size. Moreover, we provide an upper bound of the communication delay that guarantees convergence. It is revealed that the maximal delay is approximately proportional to the square root of the number of MGs. Finally, we propose a real-time implementation of the algorithm by using the feedback of frequency/voltage measurements from the physical power system. This treatment brings the following three advantages: (i) it simplifies the algorithm by reducing one order; (ii) it enables a faster response of the controller to adapt to the fast varying load demand; (iii) the impact of power loss such as line and inverter loss can be effectively eliminated.

The rest of this chapter is organized as follows. Section 7.2 formulates the power dispatch problem in hybrid MGs and proposes the synchronous algorithm. In Section 7.3.2, different types of asynchrony are introduced, and an asynchronous algorithm is proposed. The optimality of its equilibrium point and convergence is proved in Section 7.4. The real-time implementation method in hybrid MGs is introduced in Section 7.5. We confirm the performance of the controller via numerical simulations on a benchmark low voltage MG system and experiments on a dSPACE platform in Sections 7.6 and 7.7, respectively. Section 7.8 concludes the chapter with notes.

7.2 Preliminaries and Model

7.2.1 Notations and Preliminaries

Notations: A hybrid MG system is composed of a cluster of AC and DC MGs connected by lines. Each MG is treated as a bus with both generation and load. Denote AC MGs by $\mathcal{N}_{\mathrm{ac}} = \{1,2,\dots,n_{\mathrm{ac}}\}$, and DC MGs by $\mathcal{N}_{\mathrm{dc}} = \{n_{\mathrm{ac}}+1, n_{\mathrm{ac}}+2, \dots, n_{\mathrm{ac}}+n_{\mathrm{dc}}\}$. Then the set of MG buses is $\mathcal{N} = \mathcal{N}_{\mathrm{ac}} \cup \mathcal{N}_{\mathrm{dc}}$. Let $\mathcal{E} \subseteq \mathcal{N} \times \mathcal{N}$ be the set of lines, where $(i,k) \in \mathcal{E}$ if MGs i and k are connected directly. Then the overall system is modeled as a connected graph $\mathcal{G} := (\mathcal{N}, \mathcal{E})$. Denote by $N_{i,p} := \{k \mid (i,k) \in \mathcal{E}\}$ the set of neighbors of MG i over the physical graph. We also define a communication graph for MGs. Denote by $N_{i,c}$ the set of informational neighbors of MG i over the communication graph, implying MGs i,j can communicate if and only if $j \in N_{i,c}$. Denote by $N_{i,c}^2$ the set of two-hop neighbors of MG i over the communication graph. The cardinality of $N_{i,c}$ is denoted by $|N_{i,c}|$. The communication graph is also assumed to be undirected and connected, which could be different from the physical graph. Correspondingly, $N_{i,p}$ could also be different from $N_{i,c}$. Denote by L the Laplacian matrix of communication graph.

Preliminaries: In this chapter, $\mathbb{R}^n$ ($\mathbb{R}^n_+$) is the n-dimensional (nonnegative) Euclidean space. For a column vector $x \in \mathbb{R}^n$ (matrix $A \in \mathbb{R}^{m\times n}$), $x^{\mathrm{T}}(A^{\mathrm{T}})$ denotes its transpose. For vectors $x, y \in \mathbb{R}^n$, $x^{\mathrm{T}}y = \langle x, y\rangle$ denotes the inner product of x, y. $\|x\| = \sqrt{x^{\mathrm{T}}x}$ denotes the Euclidean norm of x. For a positive definite matrix G, denote the inner product $\langle x, y\rangle_G = \langle Gx, y\rangle$. Similarly, the G-matrix induced norm $\|x\|_G = \sqrt{\langle Gx, x\rangle}$. Use I to denote the identity matrix with proper dimensions. For a matrix $A = [a_{ij}]$, a_{ij} stands for the entry in the i-th row and j-th column of A. Use $\prod_{i=1}^n \Omega_i$ to denote the Cartesian product of the sets $\Omega_i, i = 1, \dots, n$. Given a collection of y_i for i in a certain set Y, y denotes the column vector $y := (y_i, i \in Y)$ with a proper dimension with y_i as its components.

For a single-valued operator $\mathcal{T} : \Omega \subset \mathbb{R}^n \to \mathbb{R}^n$, a point $x \in \Omega$ is a fixed point of $\mathcal{T}$ if $\mathcal{T}(x) \equiv x$. The set of fixed points of $\mathcal{T}$ is denoted by $Fix(\mathcal{T})$. $\mathcal{T}$ is nonexpansive if $\|\mathcal{T}(x) - \mathcal{T}(y)\| \le \|x - y\|, \forall x, y \in \Omega$. For $\alpha \in (0,1)$, $\mathcal{T}$ is called α-averaged if there exists a nonexpansive operator $\mathcal{R}$ such that $\mathcal{T} = (1-\alpha)\mathrm{Id} + \alpha\mathcal{R}$. We use $\mathcal{A}(\alpha)$ to denote the class of α-averaged operators. For $\beta \in \mathbb{R}^1_+$, $\mathcal{T}$ is called β-cocoercive if $\beta\mathcal{T} \in \mathcal{A}(\frac{1}{2})$.

7.2.2 Economic Dispatch Model

The economic power dispatch of MGs aims to achieve the power balance across MGs while minimizing the generation cost, which can be formulated as the following optimization problem

$$\min_{P_i^g} \sum_{i\in\mathcal{N}} f_i(P_i^g) \tag{7.1a}$$

$$\text{s.t.} \sum_{i\in\mathcal{N}} P_i^g = \sum_{i\in\mathcal{N}} P_i^d \tag{7.1b}$$

$$\underline{P}_i^g \le P_i^g \le \overline{P}_i^g \tag{7.1c}$$

where,

$$f_i(P_i^g) = \frac{1}{2}a_i(P_i^g)^2 + b_iP_i^g,$$

with $a_i > 0, b_i > 0$. Here, P_i^g is the power generation of MG i, and P_i^d the Load demand of MG i. The objective function (7.1a) is to minimize the total generation cost of the MGs. $\underline{P}_i^g, \overline{P}_i^g$ are the lower and upper bounds of P_i^g, respectively. Constraint (7.1b) is the power balance over MGs, and (7.1c) represents the generation limits of MGs.

For the optimization problem (7.1), we make the following assumption.

Assumption 7.1 The Slater's condition [Chapter 5.2.3] [26] of (7.1) holds.

7.3 Distributed Control Algorithms

7.3.1 Synchronous Algorithm

In this subsection, we design a synchronous algorithm to solve the problem (7.1). The Lagrangian of (7.1) is

$$\begin{aligned}\mathcal{L} = &\sum_{i\in\mathcal{N}} f_i(P_i^g) + \mu\left(\sum_{i\in\mathcal{N}} P_i^g - \sum_{i\in\mathcal{N}} P_i^d\right)\\ &+ \sum_{i\in\mathcal{N}} \gamma_i^-(\underline{P}_i^g - P_i^g) + \sum_{i\in\mathcal{N}} \gamma_i^+(P_i^g - \overline{P}_i^g)\end{aligned} \tag{7.2}$$

where $\mu, \gamma_i^-, \gamma_i^+$ are Lagrangian multipliers. Here μ is a global variable, which will be estimated by individual MGs locally.

Define the sets

$$\Omega_i := \left\{P_i^g \mid \underline{P}_i^g \le P_i^g \le \overline{P}_i^g\right\}, \quad \Omega = \prod_{i=1}^{N} \Omega_i \tag{7.3}$$

Then, we give the *s*ynchronous *d*istributed algorithm for *p*ower *d*ispatch (SDPD) algorithm. Mathematically, the update of MGi in each iteration takes the form of Krasnosel'skiĭ–Mann iteration [Chapter 5.2] [28]:

$$\tilde{\mu}_{i,k} = \mu_{i,k} + \sigma_\mu\left(-\sum_{j\in N_{i,c}}\left(\mu_{i,k} - \mu_{j,k}\right) + \sum_{j\in N_{i,c}}(z_{i,k} - z_{j,k}) + P_{i,k}^g - P_i^d\right) \tag{7.4a}$$

$$\tilde{z}_{i,k} = z_{i,k} - \sigma_z\left(2\sum_{j\in N_{i,c}}\left(\tilde{\mu}_{i,k} - \tilde{\mu}_{j,k}\right) - \sum_{j\in N_{i,c}}\left(\mu_{i,k} - \mu_{j,k}\right)\right) \tag{7.4b}$$

$$\tilde{P}_{i,k}^g = \mathcal{P}_{\Omega_i}\left(P_{i,k}^g - \sigma_g\left(f_i'(P_{i,k}^g) + 2\tilde{\mu}_{i,k} - \mu_{i,k}\right)\right) \tag{7.4c}$$

$$\mu_{i,k+1} = \mu_{i,k} + \eta\left(\tilde{\mu}_{i,k} - \mu_{i,k}\right) \tag{7.4d}$$

$$z_{i,k+1} = z_{i,k} + \eta \left(\tilde{z}_{i,k} - z_{i,k}\right) \tag{7.4e}$$

$$P^g_{i,k+1} = P^g_{i,k} + \eta \left(\tilde{P}^g_{i,k} - P^g_{i,k}\right) \tag{7.4f}$$

where $\sigma_\mu, \sigma_z, \sigma_g, \eta$ are positive constants and $\sigma_\mu, \sigma_z, \sigma_g$ are supposed to be chosen such that Φ in (7.11) is positive definite. $\tilde{\mu}_i, \mu_i$ are variables of MG i to estimate μ. It is because μ in the Lagrangian is a global variable, which needs to be estimated in a distributed way. Thus, we introduce $\tilde{\mu}_i, \mu_i$. Variables $\tilde{z}_i, z_i$ are introduced to enforce $\mu_i, i \in \mathcal{N}$ to reach consensus.

In (7.4), precisely measuring all the load demands P^d_i across the system is very challenging, if at all. Therefore, here we will provide a practical method to estimate $P^g_i - P^d_i$ instead of directly measuring P^d_i in implementation, as will be explained in Section 7.5. Moreover, later in Section 7.4, we will reveal the properties of the asynchronous version of the SDPD and show that the SDPD is simply a particular case of the distributed asynchronous algorithm. Therefore, the properties of the SDPD, such as the optimality of the equilibrium point and the convergence, are immediate consequences of the asynchronous algorithm and omitted here.

7.3.2 Asynchronous Algorithm

In the SDPD, each MG gathers information from its neighbors and then computes it locally and conveys new information to its neighbors over the communication graph. In this process, asynchrony can arise in each step. For example, individual MGs may have different sampling rates when gathering information, which results in nonidentical computation clocks accordingly. In addition, other imperfect communication situations such as time delay caused by congestion or even communication channel failures also occur from time to time in power systems, resulting in asynchrony.

In a distributed synchronous algorithm, an MG has to wait for the slowest neighbor to complete the computation by inserting a certain period of idle time. Communication delays or congestion can further lengthen the waiting time. Thus, the slowest MG and communication channel may cripple the system's stability and optimality. In contrast, the MGs with a distributed asynchronous computation can avoid waiting time and compute continuously with little idling. Even if some of its neighbors fail to update in time, the MG can use the previously stored information. That means the MG can execute an iteration without the latest information from its neighbors, enabling a more efficient and robust distributed optimization.

In this subsection, we present an *a*synchronous *d*istributed algorithm for *p*ower *d*ispatch (ASDPD) based on the SDPD. Different from the identical iteration number k used in (7.4), here each MG has its own iteration number k_i, implying that a *local clock* is used instead of the global clock. At each iteration k_i, MG i processes computation in the following way[1]:

$$\tilde{\mu}_{i,k_i} = \mu_{i,k_i} + \sigma_\mu \Bigg(-\sum_{j\in N_{i,c}} \left(\mu_{i,k_i} - \mu_{j,k_i-\tau_{ij}^{k_i}} \right) + \sum_{j\in N_{i,c}} \left(z_{i,k_i} - z_{j,k_i-\tau_{ij}^{k_i}} \right) + P^g_{i,k_i} - P^d_i \Bigg) \tag{7.5a}$$

$$\begin{aligned} \tilde{z}_{i,k_i} = z_{i,k_i} - \sigma_z \Bigg(& \sum_{j\in N_{i,c}} \left(\mu_{i,k_i} - \mu_{j,k_i-\tau_{ij}^{k_i}} \right) \\ & - \sum_{j\in N_{i,c}\cup N^2_{i,c}} 2\sigma_\mu \ell_{ij} \left(\mu_{i,k_i} - \mu_{j,k_i-\tau_{ij}^{k_i}} \right) \\ & + \sum_{j\in N_{i,c}\cup N^2_{i,c}} 2\sigma_\mu \ell_{ij} \left(z_{i,k_i} - z_{j,k_i-\tau_{ij}^{k_i}} \right) \\ & + \sum_{j\in N_{i,c}} 2\sigma_\mu \left(P^g_{i,k_i} - P^d_i \right) \Bigg) \end{aligned} \tag{7.5b}$$

$$\tilde{P}^g_{i,k_i} = \mathcal{P}_{\Omega_i} \left(P^g_{i,k_i} - \sigma_g \left(f_i'(P^g_{i,k_i}) + 2\tilde{\mu}_{i,k_i} - \mu_{i,k_i} \right) \right) \tag{7.5c}$$

$$\mu_{i,k_i+1} = \mu_{i,k_i} + \eta \left(\tilde{\mu}_{i,k_i} - \mu_{i,k_i} \right) \tag{7.5d}$$

$$z_{i,k_i+1} = z_{i,k_i} + \eta \left(\tilde{z}_{i,k_i} - z_{i,k_i} \right) \tag{7.5e}$$

$$P^g_{i,k_i+1} = P^g_{i,k_i} + \eta \left(\tilde{P}^g_{i,k_i} - P^g_{i,k_i} \right) \tag{7.5f}$$

where ℓ_{ij} is the ith row and jth column element of matrix $L^2 = L \times L$ and $\ell_{ij} \neq 0$ holds only if $j \in N_{i,c} \cup N^2_{i,c}$ [29]. $\tau_{ij}^{k_i}$ is the time difference between k_i and the time spot that MG i obtains the latest information from MG j. For example, the current iteration for MG i is $k_i = 10$, but it get the latest information from MG j at $k_i = 8$. Then, $\tau_{ij}^{k_i} = 2$ in this situation. Let $w = (\mu, z, P^g)$. Then w_{i,k_i} is the state of MG i at iteration k_i, and $w_{j,k_i-\tau_{ij}^{k_i}}$ is the latest information obtained from MG j.

Considering each MG has its local clock, we have the following distributed asynchronous algorithm.

Remark 7.1 In Algorithm 7.1, if MG i is activated, it will receive the latest information from its neighbors. Even if some neighbors are not accessible due to communication issues or other reasons, it can still execute the computation by using

1 If we omit the time delay in (7.5), it is essentially equivalent to (7.4) by substituting $\tilde{\mu}_{i,k}$ in (7.4b) and (7.4a).

Algorithm 7.1: *ASDPD*

Input: For MG i, the input is $\mu_{i,0}, z_{i,0} \in \mathbb{R}^n$, $P^g_{i,0} \in \Omega_i$.

Iteration at k_i: Supposing MG i's clock ticks at time k_i, then MG i is activated and updates its local variables as follows:

Step 1: Reading phase

Get $\mu_{j,k_i-\tau_{ij}^{k_i}}, z_{j,k_i-\tau_{ij}^{k_i}}$ from its neighbors' and two-hop neighbors' output cache, and store them to its local storage.

Step 2: Computing phase

Calculate $\tilde{\mu}_{i,k_i}, \tilde{z}_{i,k_i}$, and $\tilde{P}^g_{i,k_i}$ according to (7.5a), (7.5b), and (7.5c), respectively.

Update $\mu_{i,k_i+1}, z_{i,k_i+1}$, and P^g_{i,k_i+1} according to (7.5d), (7.5e), and (7.5f), respectively.

Step 3: Writing phase

Write $\mu_{i,k_i+1}, z_{i,k_i+1}$ to its output cache and $\mu_{i,k_i+1}, z_{i,k_i+1}, P^g_{i,k_i+1}$ to its local storage. Increase k_i to $k_i + 1$.

the previous information stored in its input cache. Despite asynchrony caused by different reasons, MG i only concerns whether the latest information comes, implying that their effect can be characterized by the time interval between two successive iterations. In this sense, our algorithm can admit different types of asynchrony.

Remark 7.2 Because the element $\ell_{ij} \neq 0$ holds only if $j \in N_{i,c} \cup N^2_{i,c}$, the ASDPD is still distributed. Communications with two-hop or multi-hop neighbors are also used in [29–31]. However, it may make the communication graph much denser. In Section 7.5, we will show that the ASDPD can be simplified based on local measurements and neighboring communication.

7.4 Optimality and Convergence Analysis

In this section, we analyze the optimality of the equilibrium point of dynamic system (7.5), as well as the convergence of Algorithm 7.1. First, we need to introduce a sequence of global iteration numbers that serve as a reference *global clock* to unify the local iterations of individual MGs in a coherent manner [32]. Then, we convert the synchronous algorithm into a fixed-point iteration problem with an averaged operator to prove the convergence. Thereby a *nonexpansive* operator is constructed, leading to the convergence results of the asynchronous algorithm. Finally, we provide an estimation of the upper bound on time delays that guarantee the convergence.

7.4.1 Virtual Global Clock

To facilitate analysis, we introduce a virtual global clock. Specifically, we arrange k_i of all MGs in the order of time and use a new number k to denote the kth iteration in the queue, which is referred to as the global clock. Figure 7.1 demonstrates this treatment by taking two MGs as an example. Suppose that, at the iteration k, the probability that MG i is activated to update its local variables follows a uniform distribution. Hence, each MG is activated with the same probability, which simplifies the convergence proof.

Then, we rewrite the algorithm (7.5) subject to the global clock, equivalently yielding

$$\begin{aligned}\tilde{\mu}_{i,k} = \mu_{i,k-\tau_i^k} + \sigma_\mu \Bigg(&-\sum_{j\in N_{i,c}} \left(\mu_{i,k-\tau_i^k} - \mu_{j,k-\tau_j^k}\right) \\ &+\sum_{j\in N_{i,c}} \left(z_{i,k-\tau_i^k} - z_{j,k-\tau_j^k}\right) + P^g_{i,k-\tau_i^k} - P_i^d \Bigg)\end{aligned} \tag{7.6a}$$

$$\begin{aligned}\tilde{z}_{i,k} = z_{i,k-\tau_i^k} - \sigma_z \Bigg(&\sum_{j\in N_{i,c}} \left(\mu_{i,k-\tau_i^k} - \mu_{j,k-\tau_j^k}\right) \\ &- \sum_{j\in N_{i,c}\cup N^2_{i,c}} 2\sigma_\mu \ell_{ij} \left(\mu_{i,k-\tau_i^k} - \mu_{j,k-\tau_j^k}\right) \\ &+ \sum_{j\in N_{i,c}\cup N^2_{i,c}} 2\sigma_\mu \ell_{ij} \left(z_{i,k-\tau_i^k} - z_{j,k-\tau_j^k}\right) \\ &+\sum_{j\in N_{i,c}} 2\sigma_\mu \left(P^g_{i,k-\tau_i^k} - P_i^d\right)\Bigg)\end{aligned} \tag{7.6b}$$

$$\tilde{P}^g_{i,k} = \mathcal{P}_{\Omega_i}\left(P^g_{i,k-\tau_i^k} - \sigma_g\left(f_i'(P^g_{i,k-\tau_i^k}) + 2\tilde{\mu}_{i,k} - \mu_{i,k-\tau_i^k}\right)\right) \tag{7.6c}$$

$$\mu_{i,k+1} = \mu_{i,k-\tau_i^k} + \eta\left(\tilde{\mu}_{i,k} - \mu_{i,k-\tau_i^k}\right) \tag{7.6d}$$

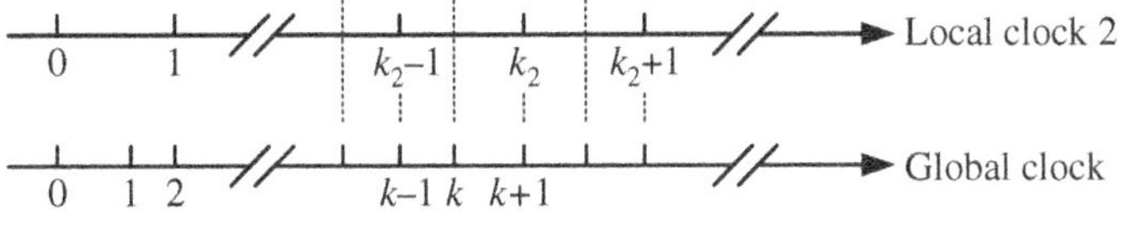

Figure 7.1 Local clocks versus global clock.

$$z_{i,k+1} = z_{i,k-\tau_i^k} + \eta \left(\tilde{z}_{i,k} - z_{i,k-\tau_i^k} \right) \tag{7.6e}$$

$$P^g_{i,k+1} = P^g_{i,k-\tau_i^k} + \eta \left(\tilde{P}^g_{i,k} - P^g_{i,k-\tau_i^k} \right) \tag{7.6f}$$

where τ_i^k is the time difference between k and time spot that MG i obtains the latest information. We also call τ_i^k the time delay if there is no confusion. It is easy to see that

$$w_{i,k-\tau_i^k} = w_{i,k-\tau_i^k+1} =, \cdots, = w_{i,k}.$$

Note that the virtual global clock is only used for convergence analysis, but not required in the implementation of the ASDPD.

7.4.2 Algorithm Reformulation

If the time delay is not considered, (7.6) is degenerated to (7.4). In this sense, the SDPD is simply a particular case of the ASDPD. Hence, we only need to analyze the properties of the ASDPD. The compact form of (7.6a)–(7.6f) without delay, i.e. (7.4a)–(7.4f), is

$$\tilde{\mu}_k = \mu_k + \sigma_\mu \left(-L \cdot \mu_k + L \cdot z_k + P^g_k - P^d \right) \tag{7.7a}$$

$$\tilde{z}_k = z_k + \sigma_z \left(-2L \cdot \tilde{\mu}_k + L \cdot \mu_k \right) \tag{7.7b}$$

$$\tilde{P}^g_k = \mathcal{P}_\Omega \left(P^g_k - \sigma_g \left(\nabla f(P^g_k) + 2\tilde{\mu}_k - \mu_k \right) \right) \tag{7.7c}$$

$$\mu_{k+1} = \mu_k + \eta \left(\tilde{\mu}_k - \mu_k \right) \tag{7.7d}$$

$$z_{k+1} = z_k + \eta \left(\tilde{z}_k - z_k \right) \tag{7.7e}$$

$$P^g_{k+1} = P^g_k + \eta \left(\tilde{P}^g_k - P^g_k \right) \tag{7.7f}$$

where $\nabla f(P^g_k)$ is the gradient of $f(P^g_k)$. The subscript k_i is replaced by a global clock k.

Next, we show that (7.7a)–(7.7f) can be converted into a fixed-point iteration problem with an averaged operator [33, 34].

Equation (7.7a) is equivalent to

$$\begin{aligned} & -L \cdot \mu_k - P^d \\ = & -P^g_k - L \cdot z_k + \sigma_\mu^{-1}(\tilde{\mu}_k - \mu_k) \\ = & -L\tilde{z}_k - \tilde{P}^g_k + \sigma_\mu^{-1}(\tilde{\mu}_k - \mu_k) \\ & + L \cdot (\tilde{z}_k - z_k) + \tilde{P}^g_k - P^g_k \end{aligned} \tag{7.8}$$

Similarly, (7.7b) is equal to

$$0 = L \cdot \tilde{\mu}_k + L \cdot (\tilde{\mu}_k - \mu_k) + \sigma_z^{-1}(\tilde{z}_k - z_k) \tag{7.9}$$

From the fact that $\mathcal{P}_\Omega(x) = (\mathrm{Id} + N_\Omega)^{-1}(x)$, (7.7c) can be rewritten as

$$\tilde{P}_k^g = (\mathrm{Id} + N_\Omega)^{-1}(P_k^g - \sigma_g(\nabla f(P_k^g) + 2\tilde{\mu}_k - \mu_k)),$$

or, equivalently,

$$-\nabla f(P_k^g) = 2\tilde{\mu}_k - \mu_k + N_\Omega(\tilde{P}_k^g) + \sigma_g^{-1}(\tilde{P}_k^g - P_k^g) \tag{7.10}$$

Then, (7.8)–(7.10) are rewritten as

$$-\begin{bmatrix} L\mu_k + P^d \\ 0 \\ \nabla f(P_k^g) \end{bmatrix} = \begin{bmatrix} -\tilde{P}_k^g - L\tilde{z}_k \\ L\tilde{\mu}_k \\ \tilde{\mu}_k + N_\Omega(\tilde{P}_k^g) \end{bmatrix} + \Phi \begin{bmatrix} \tilde{\mu}_k - \mu_k \\ \tilde{z}_k - z_k \\ \tilde{P}_k^g - P_k^g \end{bmatrix} \tag{7.11}$$

where

$$\Phi = \begin{bmatrix} \sigma_\mu^{-1} I & L & I \\ L & \sigma_z^{-1} I & 0 \\ I & 0 & \sigma_g^{-1} I \end{bmatrix} \tag{7.12}$$

Define the following two operators:

$$\mathcal{B} : \begin{bmatrix} \mu \\ z \\ P^g \end{bmatrix} \mapsto \begin{bmatrix} L\mu + P^d \\ 0 \\ \nabla f(P^g) \end{bmatrix} \tag{7.13}$$

$$\mathcal{U} : \begin{bmatrix} \mu \\ z \\ P^g \end{bmatrix} \mapsto \begin{bmatrix} -P^g - Lz \\ L\mu \\ \mu + N_\Omega(P^g) \end{bmatrix} \tag{7.14}$$

Then, (7.11) can be rewritten in a compact form of

$$-\mathcal{B}(w_k) = \mathcal{U}(\tilde{w}_k) + \Phi \cdot (\tilde{w}_k - w_k) \tag{7.15}$$

From [Lemma 5.6] [34], we know $(\mathrm{Id} + \Phi^{-1}\mathcal{U})^{-1}$ exists and is single-valued. For simplicity, we adopt the following notations:

$$\begin{aligned} w^i &= (\mu_i, z_i, P_i^g), \\ w &= (w^i), \\ \tilde{w}^i &= (\tilde{\mu}_i, \tilde{z}_i, \tilde{P}_i^g), \\ \tilde{w} &= (\tilde{w}^i). \end{aligned} \tag{7.16}$$

Then, (7.7a)–(7.7f) can be written as

$$\tilde{w}_k = \mathcal{T}(w_k) \tag{7.17}$$

$$w_{k+1} = w_k + \eta(\tilde{w}_k - w_k) \tag{7.18}$$

where the operator $\mathcal{T}$ is given by

$$\mathcal{T} = (\mathrm{Id} + \Phi^{-1}\mathcal{U})^{-1}(\mathrm{Id} - \Phi^{-1}\mathcal{B})$$

and it is straightforward to see that (7.7d)–(7.7f) are equivalent to (7.18).

Equations (7.17)–(7.18) can be further rewritten as

$$w_{k+1} = w_k + \eta(\mathcal{T}(w_k) - w_k) \tag{7.19}$$

Denote $a_{min} = \min\{a_i\}$, $a_{max} = \max\{a_i\}$, $\forall i \in \mathcal{N}$, where a_i is defined below (7.1). Denote the maximal eigenvalues of L by $\sigma_{\max}$. We have the following lemma to characterize the operator $\mathcal{T}$.

Lemma 7.1 *Take* $\zeta = \min\{\frac{1}{\sigma_{\max}^2}, \frac{a_{min}}{a_{max}^2}\}$, $\kappa > \frac{1}{2\zeta}$, *and the step sizes* $\sigma_\mu, \sigma_z, \sigma_g$ *such that* $\Phi - \kappa I$ *is positive semidefinite. Then we have the following assertions in the sense of* Φ*-induced norm:*

1) $\mathcal{T}$ *is an averaged operator, and* $\mathcal{T} \in \mathcal{A}\left(\frac{2\kappa\zeta}{4\kappa\zeta-1}\right)$.
2) There exists a nonexpansive operator $\mathcal{R}$ *such that*

$$\mathcal{T} = \left(1 - \frac{2\kappa\zeta}{4\kappa\zeta - 1}\right)\text{Id} + \frac{2\kappa\zeta}{4\kappa\zeta - 1}\mathcal{R}$$

3) Operators $\mathcal{T}$ *and* $\mathcal{R}$ *have the same fixed points, i.e.* $Fix(\mathcal{T}) = Fix(\mathcal{R})$.

Proof: **Assertion 1)**. By [Lemma 5.6] [34], we know

$$(\text{Id} + \Phi^{-1}\mathcal{U})^{-1} \in \mathcal{A}\left(\frac{1}{2}\right),$$

and

$$\text{Id} - \Phi^{-1}\mathcal{B} \in \mathcal{A}\left(\frac{1}{2\kappa\zeta}\right)$$

Then, from [Proposition 2.4] [35], we know $\mathcal{T} \in \mathcal{A}\left(\frac{2\kappa\zeta}{4\kappa\zeta-1}\right)$, which is Assertion 1).

Assertion 2). From Assertion 1) and the definition of averaged operators, there exists a nonexpansive operator $\mathcal{R}$ such that

$$\mathcal{T} = \left(1 - \frac{2\kappa\zeta}{4\kappa\zeta - 1}\right)\text{Id} + \frac{2\kappa\zeta}{4\kappa\zeta - 1}\mathcal{R} \tag{7.20}$$

Then, we have Assertion 2).

Assertion 3). Since $\mathcal{T}$ is $\frac{2\kappa\zeta}{4\kappa\zeta-1}$-averaged, $\mathcal{T}$ is also a nonexpansive operator according to [Remark 4.24] [28]. For any nonexpansive operator $\mathcal{T}$, $Fix(\mathcal{T}) \neq \emptyset$ [Theorem 4.19] [28]. Suppose x is a fixed point of $\mathcal{T}$, and then we have

$$\mathcal{T}(x) = x = \left(1 - \frac{2\kappa\zeta}{4\kappa\zeta - 1}\right)\text{Id}(x) + \frac{2\kappa\zeta}{4\kappa\zeta - 1}\mathcal{R}(x)$$

It immediately leads to

$$\frac{2\kappa\zeta}{4\kappa\zeta - 1}\text{Id}(x) = \frac{2\kappa\zeta}{4\kappa\zeta - 1}\mathcal{R}(x),$$

which is equivalent to $x = \mathcal{R}(x)$.

Similarly, suppose x is a fixed point of $\mathcal{R}$, and we have

$$\mathcal{T}(x) = \left(1 - \frac{2\kappa\zeta}{4\kappa\zeta - 1}\right)\ \mathrm{Id}(x) + \frac{2\kappa\zeta}{4\kappa\zeta - 1}\mathcal{R}(x) = x$$

Thus, Assertion 3) holds, which completes the proof. ■

So far, we have converted the ASDPD algorithm without time delay into a fixed-point iteration problem with an averaged operator (see (7.19)). Next, we will construct a nonexpansive operator $\mathcal{R}$, which allows us to prove the optimality and convergence of the ASDPD with time delay, as we will explain in the following two subsections.

7.4.3 Optimality of Equilibrium

Considering the dynamics (7.5) of the ASDPD algorithm, we give the following definition of its equilibrium point.

Definition 7.1 A point $w^* := (w_i^*, i \in \mathcal{N}) = (\mu_i^*, z_i^*, P_i^{g*})$ is an equilibrium point of dynamics (7.5) if $\lim_{k_i \to +\infty} w_{k_i} = w_i^*$ holds for all $i \in \mathcal{N}$.

Here, we present the following theorem to characterize the optimality of the equilibrium point.

Theorem 7.1 *Suppose Assumption 7.1 holds. The components P^{g*} and μ^* of the equilibrium point w^* of the dynamics* (7.5) *give the primal–dual optimal solution to the optimization problem* (7.1).

Proof: By (7.5a)–(7.5f) and Definition 7.1, we have

$$0 = -L \cdot \mu^* + L \cdot z^* + P^{g*} - P^d \tag{7.21a}$$

$$0 = L \cdot \mu^* \tag{7.21b}$$

$$-\nabla f(P^{g*}) = N_\Omega(P^{g*}) + \mu^* \tag{7.21c}$$

Then, we have

$$0 = \sum_{i \in \mathcal{N}} P_i^{g*} - \sum_{i \in \mathcal{N}} P_i^d \tag{7.22a}$$

$$\mu_i^* = \mu_j^* = \mu_0^*, \quad i, j \in \mathcal{N} \tag{7.22b}$$

$$-\nabla f(P^{g*}) = N_\Omega(P^{g*}) + \mu^* \tag{7.22b}$$

where μ_0^* is a constant. By [Theorem 3.25] [36], we know (7.22) is exactly the KKT condition of the problem (7.1). In addition, since (7.1) is a convex optimization problem and Slater's condition holds according to Assumption 7.1, P^{g*} and μ^* must be primal–dual optimal to the optimization problem (7.1), which completes the proof. ■

7.4.4 Convergence Analysis

In this subsection, we investigate the convergence of the ASDPD algorithm. The basic idea is to treat ASDPD as a randomized block-coordinate fixed-point iteration problem with delayed information by extending the operator argument presented in Section 7.4.2 to incorporate time delay. By doing so, the theory of ARock algorithms presented in [37] can be applied to establish the convergence result of the ASDPD.

Define vectors $\phi_i \in \mathbb{R}^{3n}, i \in \mathcal{N}$, which are location indicators. The jth entry of ϕ_i is denoted by $[\phi_i]_j$. Let $[\phi_i]_j = 1$ if the jth coordinate of w is also a coordinate of w_i, and $[\phi_i]_j = 0$, otherwise. Denote by φ a random variable (vector) taking values in $\{\phi_1, \phi_2, \dots, \phi_n\}$. Then $\mathbf{Prob}(\varphi = \phi_i) = \frac{1}{n}$ also follows a uniform distribution. Let φ_k be the value of φ at the kth iteration. Then, a randomized block-coordinate fixed-point iteration for (7.18) is given by

$$w_{k+1} = w_k + \eta\varphi_k \circ (\mathcal{T}(w_k) - w_k) \tag{7.23}$$

where $\circ$ is the Hadamard product. Here, we assume only one MG is activated at each iteration without loss of generality.[2]

Since (7.23) is delay-free, we modify it for admitting delayed information, which is

$$w_{k+1} = w_k + \eta\varphi_k \circ (\mathcal{T}(\hat{w}_k) - w_k) \tag{7.24}$$

where $\hat{w}_k$ is the delayed information at the iteration k. Note that, here, k is with respect to the *global clock* defined in Section 9.4. We will show that Algorithm 7.1 can be written as (7.24) if $\hat{w}_k$ is properly defined. Specifically, supposing MG i is activated at the iteration k, then $\hat{w}_k$ is defined as follows. For MG i and $j \in N_{i,c}$, replace $\mu_{i,k}$, $z_{i,k}$, and $P^g_{i,k}$ with $\mu_{i,k-\tau_i^k}$, $z_{i,k-\tau_i^k}$, and $P^g_{i,k-\tau_i^k}$, respectively. Similarly, replace $\mu_{j,k}$, $z_{j,k}$ with $\mu_{j,k-\tau_j^k}$ and $z_{j,k-\tau_j^k}$, respectively. With the random variable φ, the variables of inactivated MGs are the same as previous iterations. Then we have (7.24).

Next, we assume that the following assumption holds.

2 As mentioned previously, the global clock is virtual and only introduced to analyze the convergence. When two MGs are activated at exactly the same time, it can be simply modeled as two separate iterations. As long as we set $w_{i,k-\tau_i^k} = w_{i,k+1-\tau_i^{k+1}}, \forall i \in \mathcal{N}$, the convergence analysis can still apply.

Assumption 7.2 The time interval between arbitrary two consecutive iterations is bounded by a positive constant number χ, i.e. $0 \leq \tau_i^k \leq \chi$, $\forall i, k$.

Assumption 7.2 implies that the time delay in consideration is limited. This assumption generally holds in real-world systems if the communication does not fail permanently. With this assumption, we have the convergence result.

Theorem 7.2 *Suppose Assumptions 7.1– 7.2 hold. Take* $\zeta = \min\{\frac{1}{\sigma_{\max}^2}, \frac{a_{min}}{a_{max}^2}\}$, $\kappa > \frac{1}{2\zeta}$, *and the step sizes* $\sigma_\mu, \sigma_z, \sigma_g$ *such that* $\Phi - \kappa I$ *is positive semidefinite. Choose a proper step size* η *satisfying*

$$0 < \eta < \frac{1}{1 + 2\chi/\sqrt{n}} \frac{4\kappa\zeta - 1}{2\kappa\zeta}$$

Then, with the ASDPD algorithm, P_k^g *and* μ_k *converge to the primal–dual optimal solution to the problem* (7.1) *with probability 1.*

Proof: Combining (7.20) and (7.24), we have

$$\begin{aligned} w_{k+1} &= w_k + \eta\varphi_k \circ \left(\left(1 - \frac{2\kappa\zeta}{4\kappa\zeta - 1}\right)\hat{w}_k - w_k + \frac{2\kappa\zeta}{4\kappa\zeta - 1}\mathcal{R}(\hat{w}_k)\right) \\ &= w_k + \eta\varphi_k \circ \left(\hat{w}_k - w_k + \frac{2\kappa\zeta}{4\kappa\zeta - 1}(\mathcal{R}(\hat{w}_k) - \hat{w}_k)\right) \end{aligned} \tag{7.25}$$

With $w_{i,k-\tau_i^k} = w_{i,k-\tau_i^k+1} = \cdots = w_{i,k}$, we have $\varphi_k \circ (\hat{w}_k - w_k) = 0$. Thus, (7.25) is equivalent to

$$w_{k+1} = w_k + \frac{2\eta\kappa\zeta}{4\kappa\zeta - 1}\varphi_k \circ (\mathcal{R}(\hat{w}_k) - \hat{w}_k) \tag{7.26}$$

Invoking [37], (7.26) with the nonexpansive operator $\mathcal{R}$ is essentially a kind of the ARock algorithms introduced in [37]. Hence the convergence results given in that paper can directly be applied. Indeed, Lemma 13 and Theorem 14 of [37] indicate that the convergence of ARock is guaranteed by the condition

$$0 < \frac{2\eta\kappa\zeta}{4\kappa\zeta - 1} < \frac{1}{1 + 2\chi/\sqrt{n}} \tag{7.27}$$

Direct calculation shows that, if the step size η satisfies

$$0 < \eta < \frac{1}{1 + 2\chi/\sqrt{n}} \frac{4\kappa\zeta - 1}{2\kappa\zeta}$$

then w_k must converge to the fixed points (denoted by w_k^*) of $\mathcal{R}$ with probability 1. Recalling $Fix(\mathcal{T}) = Fix(\mathcal{R})$ and Theorem 7.1, we know P_k^{g*} and μ_k^*, as components of w_k^*, constituting the primal–dual optimal solution to the optimization problem (7.1). This completes the proof. ■

Choose $\kappa = \frac{1}{2\zeta} + \epsilon$, where $\epsilon > 0$ but sufficiently small. Then the upper bound of the valid step size η can be estimated by

$$\frac{1}{1+2\frac{\chi}{\sqrt{n}}}\frac{4\kappa\zeta - 1}{2\kappa\zeta} = \frac{1}{1+2\frac{\chi}{\sqrt{n}}}\frac{1+4\zeta\epsilon}{1+2\zeta\epsilon} \approx \frac{1}{1+2\frac{\chi}{\sqrt{n}}}$$

This result indicates $\eta < 1$. Moreover, the upper bound of the valid step size η decreases when the upper bound of time delay, χ, increases.

Given a fixed step size η and a sufficiently small $\epsilon > 0$, we have

$$\chi < \frac{\sqrt{n}(1-\eta)}{2\eta} \tag{7.28}$$

Thus, the upper bound of acceptable time delay is approximately proportional to the square root of the number of MGs, which provides a helpful insight for controller design. It should be pointed out that the upper bound of acceptable time delays is only a sufficient condition for guaranteeing the convergence but is not a necessary condition. Thereby the conservativeness is inevitable, which is also verified in both simulations and experiments.

Remark 7.3 From the convergence proof, the proposed ASDPD algorithm has two advantages compared with the works in literature. First, it is capable of simplifying and, to some extent, unifying the distributed control design considering a wide variety of asynchrony, such as non-identical sampling rates and random communication delays, which are very common in the practical operation of MGs. This approach is different from the literature [23, 24], which only addresses constant time delays. Second, the ASDPD allows using constant stepsizes, which is different from the literature [20–22, 25]. For example, in [20, 25], diminishing step sizes are required, while an extra line search needs to be performed at every iteration in [21, 22] to determine the step size. In this sense, the proposed ASDPD algorithm is easier to implemented and, hence, more appealing for practical applications.

7.5 Real-Time Implementation

This section presents a practical implementation of the ASDPD in both AC and DC MG. First, we shortly explain our motivation. Then, the variables $\tilde{z}$ and z used in the ASDPD are eliminated by utilizing the frequency/voltage measurements from the physical MGs, leading to a real-time control implementation. Finally, the optimality of the practical algorithm is theoretically justified.

7.5.1 Motivation and Main Idea

As analyzed in the previous sections, the ASDPD algorithm solves the economic dispatch problem (7.1) taking into account different kinds of asynchrony.

However, it may suffer from several limitations in practice. First, the rapid variation of renewable generations and load demand require the controller to response fast in real-time. Second, the accurate value of load demand P_i^d is difficult to know in advance. Third, there always exists power loss in MGs, which should be eliminated in real-time operation. These issues motivate us to combine the computation of economic dispatch with the real-time operation of MGs.

To consider the power loss, the problem (7.1) is modified to the following form:

$$\min_{P_i^g} \quad \sum_{i \in \mathcal{N}} f_i(P_i^g) \tag{7.29a}$$

$$\text{s.t.} \quad \sum_{i \in \mathcal{N}} P_i^g = \sum_{i \in \mathcal{N}} P_i^d + P_{\text{loss}} \tag{7.29b}$$

$$\underline{P}_i^g \leq P_i^g \leq \overline{P}_i^g \tag{7.29c}$$

where P_{loss} is the network power loss. Let P_{ij} denote the active power on line (i,j). Then $P_{ij} + P_{ji}$ is equivalent to the power loss in line (i,j), and the total network loss P_{loss} is given by

$$P_{\text{loss}} = \sum_{i \in \mathcal{N}} \sum_{j \in N_{i,p}} P_{ij}$$

The only difference between (7.1) and (7.29) appears in the equality constraints, where P_{loss} is included in (7.29), but not in (7.1). Here we need to adapt the ASDPD to solve (7.29).

Recalling (7.4) and (7.21a), the terms $\sum_{j \in N_{i,c}} (z_{i,k} - z_{j,k})$ in (7.4a) are utilized to balance the difference between $P_{i,k}^g$ and P_i^d. Denote $\delta_{ij} := z_i - z_j$, and the last three terms of (7.4a) are $P_{i,k}^g - P_i^d + \sum_{j \in N_{i,c}} \delta_{ij,k}$. From (7.21a), we know

$$0 = P_i^{g*} - P_i^d - \sum_{j \in N_{i,c}} \delta_{ij,k}^*$$

This balance equation motivates us to use the line power measurements from the physical system to replace δ_{ij} by noting that the similar power balance equations hold at each bus, i.e.

$$0 = \hat{P}_i^{g*} - P_i^d - \sum_{j \in N_{i,p}} P_{ij}^*,$$

where $\hat{P}_i^{g*}$ is the actual power generation of MG i in the steady state.[3] Thus, both $\sum_{j \in N_{i,c}} \delta_{ij,k}^*$ and $\sum_{j \in N_{i,p}} P_{ij}^*$ stand for the power difference at bus i, and δ_{ij} plays the same role as the line power P_{ij} except for the small power loss. It is a very important observation as P_{ij} are automatically given by the physical dynamics of

3 For same P_i^d, $\hat{P}_i^{g*}$ is slightly different from the computed value P_i^{g*} due to the small power loss. However, if the power loss is omitted, the two terms are essentially the same.

the power system. Hence, we only need to directly measure them as a real-time feedback to avoid complicated computation of $\tilde{z}$, z. Particularly, one can take $P^g_{i,k} - P^d_i + \sum_{j\in N_{i,c}}(z_{i,k} - z_{j,k})$ as a whole and estimate it using the measurements from the physical MGs, as we will explain later on.

7.5.2 Real-Time ASDPD

7.5.2.1 AC MGs

In AC MGs, the swing equation of bus i is given by[4]

$$M_i\dot{\omega}_i = P^g_i - P^d_i - D_i\omega_i - \sum_{j\in N_{i,p}} P_{ij},\ i \in \mathcal{N}_{ac} \tag{7.30}$$

where $M_i > 0$, $D_i > 0$ are constants and P_{ij} is the line flow from bus i to bus j. This model is suitable for both synchronous generators and inverters [27, 38, 39]. Equation (7.30) can be rewritten as

$$P^g_i - P^d_i - \sum_{j\in N_{i,p}} P_{ij} = M_i\dot{\omega}_i + D_i\omega_i,\ i \in \mathcal{N}_{ac} \tag{7.31}$$

Since the local frequency is easy to measure, one can utilize $M_i\dot{\omega}_i + D_i\omega_i$ to estimate the term $P^g_{i,k} - P^d_i + \sum_{j\in N_{i,c}}(z_{i,k} - z_{j,k})$ in (7.4a) and its delayed form in (7.5a). By doing so, the ASDPD algorithm (Algorithm 7.1) is integrated into the real-time feedback control of AC MGs.

7.5.2.2 DC Microgrids

In DC MGs, DC capacitors are used to maintain the voltage of DC buses [40]. Then, the model of DC MGs can be depicted using the following power balance equation on DC bus i (see Figure 7.2):

$$V^{dc}_i C_i \dot{V}^{dc}_i = P^g_i - P^d_i - \sum_{j\in N_{i,p}} P_{ij},\quad i \in \mathcal{N}_{dc} \tag{7.32}$$

where C_i is the shunt capacitor connected to the DC bus I and V^{dc}_i is the bus voltage. Thus, $V^{dc}_i C_i \dot{V}^{dc}_i$ can be used to estimate the term $P^g_{i,k} - P^d_i + \sum_{j\in N_{i,c}}(z_{i,k} - z_{j,k})$.

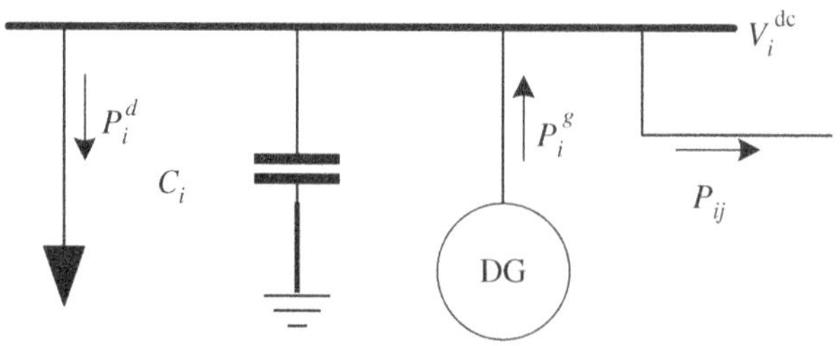

Figure 7.2 Simplified model of a DC MG.

4 In (7.30), inverter loss is also absorbed in P^d_i besides the load demand.

In this situation, we only need to measure the DC bus voltages, which is much easier to implement. By doing so, the ASDPD algorithm can be integrated to the real-time control of DC MGs.

Then, the distributed asynchronous algorithm takes the following form:

$$\tilde{\mu}_{i,k_i} = \mu_{i,k_i} + \sigma_\mu \left(-\sum_{j\in N_{i,c}} \left(\mu_{i,k_i} - \mu_{j,k_i-\tau_{ij}^{k_i}} \right) + M_i\dot{\omega}_i + D_i\omega_i \right), \quad i \in \mathcal{N}_{ac} \tag{7.33a}$$

$$\tilde{\mu}_{i,k_i} = \mu_{i,k_i} + \sigma_\mu \left(-\sum_{j\in N_{i,c}} \left(\mu_{i,k_i} - \mu_{j,k_i-\tau_{ij}^{k_i}} \right) + V_i^{dc}C_i\dot{V}_i^{dc} \right), \quad i \in \mathcal{N}_{dc} \tag{7.33b}$$

$$\tilde{P}^g_{i,k_i} = \mathcal{P}_{\Omega_i}\left(P^g_{i,k_i} - \sigma_g\left(f_i'(P^g_{i,k_i}) + 2\tilde{\mu}_{i,k_i} - \mu_{i,k_i}\right)\right) \tag{7.33c}$$

$$\mu_{i,k_i+1} = \mu_{i,k_i} + \eta\left(\tilde{\mu}_{i,k_i} - \mu_{i,k_i}\right) \tag{7.33d}$$

$$P^g_{i,k_i+1} = P^g_{i,k_i} + \eta\left(\tilde{P}^g_{i,k_i} - P^g_{i,k_i}\right) \tag{7.33e}$$

In the algorithm (7.33), only μ needs to be exchanged between immediate neighbors. Moreover, noting that the variables $\tilde{z}$, z are not necessary any longer, the controller can be further simplified. Based on (7.33), we have the *r*eal-*t*ime *as*ynchronous *d*istributed algorithm for *p*ower *d*ispatch (RTASDPD), as shown in Algorithm 7.2.

In the implementation, the RTASDPD algorithm takes the measurements from the continuous-time dynamics of the physical system as inputs. Specifically,

---

Algorithm 7.2: *RTASDPD*

Input: For MG i, the input is $\mu_{i,0} \in \mathbb{R}^n$, $P^g_{i,0} \in \Omega_i$.

Iteration at k_i: Supposing MG i's clock ticks at time k_i, then MG i is activated and updates its local variables as follows:

Step 1: Reading phase

Get $\mu_{j,k_i-\tau_{ij}^{k_i}}$ from its neighbors' output cache. For an AC MG i, measure the frequency ω_i. For a DC MG i, measure the voltage V_i.

Step 2: Computing phase

For $i \in \mathcal{N}_{ac}$, calculate $\tilde{\mu}_{i,k_i}$ and $\tilde{P}^g_{i,k_i}$ according to (7.5a) and (7.5c), respectively. For $i \in \mathcal{N}_{dc}$, calculate $\tilde{\mu}_{i,k_i}$ and $\tilde{P}^g_{i,k_i}$ according to (7.5b) and (7.5c), respectively.

Update μ_{i,k_i+1} and P^g_{i,k_i+1} according to (7.5d) and (7.5e), respectively.

Step 3: Writing phase

Write μ_{i,k_i+1} to its output cache and μ_{i,k_i+1}, P^g_{i,k_i+1} to its local storage. Increase k_i to $k_i + 1$.

The continuous-time variables ($\dot{\omega}_i$, ω_i, $\dot{V}_i^{\text{dc}}$, V_i^{dc}) are obtained from the measurements from the physical system. In practice, they also need to be discretized in the digital controller implementation. Here in (32a), (32b), we intentionally use some continuous-time notations to indicate that they are from the continuous-time physical system.

7.5.3 Control Configuration

The control diagram of the AC MG is shown in Figure 7.3, which is composed of four levels: the electric network level, the primary power control level, the asynchronous power control level, and the distributed communication level. At the electric network level, the current and voltage are measured and serve as the input of the primary power control level, where the gray circle means the

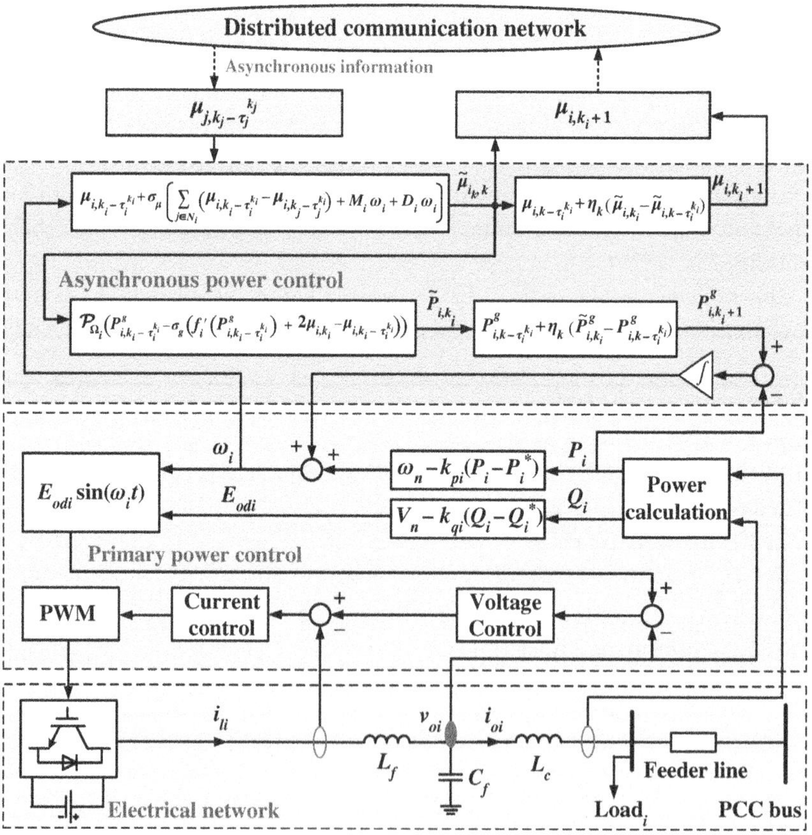

Figure 7.3 Control configuration of the proposed RTASDPD algorithm.

current measurement and gray panel is the voltage measurement. The primary power control level includes three loops, i.e. the current loop, the voltage loop, and the power loop. In the power loop, the droop control is adopted for both the active and reactive power controls, where the active power and frequency are measured and feedbacked to the asynchronous power control level. The RTAS-DPD algorithm runs iteratively based on the latest information of ω_i and $\mu_{j,k-\tau_j^k}$. From (7.33a) and (7.33d), $\mu_{i,k+1}$ is obtained and written to the output cache of MG i, which is also sent to its neighbors via the communication network. From (7.33c) and (7.33e), $P^g_{i,k+1}$ is obtained, which is sent to the primary power control level as the active power reference. The error between $P^g_{i,k+1}$ and the measured active power P_i is compensated by adding an integral term into the primary power control, which eliminates the effects of unknown power loss.

The control diagram of DC MGs is similar to that of AC MGs in Figure 7.3, where the main difference is that the DC bus voltage V_i^{dc} is measured. The details are omitted here for simplicity. The implementation of the RTASDPD is straightforward, which only needs to measure the active power and DC bus voltages from the physical system. Then, the power reference is obtained and sent to the primary control.

In the implementation, the result of each iteration is sent to the inverter as a power reference. Thus, only the latest information needs to be used to carry out the next iteration. This feature implies that the computation burden and storage requirement is very small.

7.5.4 Optimality of the Implementation

Considering the discrete dynamic (7.33), we give the following definition of its equilibrium point.

Definition 7.2 A point $\hat{w}^* = (\hat{w}_i^*, i \in \mathcal{N}) = (\hat{\mu}_i^*, \hat{P}_i^{g*}, \hat{\omega}_i^*, \hat{V}_i^{dc*})$ is an equilibrium point of the dynamic (7.33) if $\lim_{k_i \to +\infty} \hat{w}_{k_i} = \hat{w}_i^*$ holds for all i.

Then, we have that the component $(\hat{P}^{g^*}, \hat{\mu}^*)$ of the equilibrium point $\hat{w}^*$ is the primal–dual optimal solution to (7.29) in the steady state.

In the steady state, we have $\hat{\omega}_i^* = \hat{\omega}_j^* = \hat{\omega}^*, \forall i,j \in \mathcal{N}$ and $\dot{V}_i^{\mathrm{dc}} = 0, \forall i \in \mathcal{N}_{\mathrm{dc}}$. From (7.33) and Definition 7.2, we have

$$0 = \sum_{j \in N_{i,c}} \left(\hat{\mu}_i^* - \hat{\mu}_j^*\right) + D_i\hat{\omega}^*, \quad i \in \mathcal{N}_{\mathrm{ac}} \tag{7.34a}$$

$$0 = \sum_{j \in N_{i,c}} \left(\hat{\mu}_i^* - \hat{\mu}_j^*\right), \quad i \in \mathcal{N}_{\mathrm{dc}} \tag{7.34b}$$

$$\hat{P}_i^{g*} = \mathcal{P}_{\Omega_i}\left(\hat{P}_i^{g*} - \sigma_g\left(f_i'(\hat{P}_i^{g*}) + \hat{\mu}_i^*\right)\right) \tag{7.34c}$$

From (7.34a) and (7.34b), we have

$$r_1\hat{\mu}^* + D_1\hat{\omega}^* = 0 \tag{7.35a}$$

$$\vdots$$

$$r_{|\mathcal{N}_{ac}|}\hat{\mu}^* + D_{|\mathcal{N}_{ac}|}\hat{\omega}^* = 0 \tag{7.35b}$$

$$r_{|\mathcal{N}_{ac}|+1}\hat{\mu}^* = 0 \tag{7.35c}$$

$$\vdots$$

$$r_{|\mathcal{N}|}\hat{\mu}^* = 0 \tag{7.35d}$$

where r_i is the ith row of Laplacian matrix L, and $r_1 + r_2 + \cdots + r_{|\mathcal{N}|} = 0$.
Thus, we have

$$\hat{\omega}^* \sum_{i \in \mathcal{N}_{ac}} D_i = 0 \tag{7.35e}$$

It implies that $\hat{\omega}^* = 0$ due to $D_i > 0$. Then, we have

$$\hat{\mu}_i^* = \hat{\mu}_j^* = \hat{\mu}^* \tag{7.35f}$$

where μ^* is a constant. Summing (7.30) and (7.32) for all $i \in \mathcal{N}$ and recalling $P_{\text{loss}} = \sum_{i \in \mathcal{N}} \sum_{j \in N_{i,p}} P_{ij}$, we have

$$\sum_{i \in \mathcal{N}} \hat{P}_i^{g*} = \sum_{i \in \mathcal{N}} P_i^d + P_{\text{loss}} \tag{7.35g}$$

Similar to Theorem 7.1, (7.34c), (7.35f), and (7.35g) are the KKT conditions of the problem (7.29), which verifies the optimality of $(\hat{P}^{g*}, \hat{\mu}^*)$.

Remark 7.4 The RTASDPD algorithm has three main advantages:

1) Only the local frequencies of AC MG and bus voltages of DC MG need to be measured. It avoids the measurements of load demand P_i^d across the overall system, which is difficult in practice. In this sense, the proposed algorithm is much easier to implement.
2) We simplify the communication graph, where only the neighboring communication is needed. Moreover, we also simplify the controller structure. The auxiliary variables $\tilde{z}$ and z are eliminated, making the controller easier to implement.
3) In the problem (7.1), the power loss is not considered. In the real-time implementation, we measure the frequencies and voltages and intend to drive it to the nominal value. Governed by the power balance equations, the equilibrium point is the optimal solution to (7.29), which have the power loss well compensated.

Remark 7.5 From Figure 7.3, the RTASDPD algorithm is coupled with the nonlinear physical dynamics (both AC and DC MGs) and the primary control due to the feedback of measurements from the physical power system. It theoretically results in a hybrid dynamical system with both differential and difference equations. In this context, it is difficult to rigorously prove the convergence. Hence we alternatively verify the convergence by both numerical and physical experiments in various scenarios in Section 7.6 and 7.7, respectively. The experimental results evidently demonstrate that the RTASDPD works well in different conditions. However, the rigorous stability proof is an interesting topic and worthy of further investigation.

7.6 Numerical Results

7.6.1 Test System

To verify the performance of the proposed method, a 44-bus hybrid AC/DC microgrid system shown in Figure 7.4 is used for numerical simulations, which is a

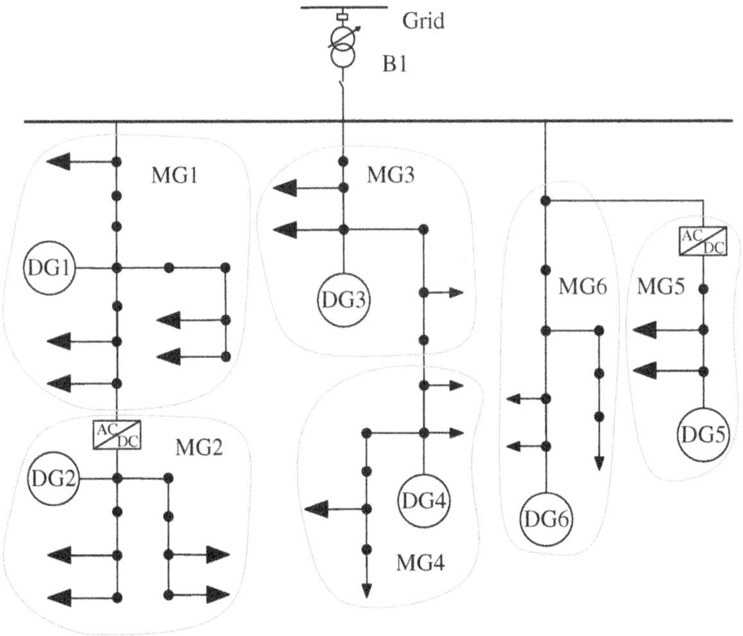

Figure 7.4 A schematic diagram of a typical 43-bus MG system.

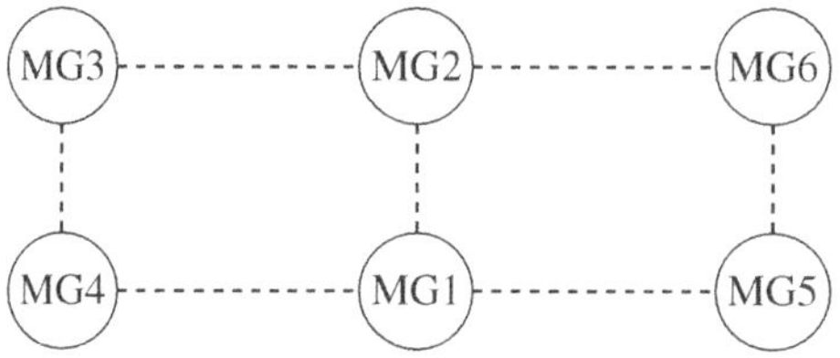

Figure 7.5 Communication graph of the system.

Table 7.1 System parameters.

DG i	1	2	3	4	5	6
a_i	0.8	1	0.65	0.75	0.9	0.85
b_i	0.01	0.01	0.014	0.012	0.01	0.01
$\overline{P}_i^g$ (kW)	85	80	90	85	80	80
$\underline{P}_i^g$ (kW)	0	0	0	0	0	0

modified benchmark of low-voltage MG systems [41, 42]. The system includes three feeders with six dispatchable MGs, where MG2 and MG5 are DC MGs while the others are AC MGs. Breaker 1 is open, which implies that the system operates in an islanded mode. All simulations are implemented in the professional power system simulation software PSCAD. The step size is set as $\eta = 0.016$, and the number of MGs $n = 6$. Then, the upper bound of time delay can be computed by (7.28), which is $\chi < 75.32$. Since the simulation period of PSCAD is set as 150 μs, the maximal acceptable time delay is 75.32 μ × 150 μs= 11.3 ms.

The simulation scenario is set as follows: (i) at $t = 2$ s, there is a 60 kW load uprush in the system, and (ii) at $t = 8$ s, there is a 30 kW load shedding. Then, each MG increases its generation to balance the power and restore the system frequency. Their initial generations are [58.93, 46.94, 66.43, 59.95, 52.06, 55.09] kW. The communication graph is undirected, as shown in Figure 7.5. Other parameters are given in Table 7.1.

7.6.2 Non-identical Sampling Rates

Individual MGs may have different sampling rates (or control periods) in practice, which could cause asynchrony and compromise the control performance. In this part, we consider the impact of non-identical sampling rates (as well as the control periods). The sampling rates of MG1–MG6 are set as 10,000, 12,000, 14,000, 16,000, 18,000, and 20,000 Hz, respectively. The local frequencies of MGs are shown in Figure 7.6.

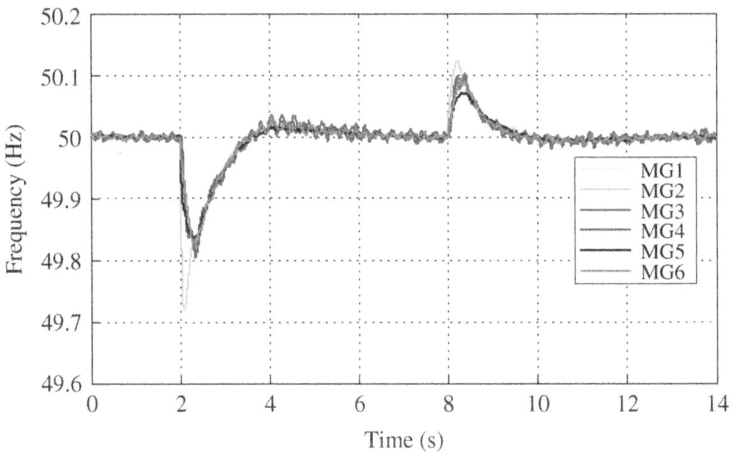

Figure 7.6 Dynamics of frequencies. For a DC MG, its frequency implies the frequency of the corresponding DC/AC inverter.

As the load change is located in MG2, the frequency nadir of MG2 is the lowest (about 0.26 Hz). The system frequency recovers in 4 s after the load change. When the load decreases, the frequency experiences an overshoot of 0.1 Hz and restores in 2 s. Voltages on the DC buses of MG2 and MG5 have a small drop when the load increases, which are illustrated in Figure 7.7. For DG2, the initial voltage is 395.3 V. When load increases, the nadir is 382.3 V, dropping about 3.28%. Then, it recovers to 393.6 V quickly, which reduces to 0.43%. When the load drops

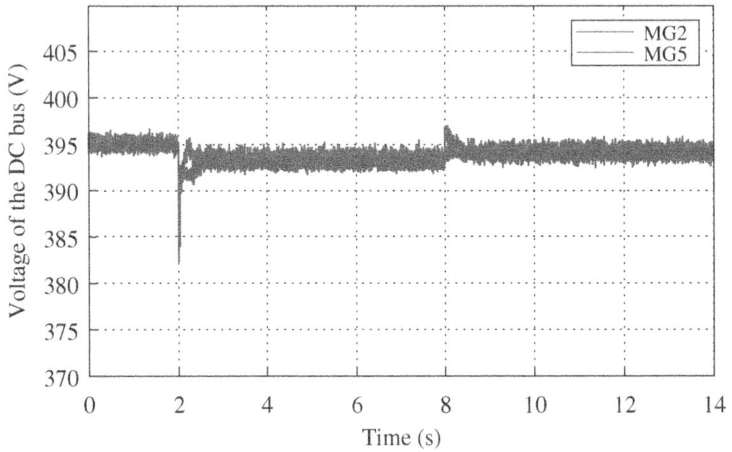

Figure 7.7 Dynamics of voltages.

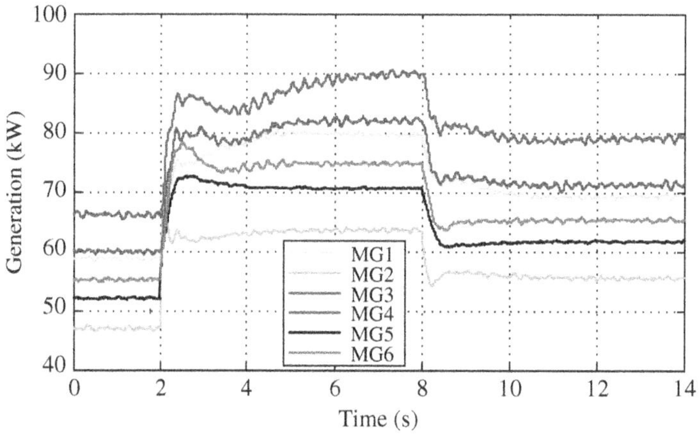

Figure 7.8 Dynamics of generations.

at $t = 8$ s, the voltage increases. The overshoot is 397.0 V, which raises 0.86%. Then, it stabilizes at 394.5 V, rising 0.21%. The voltage of DG5 also has small changes.

Dynamics of generations are given in Figure 7.8. At the end of stage one (from 2 s to 8 s), generations of MGs are [79.32, 63.60, 90, 81.82, 70.47, 75.08] kW, respectively. At the end of stage two (from 8 s to 14 s), their values are [69.50, 55.46, 79.20, 70.97, 61.86, 65.04] kW, respectively. Thus, generations are identical to those obtained by solving the centralized optimization problem (implemented by CVX). This result verifies the optimality of the proposed method. $-\mu_i$ stands for the marginal cost of MG i, whose dynamic is given in Figure 7.9. The marginal

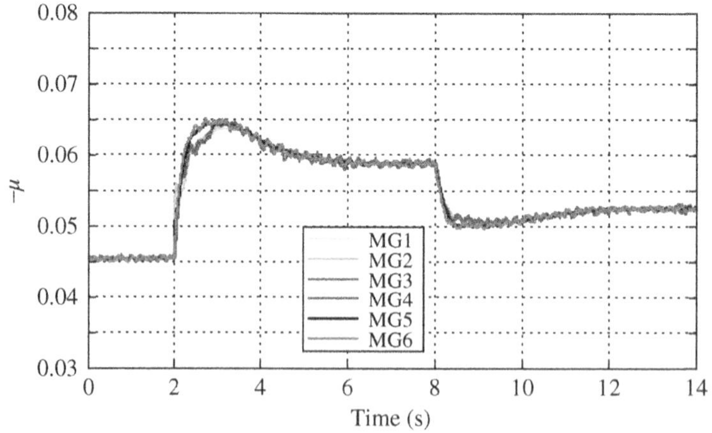

Figure 7.9 Dynamics of $-\mu$.

cost of different MGs converges to the same value when the system is stabilized, which indicates that the system operates in the optimal state.

7.6.3 Random Time Delays

In practice, time delay always exists in communication, which is usually varying up to channel situations. Therefore, the time delay is random and cannot be known in advance. In this part, we examine the impact of time-varying time delays. Initially, all the time delays in communication are set as 20 ms. Then, we intentionally increase the time delays on the channels of MG1–MG2 and MG5–MG6. Additionally, we have the time delays on these two channels varying in ranges of [100 ms, 200 ms], [200 ms, 500 ms], [500 ms, 800 ms], and [800 ms, 1000 ms], respectively, while the delays on other channels remain 20 ms. The frequency and generation dynamics of MG1 under different scenarios are shown in Figures 7.10 and 7.11, respectively. It is observed that the convergence becomes slower and the frequency overshoot becomes more considerable with the increase of time delays. However, the steady-state generations are still precisely identical to the optimal solution, which verifies the effectiveness of our controller under varying time delays. This result also reveals that the computed upper bound of the acceptable time delay is conservative.

7.6.4 Comparison with the Synchronous Algorithm

In this part, we compare the performances of the asynchronous and synchronous algorithms under imperfect communication. In the asynchronous case, the

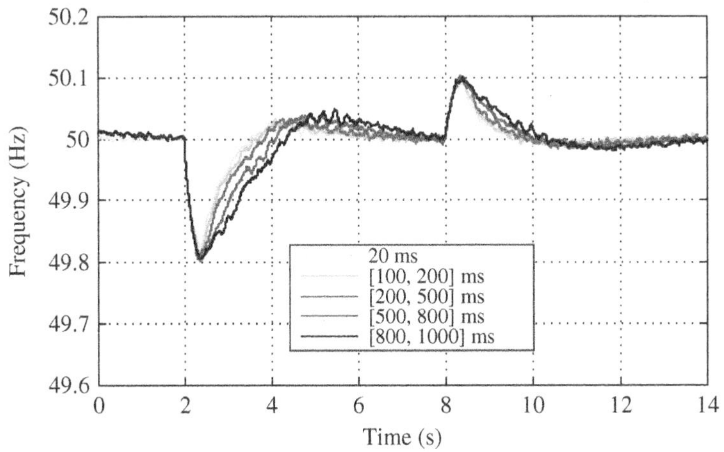

Figure 7.10 Frequencies under different/varying time delays.

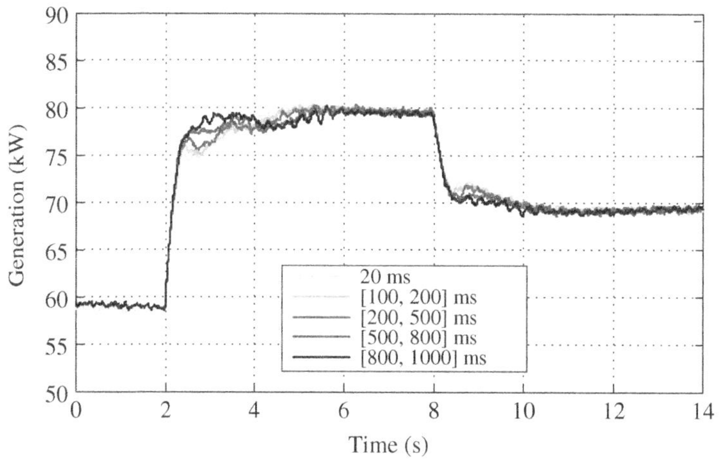

Figure 7.11 Generations under different/varying time delays.

sampling rates of MGs are set to the same as that in Section 7.6.2 Moreover, the time delay varies between [500, 800] ms. The dynamics of MG1 with two algorithms are shown in Figures 7.12 and 7.13, respectively. With the synchronous algorithm SDPD, the system remains stable after load perturbations. However, the frequency nadir and overshoot deteriorate, and the convergence becomes slower. The generation takes more time to reach the optimal solution, with considerable fluctuations. The reason is that MGs have to wait for the slowest

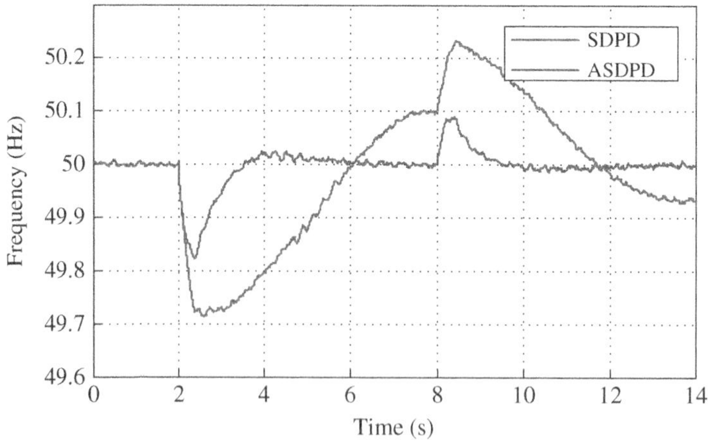

Figure 7.12 Dynamics of frequencies under synchronous and asynchronous cases.

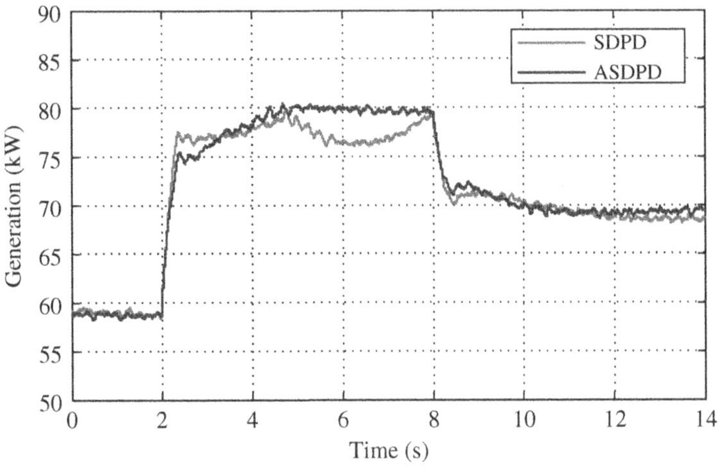

Figure 7.13 Dynamics of generations under synchronous and asynchronous cases.

one in the synchronous case. This result confirms the advantage of the proposed asynchronous algorithm.

7.7 Experimental Results

In this section, the proposed method is further verified on the experimental platform using the dSPACE RTI 1202 controller, which is presented in Figure 7.14. The experimental platform consists of three inverters, one dSPACE RTI controller, one regular load, one switchable load, and one host computer. The rated voltage is 220 V, and the maximal capacity of each inverter is 1760 W. Each inverter represents a DG (or MG). The system topology is given in Figure 7.15. Three inverter-interfaced DGs are connected to the AC bus through impedances. The breaker B0 is open, which implies that the system operates in an isolated mode. The regular load is connected at the bus of DG1, and the switchable load is at the bus of DG2. Due to the physical limit, we have no DC network. The communication topology is DG1 $\leftrightarrow$ DG2 $\leftrightarrow$ DG3 $\leftrightarrow$ DG1. Parameters in the objective functions of each inverter are $a_1 = 0.075, a_2 = 0.06, a_3 = 0.1, b_i = 0$. In the experiment, we have $\eta = 0.1, n = 3$. Then, $\chi < 7.794$. The maximal sampling and controlling period is 1 ms. Then the estimated acceptable time delay is 7.794 ms.

The experimental scenario is set as follows: (i) at $t = 20$ s, the switchable load is connected, and (ii) at $t = 60$ s, the switchable load is disconnected. Then, each

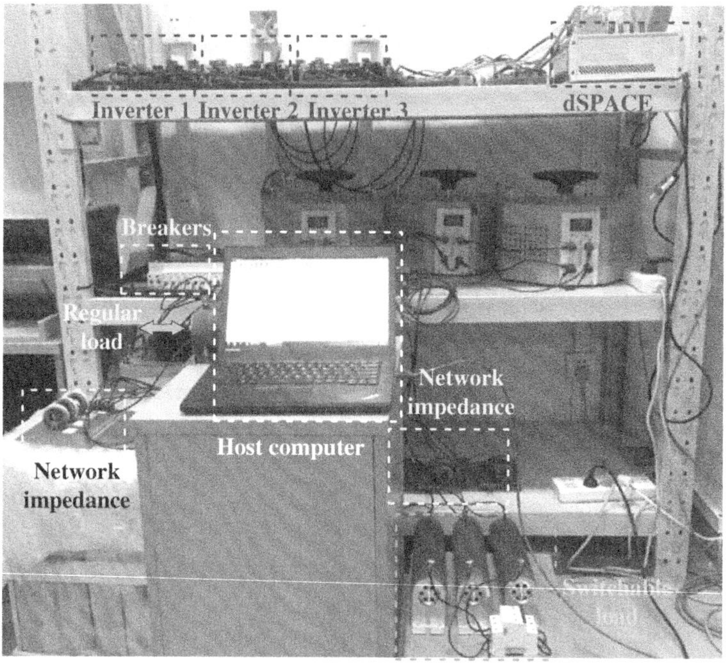

Figure 7.14 Experiment platform based on dSPACE RTI 1202 controller.

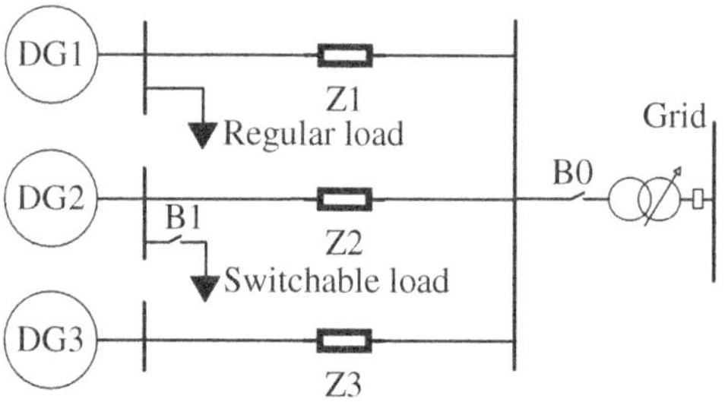

Figure 7.15 Topology of the experiment system.

DG regulates its generation to balance the power and restore system frequency. The sampling rate (and the controlling frequency) of the controller is 5000 Hz. We set a 2 ms of time delay as the baseline for each communication link. The frequency and generation dynamics are given in Figures 7.16 and 7.17, respectively. It is shown that the frequency nadir is 49.8 Hz and restores in 5 s after the load changes. From Figure 7.17, the actual power of each DG varies slightly around its computed value. The computed power is $P_1^g = 363.4$ kW, $P_2^g = 454.3$ kW, $P_3^g = 272.6$ kW after the load increase, which is optimal. This observation shows that the proposed method works as predicted in the experiment.

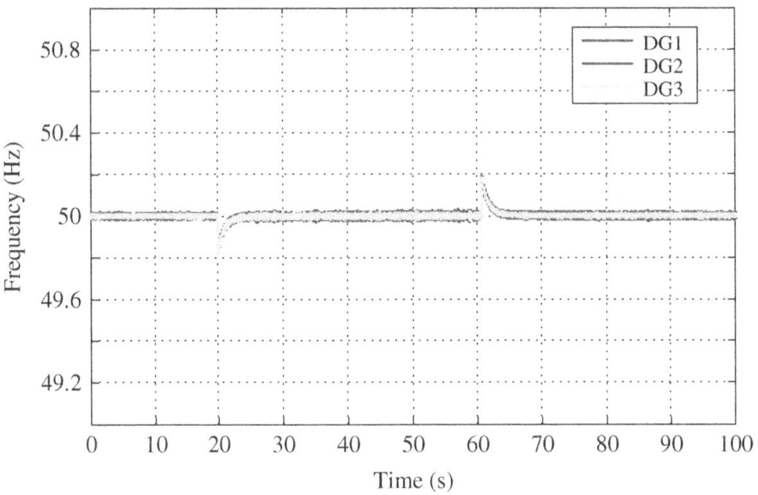

Figure 7.16 Dynamics of frequencies in the experiments. The computed power value of DG i is P^g_{i,k_i+1}.

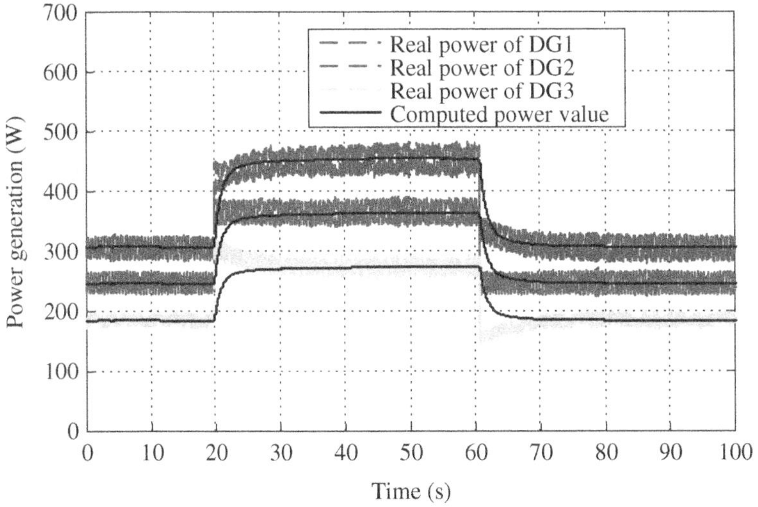

Figure 7.17 Dynamics of power generations in the experiments. The black line is the dynamics of P^g_{i,k_i+1}, and dotted lines are real power of corresponding DGs.

The values of the marginal cost $-\mu$ obtained by the asynchronous and synchronous algorithms are given in Figure 7.18. When considering different time delays, we adopt 10 and 20 ms delays, respectively. When considering different sampling rates, we decrease its value of DG3 from 5000 to 2500 and 1000 Hz, respectively. The dotted lines are results using the RTASDPD, and the solid lines are results using the SDPD. The upper part is the results under different time delays (10 and 20 ms, respectively). With the increase of time delay, the convergence speed of each algorithm is getting worse. However, the RTASDPD always converges faster than that of the SDPD under the same time delay, confirming the result of theoretical analyses.

From the lower part of Figure 7.18, we know that the marginal cost $-\mu$ converges in 20 s using the RTASDPD after the load change when the sampling rate of DG3 decreases to 2500 Hz. On the contrary, it does not converge yet even in 40 s using the synchronous algorithm SDPD. In addition, when sampling rates decrease, the convergence rate reduces for both cases. This result shows that the proposed controller converges faster when different kinds of asynchrony exist.

7.8 Conclusion and Notes

In this chapter, we have addressed the information asynchrony issue in the distributed power control of hybrid MGs to achieve an economic power dispatch (or load sharing). By introducing a virtual global clock with randomization, different kinds of asynchrony are fitted into a unified framework. Based on this, we have devised an asynchronous algorithm with proof of convergence. We have also provided an upper bound on the time delay. Furthermore, the real-time implementation method of the asynchronous distributed power control is provided in hybrid AC/DC MGs. In the implementation, by using the frequency/voltage measurement, the controller is simplified by reducing one order and can consider the power loss. Experiment results and numerical simulations confirm the superior performance of the proposed methodology.

Imperfect communication, such as packet-dropping, nonidentical sampling/controlling rates, and varying time-delays, widely exists in cyber–physical systems, which brings additional restrictions. It, however, is challenging to address diverse types of imperfect communication simultaneously. This chapter suggests an operator-splitting approach to designing and analyzing distributed optimal controllers that are robust against such imperfect communication. Here we highlight that introducing the virtual global clock allows the construction of the nonexpansive operator splitting while randomizing the clock in each iteration enables it to unify different types of imperfect communication into an asynchrony issue. The resulted asynchrony framework remarkably simplifies the convergence

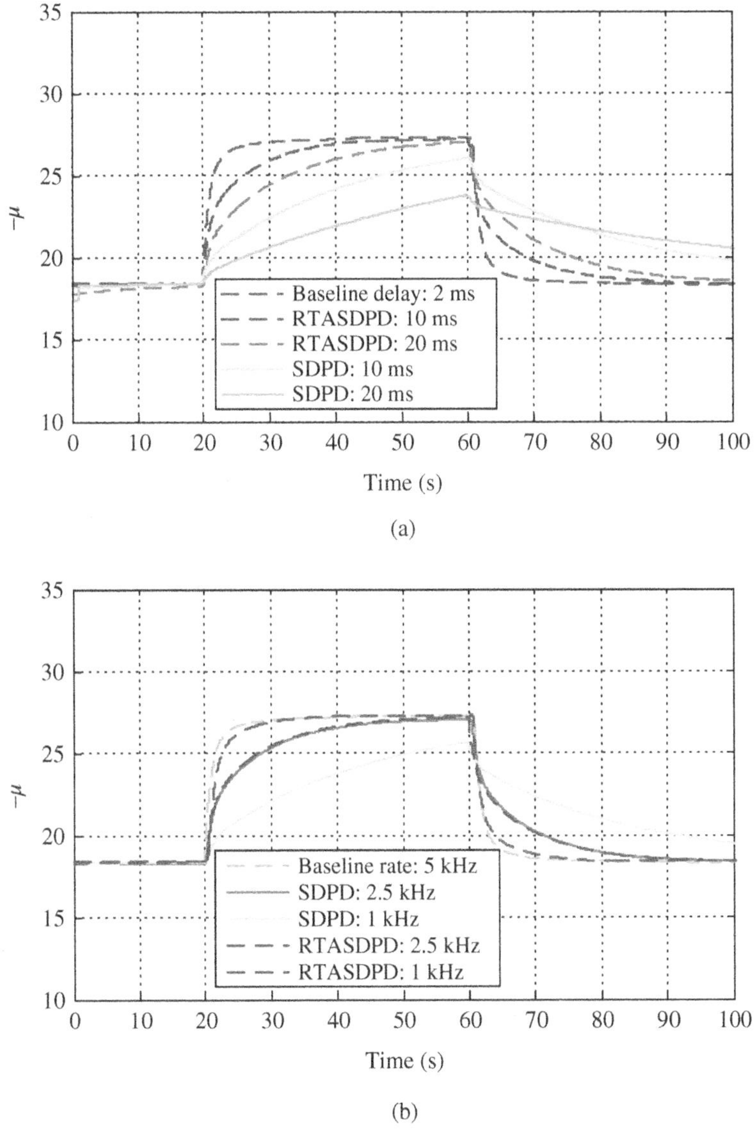

Figure 7.18 Comparison between the RTASDPD and the SDPD under different time delays and sampling rates. (a) Different time delays. (b) Different sampling rates.

analysis. It can be applied to deal with more complicated cyber and physical restrictions, as we will show in the next chapter.

Bibliography

1 Q. Xu, J. Xiao, P. Wang, and C. Wen, "A decentralized control strategy for economic operation of autonomous AC, DC, and hybrid AC/DC microgrids," *IEEE Transactions on Energy Conversion*, vol. 32, no. 4, pp. 1345–1355, 2017.

2 X. Xu, Q. Liu, C. Zhang, and Z. Zeng, "Prescribed performance controller design for DC converter system with constant power loads in DC microgrid," *IEEE Transactions on Systems, Man, and Cybernetics: Systems*, vol. 50, no. 11, pp. 4339–4348, 2020.

3 Y. Xia, W. Wei, M. Yu, X. Wang, and Y. Peng, "Power management for a hybrid AC/DC microgrid with multiple subgrids," *IEEE Transactions on Power Electronics*, vol. 33, no. 4, pp. 3520–3533, 2018.

4 F. Dörfler, J. W. Simpson-Porco, and F. Bullo, "Breaking the hierarchy: distributed control and economic optimality in microgrids," *IEEE Transactions on Control of Network Systems*, vol. 3, no. 3, pp. 241–253, 2015.

5 A. Bernstein and E. Dall'Anese, "Real-time feedback-based optimization of distribution grids: a unified approach," *IEEE Transactions on Control of Network Systems*, vol. 6, no. 3, pp. 1197–1209, 2019.

6 H. Li, G. Chen, T. Huang, Z. Dong, W. Zhu, and L. Gao, "Event-triggered distributed average consensus over directed digital networks with limited communication bandwidth," *IEEE Transactions on Cybernetics*, vol. 46, no. 12, pp. 3098–3110, 2016.

7 H. Li, C. Huang, G. Chen, X. Liao, and T. Huang, "Distributed consensus optimization in multiagent networks with time-varying directed topologies and quantized communication," *IEEE Transactions on Cybernetics*, vol. 47, no. 8, pp. 2044–2057, 2017.

8 Z. Wang, F. Liu, S. H. Low, P. Yang, and S. Mei, "Distributed load-side control: coping with variation of renewable generations," *Automatica*, vol. 109, 108556, 2019.

9 Z. Wang, F. Liu, Y. Chen, S. H. Low, and S. Mei, "Unified distributed control of stand-alone DC microgrids," *IEEE Transactions on Smart Grid*, vol. 10, no. 1, pp. 1013–1024, 2019.

10 H.-S. Ahn, B.-Y. Kim, Y.-H. Lim, B.-H. Lee, and K.-K. Oh, "Distributed coordination for optimal energy generation and distribution in cyber-physical energy networks," *IEEE Transactions on Cybernetics*, vol. 48, no. 3, pp. 941–954, 2018.

11 Y. Han, K. Zhang, H. Li, E. A. A. Coelho, and J. M. Guerrero, "MAS-based distributed coordinated control and optimization in microgrid and microgrid

clusters: a comprehensive overview," *IEEE Transactions on Power Electronics*, vol. 33, no. 8, pp. 6488–6508, 2017.

12 S. Yang, Q. Liu, and J. Wang, "Distributed optimization based on a multiagent system in the presence of communication delays," *IEEE Transactions on Systems, Man, and Cybernetics: Systems*, vol. 47, no. 5, pp. 717–728, 2017.

13 S. Magnusson, G. Qu, C. Fischione, and N. Li, "Voltage control using limited communication," *IEEE Transactions on Control of Network Systems*, vol. 6, no. 3, pp. 993–1003, 2019.

14 D. Alkano, J. M. Scherpen, and Y. Chorfi, "Asynchronous distributed control of biogas supply and multienergy demand," *IEEE Transactions on Automation Science and Engineering*, vol. 14, no. 2, pp. 558–572, 2017.

15 S. Liu, X. Wang, and P. X. Liu, "Impact of communication delays on secondary frequency control in an islanded microgrid," *IEEE Transactions on Industrial Electronics*, vol. 62, no. 4, pp. 2021–2031, 2014.

16 X. Ge, Q.-L. Han, and X.-M. Zhang, "Achieving cluster formation of multi-agent systems under aperiodic sampling and communication delays," *IEEE Transactions on Industrial Electronics*, vol. 65, no. 4, pp. 3417–3426, 2017.

17 H. Yan, X. Zhou, H. Zhang, F. Yang, and Z. Wu, "A novel sliding mode estimation for microgrid control with communication time delays," *IEEE Transactions on Smart Grid*, vol. 10, no. 2, pp. 1509–1520, 2019.

18 S. Misra, S. Bera, T. Ojha, and L. Zhou, "ENTICE: Agent-based energy trading with incomplete information in the smart grid," *Journal of Network and Computer Applications*, vol. 55, pp. 202–212, 2015.

19 S. Misra, S. Bera, T. Ojha, H. T. Mouftah, and A. Anpalagan, "ENTRUST: Energy trading under uncertainty in smart grid systems," *Computer Networks*, vol. 110, pp. 232–242, 2016.

20 J. Wu, T. Yang, D. Wu, K. Kalsi, and K. H. Johansson, "Distributed optimal dispatch of distributed energy resources over lossy communication networks," *IEEE Transactions on Smart Grid*, vol. 8, no. 6, pp. 3125–3137, 2017.

21 B. Millar and D. Jiang, "Smart grid optimization through asynchronous, distributed primal dual iterations," *IEEE Transactions on Smart Grid*, vol. 8, no. 5, pp. 2324–2331, 2017.

22 B. S. Millar and D. Jiang, "Asynchronous consensus for optimal power flow control in smart grid with zero power mismatch," *Journal of Modern Power Systems and Clean Energy*, vol. 6, no. 3, pp. 412–422, 2018.

23 C. Zhao, X. Duan, and Y. Shi, "Analysis of consensus-based economic dispatch algorithm under time delays," *IEEE Transactions on Systems, Man, and Cybernetics: Systems*, vol. 50, no. 8, pp. 2978–2988, 2020.

24 L. Ding, Q.-L. Han, L. Y. Wang, and E. Sindi, "Distributed cooperative optimal control of DC microgrids with communication delays," *IEEE Transactions on Industrial Informatics*, vol. 14, no. 9, pp. 3924–3935, 2018.

25 M. Zholbaryssov, D. Fooladivanda, and A. D. Domínguez-García, "Resilient distributed optimal generation dispatch for lossy AC microgrids," *Systems & Control Letters*, vol. 123, pp. 47–54, 2019.

26 S. Boyd and L. Vandenberghe, *Convex Optimization*. Cambridge University Press, 2004.

27 Z. Wang, F. Liu, J. Z. F. Pang, S. H. Low, and S. Mei, "Distributed optimal frequency control considering a nonlinear network-preserving model," *IEEE Transactions on Power Systems*, vol. 34, no. 1, pp. 76–86, 2019.

28 H. Bauschke and P. L. Combettes, *Convex Analysis and Monotone Operator Theory in Hilbert Spaces*. Springer, 2017.

29 T. Anderson, C.-Y. Chang, and S. Martinez, "Distributed approximate newton algorithms and weight design for constrained optimization," *Automatica*, vol. 109, 108538, 2019.

30 Z. Tang, D. J. Hill, and T. Liu, "Fast distributed reactive power control for voltage regulation in distribution networks," *IEEE Transactions on Power Systems*, vol. 34, no. 1, pp. 802–805, 2019.

31 E. Ramírez-Llanos and S. Martínez, "Distributed discrete-time optimization algorithms with applications to resource allocation in epidemics control," *Optimal Control Applications and Methods*, vol. 39, no. 1, pp. 160–180, 2018.

32 M. Cao, A. S. Morse, B. D. Anderson, *et al.*, "Agreeing asynchronously," *IEEE Transactions on Automatic Control*, vol. 53, no. 8, pp. 1826–1838, 2008.

33 P. Yi and L. Pavel, "Asynchronous distributed algorithm for seeking generalized Nash equilibria," *2018 European Control Conference*, 2018.

34 P. Yi and L. Pavel, "An operator splitting approach for distributed generalized Nash equilibria computation," *Automatica*, vol. 102, pp. 111–121, 2019.

35 P. L. Combettes and I. Yamada, "Compositions and convex combinations of averaged nonexpansive operators," *Journal of Mathematical Analysis and Applications*, vol. 425, no. 1, pp. 55–70, 2015.

36 A. P. Ruszczyński and A. Ruszczynski, *Nonlinear Optimization*, Princeton University Press, 2006.

37 Z. Peng, Y. Xu, M. Yan, and W. Yin, "ARock: an algorithmic framework for asynchronous parallel coordinate updates," *SIAM Journal on Scientific Computing*, vol. 38, no. 5, pp. A2851–A2879, 2016.

38 C. De Persis and N. Monshizadeh, "A modular design of incremental Lyapunov functions for microgrid control with power sharing," in *Control Conference (ECC), 2016 European*, pp. 1501–1506, IEEE, 2016.

39 Z. Wang, F. Liu, S. H. Low, C. Zhao, and S. Mei, "Distributed frequency control with operational constraints, Part II: network power balance," *IEEE Transactions on Smart Grid*, vol. 10, no. 1, pp. 53–64, 2019.

40 Z. Wang, W. Wu, and B. Zhang, "A distributed control method with minimum generation cost for DC microgrids," *IEEE Transactions on Energy Conversion*, vol. 31, no. 4, pp. 1462–1470, 2016.

41 S. Papathanassiou, N. Hatziargyriou, K. Strunz, *et al.*, "A Benchmark low voltage microgrid network," in *Proceedings of the CIGRE Symposium: Power Systems with Dispersed Generation*, pp. 1–8, 2005.

42 X. Wu and C. Shen, "Distributed optimal control for stability enhancement of microgrids with multiple distributed generators," *IEEE Transactions on Power Systems*, vol. 32, no. 5, pp. 4045–4059, 2017.

8

Cyber Restrictions: Imperfect Communication in Voltage Control of Active Distribution Networks

Following the same thought line of Chapter 7, this chapter stretches out to address more complicated circumstances in distributed optimal control with nonideal communication. Specifically, we consider the distributed voltage control problem of active distribution networks (ADNs) here. Due to the burst of distributed energy resources (DERs), the volatile renewable generation has brought a significant challenge to not only active power regulation but also reactive power and voltage control. In this regard, this chapter aims to develop an asynchronous distributed voltage control (ASDVC) based on the partial primal–dual (PPD) gradient algorithm. Therein one has to carefully address asynchrony issues in communication, as renewable generations can vary fairly fast. Similar to Chapter 7, different types of asynchrony due to imperfect communication or other practical limits, such as random time delays and nonidentical sampling/computation rates, are formulated into a unified asynchrony framework. However, unlike that work, this chapter additionally considered both active and reactive powers of DERs. It results in complex coupling constraints twisted with the asynchrony issue, which are much more challenging albeit more practical for the distributed voltage control design. Moreover, the strong convexity of the objective function, as assumed in most literature, may not hold for voltage control, further complicating the problem. In this chapter, we show that the operator splitting based framework presented in Chapter 7 can be extended to deal with such a problem. Again, we devise an asynchronous algorithm for the distributed voltage control by exploiting the unique structure of the voltage optimization problem. Then, we equivalently convert it into a fixed-point problem by leveraging the operator splitting technique, which leads to a simple but concise convergence proof. We also present an online implementation to adapt the controller to time-varying environments, which is appealing to practical applications. Similar to Chapter 7, this chapter omits the fast internal dynamics of DERs. Hence this chapter also mainly involves feedback-based optimization, where the measurements of

Merging Optimization and Control in Power Systems: Physical and Cyber Restrictions in Distributed Frequency Control and Beyond, First Edition. Feng Liu, Zhaojian Wang, Changhong Zhao, and Peng Yang.

voltages, active powers, and reactive powers are utilized to improve the control performance while relaxing communication requirements.

8.1 Background

With the proliferation of DERs, such as small hydro plants, photovoltaics (PVs), and energy storage systems, voltage regulation in ADNs is facing significant challenges. On the one hand, the voltage quality remarkably degrades, e.g. the voltage may fluctuate rapidly due to the variation of renewable generations, and overvoltage exists at the buses DERs connected. On the other hand, many DERs, such as some small hydro plants [1] and inverter-integrated DERs [2], have great potential for voltage regulation by appropriately managing their active or reactive power outputs. Beyond the capability of traditional voltage regulation schemes, these challenges call for a new voltage control paradigm.

The voltage control in an ADN aims to minimize the voltage mismatch by regulating active or reactive power outputs of controllable DERs. Generally speaking, one can regard it as a sort of optimal power flow problem, where the branch power flow model is usually utilized [3]. Similar topics have been studied extensively in the literature, which can roughly be categorized into three classes in terms of communication requirements: centralized control, local control, and distributed control. In the centralized voltage control, a global optimization problem is formulated and solved by a central controller to determine optimal set points for the overall system [4, 5]. In this case, the central controller collects all the required information and communicates with all DERs. Thus, it suffers from the single-point failure issue and heavy computation burden when the number of DERs is extremely large. As for the local voltage control, one utilizes locally available information such as bus voltage magnitude to design the controller [2]. Remarkably, the linearized power flow model is extensively used in the problem formulation [6–8]. Because it uses only local information, the response is rapid. However, the control objective associated with the voltage deviation requires a specific form, i.e. the admittance-matrix-induced norm instead of the commonly used Euclidean norm in practice, which restricts its application. The distributed voltage control can avoid the disadvantages of centralized and fully local control to some extent [9]. Compared with the centralized control, the distributed voltage control requires the communication only between immediate neighbors [10–14] or two-hop neighbors [15], which avoids the notorious problem of single-point failure. The controller works as long as the communication graph is connected. Compared with the local control, the control objective associated with the voltage deviation can be the commonly used Euclidean norm, which also relaxes the limitation to the objective functions.

In most of the aforementioned literature, the distributed voltage control is synchronous, i.e. each bus is required to wait until they receive information from all of their neighbors and then perform a control action simultaneously at the same pace. Unfortunately, it is not the case in practice, since asynchrony widely exists in power systems, such as communication time delays and different sampling or computation rates. When a synchronous algorithm is adopted, the slowest bus and communication channel can cripple the system [16]. To address this problem, ASDVC methods are investigated, considering asynchronous updates [11, 13], time delays [14], event-triggered communications [17], etc. Although these works are very inspiring, they still have some limitations, e.g. communication delays cannot be considered in [11, 13], reference [14] only designs a synchronous algorithm intuitively without a theoretic guarantee, and global communication is required in [17].

In this chapter, we design an ASDVC method in ADNs. Each bus can perform its active/reactive power updates asynchronously. We prove that the algorithm converges to an optimal solution to the optimal voltage regulation problem. Salient features include two aspects:

1) **Practicality improvement**: in this work, we address several important practical issues in designing distributed controllers for ADNs to improve the practicality of the controller. On the one hand, various types of asynchrony are considered, such as communication delays and different sampling/controlling rates, leading to a unified framework for convergence analysis. This is different from [11, 13, 14], which only designs synchronous algorithm [14] or only considers asynchronous iterations and assumes no communication delay [11, 13]. On the other hand, in contrast to [2, 6–15] only controlling reactive power, we consider the regulation of both active and reactive powers of DERs with coupling constraints. This is crucially important in ADNs due to the following: (i) many kinds of DERs such as small hydro plants have no capability of reactive power regulation; (ii) the ADNs have comparable resistance and reactance, which implies that the impact of active power on voltage is not neglectable; and (iii) the constraints of active and reactive power may be coupled with each other, i.e. the output limits of inverters. These considerations allow a more practical distributed voltage control for ADNs, which, however, makes the problem much more difficult to solve. Here we present a systematic method to address this problem.
2) **Solution algorithm**: in the controller design, the PPD gradient algorithm is utilized, which is formulated as the form of the Krasnosel'skiĭ–Mann iteration. It has the following three advantages. First, it relaxes the assumption on the objective function. Here we only require that the objective function is *convex* and has a Lipschitzian gradient, which is weaker than the assumption

commonly used in most of the literature (where *strong convexity* is required) [6–8, 13–15, 18]. Consequently, the proposed method turns out to be more applicable as many regulation cost functions are not strongly convex, such as demand response with a piecewise linear cost function. Second, dual variables need not be updated simultaneously, which makes the algorithm easier to apply. It is a significant improvement compared with [19], where dual variables of different agents are required to be updated simultaneously in the asynchronous distributed primal–dual algorithm. Therefore, our method can be applied to address a broader class of asynchronous distributed optimization problems in practice. Third, we convert the algorithm into a fixed-point problem by employing the operator splitting method. This conversion leads to a concise convergence proof that is not presented in the literature [2, 6–15].

The rest of this chapter is organized as follows. In Section 8.2, we introduce some preliminaries and the model of ADNs. Section 8.3 formulates the optimal voltage control problem. The asynchronous controller is derived and analyzed in Section 8.4. In Section 8.5, convergence and optimality of the equilibrium are proved. Section 8.6 introduces the implementation of the proposed method. We confirm the performance of controllers via simulations on an 8-bus system and the IEEE 123-bus system in Section 8.7. Finally, Section 8.8 concludes this chapter with notes.

8.2 Preliminaries and System Model

8.2.1 Note and Preliminaries

Throughout this chapter, we adopt the following notations. $\mathbb{R}^n$ represents the n-dimensional Euclidean space. For a column vector $x \in \mathbb{R}^n$ (matrix $A \in \mathbb{R}^{m \times n}$), its transpose is denoted by $x^{\mathrm{T}}(A^{\mathrm{T}})$. For vectors $x, y \in \mathbb{R}^n$, $x^{\mathrm{T}}y = \langle x, y \rangle$ is the inner product of x, y. $\|x\| = \sqrt{x^{\mathrm{T}}x}$ is the norm induced by the inner product. For a positive definite matrix G, the inner product induced by G is denoted by $\langle x, y \rangle_G = \langle Gx, y \rangle$. Similarly, the G induced norm $\|x\|_G = \sqrt{\langle Gx, x \rangle}$. The identity matrix with dimension n is denoted by I_n. Sometimes, we also omit n to represent the identity matrix with proper dimension. For a matrix $A = [a_{ij}]$, a_{ij} is the entry in the ith row and jth column of A. The Cartesian product of the sets $\Omega_i, i = 1, \dots, n$ is denoted by $\prod_{i=1}^n \Omega_i$. Given a collection of y_i for i in a certain set Y, the vector composed of y_i is defined as $\mathbf{y} = \mathrm{col}(y_i) := (y_1, y_2, \dots, y_n)^{\mathrm{T}}$. For a single-valued operator $\mathcal{T} : \Omega \subset \mathbb{R}^n \to \mathbb{R}^n$, a point $x^* \in \Omega$ is a fixed point of $\mathcal{T}$ if $\mathcal{T}(x^*) \equiv x^*$.

8.2.2 System Modeling

Consider a radial ADN with $(n+1)$ buses collected in the set $\mathcal{N}_0 := \{0\} \cup \mathcal{N}$, where $\mathcal{N} := \{1, \ldots, n\}$ and bus 0 is the substation bus (slack bus) with a fixed voltage U_0. The set of lines is denoted by $\mathcal{E} := \{(i,j)\} \subset \mathcal{N} \times \mathcal{N}$. Due to the tree topology of ADNs, the cardinality of $\mathcal{E}$ satisfies $|\mathcal{E}| = n$. The set of immediate neighbors and two-hop neighbors of bus j is denoted by N_j and N_j^2, respectively.

For bus j, its voltage magnitude is denoted by U_j. The active and reactive power generations are denoted by p_j and q_j, respectively, which are controllable. p_j^c and q_j^c are active and reactive power loads, which are uncontrollable. For line $(i,j) \in \mathcal{E}$, r_{ij} and x_{ij} are its line resistance and reactance. The active and reactive powers from bus i to j are denoted by P_{ij} and Q_{ij}, respectively. From references [3, 6], the linearized power flow equations are given by

$$P_{ij} + p_j - p_j^c = \sum_{k \in N_j} P_{jk} \tag{8.1a}$$

$$Q_{ij} + q_j - q_j^c = \sum_{k \in N_j} Q_{jk} \tag{8.1b}$$

$$U_i^2 - U_j^2 = r_{ij}P_{ij} + x_{ij}Q_{ij} \tag{8.1c}$$

It has been reported that the relative error of the linearization is fairly small, at the order of 1%; hence it is valid in most scenarios [6]. The incidence matrix of the network $(\mathcal{N}_0, \mathcal{E})$ is denoted by $\mathcal{M} \in \mathbb{R}^{(n+1)\times n}$. The first row of $\mathcal{M}$ is denoted by $\mathbf{m}_0^{\mathrm{T}}$, while the rest of the matrix is denoted by $\mathbf{M}$. Introduce a new variable $V_i := \frac{U_i^2}{2}$, and then the compact form of (8.1) can be written as

$$\mathbf{MP} = \mathbf{p} - \mathbf{p}^c \tag{8.2a}$$

$$\mathbf{MQ} = \mathbf{q} - \mathbf{q}^c \tag{8.2b}$$

$$[\mathbf{m}_0,\ \mathbf{M}^{\mathrm{T}}] \cdot [V_0,\ \mathbf{V}^{\mathrm{T}}]^{\mathrm{T}} = \mathrm{diag}(\mathbf{r})\mathbf{P} + \mathrm{diag}(\mathbf{x})\mathbf{Q} \tag{8.2c}$$

where diag($\mathbf{r}$) is the diagonal matrix composed of r_{ij} and similar is diag($\mathbf{x}$). Because the network is connected, the rank of $\mathcal{M}$ is n. Thus, $\mathbf{M}$ is of full rank and invertible [6]. By multiplying (8.2a) and (8.2b) with $\mathbf{M}^{-1}$, replacing P, Q in (8.2c) with the resulting equations, and solving (8.2c) for $\mathbf{V}$, we obtain

$$\mathbf{V} = \mathbf{Rp} + \mathbf{Xq} - \mathbf{M}^{-\mathrm{T}}\mathbf{m}_0 V_0 - \mathbf{Rp}^c - \mathbf{Xq}^c \tag{8.3}$$

where

$$\mathbf{R} = \mathbf{M}^{-\mathrm{T}}\ \mathrm{diag}(\mathbf{r})\mathbf{M}^{-1};$$

$$\mathbf{X} = \mathbf{M}^{-\mathrm{T}}\ \mathrm{diag}(\mathbf{x})\mathbf{M}^{-1}$$

Here **R** and **X** are all symmetric positive definite matrices. By Kekatos *et al.* [18] and Hale *et al.* [20], we know

$$-\mathbf{M}^{-\mathrm{T}}\mathbf{m}_0 = \mathbf{1}_n$$

The inverse of **X** is denoted by

$$\mathbf{B} = \mathbf{M}\mathrm{diag}(\mathbf{x}^{-1})\mathbf{M}^{\mathrm{T}}$$

which is also positive definite. It is proved in [Theorem 2] [8] that

$$\mathbf{B} = \mathbf{L} + \mathrm{diag}\left(\frac{1}{x_{0j}}\right)$$

where **L** is the weighted Laplacian matrix of the subtree (i.e. without bus 0) and x_{0j} is the reactance of the line connected to bus 0. If bus j is not connected to the bus 0 directly, then we let $x_{0j} = \infty$.

If lines have an identical resistance–reactance ratio, i.e. there exists a uniform constant $K = \frac{r_{ij}}{x_{ij}}, \forall(i,j) \in \mathcal{E}$, the network is called homogeneous. For a homogeneous network, we have $\mathbf{R} = K\mathbf{X}$ and $\mathbf{B} \cdot \mathbf{R} = K$. In the analysis of this chapter, it is assumed that the ADN is homogeneous, which is similar to the literature [11, 15]. In the simulations, however, we use heterogeneous networks to verify the performance of the controller in a more practical fashion.

8.3 Problem Formulation

The optimal voltage control problem is solved by the following optimization problem:

$$\min_{\mathbf{V},\mathbf{p},\mathbf{q}} \quad f = \frac{1}{2}\|\mathbf{V} - \mathbf{V}^{\mathrm{o}}\|^2 + \sum_{j\in\mathcal{N}} g_j(p_j, q_j) \tag{8.4a}$$

$$\text{s.t.} \quad \mathbf{BV} = K\mathbf{p} + \mathbf{q} + \boldsymbol{\varpi}^S \tag{8.4b}$$

$$\underline{p}_j \le p_j \le \overline{p}_j, \quad \forall j \tag{8.4c}$$

$$\underline{q}_j \le q_j \le \overline{q}_j, \quad \forall j \tag{8.4d}$$

$$0 \le p_j^2 + q_j^2 \le s_j^2, \quad \forall j \tag{8.4e}$$

where

$$\boldsymbol{\varpi}^S = \mathbf{B}(-\mathbf{M}^{-\mathrm{T}}\mathbf{m}_0 V_0 - \mathbf{R}\mathbf{p}^c - \mathbf{X}\mathbf{q}^c)$$

Here, $\mathbf{V}^{\mathrm{o}} = 0.5 \times \mathbf{1}_n$ is the desired voltage profile; $\underline{p}_j, \overline{p}_j$ is the lower and upper bounds of p_j; $\underline{q}_j, \overline{q}_j$ is the lower and upper bounds of $q_j \cdot s_j$ is the apparent power capability of the inverter.

Considering the optimization problem (8.4), we make an assumption.

Assumption 8.1 In (8.4), the function $g_i(x)$ is convex, and $\nabla g_i(x)$ is ϑ-Lipschitzian, i.e. there exists some $\vartheta > 0$ such that

$$\|\nabla g_i(x_1) - \nabla g_i(x_2)\| \le \vartheta \|x_1 - x_2\|, \forall x_1, x_2$$

For each bus j, the feasible region is defined as

$$\Omega_j = \{(p_j, q_j) \mid p_j, q_j \text{ satisfy } (8.4c), (8.4d), (8.4e)\}$$

The Lagrangian of (8.4) is

$$\underset{(\mathbf{p},\mathbf{q})\in\Omega}{\mathcal{L}(\mathbf{V}, \mathbf{p}, \mathbf{q}, \lambda)} = \frac{1}{2}\|\mathbf{V} - \mathbf{V}^*\|^2 + \sum_{j\in\mathcal{N}} g_j(p_j, q_j) \tag{8.5}$$

$$+\lambda^{\mathrm{T}}(\mathbf{BV} - K\mathbf{p} - \mathbf{q} - \boldsymbol{\varpi}^S) \tag{8.6}$$

where $\Omega = \prod_{j\in\mathcal{N}} \Omega_j$.

Remark 8.1 ***(Objective Function)*** The term $g_j(p_j, q_j)$ in the objective function is more general than the works in literature [6–8, 13–15, 18], which is only required to be convex and has a Lipschitzian gradient instead of being strongly convex. This allows to include common practical objectives such as the demand response with a piecewise linear cost function. In addition, if the objective function is formulated by the **B**-induced norm, i.e. $f = \frac{1}{2}\|\mathbf{V} - \mathbf{V}^o\|_{\mathbf{B}}^2$, we can design a local controller, as done in [6, 7, 11].

In the problem (8.4), we consider the regulation of both active and reactive powers of DERs. The main motivations are twofold: first, some DERs such as many small hydro plants have no capability of reactive power regulation, and second, the ADNs have comparable resistance and reactance. Thus, regulating both active and reactive powers turns to be necessary for the voltage control of ADNs.

8.4 Asynchronous Voltage Control

In the asynchronous controller design, the PPD algorithm is adopted.[1] The continuous-time form of the PPD algorithm is previously introduced in [21] to design the primary frequency controller in power systems. Reference [22] provides its general form and gives a rigorous proof of the convergence. It is further developed in [23–25] to design a secondary frequency controller. The discrete-time form of the PPD algorithm is introduced in [14], which is utilized to design distributed voltage controllers. However, Liu *et al.* [14] do not consider asynchrony in the design.

---

1 The detailed introduction of PPD algorithm is given in Appendix A.2.3.

By the PPD algorithm, the problem (8.7) with the decision variable $\mathbf{V}$ can be solved in a closed-form by

$$\mathbf{V} = \arg\min_{\mathbf{V}} \mathcal{L}(\mathbf{V}, \mathbf{p}, \mathbf{q}, \lambda) = -\mathbf{B}^{\mathrm{T}} \lambda + \mathbf{V}^{o} \tag{8.7}$$

Thus, as long as λ is obtained, $\mathbf{V}$ can be directly figured out.

Let $\boldsymbol{\varpi}^a = \boldsymbol{\varpi}^s - \mathbf{B}\mathbf{V}^o$. Each bus has its own iteration number t_j, implying that a *local clock* is adopted. Then, various types of asynchrony can be considered in the time interval between two consecutive iterations. At t_j, bus j computes in the following way, which serves as a Krasnosel'skiĭ–Mann iteration:

$$\begin{bmatrix} \tilde{p}_{j,t_j} \\ \tilde{q}_{j,t_j} \end{bmatrix} = \mathcal{P}_{\Omega_j}\left[\begin{pmatrix} p_{j,t_j} \\ q_{j,t_j} \end{pmatrix} - \alpha_{pq}\begin{pmatrix} \frac{\partial g_j}{\partial p_j}(p_{j,t_j}, q_{j,t_j}) - K\lambda_{j,t_j} \\ \frac{\partial g_j}{\partial q_j}(p_{j,t_j}, q_{j,t_j}) - \lambda_{j,t_j} \end{pmatrix}\right] \tag{8.8a}$$

$$\begin{aligned} \tilde{\lambda}_{j,t_j} = \lambda_{j,t_j} + \alpha_\lambda \Bigg(& -\sum_{k\in N_j \cup N_j^2} \tilde{B}_{jk} \lambda_{k,t_j-\tau_{jk,t_j}} \\ & - 2K\tilde{p}_{j,t_j} - 2\tilde{q}_{j,t_j} + Kp_{j,t_j} + q_{j,t_j} - \varpi_j^a \Bigg) \end{aligned} \tag{8.8b}$$

$$\lambda_{j,t_j+1} = \lambda_{j,t_j} + \eta(\tilde{\lambda}_{j,t_j} - \lambda_{j,t_j}) \tag{8.8c}$$

$$p_{j,t_j+1} = p_{j,t_j} + \eta(\tilde{p}_{j,t_j} - p_{j,t_j}) \tag{8.8d}$$

$$q_{j,t_j+1} = q_{j,t_j} + \eta(\tilde{q}_{j,t_j} - q_{j,t_j}) \tag{8.8e}$$

$$V_{j,t_j+1} = -\sum_{k\in \mathcal{N}_j} B_{jk}\lambda_{j,t_j+1} + V_j^o \tag{8.8f}$$

where ϖ_j^a is the jth component of $\boldsymbol{\varpi}^a$ and step sizes $\eta, \alpha_{pq}, \alpha_\lambda > 0$. τ_{jk,t_j} is the time difference between t_j and time instant when the bus j obtains the latest information from bus k. For example, the current iteration for bus j is $t_j = 12$, but it receives the latest information from bus k at $t_j = 8$. Then, $\tau_{jk,t_j} = 4$. $\tilde{B}_{jk}$ is the jth row and kth column element of matrix $\tilde{\mathbf{B}} = \mathbf{B}^2$. As $\mathbf{B}$ has the same sparse structure with the Laplacian of the subtree, the matrix $\mathbf{B}^2$ has nonzero entries matching the immediate neighbors and two-hop neighbors of each bus. It implies that each bus only needs the information of its one- and two-hop neighbors to compute the variable $\tilde{\lambda}_{j,t_j}$. In this context, the ASDVC algorithm based on (8.8) can be devised, as given in Algorithm 8.1.

Remark 8.2 ***(Asynchronous update)*** As proved in [19], the asynchronous distributed primal–dual algorithm cannot guarantee the convergence if dual variables are not updated simultaneously. In ASDVC, there is no need to update λ simultaneously. We use the information of neighbors and two-hop neighbors for a trade-off.

Algorithm 8.1: *ASDVC*

Input: For bus j, the input is $(p_{j,0}, q_{j,0}) \in \Omega_j$, $\lambda_{j,0} \in \mathbb{R}$.
Iteration at t_j: Supposing the clock of bus j ticks at time t_j, then bus j is activated and updates its local variables as follows:

Step 1: Reading phase

Get $\lambda_{k,t_j-\tau_{jk,t_j}}$, $k \in N_j \cup N_j^2$ from output cache of its neighbors and two-hop neighbors, and store them to local storage.

Step 2: Computing phase

Calculate $\tilde{p}_{j,t_j}$, $\tilde{q}_{j,t_j}$, and $\tilde{\lambda}_{j,t_j}$ according to (8.8a) and (8.8b).

Update λ_{j,t_j+1}, p_{j,t_j+1}, q_{j,t_j+1}, and V_{j,t_j+1} according to (8.8c) – (8.8f), respectively.

Step 3: Writing phase

Write λ_{j,t_j+1} to its output cache and p_{j,t_j+1}, q_{j,t_j+1}, V_{j,t_j+1} to its local storage. Increase t_j to $t_j + 1$.

8.5 Optimality and Convergence

In this section, we formulate the algorithm (8.8) into a fixed-point iteration problem using the operator splitting method and prove its convergence and optimality of the equilibrium. Following the line in Chapter 7, we introduce a **virtual global clock** to substitute the local clocks of individual buses in ASDVC. The main idea is to queue t_j of all buses in the order of real time and use a new number t to represent the tth iteration in the queue, denoted by $t \in T$. Take two local clocks as an example. Suppose the local clocks to be $T_1 = \{1, 3, 6, \dots\}$ and $T_2 = \{2, 4, 5, \dots\}$, and then the global clock is $T = \{1, 2, 3, 4, 5, 6, \dots\}$. In the global clock, the probability that bus j is activated to update its local variables is assumed to follow a uniform distribution, i.e. each bus is activated with the same probability.

Then, taking into account asynchrony, the algorithm (8.8) under the global clock can be rewritten as

$$\begin{bmatrix} \tilde{p}_{j,t} \\ \tilde{q}_{j,t} \end{bmatrix} = \mathcal{P}_{\Omega_j}\left[\begin{pmatrix} p_{j,t-\tau_j^t} \\ q_{j,t-\tau_j^t} \end{pmatrix} - \alpha_{pq}\begin{pmatrix} \frac{\partial g_j}{\partial p_j}(p_{j,t-\tau_j^t}, q_{j,t-\tau_j^t}) - K\lambda_{j,t-\tau_j^t} \\ \frac{\partial g_j}{\partial q_j}(p_{j,t-\tau_j^t}, q_{j,t-\tau_j^t}) - \lambda_{j,t-\tau_j^t} \end{pmatrix}\right] \tag{8.9a}$$

$$\begin{aligned} \tilde{\lambda}_{j,t} = \lambda_{j,t-\tau_j^t} + \alpha_\lambda \Bigg(& - \sum_{k \in N_j \cup N_j^2} \tilde{B}_{jk}\lambda_{k,t-\tau_k^{\mathrm{T}}} \\ & - 2K\tilde{p}_{j,t-\tau_j^t} - 2\tilde{q}_{j,t-\tau_j^t} + Kp_{j,t-\tau_j^t} + q_{j,t-\tau_j^t} - \varpi_j^a \Bigg) \end{aligned} \tag{8.9b}$$

$$\lambda_{j,t+1} = \lambda_{j,t-\tau_j^t} + \eta(\tilde{\lambda}_{j,t} - \lambda_{j,t-\tau_j^t}) \tag{8.9c}$$

$$p_{j,t+1} = p_{j,t-\tau_j^t} + \eta(\tilde{p}_{j,t} - p_{j,t-\tau_j^t}) \tag{8.9d}$$

$$q_{j,t+1} = q_{j,t-\tau_j^t} + \eta(\tilde{q}_{j,t} - q_{j,t-\tau_j^t}) \tag{8.9e}$$

$$V_{j,t+1} = -\sum_{k \in \mathcal{N}_j} B_{jk}\lambda_{j,t+1} + V_j^o \tag{8.9f}$$

where $t \in T$. τ_j^t is the time difference between t and time spot when bus j obtains the latest information. We also simply call τ_j^t the time delay if there is no confusion. Note that the global clock is only used for analysis but does not appear in the implementation.

8.5.1 Algorithm Reformulation

Let $z_j = \text{col}(p_j, q_j)$ and $\tilde{z}_j = \text{col}(\tilde{p}_j, \tilde{q}_j)$. If the asynchrony is not considered and all clocks are synchronous, the compact form of (8.9) can be obtained, denoted by synchronous distributed voltage control (SDVC). Because $\mathbf{V}$ is absent from the iteration process of other variables and determined by λ, it is omitted here:

$$\tilde{\mathbf{z}}_t = \mathcal{P}_\Omega\left(\mathbf{z}_t - \alpha_{pq}(\nabla_{z_t} g(z_t) - \text{col}(K\lambda_t, \lambda_t))\right) \tag{8.10a}$$

$$\tilde{\lambda}_t = \lambda_t + \alpha_\lambda\left(-\mathbf{B}^2\lambda_t - 2(K \cdot I_n, I_n)\tilde{\mathbf{z}}_t + (K \cdot I_n, I_n)\mathbf{z}_t - \boldsymbol{\varpi}^a\right) \tag{8.10b}$$

$$\mathbf{z}_{t+1} = \mathbf{z}_t + \eta(\tilde{\mathbf{z}}_t - \mathbf{z}_t) \tag{8.10c}$$

$$\lambda_{t+1} = \lambda_t + \eta(\tilde{\lambda}_t - \lambda_t) \tag{8.10d}$$

In the rest of the chapter, let $F(\mathbf{z}) = \nabla_{\mathbf{z}} g(\mathbf{z})$. Noticing the fact that

$$\mathcal{P}_\Omega(x) = (\text{Id} + N_\Omega)^{-1}(x)$$

and

$$N_\Omega(x) = \alpha_{pq} N_\Omega(x), \alpha_{pq} > 0$$

Eq. (8.10a) is equivalent to

$$\begin{aligned}
&\tilde{\mathbf{z}}_t = (\text{Id} + N_\Omega)^{-1}\left(\mathbf{z}_t - \alpha_{pq}(F(\mathbf{z}_t) - \text{col}(K\lambda_t, \lambda_t))\right)\\
\Rightarrow\ &\tilde{\mathbf{z}}_t + N_\Omega(\tilde{\mathbf{z}}_t) = \mathbf{z}_t - \alpha_{pq}(F(\mathbf{z}_t) - \text{col}(K\lambda_t, \lambda_t)\\
\Rightarrow\ &-F(\mathbf{z}_t) = N_\Omega(\tilde{\mathbf{z}}_t) + \alpha_{pq}^{-1}(\tilde{\mathbf{z}}_t - \mathbf{z}_t) - \text{col}(K\tilde{\lambda}_t, \tilde{\lambda}_t)\\
&\qquad + \text{col}(K\tilde{\lambda}_t, \tilde{\lambda}_t) - \text{col}(K\lambda_t, \lambda_t)
\end{aligned} \tag{8.11}$$

Then, (8.10a)–(8.10b) are equivalent to

$$\begin{aligned}
-F(\mathbf{z}_t) &= N_\Omega(\tilde{\mathbf{z}}_t) - \text{col}(K\tilde{\lambda}_t, \tilde{\lambda}_t)\\
&\quad + \alpha_{pq}^{-1}(\tilde{\mathbf{z}}_t - \mathbf{z}_t) + (K \cdot I_n, I_n)^{\mathrm{T}}(\tilde{\lambda}_t - \lambda_t)
\end{aligned} \tag{8.12a}$$

$$-\varpi^a - \mathbf{B}^2\lambda_t = (K \cdot I_n, I_n)\tilde{\mathbf{z}}_t + (K \cdot I_n, I_n)(\tilde{\mathbf{z}}_t - \mathbf{z}_t) + \alpha_\lambda^{-1}(\tilde{\lambda}_t - \lambda_t) \tag{8.12b}$$

To formulate (8.12) into a fixed-point iteration problem, two operators are defined.

$$C : \begin{bmatrix} \mathbf{z} \\ \lambda \end{bmatrix} \mapsto \begin{bmatrix} F(\mathbf{z}) \\ \varpi^a + \mathbf{B}^2\lambda \end{bmatrix} \tag{8.13a}$$

$$\mathcal{D} : \begin{bmatrix} \mathbf{z} \\ \lambda \end{bmatrix} \mapsto \begin{bmatrix} N_\Omega(\mathbf{z}) - \mathrm{col}(K\lambda, \lambda) \\ (K \cdot I_n, I_n)\mathbf{z} \end{bmatrix} \tag{8.13b}$$

Let $\mathbf{w}_t = \mathrm{col}(\mathbf{z}_t, \lambda_t)$ and $\tilde{\mathbf{w}}_t = \mathrm{col}(\tilde{\mathbf{z}}_t, \tilde{\lambda}_t)$. Then, (8.12) can be rewritten as

$$-C(\mathbf{w}_t) = \mathcal{D}(\tilde{\mathbf{w}}_t) + \Gamma \cdot (\tilde{\mathbf{w}}_t - \mathbf{w}_t) \tag{8.14}$$

where

$$\Gamma := \begin{bmatrix} \alpha_{pq}^{-1} I_{2n} & (K \cdot I_n, I_n)^{\mathrm{T}} \\ (K \cdot I_n, I_n) & \alpha_\lambda^{-1} I_n \end{bmatrix} \tag{8.15}$$

Here, $\alpha_{pq}, \alpha_\lambda$ are chosen to make Γ positive definite.

The maximal eigenvalue of $\mathbf{B}$ is denoted by $\sigma_{\max}$, then the 2-norm of $\|\mathbf{B}\| = \sigma_{\max}$ and $\|\mathbf{B}^2\| = \sigma_{\max}^2$ [Proposition 5.2.7, 5.2.8] [26]. We have following results.

Lemma 8.1 *Operators C and $\mathcal{D}$ defined in* (8.13a) *and* (8.13b) *have the following properties:*

1) *Operator C is β-cocoercive under the 2-norm with $0 < \beta \leq \min\{\frac{1}{\sigma_{\max}^2}, \frac{1}{\vartheta}\}$.*
2) *Operator $\mathcal{D}$ is maximally monotone.*
3) *$\Gamma^{-1}\mathcal{D}$ is maximally monotone under the Γ-induced norm.*
4) *$(\mathrm{Id} + \Gamma^{-1}\mathcal{D})^{-1}$ exists and is firmly nonexpansive.*

Proof:

Assertion 1). According to the definition of C and the definition of β-cocoercive, it suffices to prove

$$\langle C(\mathbf{w}_1) - C(\mathbf{w}_2), \mathbf{w}_1 - \mathbf{w}_2 \rangle \geq \beta \|C(\mathbf{w}_1) - C(\mathbf{w}_2)\|^2$$

or, equivalently,

$$\left(F(\mathbf{z}_1) - F(\mathbf{z}_2)\right)^{\mathrm{T}}(\mathbf{z}_1 - \mathbf{z}_2) + (\lambda_1 - \lambda_2)^{\mathrm{T}}\mathbf{B}^2(\lambda_1 - \lambda_2) \geq \beta\left(\left\|\mathbf{B}^2\lambda_1 - \mathbf{B}^2\lambda_2\right\|^2 + \|F(\mathbf{z}_1) - F(\mathbf{z}_2)\|^2\right) \tag{8.16}$$

Note that $\varpi^a + \mathbf{B}^2\lambda$ is the gradient of the function

$$\hat{f}(\lambda) = \frac{1}{2}\lambda^{\mathrm{T}}\mathbf{B}^2\lambda + \lambda^{\mathrm{T}}\varpi^a$$

As $\nabla^2 \hat{f}(\lambda) = \mathbf{B}^2 > 0, \hat{f}(\lambda)$ is a convex function. For its gradient, we have

$$\begin{aligned}\left\|\mathbf{B}^2(\lambda_1 - \lambda_2)\right\| &\leq \left\|\mathbf{B}^2\right\| \left\|\lambda_1 - \lambda_2\right\| \\ &= \sigma_{\max}^2 \left\|\lambda_1 - \lambda_2\right\| \end{aligned} \tag{8.17}$$

Thus, $\nabla \hat{f}(\lambda)$ is $\sigma_{\max}^2$-Lipschitzian. It follows that $\nabla \hat{f}(\lambda) = \boldsymbol{\varpi}^a + \mathrm{B}^2\lambda$ is $\frac{1}{\sigma_{\max}^2}$-cocoercive [Corollary 18.16] [27], i.e.

$$(\lambda_1 - \lambda_2)^{\mathrm{T}} \mathbf{B}^2 (\lambda_1 - \lambda_2) \geq \frac{1}{\sigma_{\max}^2} \left\|\mathbf{B}^2 \lambda_1 - \mathbf{B}^2 \lambda_2\right\|^2 \tag{8.18}$$

Moreover, F is $\frac{1}{\vartheta}$-cocoercive, i.e.

$$\left(F(\mathbf{z}_1) - F(\mathbf{z}_2)\right)^{\mathrm{T}} (\mathbf{z}_1 - \mathbf{z}_2) \geq \frac{1}{\vartheta} \left\|F(\mathbf{z}_1) - F(\mathbf{z}_2)\right\|^2 \tag{8.19}$$

Combining (8.18) and (8.19) and taking $0 < \beta \leq \min\{\frac{1}{\sigma_{\max}^2}, \frac{1}{\vartheta}\}$, we can get the first assertion.

Assertion 2). The operator $\mathcal{D}$ can be decomposed into

$$\begin{aligned}\mathcal{D} &= \begin{bmatrix} 0 & -(KI_n, I_n)^{\mathrm{T}} \\ (KI_n, I_n) & 0 \end{bmatrix} \begin{bmatrix} \mathbf{z} \\ \lambda \end{bmatrix} + \begin{bmatrix} N_\Omega(\mathbf{z}) \\ 0 \end{bmatrix} \\ &= \mathcal{D}_1 + \mathcal{D}_2 \end{aligned} \tag{8.20}$$

As $\mathcal{D}_1$ is a skew-symmetric matrix, $\mathcal{D}_1$ is maximally monotone [Example 20.30] [27]. Moreover, $N_\Omega(\mathbf{z})$ and 0 are all maximally monotone [Example 20.41] [27], so $\mathcal{D}_2$ is also maximally monotone. Thus, $\mathcal{D} = \mathcal{D}_1 + \mathcal{D}_2$ is maximally monotone.

Assertion 3). As Γ is symmetric positive definite and $\mathcal{D}$ is maximally monotone, it is easy to prove that $\Gamma^{-1}\mathcal{D}$ is maximally monotone by the similar analysis in Lemma 5.6 of [28].

Assertion 4). As $\Gamma^{-1}\mathcal{D}$ is maximally monotone, $(\mathrm{Id} + \Gamma^{-1}\mathcal{D})^{-1}$ exists and is firmly nonexpansive by [Proposition 23.7] [27]. ■

By the last assertion of Lemma 8.1, (8.14) is equivalent to

$$\tilde{\mathbf{w}}_t = (\mathrm{Id} + \Gamma^{-1}\mathcal{D})^{-1}(\mathrm{Id} - \Gamma^{-1}\mathcal{C})\mathbf{w}_t \tag{8.21a}$$

$$\mathbf{w}_{t+1} = \mathbf{w}_t + \eta(\tilde{\mathbf{w}}_t - \mathbf{w}_t) \tag{8.21b}$$

Let

$$\begin{aligned} \mathcal{S}_1 &= (\mathrm{Id} + \Gamma^{-1}\mathcal{D})^{-1}, \\ \mathcal{S}_2 &= (\mathrm{Id} - \Gamma^{-1}\mathcal{C}), \\ \mathcal{S} &= \mathcal{S}_1\mathcal{S}_2, \end{aligned} \tag{8.22}$$

and then we have following results.

Lemma 8.2 *Let* $0 < \beta \le \min\{\frac{1}{\sigma_{\max}^2}, \frac{1}{\vartheta}\}$, $\kappa > \frac{1}{2\beta}$, *and the step sizes* $\alpha_{pq}, \alpha_\lambda$ *such that the matrix* $\Gamma - \kappa I$ *is positive semi-definite.*[2] *The following results are true in the sense of the* Γ*-induced norm* $\|\cdot\|_\Gamma$*:*

1) S_1 *is a* $\frac{1}{2}$*-averaged operator, i.e.* $S_1 \in \mathcal{A}\left(\frac{1}{2}\right)$.

2) S_2 *is a* $\frac{1}{2\beta\kappa}$*-averaged operator, i.e.* $S_2 \in \mathcal{A}\left(\frac{1}{2\beta\kappa}\right)$.

3) S *is a* $\frac{2\kappa\beta}{4\kappa\beta-1}$*-averaged operator, i.e.* $S \in \mathcal{A}\left(\frac{2\kappa\beta}{4\kappa\beta-1}\right)$.

Proof:

Assertion 1). From the Assertion 4) of Lemma 8.1, $S_1 = (\mathrm{Id} + \Gamma^{-1}\mathcal{D})^{-1}$ is firmly nonexpansive, implying $S_1 \in \mathcal{A}(\frac{1}{2})$.

Assertion 2). First, we prove that $\Gamma^{-1}C$ is $\beta\kappa$-cocoercive, i.e.

$$\left\langle \Gamma^{-1}C(\mathbf{w}_1) - \Gamma^{-1}C(\mathbf{w}_2), \mathbf{w}_1 - \mathbf{w}_2 \right\rangle_\Gamma \tag{8.23}$$

$$\ge \beta\kappa \left\|\Gamma^{-1}C(\mathbf{w}_1) - \Gamma^{-1}C(\mathbf{w}_2)\right\|_\Gamma^2 \tag{8.24}$$

The maximal and minimal eigenvalues of Γ are denoted by $\delta_{\max}$ and $\delta_{\min}$, respectively, and we have

$$\delta_{\max} \ge \delta_{\min} \ge \kappa > 0.$$

From [Proposition 5.2.7, 5.2.8] [26], the Euclidean norms of Γ and Γ^{-1} are

$$\|\Gamma\|_2 = \delta_{\max}$$

and

$$\|\Gamma^{-1}\|_2 = \frac{1}{\delta_{\min}},$$

respectively.

For the right-hand side of (8.23), we have

$$\begin{aligned}
&\beta\kappa \left\|\Gamma^{-1}C(\mathbf{w}_1) - \Gamma^{-1}C(\mathbf{w}_2)\right\|_\Gamma^2 \\
&= \beta\kappa \left\|C(\mathbf{w}_1) - C(\mathbf{w}_2)\right\|_{\Gamma^{-1}}^2 \\
&= \beta\kappa (C(\mathbf{w}_1) - C(\mathbf{w}_2))^{\mathrm{T}}\Gamma^{-1}(C(\mathbf{w}_1) - C(\mathbf{w}_2)) \\
&\le \beta\kappa \|\Gamma^{-1}\|_2 \left\|C(\mathbf{w}_1) - C(\mathbf{w}_2)\right\|_2^2 \\
&\le \beta\kappa \cdot \frac{1}{\kappa} \left\|C(\mathbf{w}_1) - C(\mathbf{w}_2)\right\|_2^2
\end{aligned} \tag{8.25}$$

2 If one hopes to make Γ positive definite and $\Gamma - \kappa * I$ positive semidefinite, it is sufficient to have $\Gamma - \kappa * I$ being strictly diagonally dominant. This is easy to realize as long as $0 < \alpha_{pq} < \frac{1}{K+1} + \kappa, 0 < \alpha_\lambda < \frac{1}{K+1} + \kappa$. Thus, such α_{pq}, α_λ always exist.

where the first "$\leq$" is due to the Cauchy–Schwarz inequality and the second is due to $\delta_{\min} \geq \kappa$.

For the left part of (8.23), we have

$$\begin{aligned}
&\left\langle \Gamma^{-1}C(\mathbf{w}_1) - \Gamma^{-1}C(\mathbf{w}_2), \mathbf{w}_1 - \mathbf{w}_2 \right\rangle_\Gamma \\
&\quad = \langle C(\mathbf{w}_1) - C(\mathbf{w}_2), \mathbf{w}_1 - \mathbf{w}_2 \rangle \\
&\quad \geq \beta \left\| C(\mathbf{w}_1) - C(\mathbf{w}_2) \right\|_2^2
\end{aligned} \tag{8.26}$$

where the inequality is from Assertion 1) of Lemma 8.1. From (8.25) and (8.26), we have (8.23).

As $\Gamma^{-1}C$ is $\beta\kappa$-cocoercive, we have $\beta\kappa\Gamma^{-1}C \in \mathcal{A}(\frac{1}{2})$. That is to say, there is a nonexpansive operator $\tilde{S}$ such that

$$\beta\kappa\Gamma^{-1}C = \frac{1}{2}\,\mathrm{Id} + \frac{1}{2}\tilde{S},$$

or equivalently,

$$\Gamma^{-1}C = \frac{1}{2\beta\kappa}\,\mathrm{Id} + \frac{1}{2\beta\kappa}\tilde{S}$$

Therefore we have

$$S_2 = \mathrm{Id} - \Gamma^{-1}C = \left(1 - \frac{1}{2\beta\kappa}\right)\mathrm{Id} - \frac{1}{2\beta\kappa}\tilde{S} \tag{8.27}$$

As $0 < \frac{1}{2\beta\kappa} < 1$ and $-\tilde{S}$ are also nonexpansive, we have $S_2 \in \mathcal{A}(\frac{1}{2\beta\kappa})$.

Assertion 3). From [Proposition 2.4] [29], $S = S_1S_2$ is a a-averaged operator with $a = \frac{a_1+a_2-2a_1a_2}{1-a_1a_2}$, if S_1 is a_1-averaged and S_2 is a_2-averaged. As $S_1 \in \mathcal{A}\left(\frac{1}{2\beta\kappa}\right)$ and $S_2 \in \mathcal{A}\left(\frac{1}{2}\right)$, we have $S \in \mathcal{A}\left(\frac{2\kappa\beta}{4\kappa\beta-1}\right)$. It completes the proof. ■

By the definition of the averaged operator and assertion 3) of Lemma 8.2, there exists a nonexpansive operator $\mathcal{T}$ such that

$$S = \left(1 - \frac{2\kappa\beta}{4\kappa\beta - 1}\right)\mathrm{Id} + \frac{2\kappa\beta}{4\kappa\beta - 1}\mathcal{T} \tag{8.28}$$

Apparently, operators S and $\mathcal{T}$ have the same fixed points, i.e. $Fix(S) = Fix(\mathcal{T})$.

So far, we have converted the asynchronous algorithm into a fixed-point iteration problem with an averaged operator. Moreover, a nonexpansive operator $\mathcal{T}$ is constructed, which paves the way to the convergence proof of ASDVC.

8.5.2 Optimality of Equilibrium

The definition of the equilibrium point of ASDVC is introduced as follows.

Definition 8.1 A point $\mathbf{w}^* = \mathrm{col}(w_j^*) = \mathrm{col}(x_j^*, \lambda_j^*)$ is an equilibrium point of system (8.8) if $\lim_{t_j\to\infty} w_{t_j} = w_j^*$, holds for all $j \in \mathcal{N}$.

Now, we give the KKT condition of the optimization problem (8.4) [Theorem 3.25] [30]:

$$0 = (\mathbf{V} - \mathbf{V}^{o}) + \mathbf{B}^{T}\lambda \tag{8.29a}$$

$$0 = \nabla_{\mathbf{z}} g(\mathbf{z}) - \mathrm{col}(K\lambda, \lambda) + N_{\Omega}(\mathbf{z}) \tag{8.29b}$$

$$0 = \boldsymbol{\varpi}^{s} - \mathbf{B}\mathbf{V} + (K \cdot I, I)\mathbf{z} \tag{8.29c}$$

The voltage at the equilibrium point is defined as $\mathbf{V}^{*} := -\mathbf{B}^{T}\lambda^{*} + \mathbf{V}^{o}$, and we have the following theorem.

Theorem 8.1 *The point* $(\mathbf{V}^{*}, \mathbf{z}^{*}, \lambda^{*})$ *satisfies the KKT condition* (8.29), *i.e. it is the primal–dual optimal solution to the optimization problem* (8.4).

Proof: By Definition 8.1, we know

$$w_j^{*} = \lim_{t_j \to \infty} w_{j,t_j-1} = \lim_{t_j \to \infty} w_{j,t_j} = \lim_{t_j \to \infty} w_{j,t_j+1} = \lim_{t_j \to \infty} \tilde{w}_{j,t_j}$$

From (8.12) and

$$\mathbf{V}_t = -\mathbf{B}^{T}\lambda_t + \mathbf{V}^{o}$$

$$\boldsymbol{\varpi}^{a} = \boldsymbol{\varpi}^{s} - \mathbf{B}\mathbf{V}^{o}$$

we have

$$-(\mathbf{V}^{*} - \mathbf{V}^{o}) = \mathbf{B}^{T}\lambda^{*} \tag{8.30a}$$

$$-\nabla_{\mathbf{z}^{*}} g(\mathbf{z}^{*}) = N_{\Omega}(\mathbf{z}^{*}) - \mathrm{col}(K\lambda^{*}, \lambda^{*}) \tag{8.30b}$$

$$-\boldsymbol{\varpi}^{s} = -\mathbf{B}\mathbf{V}^{*} + (K \cdot I, I)\mathbf{z}^{*} \tag{8.30c}$$

Comparing (8.29) with (8.30), we know $(\mathbf{V}^{*}, \mathbf{z}^{*}, \lambda^{*})$ satisfies the KKT condition, which completes the proof. ■

8.5.3 Convergence Analysis

In this subsection, we justify the convergence of the ASDVC algorithm. We first treat the ASDVC as a randomized block-coordinate fixed-point iteration problem with delayed information. Then, the results in [31] can be applied.

To consider stochastic asynchrony due to different reasons, we randomize the virtual global clock in each iteration. Similar to the previous chapter, we take vectors $\psi_j \in \mathbb{R}^{3n}, j \in \mathcal{N}$, as location indicators, and the ith entry of ψ_j is denoted by $[\psi_j]_i$. If the ith coordinate of $\mathbf{w}$ is also a coordinate of w_j, then $[\psi_j]_i = 1$. Otherwise, $[\psi_j]_i = 0$. A random variable (vector) taking values in $\psi_j, j \in \mathcal{N}$ is denoted by ξ. The probability $\mathbf{Prob}(\xi = \psi_j) = 1/n$ also follows a uniform distribution. Let ξ_t be the value of ξ at the tth iteration. Then, a randomized block-coordinate fixed-point iteration for (8.21) is given by

$$\mathbf{w}_{t+1} = \mathbf{w}_t + \eta\xi_t \circ (S(\mathbf{w}_t) - \mathbf{w}_t) \tag{8.31}$$

where $\circ$ is the Hadamard product. In (8.31), only one bus j is activated at each iteration.

Since (8.31) is delay-free, we further modify it for considering delayed information, which is

$$\mathbf{w}_{t+1} = \mathbf{w}_t + \eta\xi_t \circ (S(\hat{\mathbf{w}}_t) - \mathbf{w}_t) \tag{8.32}$$

where $\hat{\mathbf{w}}_t$ is the information with delay at iteration t.

Here we claim that Algorithm 8.1 can be written as (8.32) if $\hat{\mathbf{w}}_t$ is properly defined, as we explain. Supposing bus j is activated at the iteration t, then $\hat{\mathbf{w}}_t$ is defined as follows. For bus j, replace p_{j,t_j}, q_{j,t_j}, and λ_{j,t_j} with p_{j,t_j}, q_{j,t_j}, and λ_{j,t_j}. Similarly, replace λ_{k,t_k} with $\lambda_{k,t_k-\tau_k^{t_k}}$ from its neighbors and two-hop neighbors. For inactivated buses, their state values keep unchanged.

Before proving the convergence, we make an assumption.

Assumption 8.2 The maximal delay between two consecutive iterations is bounded by χ, i.e. $\tau_j^t \leq \chi, \forall t, \forall j$.

With the assumption, we can establish the following theorem to justify the convergence of the ASDVC algorithm.

Theorem 8.2 *Suppose Assumptions 8.1 and 8.2 hold. Take* $0 < \beta \leq \min\{\frac{1}{\sigma_{\max}^2}, \frac{1}{\vartheta}\}$, $\kappa > \frac{1}{2\beta}$ *and the step sizes* $\alpha_{pq}, \alpha_\lambda$ *such that* $\Gamma - \kappa I$ *is positive semidefinite. Choose* $0 < \eta < \frac{1}{1+2\chi/\sqrt{n}}\frac{4\kappa\beta-1}{2\kappa\beta}$. *Then, with the ASDVC,* $\mathbf{w}_t$ *converges to the point* $\mathbf{w}^*$ *defined in Definition 8.1 with probability 1.*

Proof: Combining (8.28) and (8.32), we have

$$\begin{aligned}\mathbf{w}_{t+1} &= \mathbf{w}_t + \eta\xi_t \circ \left(\left(1 - \frac{2\kappa\beta}{4\kappa\beta - 1}\right)\hat{\mathbf{w}}_t - \mathbf{w}_t + \frac{2\kappa\beta}{4\kappa\beta - 1}\mathcal{T}(\hat{\mathbf{w}}_t)\right)\\ &= \mathbf{w}_t + \eta\xi_t \circ \left(\hat{\mathbf{w}}_t - \mathbf{w}_t + \frac{2\kappa\beta}{4\kappa\beta - 1}(\mathcal{T}(\hat{\mathbf{w}}_t) - \hat{\mathbf{w}}_t)\right)\end{aligned} \tag{8.33}$$

By noting $w_{j,t-\tau_j^t} = w_{j,t-\tau_j^T+1} =, \cdots, = w_{j,t}$, we have $\xi_k \circ \left(\hat{\mathbf{w}}_t - \mathbf{w}_t\right) = 0$. Thus, (8.33) is equivalent to

$$\mathbf{w}_{t+1} = \mathbf{w}_t + \frac{2\eta\kappa\beta}{4\kappa\beta - 1}\xi_t \circ \left(\mathcal{T}(\hat{\mathbf{w}}_t) - \hat{\mathbf{w}}_t\right) \tag{8.34}$$

In fact, (8.34) has the form of the ARock algorithms presented in [31]. In Lemma 13 and Theorem 14 of [31], it is proved that $\mathbf{w}_t$ generated by (8.34) is bounded. Moreover, if η satisfies $0 < \eta < \frac{1}{1+2\chi/\sqrt{n}}\frac{4\kappa\beta-1}{2\kappa\beta}$, $\mathbf{w}_t$ converges to a random variable that takes value in the fixed points of $\mathcal{T}$ with probability 1, denoted by $\mathbf{w}^*$. Recall $Fix(\mathcal{S}) = Fix(\mathcal{T})$ and Theorem 8.1, and we know that $\mathbf{w}^*$ satisfies the KKT condition (8.29), i.e. the equilibrium in Definition 8.1. This completes the proof. ∎

As $\mathcal{S}$ is $\frac{2\kappa\beta}{4\kappa\beta-1}$-averaged, it is also a nonexpansive operator [Remark 4.24] [27]. Then, in Theorem 8.2, we can also use $\mathcal{S}$ instead of $\mathcal{T}$ to prove the convergence of the ASDVC. In this situation, the bound of η is $0 < \eta < \frac{1}{1+2\chi/\sqrt{n}}$. Since $\kappa > \frac{1}{2\beta}$, we have $\frac{4\kappa\beta-1}{2\kappa\beta} > 1$. This implies that the operator $\mathcal{T}$ can increase the upper bound of η compared with $\mathcal{S}$.

The proof also indicates that the maximal time delay χ affects the upper bound of the step size η. Only when $0 < \eta < \frac{1}{1+2\chi/\sqrt{n}}\frac{4\kappa\beta-1}{2\kappa\beta}$, the convergence can be guaranteed. It also makes sense in practice to restrict the upper bound of time delay. If the time delay is too large, one should check whether or not the physical communication channel is working properly.

8.6 Implementation

8.6.1 Communication Graph

Although the information of two-hop neighbors is utilized in the algorithm (8.8), the communication graph still can be fully distributed, as we explain. In (8.8b), the information of two-hop neighbors is needed to obtain $\tilde{B}_{jk}$ and $\lambda_{k,t_j-\tau_{jk,t_j}}, k \in N_j^2$. For $\tilde{B}_{jk}$, it can be obtained from two neighboring communications. In a normal situation, the topology of an ADN does not change frequently, so $\tilde{B}_{jk}$ can be obtained in advance. For $\lambda_{k,t_j-\tau_{jk,t_j}}, k \in N_j^2$, it also can be obtained from twice neighboring communications, which is illustrated in Figure 8.1.

In this situation, the time delay may be longer. Because this is an asynchronous algorithm, we can simply treat $\tau_{jk,t_j} + \tau_{ij,t_j}$ as one delay τ_{ik,t_j} if there is no confusion. Then, one-step and two-step communication delays can be formulated into a unified framework. In this way, only neighboring communications need to be considered in the asynchronous algorithm.

Figure 8.1 Two-step communications.

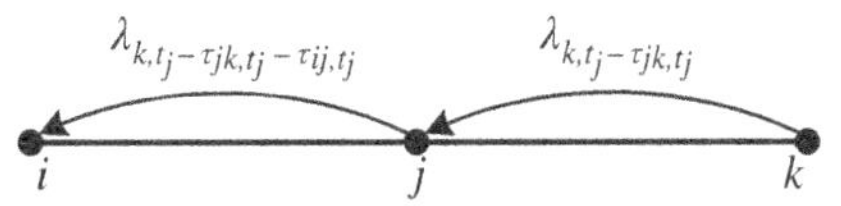

8.6.2 Online Implementation

In the ASDVC, ϖ_j^a is assumed to be available for every bus j. By its definition, ϖ_j^a is determined by almost all the power injections across the network, which implies that a centralized coordinator is needed to collect all information and compute ϖ_j^a. However, it is a very challenging task in practice as the system states vary rapidly due to the fast variation of renewable generations and loads. In this regard, additional design is required for facilitating the online implementation.

Fortunately, in (8.4b), $\boldsymbol{\varpi}^s$ can be figured out locally without the centralized coordinator if one can directly measure the local voltage and active and reactive power injections from the physical power network and get the neighbors' voltages through communication. It is the same with $\boldsymbol{\varpi}^a$ due to $\boldsymbol{\varpi}^a = \boldsymbol{\varpi}^s - \mathbf{B}\mathbf{V}^o$. The set of buses connected directly to bus 0 is denoted by N_0. Let $\mathbf{V}^o = 0.5 \times \mathbf{1}$, and we have

$$\mathbf{B}\mathbf{V}^o = \begin{cases} 0, & j \notin N_0 \\ \frac{1}{2x_{0j}}, & j \in N_0 \end{cases}$$

Then, ϖ_j^a in the ASDVC algorithm can be obtained by

$$\varpi_j^a = \begin{cases} -\sum\limits_{k\in N_j} B_{jk} V_k^m + K p_j^m + q_j^m, & j \notin N_0 \\ -\frac{1}{2x_{0j}} - \sum\limits_{k\in N_j} B_{jk} V_k^m + K p_j^m + q_j^m, & j \in N_0 \end{cases} \tag{8.35}$$

where p_j^m, q_j^m are the injected active and reactive powers measured locally and V_k^m is the square of the measured voltage of the neighbor.

In (8.35), only communications between one-hoc neighbors are needed. We can also use $p_{j,t}, q_{j,t}$ instead of p_j^m, q_j^m in inverter-integrated DERs, since $p_{j,t} \approx p_j^m$ and $q_{j,t} \approx q_j^m$ as the response of inverters are very fast. The voltage measurements contain the latest system information, which makes the ASDVC be able to track the time-varying operating conditions.

Remark 8.3 Algorithm 8.1 is mainly designed for the distributed optimal voltage control of an ADN in normal operations, which has a limitation in the scenarios of topology changing. Because information of two-hop neighbors is utilized, it is not fully distributed. If the ADN topology changes rapidly, $\tilde{B}_{jk}$ should be broadcast to each bus after each change of topology. This may slow down the response speed.

8.7 Case Studies

In this section, simulation results are presented to demonstrate the effectiveness of the proposed voltage control methods. To this end, an 8-bus feeder system [15] and the IEEE 123-bus feeder system are utilized for test.

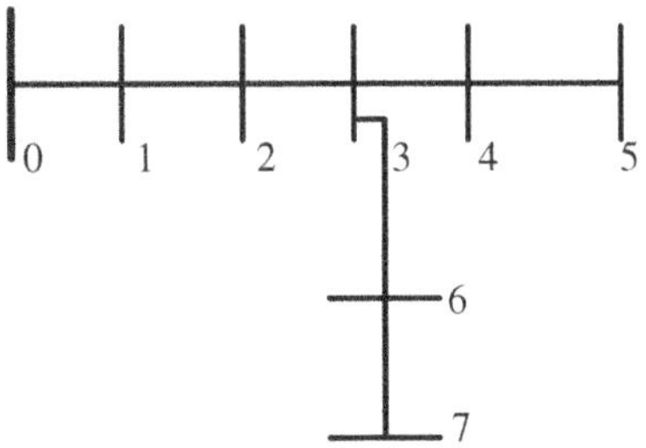

Figure 8.2 The graph of the 8-bus distribution network.

8.7.1 8-Bus Feeder System

The ADN with 8-bus feeder is illustrated in Figure 8.2 [15], where bus 0 is the substation. The impedance of each line segment is identical, which is $0.9216 + j0.4608\ \Omega$ with $K = 2$. All buses except bus 0 have DERs with power limits as

$$\overline{\mathbf{p}} = -\underline{\mathbf{p}} = [90, 100, 0, 120, 170, 90, 70]\text{kW};$$
$$\overline{\mathbf{q}} = -\underline{\mathbf{q}} = [100, 100, 110, 100, 130, 100, 120]\text{kVar}.$$

The capacity limit of inverter at each bus is $0.9 * \sqrt{\overline{\mathbf{p}}^2 + \overline{\mathbf{q}}^2}$kVA.

First, CPLEX is used to obtain the optimal solution, which gives

$$\mathbf{V}^* = [0.9934, 1.0063, 1.0083, 1.0282, 0.9492, 1.0073, 0.9987]\ \text{p.u.};$$
$$\mathbf{p}^* = [63.08, 65.41, 0, 120, 170, 70.57, 70]\ \text{kW};$$
$$\mathbf{q}^* = [31.53, 32.71, 63, 73.24, 90.54, 35.28, 41.29]\ \text{kVar}.$$

Then, the proposed SDVC and ASDVC algorithms are utilized to solve the voltage control problem. In the SDVC, no time delay is considered. In the ASDVC, random time delays are included, where the maximal delay is set as 5 iterations. We compare the performance by showing how the relative error $\|\mathbf{w} - \mathbf{w}^*\|_2^2 / \|\mathbf{w}^*\|_2^2$ is evolving with the number of average iterations, as illustrated in Figure 8.3. It is observed that the ASDVC algorithm can achieve almost the same convergence as the SDVC, both taking about 50 iterations. The ASDVC is even slightly better than the SDVC in the average number of iterations.

In power systems, many regulation cost functions are not strongly convex, such as demand response. For example, the regulation cost of controllable load may be $k|p|$, where $k > 0$ is the price and p is the controllable load power. The cost function is not strongly convex, which needs to be carefully addressed. To make its gradient Lipschitzian, we add a small perturbation at $p = 0$. Without loss of generality, take $k = 1$, and then the $g_j(p_j, q_j)$ takes the form:

$$g_j(p_j, q_j) = \begin{cases} |p| + \frac{1}{2}q_j^2, & p < -0.001 \text{ or } p > 0.001 \\ 500p^2 + \frac{1}{2} \times 0.001 + \frac{1}{2}q_j^2, & \text{otherwise} \end{cases}$$

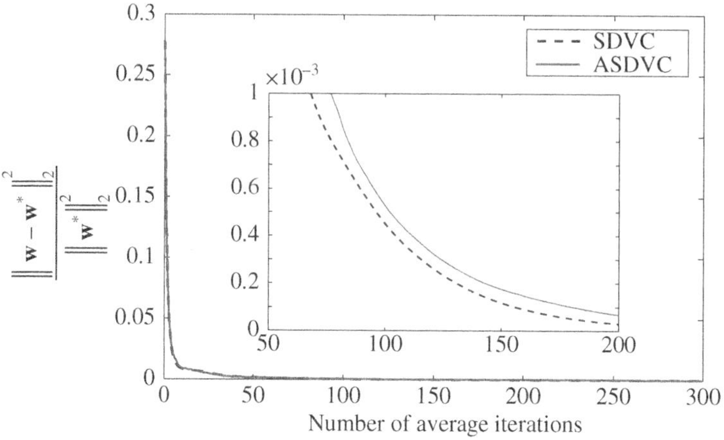

Figure 8.3 Comparison of algorithm convergence in terms of number of average iterations for SDVC and ASDVC.

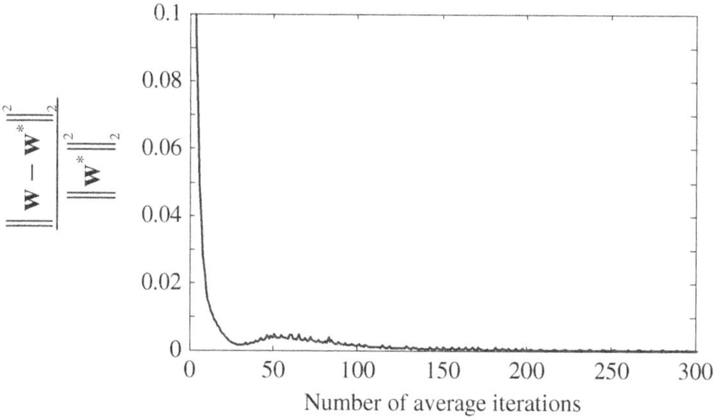

Figure 8.4 Algorithm convergence with non-strongly objective function.

It can be verified that $g_j(p_j, q_j)$ is convex and its gradient is Lipschitzian. The performance of the algorithm is shown in Figure 8.4. The convergence rate is slower than that with strongly convex ones, which takes about 150 iterations. The result is the same as that obtained by the centralized method, which validates that the algorithm can address the non-strongly convex cost.

The convergence of the algorithm with different time delays and η is given in Figure 8.5. In the upper part, the $\eta = 0.31$ and three different types of delays are considered. The algorithm can converge smoothly with 2 iterations. When it increases to 5 iterations, there is oscillation. When it increases to 10 iterations,

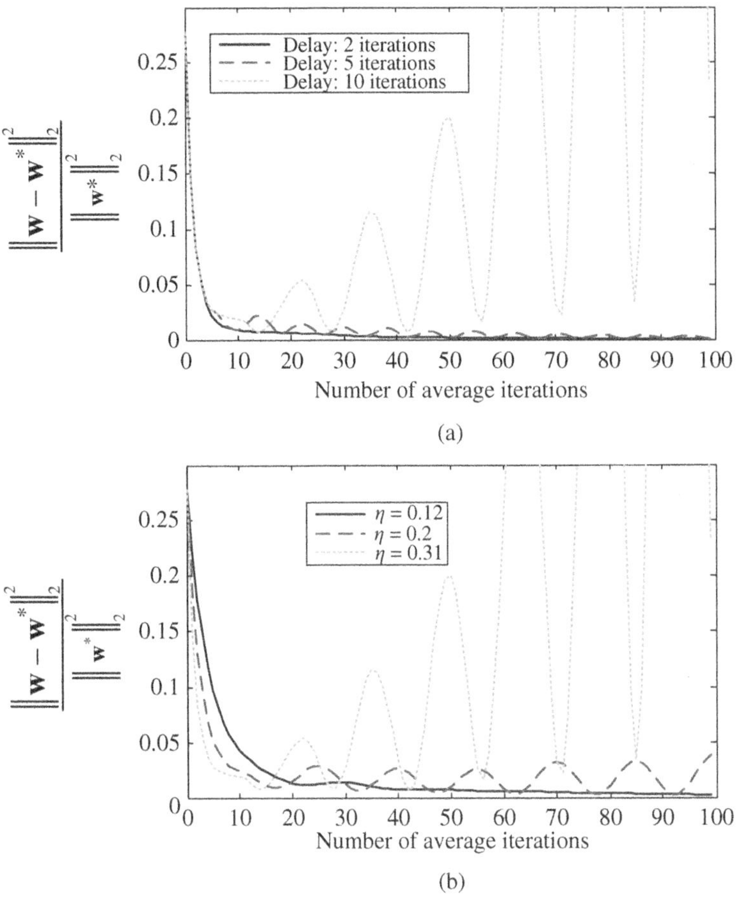

Figure 8.5 Convergence with different time (a) delays and (b) η. The time delay here exists in the communication channel with bus 1.

the algorithm does not converge. In the lower part, the time delay is 10 iterations. When η decreases to 0.12, the algorithm converges. This verifies that with smaller χ, it will allow larger η and so will achieve a faster convergence rate.

In distribution networks, **R**, **X** are comparable. Thus, the impact of active power cannot be neglected. We also demonstrate this by simulation results. The upper bound of active power is modified to

$$\overline{\mathbf{p}} = [135, 150, 165, 180, 255, 135, 105] \text{ kW}$$

and $\overline{\mathbf{p}}$ does not change. In the regulation with active power, the upper and lower bounds of active and reactive powers are $\overline{\mathbf{p}} = -\underline{\mathbf{p}}$ and $\overline{\mathbf{q}} = -\underline{\mathbf{q}}$, respectively. In the

Table 8.1 Comparison between the regulations with and without active power.

	Regulation with active power			Regulation without active power		
Bus_i	U (p.u.)	p (kW)	q (kVar)	U (p.u.)	p (kW)	q (kVar)
1	1	80.50	39.01	1.13	135	−68.10
2	1	59.29	21.41	1.25	150	−61.85
3	1	36.58	14.85	1.32	165	−68.03
4	1	59.39	17.20	1.35	180	−44.10
5	0.99	230.40	115.21	1.31	255	−36.53
6	1	66.23	27.55	1.31	135	−68.10
7	1	72.98	54.04	1.28	105	−97.82
$\|\mathbf{U}-1\|_2$		0.0078			0.7588	

regulation without active power, the upper and lower bounds of active and reactive powers are $\overline{\mathbf{p}} = \underline{\mathbf{p}}$ and $\overline{\mathbf{q}} = -\underline{\mathbf{q}}$, respectively, i.e. the active power cannot be regulated. Then the results of comparison between regulations with and without active power are given in Table 8.1.

It is shown in Table 8.1 that the voltage difference in the case with active power is $\|\mathbf{V}-1\|_2 = 0.0078$. If the active power is not regulated, the voltage difference is $\|\mathbf{V}-1\|_2 = 0.7588$. This verifies that controlling both active and reactive powers can achieve a better performance than reactive power alone. In this case, the voltage difference reduces by 98.97%.

8.7.2 IEEE 123-Bus Feeder System

In addition to the 8-bus feeder, we also test the proposed method on the IEEE 123-bus system to show the scalability and practicability [6], which is illustrated in Figure 8.6. It should be noted that the IEEE 123-bus system is not homogeneous, where the $\frac{r_{ij}}{x_{ij}}$ ranges from 0.42 to 2.02. In the implementation of the controller, we simply take $K = 1$ and examine the robustness of the proposed control. All the simulations are carried out in the Matlab R2017b simulator, and OpenDSS is used for solving the nonlinear power flow (Table 8.1).

Each bus is equipped with PVs, offering limited flexible active and reactive power supplies to the feeder. The profiles of residential load and solar generation come from real data. The minute-sampled profiles of active and reactive loads are from an online data repository [32], and we use the data on 13 July 2010. The minute-sampled profile of solar generation, collected in a city in Utah, USA,

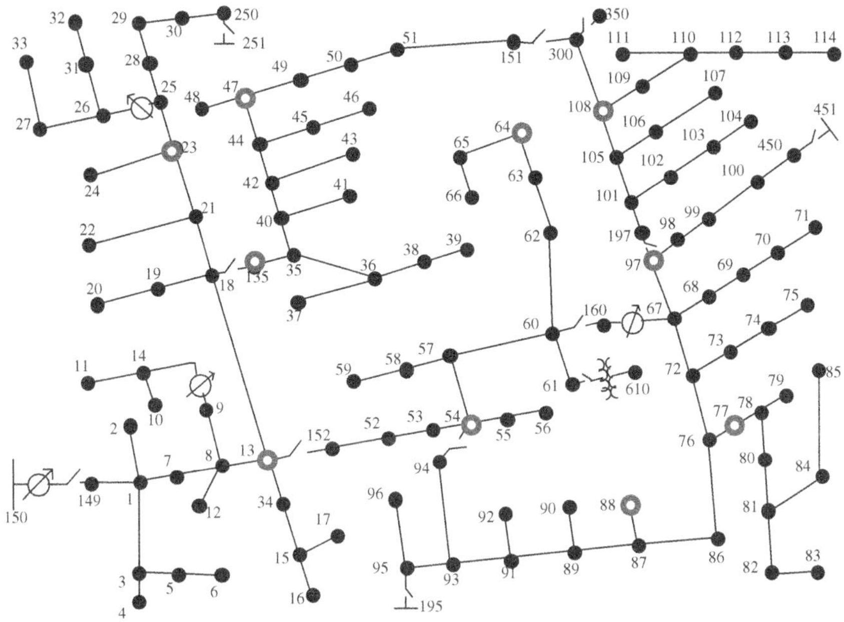

Figure 8.6 IEEE 123-bus system.

is from [33], and we use the data on 14 July 2010. The profiles of active and reactive loads and solar generation are given in Figure 8.7, where the curve is the active power (kW), the light grey curve is the reactive power (kVar), and the dotted line is the solar generation (kW). In the simulation, the tap positions of voltage regulators are kept constant to better capture the performance of the proposed method. The voltage at the substation of the feeder head is set as 1 p.u., and the value of $\mathbf{V}^0$ is $0.5 \times \mathbf{1}$ p.u. The capacity limit of each bus is 100 kVA. The upper limit of active power is the instantaneous generation of the PV, and the reactive power limit is determined correspondingly. Buses 13, 23, 135, 47, 54, 64, 88, 77, 97, and 108 are equipped with small hydro plants with the active power limits as 300 kW, which are marked as the light grey circle in Figure 8.6. When load and solar generation change, the method in Section 8.6.2 is utilized for the online implementation. A quasi-static operating condition is adopted in each minute, and the proposed controller runs in each iteration updated every 0.2 s (a total of 300 iterations per minute).

We compare the performances of the SDVC and the ASDVC under random time delays. The maximal time delay is 5 s. The profiles of daily network-wide voltage errors with the ASDVC and the SDVC are given in Figure 8.8.It is observed that the

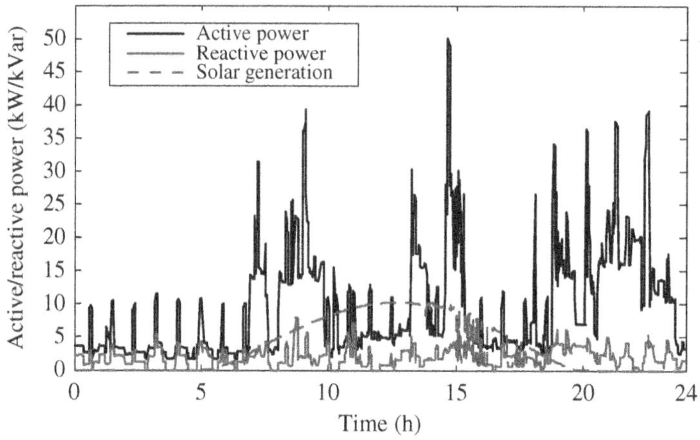

Figure 8.7 Active and reactive loads and solar generation through the 24 h.

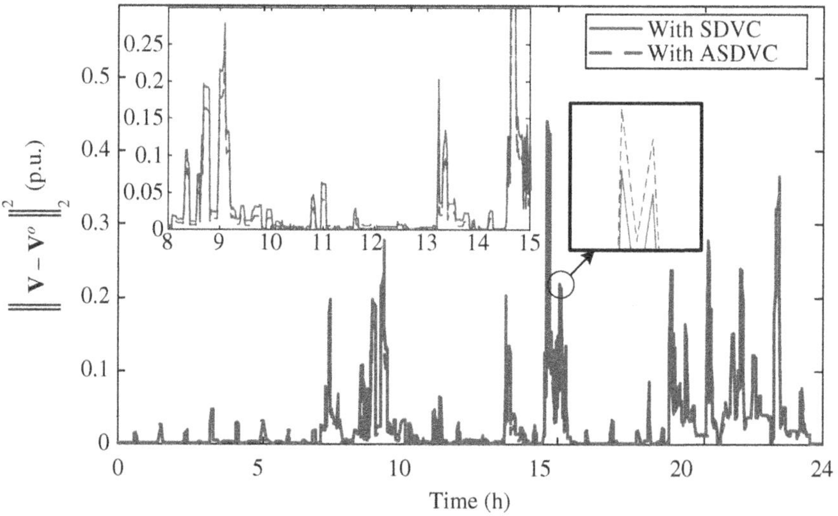

Figure 8.8 Daily voltage mismatch under the SDVC and the ASDVC in the presence of random time delays. If the SDVC is adopted, every bus has to wait for the slowest one to proceed with the next iteration. In contrast, every bus can update as long as new information is received in the ASDVC.

voltage deviation with the SDVC is more significant than that with the ASDVC at most time spots. The reason is that each bus under SDVC has to wait for the slowest neighbor before carrying out the next step algorithm. Hence it cannot accurately track the rapid changes. In contrast, the ASDVC does not suffer from idling time.

This result demonstrates that the ASDVC has better performance in time-varying environments with time delays.

It is also observed that, at some moments, the voltage deviations with the SDVC are smaller than that with ASDVC, as shown in the zoomed-in image in the circle. This phenomenon may be caused by two reasons: (i) The objective function (8.4) is composed of two parts. In Figure 8.8, we only give the voltage deviation, which does not completely reflect the performance of the algorithm in minimizing the objective function. (ii) The convergence of the algorithm guarantees that $\|\mathbf{w}_t - \mathbf{w}^*\|$ reduces, which is slightly different from the voltage deviation.

8.8 Conclusion and Notes

In this chapter, we have developed an asynchronous distributed control method to regulate the voltage in ADNs by using both the active and reactive powers of DERs. The PPD gradient algorithm is utilized to design the controller. The operator splitting theory is applied to prove the convergence and optimality of the equilibrium point. Finally, numerical tests with real data show that the voltage deviation can be effectively reduced by using ASDVC. Furthermore, simulations under random time delays show that the asynchronous algorithm has better performance in time-varying environments. In the theoretic analysis, we assume the ADN is three-phase symmetric and homogeneous. How to eliminate these restrictions is worthy of investigating, although the presented methodology should be applicable in theory.

To unify different types of imperfect communication into the same asynchrony framework, we introduce a virtual global clock and randomize it in each iteration. Then we construct non-expansive operators to convert the convergence problem into a fixed-point problem. The methodology is similar to what we have done in Chapter 7 except that the voltage control of ADN turns to be much more challenging due to the complicated physical constraints created by the couplings between active and reactive power. Nevertheless, the theoretical analysis and numerical experiments show that the methodology applies even in complicated circumstances. This chapter also demonstrates that the operator splitting theory can conveniently combine with other techniques, such as primal–dual algorithms and Krasnosel'skiĭ–Mann iteration, to deal with a wide variety of asynchronous distributed control/optimization problems considering both cyber and physical restrictions. Another interesting trick worthy of mentioning is that the utilization of measurement feedback avoids the need of two-hoc communication, which is helpful in improving control performance and reducing communication requirements.

Bibliography

1 G. Ardizzon, G. Cavazzini, and G. Pavesi, "A new generation of small hydro and pumped-hydro power plants: advances and future challenges," *Renewable and Sustainable Energy Reviews*, vol. 31, pp. 746–761, 2014.

2 K. Turitsyn, P. Sulc, S. Backhaus, and M. Chertkov, "Options for control of reactive power by distributed photovoltaic generators," *Proceedings of the IEEE*, vol. 99, no. 6, pp. 1063–1073, 2011.

3 M. E. Baran and F. F. Wu,"Optimal capacitor placement on radial distribution systems," *IEEE Transactions on Power Delivery*, vol. 4, no. 1, pp. 725–734, 1989.

4 M. Farivar, R. Neal, C. Clarke, and S. Low, "Optimal inverter VAR control in distribution systems with high PV penetration," in *Power and Energy Society General Meeting, 2012 IEEE*, pp. 1–7, IEEE, 2012.

5 V. Kekatos, G. Wang, A. J. Conejo, and G. B. Giannakis, "Stochastic reactive power management in microgrids with renewables," *IEEE Transactions on Power Systems*, vol. 30, no. 6, pp. 3386–3395, 2015.

6 H. Zhu and H. J. Liu, "Fast local voltage control under limited reactive power: optimality and stability analysis," *IEEE Transactions on Power Systems*, vol. 31, no. 5, pp. 3794–3803, 2016.

7 H. J. Liu, W. Shi, and H. Zhu, "Decentralized dynamic optimization for power network voltage control," *IEEE Transactions on Signal and Information Processing Over Networks*, vol. 3, no. 3, pp. 568–579, 2017.

8 X. Zhou, L. Chen, M. Farivar, Z. Liu, and S. Low, "Reverse and forward engineering of local voltage control in distribution networks," *IEEE Transactions on Automatic Control*, vol. 66, no. 3, pp. 1116–1128, 2020.

9 K. E. Antoniadou-Plytaria, I. N. Kouveliotis-Lysikatos, P. S. Georgilakis, and N. D. Hatziargyriou, "Distributed and decentralized voltage control of smart distribution networks: models, methods, and future research," *IEEE Transactions on Smart Grid*, vol. 8, no. 6, pp. 2999–3008, 2017.

10 P. Šulc, S. Backhaus, and M. Chertkov, "Optimal distributed control of reactive power via the alternating direction method of multipliers," *IEEE Transactions on Energy Conversion*, vol. 29, no. 4, pp. 968–977, 2014.

11 S. Bolognani, R. Carli, G. Cavraro, and S. Zampieri, "Distributed reactive power feedback control for voltage regulation and loss minimization," *IEEE Transactions on Automatic Control*, vol. 60, no. 4, pp. 966–981, 2015.

12 B. Zhang, A. Y. Lam, A. D. Domínguez-García, and D. Tse, "An optimal and distributed method for voltage regulation in power distribution systems," *IEEE Transactions on Power Systems*, vol. 30, no. 4, pp. 1714–1726, 2015.

13 H. J. Liu, W. Shi, and H. Zhu, "Distributed voltage control in distribution networks: online and robust implementations," *IEEE Transactions on Smart Grid*, vol. 9, no. 6, pp. 6106–6117, 2018.

14 H. J. Liu, W. Shi, and H. Zhu, "Hybrid voltage control in distribution networks under limited communication rates," *IEEE Transactions on Smart Grid*, vol. 10, no. 3, pp. 2416–2427, 2018.

15 Z. Tang, D. J. Hill, and T. Liu, "Fast distributed reactive power control for voltage regulation in distribution networks," *IEEE Transactions on Power Systems*, vol. 34, no. 1, pp. 802–805, 2019.

16 P. Yi and L. Pavel, "Asynchronous distributed algorithm for seeking generalized Nash equilibria," *2018 European Control Conference*, 2018.

17 S. Magnusson, C. Fischione, and N. Li, "Optimal voltage control using event triggered communication," in *Proceedings of the Tenth ACM International Conference on Future Energy Systems*, pp. 343–354, 2019.

18 V. Kekatos, L. Zhang, G. B. Giannakis, and R. Baldick, "Voltage regulation algorithms for multiphase power distribution grids," *IEEE Transactions on Power Systems*, vol. 31, no. 5, pp. 3913–3923, 2016.

19 M. T. Hale, A. Nedić, and M. Egerstedt, "Asynchronous multiagent primal-dual optimization," *IEEE Transactions on Automatic Control*, vol. 62, no. 9, pp. 4421–4435, 2017.

20 V. Kekatos, L. Zhang, G. B. Giannakis, and R. Baldick, "Fast localized voltage regulation in single-phase distribution grids," in *Smart Grid Communications (SmartGridComm), 2015 IEEE International Conference on*, pp. 725–730, IEEE, 2015.

21 C. Zhao, U. Topcu, N. Li, and S. H. Low, "Design and stability of load-side primary frequency control in power systems," *IEEE Transactions on Automatic Control*, vol. 59, no. 5, pp. 1177–1189, 2014.

22 N. Li, C. Zhao, and L. Chen, "Connecting automatic generation control and economic dispatch from an optimization view," *IEEE Transactions on Control of Network Systems*, vol. 3, no. 3, pp. 254–264, 2016.

23 E. Mallada, C. Zhao, and S. Low, "Optimal load-side control for frequency regulation in smart grids," *IEEE Transactions on Automatic Control*, vol. 62, no. 12, pp. 6294–6309, 2017.

24 Z. Wang, F. Liu, S. H. Low, C. Zhao, and S. Mei, "Distributed frequency control with operational constraints, Part I: per-node power balance," *IEEE Transactions on Smart Grid*, vol. 10, no. 1, pp. 40–52, 2019.

25 Z. Wang, F. Liu, S. H. Low, C. Zhao, and S. Mei, "Distributed frequency control with operational constraints, Part II: network power balance," *IEEE Transactions on Smart Grid*, vol. 10, pp. 53–64, 2019.

26 C. D. Meyer, *Matrix Analysis and Applied Linear Algebra*, vol. 71. SIAM, 2000.

27 H. Bauschke and P. L. Combettes, *Convex Analysis and Monotone Operator Theory in Hilbert Spaces*. Springer, 2017.

28 P. Yi and L. Pavel, "An operator splitting approach for distributed generalized Nash equilibria computation," *Automatica*, vol. 102, pp. 111–121, 2019.

29 P. L. Combettes and I. Yamada, "Compositions and convex combinations of averaged nonexpansive operators," *Journal of Mathematical Analysis and Applications*, vol. 425, no. 1, pp. 55–70, 2015.

30 A. P. Ruszczyński and A. Ruszczynski, *Nonlinear Optimization*, vol. 13. Princeton University Press, 2006.

31 Z. Peng, Y. Xu, M. Yan, and W. Yin, "ARock: An algorithmic framework for asynchronous parallel coordinate updates," *SIAM Journal on Scientific Computing*, vol. 38, no. 5, pp. A2851–A2879, 2016.

32 UCI, "Individual household electric power consumption data set." https://archive.ics.uci.edu/ml/datasets/individual+household+electric+power+consumption, 2012.

33 NREL, "SOLRMAP Utah Geological Survey." https://midcdmz.nrel.gov/usep_cedar/, 2018.

9

Robustness and Adaptability: Unknown Disturbances in Load-Side Frequency Control

So far we have addressed how to deal with cyber and physical restrictions arising from practical operational requirements. However, in real-world complicated environments, the control system may suffer from unknown external disturbances, partial control coverage, heterogeneous control configurations, etc. It, therefore, is desired to endow the control system with strong robustness and adaptability. This chapter focuses on the issue of hedging against unknown disturbances.

In the previous chapters, the change of power/load is simply modeled as a step change. It is sufficient for a power system with slowly varying load and generation. However, with the constantly increasing penetration of renewable generation, the power mismatch across the power system becomes fast-time varying and is always unknown. In this case, the design of frequency control turns to be much more sophisticated. This chapter aims to address the distributed load-side frequency control problem in a multi-area power system taking into account of unknown time-varying power imbalance. Particularly, fast controllable loads are utilized to restore system's frequency. To this end, the imbalanced power is decomposed into three components: a known steady part, an unknown low-frequency variation, and a high-frequency residual. The known steady part is the prediction of power imbalance. The variation part may result from the fluctuation of renewable generation sources, electric vehicle charging, load variation, etc., which is usually unknown. The high-frequency residual is also unknown and treated as an external disturbance. Correspondingly, we address the following subproblems in three timescales (i) to allocate the steady part of power imbalance economically in a distributed manner, (ii) to mitigate the effect of unknown low-frequency power variation locally, and (iii) to effectively attenuate unknown high-frequency disturbances.

Merging Optimization and Control in Power Systems: Physical and Cyber Restrictions in Distributed Frequency Control and Beyond, First Edition. Feng Liu, Zhaojian Wang, Changhong Zhao, and Peng Yang.

9.1 Background

In power system operation, frequency deviation is usually a consequence of power mismatch due to unexpected disturbances, such as sudden load leaping/dropping or generator tripping. Therefore, frequency control plays a core role in underpinning the functional operation of a power system. The frequency control literature can be roughly divided into two categories in terms of the power imbalance modeling: (i) constant power imbalance [1–7] and (ii) time-varying power imbalance [8–10]. In the first category, a step change of load/generation is considered. Then, generators and/or controllable loads are utilized to eliminate the power imbalance and restore the nominal frequency. In [1], an optimal load-side control problem is formulated, and a primary frequency controller is derived to rebalance the system after a step power change by using controllable loads. It is further extended in [2] to the scope of secondary frequency control, i.e. restoring the nominal frequency. The design approach is generalized in [4], where the model requirement is relaxed by introducing a passivity condition to guarantee asymptotic stability of the closed-loop system. References [5, 6] further consider both steady-state and transient operational constraints in distributed optimal frequency control. In [3], a nonlinear network-preserving model is considered, and only limited control coverage is needed to implement the distributed optimal frequency control. A different type of disturbance is considered in [7], where the secondary frequency controller may experience malicious attacks. In this regard, an attack detection procedure is added to the distributed frequency controller.

In the second category, the power imbalance is no longer constant, creating a much greater challenge to controller design and stability analysis. In [8], the power variation is modeled as the output of a *known* exosystem. Then an internal model controller is designed to compensate for the time-varying power imbalance. The idea of combining distributed control with internal model control is appealing and attracts a lot of attention. In [9], a centralized controller is proposed to track the power imbalance and maintain the system frequency within a desired range in the presence of slowly varying power imbalance. In [10], measurement noise is considered in frequency control, and a leaky integral controller is proposed that can strike an acceptable trade-off between performance and robustness.

To sum up, in most of the literature, power disturbance is modeled as a step change. The time-varying power disturbance is usually regarded as the output of a known exosystem for simplicity. However, such models are not realistic enough for practical power system operation, especially when a large amount of renewable generation and electric vehicles are integrated. In such a situation, the power imbalance is always time-varying and unknown, which should be carefully considered in the design of distributed frequency control.

In power systems, the power imbalance can be decomposed into three parts: a known constant part, an unknown low-frequency time-varying part, and a high-frequency residual. The first one is usually obtained by prediction while the latter two are fluctuations around the prediction. Offset error of prediction can also be considered in the unknown time-varying part. This decomposition suggests a compositional way to deal with time-varying disturbances, as presented in this chapter. First, a distributed control is proposed based on the consensus method to balance the known constant part economically, which resolves a slow timescale operation problem. Second, a decentralized supplementary controller based on the adaptive internal model control is proposed to mitigate the effect of the unknown low-frequency variation at a faster timescale. Third, we show that the proposed controller can well attenuate the impact of high-frequency residual. Finally, we confirm the performance of the proposed controller via various simulations with real data.

In the remainder of this chapter is organized as follows. The network and power imbalance models are formulated in Section 9.2, and the controller is proposed based on adaptive internal model control in Section 9.3. Then, the equilibrium of the closed-loop system is characterized with a proof of asymptotic stability in Section 9.4. The robustness of the proposed controller under uncertainties is also analyzed in Section 9.5. The control performance is verified via simulations with real data in Section 9.6. Finally, Section 9.7 concludes this chapter with notes.

9.2 Problem Formulation

9.2.1 Power Network

A large power-grid network is usually composed of multiple control areas, which are interconnected through tie lines. For simplicity, we treat each control area as a node with an aggregate controllable load and an aggregate uncontrollable power injection.[1] Then the power network can be modeled as a graph $\mathcal{G} := (\mathcal{N}, \mathcal{E})$ where $\mathcal{N} = \{1,2, \dots n\}$ is the set of nodes (control areas) and $\mathcal{E} \subseteq \mathcal{N} \times \mathcal{N}$ is the set of edges (tie lines). If a pair of nodes i and j are connected by a tie line directly, we denote the tie line by $(i,j) \in \mathcal{E}$. $\mathcal{G}$ is directed with arbitrary orientation, and we use $(i,j) \in \mathcal{E}$ or $i \rightarrow j$ interchangeably to denote a directed edge from i to j. Without loss of generality, we assume $\mathcal{G}$ is connected.

Besides the graph of the physical power network, we also need to consider its communication network, which is modeled by another graph $\mathcal{H}$ whose nodes

1 In our study, all controllable loads in the same area are aggregated into one controllable load. It is the same for the aggregate uncontrollable power injection. This simplification is practically reasonable when dealing with the frequency control problem in power systems [11].

are the same set $\mathcal{N}$ of graph $\mathcal{G}$ with possibly a different set of edges. An edge in $\mathcal{H}$ means that the two endpoints of the edge can communicate with each other directly. In this chapter, we assume $\mathcal{H}$ is always connected. The set of the neighbors of node j in the communication graph $\mathcal{H}$ is denoted by N_{cj}. The Laplacian matrix of $\mathcal{H}$ is denoted by L.

A second-order linearized power system dynamic model is adopted to describe the frequency dynamics of each node. We assume that the tie lines are lossless so that the DC power flow model can be used, which is a common assumption when considering a high-voltage transmission system. Then for each node $j \in \mathcal{N}$, we have

$$\dot{\theta}_j = \omega_j \tag{9.1a}$$

$$M_j\dot{\omega}_j = P_j^{in} - P_j^l - D_j\omega_j + \sum_{i:i\to j} B_{ij}(\theta_i - \theta_j) - \sum_{k:j\to k} B_{jk}(\theta_j - \theta_k) \tag{9.1b}$$

where θ_j represents the rotor angle at node j, ω_j the frequency deviation, P_j^{in} is the uncontrollable power injection, and P_j^l is the controllable load. $M_j > 0$, $D_j > 0$ are inertia and damping constants, respectively. $B_{jk} > 0$ are line parameters that depend on the reactances of line $(j, k) \in \mathcal{E}$. In fact, $P_{ij} := B_{ij}(\theta_i - \theta_j)$ represents the line flow from node i to j in the sense of DC power flow.

9.2.2 Power Imbalance

Let P_j^{in} to denote the imbalanced power in the system. It can be decomposed into two parts: a constant part and a variation part. That is

$$P_j^{in}(t) = \overline{P}_j^{in} + \tilde{q}_j(t) \tag{9.2}$$

where $\overline{P}_j^{in}$ is the known constant part, which could be the prediction of renewable generations and/or loads. $\tilde{q}_j(t)$ is the variation part, which is assumed unknown.[2]

The known constant part is easy to handle, while the variation part is nontrivial. The main idea is to further decompose it into the sum of a series of sinusoidal functions, whose parameters are unknown. Then an adaptive internal model control can be utilized to asymptotically trace these sinusoidal components and then eliminate their effects on the system frequency.

---

2 As $\overline{P}_j^{in}$ may not be accurate, the offset error of prediction is included in the component of $\tilde{q}_j(t)$. We abuse the term $\tilde{q}_j(t)$ "variation part" for simplicity.

In light of [12–15], we approximate the variation of renewable generation and load demands by a superposition of a few sinusoidal functions. Specifically, we decompose the time-varying power imbalance $\tilde{q}_j(t)$ at node j into

$$\tilde{q}_j(t) := q_{j0} + \sum_{k=1}^{s_j} q_{jk} \sin(a_{jk} \cdot t + \phi_{jk}) + w_j(t) \tag{9.3}$$

where q_{j0} is the prediction offset error (which is an unknown constant). The second term models the variation part, which is a superposition of s_j sinusoidal functions. Their amplitudes q_{jk}, frequencies $a_{jk} > 0$, and initial phases ϕ_{jk} are all unknown but belong to known bounded intervals. Here we consider only a few components of low-frequency power fluctuations, as the remaining high-frequency residuals, denoted by $w_j(t)$, are usually quite small. So we treat $w_j(t)$ as an external disturbance and do not consider its detailed model in our work but simply assume that it belongs to the $\mathcal{L}_2^{\mathrm{T}}$ space, i.e. for any $w_j(t)$ ($j \in \mathcal{N}$), $\int_0^T ||w_j(t)||^2 \mathrm{d}t < +\infty$ holds for all $0 < T < +\infty$.

9.2.3 Equivalent Transformation of Power Imbalance

We further investigate the dominant part in $\tilde{q}_j(t)$. Denote

$$q_j(t) := q_{j0} + \sum_{k=1}^{s_j} q_{jk} \sin(a_{jk} \cdot t + \phi_{jk}) \tag{9.4}$$

Then we show that $q_j(t)$ can be expressed as the output of an exosystem. To this end, define

$$\begin{aligned} \lambda_{j1} &:= q_j(t) \\ \lambda_{jk} &:= \left(\frac{\mathrm{d}}{\mathrm{d}t}\right)^{k-1} q_j(t) \quad (2 \le k \le \bar{s}_j) \end{aligned} \tag{9.5}$$

where $\bar{s}_j := 2s_j + 1$. Then $q_j(t)$ is just the output of the following dynamic system [16, 17]:

$$\dot{\lambda}_j = A_j(\alpha_j)\lambda_j \tag{9.6a}$$

$$q_j(t) = [1 \ \mathbf{0}_{1\times 2s_j}] \cdot \lambda_j \tag{9.6b}$$

where

$$\begin{aligned} \lambda_j &:= [\lambda_{j1}, \dots, \lambda_{j,\bar{s}_j}]^{\mathrm{T}} \\ A_j(\alpha_j) &:= \begin{bmatrix} \mathbf{0}_{2s_j\times 1} & I_{2s_j} \\ 0 & \alpha_{j1}, 0, \dots, \alpha_{js_j}, 0 \end{bmatrix} \end{aligned} \tag{9.7}$$

with

$$\alpha_{j1} = -\prod_{l=1}^{s_j} a_{jl}^2$$

$$\alpha_{j2} = -\sum_{k=1}^{s_j}\prod_{\substack{l=1\\ l\neq k}}^{s_j} a_{jl}^2$$

$$\vdots$$

$$\alpha_{js_j} = -\sum_{l=1}^{s_j} a_{jl}^2$$

where the parameters a_{jl} are defined in (9.4).

To facilitate the controller design, a transformation is constructed as follows. Let $R_j := [r_{i1}, \dots, r_{i,\bar{s}_j-1}, 1]$, such that all the roots of polynomial $\tau^{\bar{s}_j-1} + r_{i,\bar{s}_j-1}\tau^{\bar{s}_j-2} + \cdots + r_{i2}\tau + r_{i1}$ have negative real parts. Then define a vector as

$$\tilde{A}_j(\alpha_j) := R_j(I_{\bar{s}_j} + A_j(\alpha_j)),$$

and construct the following matrix

$$O_j(\alpha_j) := \left[\tilde{A}_j^{\mathrm{T}}(\alpha_j), \dots, \left(\tilde{A}_j(\alpha_j)A_j^{\bar{s}_j-1}(\alpha_j)\right)^{\mathrm{T}}\right]^{\mathrm{T}}.$$

In [18], it is proved that $O_j(\alpha_j)$ is nonsingular. Moreover

$$O_j^{-1}(\alpha_j)A_j(\alpha_j)O_j(\alpha_j) = A_j(\alpha_j)$$

Let $\varphi_j := O_j^{-1}(\alpha_j)\lambda_j$. Then we have

$$\dot{\varphi}_j = A_j(\alpha_j)\varphi_j \tag{9.8a}$$

$$q_j(t) = \tilde{A}_j(\alpha_j)\varphi_j \tag{9.8b}$$

So far, $q_j(t)$ is written as the output of a new exosystem (9.8a). However, the elements in $A_j(\alpha_j)$ and $\tilde{A}_j(\alpha_j)$ are still unknown. According to the definition of $q_j(t)$ and the boundedness of q_{jk}, a_{jk}, we know λ_j is also bounded. Again, φ_j is bounded due to the nonsingularity of $O_j^{-1}(\alpha_j)$.

From (9.2), (9.3), and (9.4), $P_j^{in}(t)$ is composed of three parts: $\overline{P}_j^{in}$, $q_j(t)$, and $w_j(t)$. We will address them in different ways one by one, giving rise to the following three subproblems:

- **P1**: balancing $\overline{P}_j^{in}$ economically across the physical network in a distributed way over the cyber network.
- **P2**: coping with the variation of $q_j(t)$ locally.
- **P3**: attenuating the negative impact of external disturbance $w_j(t)$.

Remark 9.1 ***(Timescales)*** The above three subproblems can be interpreted from the perspective of multi-timescale nature of power systems. **P1** is considered in the long-term operation, i.e. the system should operate economically in steady state, where the timescale is from several minutes or longer. **P2** belongs to the short-term control problem with a timescale of several seconds, where the low-frequency variation should be eliminated by designing a proper controller. The timescale of **P3** is even faster than that of **P2**, where the controller cannot track the high-frequency disturbance accurately. In this situation, we need to attenuate its negative impact via controller design. Thus, we resolve the distributed frequency control problem under time-varying power imbalance systematically in three different timescales, which coincides with subproblems **P1–P3**.

9.3 Controller Design

In this section, the known steady-state part $\overline{P}_j^{in}$ is optimally balanced across all areas using a consensus-based distributed control, which resolves **P1**. Then the effect of variation part $q_j(t)$ is eliminated locally by using a supplementary internal model controller, resolving **P2**. In terms of **P3**, here we do not design a specific controller to deal with $w_j(t)$. Instead, we show that the proposed controller can effectively attenuate $w_j(t)$, as discussed in Section 9.5.

9.3.1 Controller for Known $\overline{P}_j^{in}$

Following the idea of optimization-guided control, first we formulate an optimization model for the optimal load control problem:

$$\text{OLC:} \quad \min_{P_j^l} \quad \sum_{j \in \mathcal{N}} \frac{1}{2} \beta_j \cdot \left(P_j^l\right)^2 \tag{9.9a}$$

$$\text{s. t.} \quad \sum_{j \in \mathcal{N}} \overline{P}_j^{in} = \sum_{j \in \mathcal{N}} P_j^l \tag{9.9b}$$

where $\beta_j > 0$ is a constant. The control goal of each area is to minimize the regulation cost of the controllable load, which is in a quadratic form [8]. (9.9b) is the power balance constraint. Suppose for $\tilde{q}_j(t) = 0$. We design a consensus-based controller [8]:

$$P_j^l = \frac{\mu_j}{\beta_j} \tag{9.10a}$$

$$\dot{\mu}_j = -\sum_{k \in N_{cj}} (\mu_j - \mu_k) + \frac{\omega_j}{\beta_j} \tag{9.10b}$$

In (9.10a), μ_j are the consensus variables, and $-\mu_j$ is for the marginal costs of individual controllable loads. In the steady state, all μ_j should converge to an identical value for all controllable loads when ω_j converges to zero.

This simple controller can restore the frequency and minimize the regulation cost of the controllable loads when $\tilde{q}_j(t) = 0$. However, a fast time-varying $\tilde{q}_j(t)$ may undermine the controller. Next we construct an adaptive internal model controller to deal with $\tilde{q}_j(t)$.

9.3.2 Controller for Time-Varying Power Imbalance

In this subsection, an adaptive internal model control is supplemented to mitigate $q_j(t)$, which is given by

$$P_j^l = \frac{\mu_j}{\beta_j} + [d_j\omega_j + \tilde{A}_j(\hat{\alpha}_j)\zeta_j] \tag{9.11a}$$

$$\dot{\mu}_j = -\sum_{k\in N_{cj}} (\mu_j - \mu_k) + \omega_j/\beta_j \tag{9.11b}$$

$$\begin{aligned}\dot{\eta}_j &= -\eta_j + \overline{P}_j^{in} - P_j^l - D_j\omega_j \\ &\quad + \sum_{i:i\to j} B_{ij}(\theta_i - \theta_j) - \sum_{k:j\to k} B_{jk}(\theta_j - \theta_k)\end{aligned} \tag{9.11c}$$

$$\dot{\zeta}_j = A_j(\hat{\alpha}_j)\zeta_j - G_j(\eta_j + R_j\zeta_j) \tag{9.11d}$$

$$\dot{\hat{\alpha}}_j = -k_\alpha \Lambda_j(\zeta_j)(\eta_j + R_j\zeta_j) \tag{9.11e}$$

where $k_\alpha > 0, \gamma > 0$ are constant coefficients and

$$G_j = [\ \mathbf{0}_{1\times(\bar{s}_j-2)},\ 1,\ \gamma\ ]^{\mathrm{T}},$$

$$\Lambda_j(\zeta_j) = [\ \zeta_{j2},\ \zeta_{j4},\ \dots,\ \zeta_{j,\bar{s}_j-1}\ ]^{\mathrm{T}}.$$

Here, (9.11b) is the same as (9.10b), which is used to achieve the consensus of μ_j and restore the system frequency. Dynamics of $\eta_j, \zeta_j, \hat{\alpha}_j$ are derived from the adaptive internal model. Comparing (9.11c) and (9.1b), we have

$$\dot{\eta}_j = -\eta_j + M_j\dot{\omega}_j - \tilde{q}_j(t)$$

which implies that η_j is intended to estimate unknown $\tilde{q}_j(t)$ by using local frequency measurements. ζ_j reproduces the dynamics of φ_j in (9.8a). $\hat{\alpha}_j$ is the estimation of α_j. It should be noted that $\tilde{A}_j(\hat{\alpha}_j)\zeta_j$ in (9.11a) are the estimated values of $\tilde{A}_j(\alpha_j)\varphi_j$, i.e. q_j, in (9.8b). It will be proved in Section 9.4 that $\hat{\alpha}_j = \alpha_j$ and $\zeta_j = \varphi_j$ hold in a steady state, leading to $q_j = \tilde{A}_j(\hat{\alpha}_j)\zeta_j$.

In the controller (9.11a), μ_j/β_j allocates $\overline{P}_j^{in}$ economically; $\tilde{A}_j(\hat{\alpha}_j)\zeta_j$ is the output of the internal model, which is used to eliminate $q_j(t)$ asymptotically; and $d_j\omega_j$ is used to guarantee the stability and enhance the robustness of the controller. It is illustrated in Section 9.6 that a low-order internal model control suffices to trace and compensate for the power variation fairly well.

9.3.3 Closed-Loop Dynamics

Combining (9.1) with (9.11) and omitting $w_j(t)$, we obtain a closed-loop system. Since only angle difference between two areas is concerned, we use $\tilde{\theta}_{ij} := \theta_i - \theta_j$ as the new state variable and then perform the following transformation:

$$\tilde{\eta}_j := R_j\varphi_j + \eta_j \tag{9.12}$$

$$\tilde{\zeta}_j := \zeta_j - \varphi_j \tag{9.13}$$

$$\tilde{\alpha}_j := \hat{\alpha}_j - \alpha_j \tag{9.14}$$

Their time derivatives are

$$\begin{aligned}
\dot{\tilde{\eta}}_j &= R_j\dot{\varphi}_j + \dot{\eta}_j \\
&= R_jA_j(\alpha_j)\varphi_j - \eta_j + \overline{P}_j^{in} + \sum_{i:i\to j} B_{ij}\tilde{\theta}_{ij} \\
&\quad - \sum_{k:j\to k} B_{jk}\tilde{\theta}_{jk} - D_j\omega_j - \left(\frac{\mu_j}{\beta_j} + d_j\omega_j + \tilde{A}_j(\hat{\alpha}_j)\zeta_j\right) \\
&= -\tilde{\eta}_j + \tilde{A}_j(\alpha_j)\varphi_j - \tilde{A}_j(\hat{\alpha}_j)\zeta_j + \overline{P}_j^{in} + \sum_{i:i\to j} B_{ij}\tilde{\theta}_{ij} \\
&\quad - \sum_{k:j\to k} B_{jk}\tilde{\theta}_{jk} - D_j\omega_j - \frac{\mu_j}{\beta_j} - d_j\omega_j
\end{aligned} \tag{9.15a}$$

$$\begin{aligned}
\dot{\tilde{\zeta}}_j &= \dot{\zeta}_j - \dot{\varphi}_j = A_j(\hat{\alpha}_j)\zeta_j - G_j(\eta_j + R_j\zeta_j) - A_j(\alpha_j)\varphi_j \\
&= (A_j(\hat{\alpha}_j) - A_j(\alpha_j))\zeta_j + A_j(\alpha_j)(\zeta_j - \varphi_j) \\
&\quad - G_j(\eta_j + R_j\zeta_j - R_j\varphi_j + R_j\varphi_j) \\
&= (A_j(\hat{\alpha}_j) - A_j(\alpha_j))\zeta_j + (A_j(\alpha_j) - G_jR_j)\tilde{\zeta}_j - G_j\tilde{\eta}_j
\end{aligned} \tag{9.15b}$$

$$\begin{aligned}
\dot{\tilde{\alpha}}_j &= \dot{\hat{\alpha}}_j - \dot{\alpha}_j = -k_\alpha\Lambda_j(\zeta_j)(\eta_j + R_j\zeta_j) \\
&= -k_\alpha\Lambda_j(\zeta_j)(\tilde{\eta}_j + R_j\tilde{\zeta}_j)
\end{aligned} \tag{9.15c}$$

Define

$$\rho_{qj} := \tilde{A}_j(\alpha_j)\varphi_j - \tilde{A}_j(\hat{\alpha}_j)\zeta_j$$

Then the closed-loop system is converted into

$$\dot{\tilde{\theta}}_{ij} = \omega_i - \omega_j \tag{9.16a}$$

$$\dot{\omega}_j = \frac{1}{M_j}\Bigg(\overline{P}_j^{in} + \rho_q + \sum_{i:i\to j} B_{ij}\tilde{\theta}_{ij} - \sum_{k:j\to k} B_{jk}\tilde{\theta}_{jk} - \frac{\mu_j}{\beta_j} - d_j\omega_j - D_j\omega_j\Bigg) \tag{9.16b}$$

$$\dot{\mu}_j = -\sum_{k\in N_{cj}} (\mu_j - \mu_k) + \frac{\omega_j}{\beta_j} \tag{9.16c}$$

$$\dot{\tilde{\eta}}_j = -\tilde{\eta}_j + \overline{P}_j^{in} + \rho_q + \sum_{i:i\to j} B_{ij}\tilde{\theta}_{ij} - \sum_{k:j\to k} B_{jk}\tilde{\theta}_{jk} - D_j\omega_j - \frac{\mu_j}{\beta_j} - d_j\omega_j \tag{9.16d}$$

$$\dot{\tilde{\zeta}}_j = (A_j(\hat{\alpha}_j) - A_j(\alpha_j))\zeta_j + (A_j(\alpha_j) - G_jR_j)\tilde{\zeta}_j - G_j\tilde{\eta}_j \tag{9.16e}$$

$$\dot{\tilde{\alpha}}_j = -k_\alpha \Lambda_j(\zeta_j)(\tilde{\eta}_j + R_j\tilde{\zeta}_j) \tag{9.16f}$$

The new closed-loop system (9.16) is equivalent to the original system (9.1) and (9.11). We hence alternatively analyze the equilibrium point and stability of the equivalent system (9.16).

9.4 Equilibrium and Stability Analysis

In this section, we analyze the equilibrium and stability of the closed-loop system (9.16) when the noise $w_j(t)$ is *not* considered.

9.4.1 Equilibrium

First we define the equilibrium point of the closed-loop system (9.16).

Definition 9.1 A point $(\tilde{\theta}^*, \omega^*, \mu^*, \tilde{\eta}^*, \tilde{\zeta}^*, \tilde{\alpha}^*)$[3] is an *equilibrium point* or an *equilibrium* of the closed-loop system (9.16) if the right-hand side of (9.16) vanishes at $(\tilde{\theta}^*, \omega^*, \mu^*, \tilde{\eta}^*, \tilde{\zeta}^*, \tilde{\alpha}^*)$.

The next theorem shows that the two subproblems **P1** and **P2** are solved simultaneously at the equilibrium point.

Theorem 9.1 *At the equilibrium point, x^*, of the closed-loop dynamics (9.16), the following assertions are true:*

3 Given a collection of y_i for i in a certain set Y, y denotes the column vector $y := (y_i, i \in Y)$ of a proper dimension with y_i as its components.

1) $\tilde{\eta}_j^* = \tilde{\zeta}_j^* = \tilde{\alpha}_j^* = 0$, *which implies that* $q_j(t)$ *is accurately estimated.*
2) *System frequency recovers to its nominal value, i.e.* $\omega_j^* = 0$ *for all* $j \in \mathcal{N}$.
3) *The marginal controllable load costs satisfy* $\mu_j^* = \mu_k^*$ *for all* $j, k \in \mathcal{N}$.

Proof: Considering an equilibrium point, we have

$$0 = \omega_i^* - \omega_j^* \tag{9.17a}$$

$$0 = \overline{P}_j^{in} + \rho_q^* + \sum_{i:i\to j} B_{ij}\tilde{\theta}_{ij}^* - \sum_{k:j\to k} B_{jk}\tilde{\theta}_{jk}^* - \frac{\mu_j^*}{\beta_j} - (d_j + D_j)\omega_j^* \tag{9.17b}$$

$$0 = -\sum_{k\in N_{cj}} (\mu_j^* - \mu_k^*) + \frac{\omega_j^*}{\beta_j} \tag{9.17c}$$

$$0 = -\tilde{\eta}_j^* + \overline{P}_j^{in} + \rho_q^* + \sum_{i:i\to j} B_{ij}\tilde{\theta}_{ij}^* - \sum_{k:j\to k} B_{jk}\tilde{\theta}_{jk}^* - \frac{\mu_j^*}{\beta_j} - (d_j + D_j)\omega_j^* \tag{9.17d}$$

$$0 = (A_j(\hat{\alpha}_j^*) - A_j(\alpha_j))\zeta_j + (A_j(\alpha_j) - G_jR_j)\tilde{\zeta}_j^* - G_j\tilde{\eta}_j^* \tag{9.17e}$$

$$0 = -k_\alpha \Lambda_j(\zeta_j)(\tilde{\eta}_j^* + R_j\tilde{\zeta}_j^*) \tag{9.17f}$$

We have $\tilde{\eta}_j^* = 0$ due to (9.17d) and (9.17b). Then (9.7) directly yields

$$A_j(\hat{\alpha}_j^*) - A_j(\alpha_j) = \begin{bmatrix} \mathbf{0}_{2s_j\times 1} & \mathbf{0}_{2s_j} \\ 0 & \tilde{\alpha}_{j1}, 0, \ldots, \tilde{\alpha}_{js_j}, 0 \end{bmatrix} \tag{9.18}$$

and

$$\begin{aligned} A_j(\alpha_j) - G_jR_j &= \begin{bmatrix} \mathbf{0}_{2s_j\times 1} & I_{2s_j} \\ 0 & \alpha_{j1}, 0, \ldots, \alpha_{js_j}, 0 \end{bmatrix} - [\mathbf{0}^{\mathrm{T}}, 1, \gamma]^{\mathrm{T}} \cdot [r_{j1}, \ldots, r_{j,\bar{s}_j-1}, 1] \\ &= \begin{bmatrix} \mathbf{0}_{\bar{s}_j-2,1} & I_{\bar{s}_j-2} & \mathbf{0}_{\bar{s}_j-2,1} \\ -r_{j1} & -r_{j2}, \ldots, -r_{j,\bar{s}_j-1} & 0 \\ -\gamma r_{i1} & \alpha_{j1} - \gamma r_{j2}, -\gamma r_{j3}, \ldots, -\gamma r_{j,\bar{s}_j-1} & -\gamma \end{bmatrix} \end{aligned} \tag{9.19}$$

Then the first $(\bar{s}_j - 1)$ dimension of (9.17e) is rewritten as

$$\underbrace{\begin{bmatrix} \mathbf{0}_{\bar{s}_j-2,1} & I_{\bar{s}_j-2} \\ -r_{j1} & -r_{j2}, \dots, -r_{j,\bar{s}_j-1} \end{bmatrix}}_{\Psi} \begin{bmatrix} \tilde{\zeta}^*_{j1} \\ \vdots \\ \tilde{\zeta}^*_{j,\bar{s}_j-1} \end{bmatrix} = \mathbf{0} \tag{9.20}$$

By noting the first matrix in (9.20), denoted by Ψ, is nonsingular, we have

$$\left[\tilde{\zeta}^*_{j1}, \dots, \tilde{\zeta}^*_{j,\bar{s}_j-1}\right]^{\mathrm{T}} = \mathbf{0}$$

Denote

$$\tilde{\alpha}_j := [\tilde{\alpha}_{j1}, \dots, \tilde{\alpha}_{j,s_j}]^{\mathrm{T}}$$

Then the $\bar{s}_j$th dimension of (9.17e) together with (9.17f) yields

$$\Lambda_j^{\mathrm{T}}(\zeta_j)\tilde{\alpha}_j^* - \gamma\tilde{\zeta}^*_{j,\bar{s}_j} \equiv 0$$
$$\Lambda_j(\zeta_j)\tilde{\zeta}^*_{j,\bar{s}_j} \equiv \mathbf{0}$$

This implies $\tilde{\zeta}^*_{j,\bar{s}_j} = 0$ and $\tilde{\alpha}_j^* = \mathbf{0}$. The first assertion is proved.
From the first assertion, we have

$$p_{qj}^* = -\tilde{A}_j(\hat{\alpha}_j^*)\zeta_j + \tilde{A}_j(\alpha_j)\varphi_j = 0 \tag{9.21}$$

From (9.17a), we have $\omega_i^* = \omega_j^* = \omega_0$, with a constant ω_0. Considering the compact form of (9.17c), we have

$$-L\mu^* + \omega_0 \cdot \beta^{-1} = 0 \tag{9.22}$$

where $\beta^{-1} := [\beta_1^{-1}, \dots, \beta_n^{-1}]^{\mathrm{T}}$. Multiply $\mathbf{1}^{\mathrm{T}}$ on both sides of (9.22), and we have

$$-\mathbf{1}^{\mathrm{T}} \cdot L\mu^* + \omega_0\mathbf{1}^{\mathrm{T}} \cdot \beta^{-1} = 0 = \omega_0(\beta_1^{-1} + \dots + \beta_n^{-1}) \tag{9.23}$$

where $\mathbf{1}$ is a vector with all elements as 1 and the second equation is due to $\mathbf{1}^{\mathrm{T}} \cdot L = 0$. Thus we have $\omega_0 = 0$ due to $\beta_j > 0, \forall j$, which is the second assertion.
From (9.22), we have $L\mu^* = 0$. Equivalently, $\mu^* = \mu_0 \cdot \mathbf{1}$ with a constant μ_0, implying the third assertion. Then the proof completes. ∎

In fact, the equilibrium $(\tilde{\theta}^*, \omega^*, \mu^*, \tilde{\eta}^*, \tilde{\zeta}^*, \tilde{\alpha}^*)$ is unique, with $\tilde{\theta}^*$ being unique up to reference angles θ_0. As the optimization problem (9.9) is with a strongly convex objective function and linear constraints, its solution P_j^l is unique. Then, μ_j^* is unique by (9.10a). In Theorem 9.1, we have proved that $\omega_j^* = \tilde{\eta}_j^* = \tilde{\zeta}_j^* = \tilde{\alpha}_j^* = 0$, which are also unique. If the angle of the reference node is set as a constant θ_0, $\tilde{\theta}^*$ is also unique (see [[6], Theorem 2]). Thus, the equilibrium point of (9.16) is unique.
From the first assertion and invoking (9.12), we have $\zeta_j^* = \varphi_j, \hat{\alpha}_j^* = \alpha_j$, implying the variation $q_j(t)$ is accurately eliminated. Then the subproblem **P2** is solved. From the third assertion, **P1** is solved. Therefore, the two subproblems **P1** and **P2** are solved simultaneously.

9.4.2 Asymptotic Stability

In this subsection, we prove the asymptotic stability of the closed-loop system (9.16) when the noise $w_j(t)$ is not considered. We start with transforming it into an equivalent form.

Denote

$$\hat{\eta}_j := \tilde{\eta}_j - M_j\omega_j$$
$$v_j := [\hat{\eta}_j,\ \tilde{\zeta}_j,\ \tilde{\alpha}_j]^{\mathrm{T}}$$

Then (9.16) can be rewritten as

$$\dot{\tilde{\theta}}_{ij} = \omega_i - \omega_j \tag{9.24a}$$

$$\dot{\omega}_j = \frac{1}{M_j}\left(\overline{P}_j^{in} + \rho_{qj} + \sum_{i:i\to j} B_{ij}\tilde{\theta}_{ij} - \sum_{k:j\to k} B_{jk}\tilde{\theta}_{jk} - \frac{\mu_j}{\beta_j} - d_j\omega_j - D_j\omega_j\right) \tag{9.24b}$$

$$\dot{\mu}_j = -\sum_{k\in N_{cj}}(\mu_j - \mu_k) + \frac{\omega_j}{\beta_j} \tag{9.24c}$$

$$\dot{v}_j = \phi_j(v_j, \omega_j) \tag{9.24d}$$

where

$$\phi_j(v_j,\omega_j) = \begin{bmatrix} -\hat{\eta}_j - M_j\omega_j \\ \begin{pmatrix} (A_j(\hat{\alpha}_j) - A_j(\alpha_j))\zeta_j - G_j(\hat{\eta}_j + M_j\omega_j) \\ +(A_j(\alpha_j) - G_jR_j)\tilde{\zeta}_j \end{pmatrix} \\ -k_\alpha\Lambda_j(\zeta_j)(\hat{\eta}_j + M_j\omega_j + R_j\tilde{\zeta}_j) \end{bmatrix}$$

It is obvious that if (9.16) is stable, (9.24) is also stable. Thus, we turn to prove the stability of (9.24).

Consider the subsystem v_j, we have the following Lemma.

Lemma 9.1 *Consider the subsystem (9.24d), and let $\omega_j \equiv 0$. Then for each $j \in N$, there exists a C^1 function $U_j(t, v_j)$ such that*

$$\begin{aligned} &\underline{U}_j(v_j) \le U_j(t, v_j) \le \overline{U}_j(v_j) \\ &\frac{\partial U_j(t,v_j)}{\partial t} + \frac{\partial U_j(t,v_j)}{\partial v_j}\phi_j(v_j, 0) \le -\left\|v_j\right\|^2 \\ &\left\|\frac{\partial U_j(t,v_j)}{\partial v_j}\right\| \le b_{j0}(\left\|v_j\right\| + \left\|v_j\right\|^3) \end{aligned} \tag{9.25}$$

for some constant $b_{j0} > 0$ and positive definite and radially unbounded functions $\underline{U}_j(v_j)$, $\overline{U}_j(v_j)$.

The proof of Lemma 9.1 is similar to [[17], Lemma 3], which is omitted here.

Before giving the stability result, we first study the two Euclidean norms: $\left\|\rho_{qj}\right\|$ and $\left\|\frac{\partial U_j(t,v_j)}{\partial v_j}\left(\phi_j(v_j,\omega_j)-\phi_j(v_j,0)\right)\right\|$.

For ρ_{qj}, we have

$$\begin{aligned}\left\|\rho_{qj}\right\| &= \left\|\tilde{A}_j(\alpha_j)\varphi_j - \tilde{A}_j(\hat{\alpha}_j)(\tilde{\zeta}_j+\varphi_j)\right\| \\ &\le \left\|R_j\hat{A}_j(\tilde{\alpha}_j)\varphi_j\right\| + \left\|\tilde{A}_j(\tilde{\alpha}_j+\alpha_j)\tilde{\zeta}_j\right\| \\ &\le \left\|R_j\right\|\left\|\hat{A}_j(\tilde{\alpha}_j)\right\|\left\|\varphi_j\right\| + \left\|R_j\tilde{\zeta}_j + R_jA_j(\tilde{\alpha}_j+\alpha_j)\tilde{\zeta}_j\right\| \\ &\le \left\|R_j\right\|\left\|\hat{A}_j(\tilde{\alpha}_j)\right\|\left\|\varphi_j\right\| \qquad (9.26)\\ &\quad + \left\|R_j\tilde{\zeta}_j\right\| + \left\|R_jA_j(\alpha_j)\tilde{\zeta}_j\right\| + \left\|R_j\bar{A}_j(\tilde{\alpha}_j)\tilde{\zeta}_j\right\| \\ &\le c_2\left(\left\|v_j\right\| + \left\|v_j\right\|^2\right) \qquad (9.27)\end{aligned}$$

where

$$\hat{A}_j(\tilde{\alpha}_j) = \begin{bmatrix}\tilde{\alpha}_{j1} & 0 & \tilde{\alpha}_{j2} & 0 & \cdots & \tilde{\alpha}_{js_j} & 0\end{bmatrix}$$

$$\bar{A}_j(\tilde{\alpha}_j) = \begin{bmatrix}\mathbf{0}_{(\bar{s}_j-1)\times 1} & \mathbf{0}_{\bar{s}_j-1,\bar{s}_j-1} \\ \tilde{\alpha}_{j1} & 0,\tilde{\alpha}_{j2},0,\ldots,\tilde{\alpha}_{js_j},0\end{bmatrix}$$

$$c_2 \ge \left\|R_j\right\|\left\|\hat{A}_j(\tilde{\alpha}_j)\right\| + \left\|R_j\right\| + \left\|R_jA_j(\alpha_j)\right\|, \forall j \in N$$

The last "$\le$" is due to the boundedness of ϕ_j.

Define a set Ω_v as

$$\Omega_v := \left\{ v \mid \sum_{j\in\mathcal{N}} U_j(t,v_j) \le \tilde{c} \right\}$$

Since $U_j(t,v_j)$ is radially unbounded, there exists a constant $\bar{c}$ such that $\left\|v_j(t)\right\| \le \bar{c}$ for any $v \in \Omega_v$. In Ω_v, we have

$$\left\|\rho_{qj}\right\| = \left\|\tilde{A}_j(\alpha_j)\phi_j - \tilde{A}_j(\hat{\alpha}_j)\zeta_j\right\| \le c_3\left\|v_j\right\| \qquad (9.28)$$

for a suitable $c_3 > 0$ (defined in (9.32)).

Similarly,

$$\begin{aligned}\left\|\phi_j(v_j,\omega_j)-\phi_j(v_j,0)\right\| &= \left\|\begin{bmatrix}-M_j\omega_j \\ -G_jM_j\omega_j \\ -k_\alpha\Lambda_j(\zeta_j)M_j\omega_j\end{bmatrix}\right\| \\ &\le \left(\left\|M_j\right\| + \left\|G_j\right\|\left\|M_j\right\| + k_\alpha\left\|M_j\right\|\left\|v_j\right\|\right)\left\|\omega_j\right\| \\ &\le c_3\left\|\omega_j\right\| \qquad (9.29)\end{aligned}$$

From Lemma 9.1, we have

$$\left\| \frac{\partial U_j(t, v_j)}{\partial v_j} \right\| \leq c_3 \left\| v_j \right\| \tag{9.30}$$

Combining (9.29) and (9.30) yields

$$\left\| \frac{\partial U_j(t, v_j)}{\partial v_j} \left(\phi_j(v_j, \omega_j) - \phi_j(v_j, 0) \right) \right\| \leq \frac{1}{2} \left\| v_j \right\|^2 + \frac{1}{2} c_3^4 \left\| \omega_j \right\|^2 \tag{9.31}$$

where

$$c_3 \geq \max \left\{ 1,\ c_2(1 + \bar{c}),\ b_{j0}(1 + \bar{c}^2), \right.$$
$$\left. \left\| M_j \right\| + \left\| G_j \right\| \left\| M_j \right\| + k_\alpha \left\| M_j \right\| \bar{c} \right\},\ \forall j \in \mathcal{N} \tag{9.32}$$

To continue the analysis, we make the following assumption.

Assumption 9.1 The control parameter d_j satisfies

$$d_j > \max \left\{ \frac{1 + 2c_3^6}{2} - D_j,\ \frac{2c_3^2 + 1}{4c_3^2} + \frac{2c_3^6 - c_3^4}{2c_3^2 - 2} + 2c_3^2 - D_j \right\} \tag{9.33}$$

Assumption 9.1 is easy to satisfy by letting d_j be large enough. Denote the state variables of (9.24) as $x = \left[\tilde{\theta}^{\mathrm{T}}, \omega^{\mathrm{T}}, \mu^{\mathrm{T}}, v^{\mathrm{T}} \right]^{\mathrm{T}}$ and $x_1 = \left[\tilde{\theta}^{\mathrm{T}}, \omega^{\mathrm{T}}, \mu^{\mathrm{T}} \right]^{\mathrm{T}}$. Similar to Definition 9.1, we have

Definition 9.2 A point x^* is an *equilibrium point* of the closed-loop system (9.24) if the right-hand side of (9.24) vanishes at x^*.

Define a Lyapunov candidate function as

$$V(t, x_1, v) = \frac{1}{2c_3^2} V_1 + V_2 \tag{9.34}$$

where

$$V_1 = \frac{1}{2}(x_1 - x_1^*)^{\mathrm{T}} \Gamma (x_1 - x_1^*) \tag{9.35}$$

with $\Gamma := \operatorname{diag}\left(B, M, I_n\right)$,

$$V_2 = \sum_{j \in \mathcal{N}} U_j(t, v_j) \tag{9.36}$$

From Lemmas 9.1 and (9.35), there are positive definite and radially unbounded functions $\underline{V}(x_1, v), \overline{V}(x_1, v)$ such that $\underline{V}(x_1, v) \leq V(t, x_1, v) \leq \overline{V}(x_1, v)$. Define a set as

$$\Omega_{\overline{V}} = \left\{ (x_1, v) \mid \overline{V}(x_1, v) \leq \tilde{c} \right\}$$

Then, for any $(x_1, v) \in \Omega_{\overline{V}}$, there must be $v \in \Omega_v$ and $\left\|v_j(t)\right\| \leq \bar{c}$.
Now we are ready to establish the stability result.

Theorem 9.2 *Suppose Assumption 9.1 holds. Then every trajectory of (9.24) $x(t)$ starting from $\Omega_{\overline{V}}$ converges to x^* asymptotically.*

Proof: Define the following function:

$$h(x_1) = \begin{bmatrix} BC^{\mathrm{T}}\omega \\ \overline{P}^{in} - \beta^{-1}\mu - (D+d)\omega - CB\tilde{\theta} \\ -L\mu + \beta^{-1}\omega \end{bmatrix} \tag{9.37}$$

The derivative of V_1 is

$$\dot{V}_1 = (x_1 - x_1^*)^{\mathrm{T}} h(x_1) + \sum_{j\in\mathcal{N}} \omega_j \left(\tilde{A}_j(\alpha_j)\varphi_j - \tilde{A}_j(\hat{\alpha}_j)\zeta_j\right) \tag{9.38}$$

The first part of $\dot{V}_1$ is

$$\begin{aligned} &(x_1 - x_1^*)^{\mathrm{T}} h(x_1) \\ &= \int_0^1 (x_1 - x_1^*)^{\mathrm{T}} \frac{\partial}{\partial y} h(y(s))(x_1 - x_1^*) ds + (x_1 - x_1^*)^{\mathrm{T}} h(x_1^*) \\ &\leq \frac{1}{2}\int_0^1 (x_1 - x_1^*)^{\mathrm{T}} \left[\frac{\partial^{\mathrm{T}}}{\partial y} h(y(s)) + \frac{\partial}{\partial y} h(y(s))\right] (x_1 - x_1^*) ds \\ &= \int_0^1 (x_1 - x_1^*)^{\mathrm{T}} \left[H(y(s))\right] (x_1 - x_1^*) ds \end{aligned} \tag{9.39}$$

where $y(s) = x_1^* + s(x_1 - x_1^*)$.
The second equation is from the fact that

$$h(x_1) - h(x_1^*) = \int_0^1 \frac{\partial}{\partial y} h(y(s))(x_1 - x_1^*) ds$$

The inequality is due to either $h(x_1^*) = 0$ or $h(x_1^*) < 0$, $x_1 \geq 0$, i.e. $(x_1 - x_1^*)^{\mathrm{T}} h(x_1^*) \leq 0$. Direct calculation gives

$$\frac{\partial h(x_1)}{\partial x_1} = \begin{bmatrix} 0 & BC^{\mathrm{T}} & 0 \\ -CB & -(D+d) & -\beta^{-1} \\ 0 & \beta^{-1} & -L \end{bmatrix} \tag{9.40}$$

where $D = \mathrm{diag}(D_i)$, $d = \mathrm{diag}(d_i)$, and C is the incidence matrix of the communication graph.
Finally, H in (9.39) is

$$H = \frac{1}{2}\left[\frac{\partial^{\mathrm{T}}}{\partial x_1} h(x_1) + \frac{\partial}{\partial x_1} h(x_1)\right] = \begin{bmatrix} 0 & 0 & 0 \\ 0 & -(D+d) & 0 \\ 0 & 0 & -L \end{bmatrix}$$

The second part of $\dot{V}_1$ is

$$\begin{aligned}\sum_{j\in\mathcal{N}}\omega_j\left(\tilde{A}_j(\alpha_j)\phi_j-\tilde{A}_j(\hat{\alpha}_j)\zeta_j\right)\\ &\leq\frac{1}{2}\|\omega\|^2+\frac{1}{2}\sum_{j\in\mathcal{N}}(\tilde{A}_j(\alpha_j)\varphi_j-\tilde{A}_j(\hat{\alpha}_j)\zeta_j)^2\\ &\leq\frac{1}{2}\|\omega\|^2+\frac{1}{2}c_3^2\,\|v\|\end{aligned}\tag{9.41}$$

where the last inequality is due to (9.28).

Thus, the following inequality holds:

$$\dot{V}_1\leq\int_0^1(x_1-x_1^*)^{\mathrm{T}}\,[H(y(s))]\,(x_1-x_1^*)\mathrm{d}s+\frac{1}{2}\|\omega\|^2+\frac{1}{2}c_3^2\,\|v\|\tag{9.42}$$

The derivative of V_2 is given by

$$\begin{aligned}\dot{V}_2&=\sum_{j\in\mathcal{N}}\left(\frac{\partial U_j(t,v_j)}{\partial t}+\frac{\partial U_j(t,v_j)}{\partial v_j}\phi_j(v_j,\omega_j)\right)\\ &=\sum_{j\in\mathcal{N}}\left(\frac{\partial U_j(t,v_j)}{\partial t}+\frac{\partial U_j(t,v_j)}{\partial v_j}\phi_j(v_j,0)\right)\\ &\quad+\sum_{j\in\mathcal{N}}\left(\frac{\partial U_j(t,v_j)}{\partial v_j}\left(\phi_j(v_j,\omega_j)-\phi_j(v_j,0)\right)\right)\\ &\leq-\|v\|^2+\frac{1}{2}\|v\|^2+\frac{1}{2}c_3^4\|\omega\|^2\\ &=-\frac{1}{2}\|v\|^2+\frac{1}{2}c_3^4\|\omega\|^2\end{aligned}\tag{9.43}$$

where the inequality is due to Lemma 9.1 and (9.31).

In $\Omega_{\overline{V}}$, the derivative of V is

$$\begin{aligned}\dot{V}&\leq\frac{1}{2c_3^2}\int_0^1(x_1-x_1^*)^{\mathrm{T}}\,[H(y(s))]\,(x_1-x_1^*)\mathrm{d}s+\frac{1}{4c_3^2}\|\omega\|^2\\ &\quad+\frac{1}{4c_3^2}\sum_{j\in N}(\tilde{A}_j(\alpha_j)\phi_j-\tilde{A}_j(\hat{\alpha}_j)\zeta_j)^2-\frac{1}{2}\|v\|^2+\frac{1}{2}c_3^4\|\omega\|^2\\ &\leq-\frac{1}{4}\|v\|^2+\frac{1}{2c_3^2}\int_0^1(x_1-x_1^*)^{\mathrm{T}}\,[H(y(s))]\,(x_1-x_1^*)\mathrm{d}s\\ &\quad+\frac{1+2c_3^6}{4c_3^2}\|\omega\|^2\end{aligned}\tag{9.44}$$

Define a matrix $\tilde{H}$ as

$$\tilde{H}:=\begin{bmatrix}0&0&0\\0&-(D+d)+\frac{1+2c_3^6}{2}I_n&0\\0&0&-L\end{bmatrix}$$

Then we have

$$\dot{V} \leq -\frac{1}{4}\|v\|^2 + \frac{1}{2c_3^2}\int_0^1 (x_1 - x_1^*)^{\mathrm{T}}\, \tilde{H}\, (x_1 - x_1^*)ds. \tag{9.45}$$

Clearly, $\tilde{H} \leq 0$ holds if

$$-(D+d) + \frac{1+2c_3^6}{2}I_n < 0 \tag{9.46}$$

where I_n is an n-dimensional identity matrix. Indeed, Assumption 9.1 guarantees that the inequality (9.46) holds.

By LaSalle's invariance principle, the trajectory $x(t)$ converges to the largest invariant subset of

$$W_1 = \{x | v^* = 0, \omega = \omega^* = 0, \mu = \mu^*\}$$

Next we prove that the convergence is to an equilibrium point. Since $\omega = \omega^*$ are constants, $\tilde{\theta} = C^{\mathrm{T}}\omega^*$ are also constants. Then by [[19], Corollary 4.1], $x(t)$ converges to its equilibrium point x^* asymptotically. ■

9.5 Robustness Analysis

9.5.1 Robustness Against Uncertain Parameters

In the controller (9.11), the exact value of D_j is difficult to know and may even change. However, we claim that the inaccuracy of D_j does not influence the equilibrium point of the closed-loop system (9.16) and its stability, as we explain.

We first consider the equilibrium point. Suppose the estimation of D_j is $\hat{D}_j$ and the estimation error is $\Delta D_j := \hat{D}_j - D_j$. As $D_j > 0$, we assume its estimation $\hat{D}_j > 0$. Then (9.16d) can be rewritten as

$$\begin{aligned}\dot{\tilde{\eta}}_j = -\tilde{\eta}_j + \overline{P}_j^{in} + p_{qj} + \sum_{i:i\to j} B_{ij}\tilde{\theta}_{ij} - \sum_{k:j\to k} B_{jk}\tilde{\theta}_{jk}\\ -D_j\omega_j - \Delta D_j\omega_j - \tfrac{\mu_j}{\beta_j} - d_j\omega_j\end{aligned} \tag{9.47}$$

Since ω_j vanishes at equilibrium, ΔD_j does not influence the equilibrium point of the closed-loop system (9.16a)–(9.16c), (9.47), and (9.16e)–(9.16f).

Next, we discuss the stability under parameter uncertainty. When ΔD_j is considered, (9.24d) is rewritten as

$$\dot{v}_j = \begin{bmatrix} -\hat{\eta}_j - (M_j + \Delta D_j)\omega_j \\ \begin{pmatrix} (A_j(\hat{\alpha}_j) - A_j(\alpha_j))\zeta_j - G_j(\hat{\eta}_j + M_j\omega_j) \\ +(A_j(\alpha_j) - G_jR_j)\tilde{\zeta}_j \end{pmatrix} \\ -k_\alpha\Lambda_j(\zeta_j)(\hat{\eta}_j + M_j\omega_j + R_j\tilde{\zeta}_j) \end{bmatrix} \tag{9.48}$$

Suppose $\tilde{x}(t)$ is the state variable of (9.24a)–(9.24c) and (9.48) and $\tilde{x}^*$ is the equilibrium point. To continue the analysis, we introduce an additional assumption.

Assumption 9.2 The parameter d_j satisfies (9.33), where c_3 is given by

$$c_3 \geq \max \left\{1,\ c_2(1+\bar{c}),\ b_{j0}(1+\bar{c}^2),\right.$$
$$\left. \left\|M_j + \Delta D_j\right\| + \left\|G_j\right\| \left\|M_j\right\| + k_\alpha \left\|M_j\right\| \bar{c}\right\}$$

With Assumption 9.2, we have the following result.

Corollary 9.1 Suppose Assumption 9.2 holds, every trajectory $\tilde{x}(t)$ of (9.16a)–(9.16c), (9.47), and (9.16e)–(9.16f) starting from $\Omega_{\overline{V}}$ converges to the equilibrium point $\tilde{x}^*$ asymptotically.

Note that one can always choose a large enough d_j in controller design. Hence Corollary 9.1 can be easily proved following the line of proving Theorem 9.2, which is omitted here.

In summary, the unknown parameter D_j does not influence the equilibrium point and its stability, indicating that our controller is robust against parameter uncertainty.

9.5.2 Robustness Against Unknown Disturbances

To attenuate the effect of $w_j(t)$, one needs to guarantee that, for a given constant $\gamma > 0$, the robustness performance index $\left\|\omega_j(t)\right\|^2 \leq \gamma \left\|w_j(t)\right\|^2$ holds [[20], Chapter 16], [21]. It means that, for a bounded external disturbance $w_j(t)$, the frequency deviation is always bounded by a number γ. A smaller γ results in a better attenuation performance. The lower bound of γ (if it exists) is referred to as the L_2 gain of the system.

When considering $w_j(t)$, the closed-loop system is

$$\dot{\tilde{\theta}}_{ij} = \omega_i - \omega_j \tag{9.49a}$$

$$\dot{\omega}_j = \frac{1}{M_j}\left(\overline{P}_j^{in} + w_j(t) + \rho_{qj} + \sum_{i:i\to j} B_{ij}\tilde{\theta}_{ij} - \sum_{k:j\to k} B_{jk}\tilde{\theta}_{jk}\right.$$
$$\left. -\frac{\mu_j}{\beta_j} - d_j\omega_j - D_j\omega_j\right) \tag{9.49b}$$

$$\dot{\mu}_j = -\sum_{k\in N_{cj}} (\mu_j - \mu_k) + \frac{\omega_j}{\beta_j} \tag{9.49c}$$

$$\dot{v}_j = \phi_j(v_j, \omega_j, w_j) \tag{9.49d}$$

where

$$\tilde{\phi}_j = \begin{bmatrix} -\hat{\eta}_j - M_j\omega_j - w_j(t) \\ \begin{pmatrix} (A_j(\hat{\alpha}_j) - A_j(\alpha_j))\zeta_j - G_j(\hat{\eta}_j + M_j\omega_j) \\ +(A_j(\alpha_j) - G_jR_j)\tilde{\zeta}_j \end{pmatrix} \\ -k_\alpha \Lambda_j(\zeta_j)(\hat{\eta}_j + M_j\omega_j + R_j\tilde{\zeta}_j) \end{bmatrix}$$

By an analysis similar to (9.29), we have

$$\left\|\phi_j(v_j, \omega_j, w_j) - \phi_j(v_j, 0,0)\right\| \leq c_3 \left\|\omega_j\right\| + \left\|w_j\right\| \tag{9.50}$$

where c_3 is the same as that in (9.32). Then

$$\begin{aligned} &\left\|\frac{\partial U_j(t,v_j)}{\partial v_j}\left(\phi_j(v_j, \omega_j, w_j) - \phi_j(v_j, 0,0)\right)\right\| \\ &\quad \leq \frac{1}{2}\left\|v_j\right\|^2 + \frac{2c_3^6 - c_3^4}{2c_3^2-2}\left\|\omega_j\right\|^2 + \frac{1}{2}\left\|w_j\right\|^2 \end{aligned} \tag{9.51}$$

Using V_1, V_2 defined in (9.35) and (9.36) again, we have

$$\begin{aligned} \dot{V}_1 \leq &\int_0^1 (x_1 - x_1^*)^{\mathrm{T}} \left[H(y(s))\right] (x_1 - x_1^*)\mathrm{d}s + \frac{1}{2}c_3^2\|v\|^2 \\ &+\frac{1}{2}\|\omega\|^2 + \frac{1}{4c_3^2}\|\omega\|^2 + c_3^2\|w\|^2 \end{aligned} \tag{9.52}$$

and

$$\begin{aligned} \dot{V}_2 &\leq -\|v\|^2 + \frac{1}{2}\left\|v_j\right\|^2 + \frac{2c_3^6 - c_3^4}{2c_3^2 - 2}\|\omega\|^2 + \frac{1}{2}\|w\|^2 \\ &= -\frac{1}{2}\|v\|^2 + \frac{2c_3^6 - c_3^4}{2c_3^2 - 2}\|\omega\|^2 + \frac{1}{2}\|w\|^2 \end{aligned} \tag{9.53}$$

Using the same Lyapunov function as in (9.34) gives

$$\begin{aligned} \dot{V} \leq &-\frac{1}{4}\|v\|^2 - \frac{1}{2c_3^2}(\mu - \mu^*)^{\mathrm{T}}L(\mu - \mu^*) + \|w\|^2 \\ &-\frac{1}{2c_3^2}\omega^{\mathrm{T}}\left(D + d - \frac{2c_3^2 + 1}{4c_3^2}I_n - \frac{2c_3^6 - c_3^4}{2c_3^2 - 2}I_n\right)\omega \end{aligned} \tag{9.54}$$

It immediately leads to

$$\left\|\omega_j\right\|^2 \leq \gamma\left\|w_j\right\|^2 \tag{9.55}$$

where

$$\frac{1}{\gamma} = \min\left\{\frac{1}{2c_3^2}\left(D_j + d_j - \frac{2c_3^2 + 1}{4c_3^2} - \frac{2c_3^6 - c_3^4}{2c_3^2 - 2}\right)\right\}, \quad \forall j \in N \tag{9.56}$$

Assumption 9.1 guarantees $\gamma < 1$.

Inequalities (9.55) and (9.56) indicate that the controller is robust to $w_j(t)$ with the L_2-gain $\gamma < 1$. In practice, the amplitudes of $w_j(t)$ are usually quite small.

As a consequence, the deviation of ω_j is also small. According to (9.56), a larger d_j is helpful to enhance the attenuation performance.

The analysis above shows that the controller is robust in terms of uncertain parameter D_j and unknown disturbance $w_j(t)$. Therefore, the subproblem **P3** is resolved.

9.6 Case Studies

9.6.1 System Configuration

To verify the performance of the proposed controller, the New England 39-bus system with 10 generators as shown in Figure 9.1 is used for test. All simulations are conducted using the industrial professional electromagnetic transient simulator PSCAD 4.2.

We add four (aggregate) controllable loads to the system by connecting them at buses 32, 36, 38, and 39, respectively. Their initial values are set as [74.1, 52.7, 52.7, 105.4] MW. Then the system is divided into four control areas, as shown in Figure 9.1. Each area contains a controllable load. The communication graph is

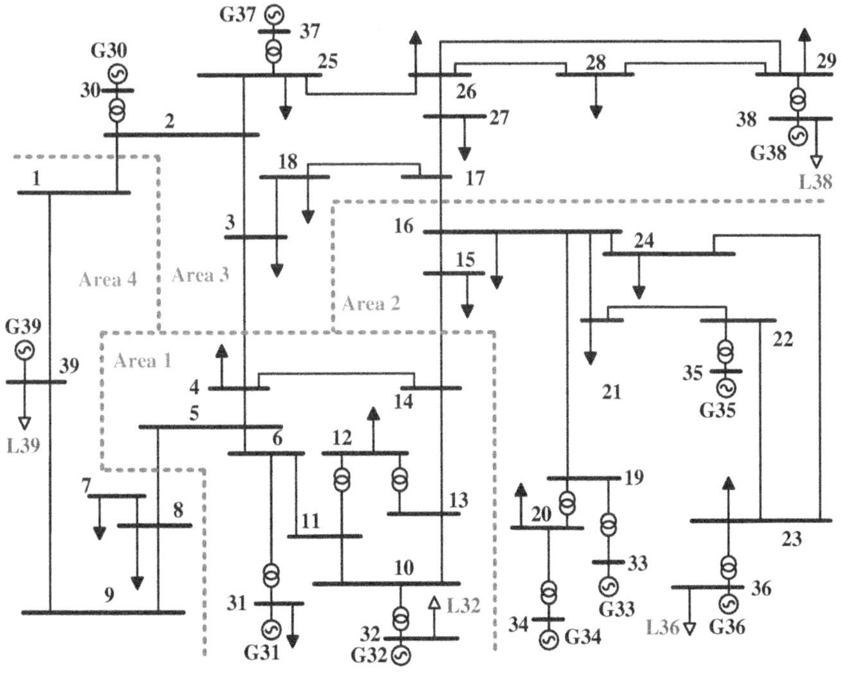

Figure 9.1 The New England 39-bus system.

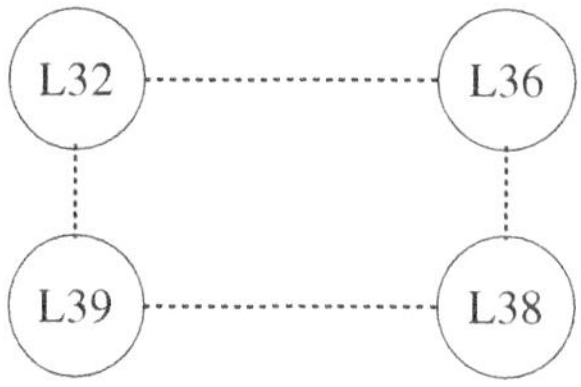

Figure 9.2 Communication graph.

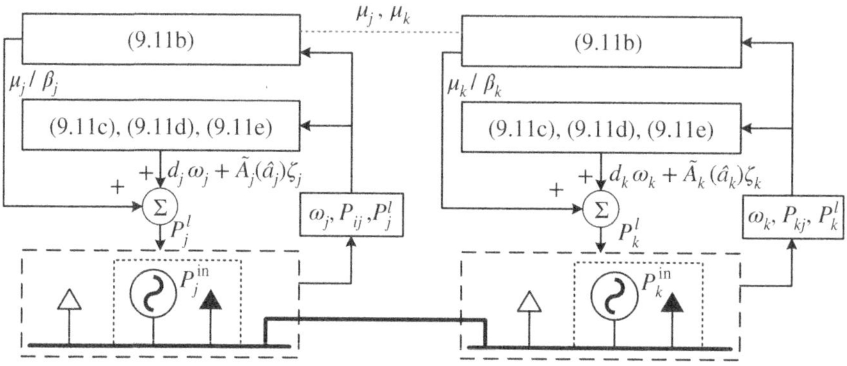

Figure 9.3 Diagram of the closed-loop system.

Table 9.1 Parameters used in the controller (9.11).

Area i	1	2	3	4
β	1	0.8	0.8	0.4
d	1000	1000	1000	1000
D	50	50	50	80
γ	1	1	1	1
k_α	10	10	10	10

undirected and connected among the controllable loads. Specifically, in the case studies, the communication topology is set as $L32 \leftrightarrow L36 \leftrightarrow L38 \leftrightarrow L39 \leftrightarrow L32$, as shown in Figure 9.2.

The closed-loop system diagram is shown in Figure 9.3. We need measure local frequency, controllable load, and tie-line power flows to compute control demands. Only μ is exchanged between neighbors (dotted line below μ_j, μ_k).

In our tests, two cases are studied based on different data sets: (i) the self-generated data in PSCAD and (ii) the real data of an offshore wind farm. The variation in the first case is faster than the latter. Parameters used in the controller (9.11) are given in Table 9.1. The values of B_{ij} are given in Table 9.2.

Table 9.2 Parameters used in the controller (9.11).

Line	(1, 2)	(1, 3)	(1, 4)	(2, 3)	(2, 4)
B_{ij}	46	47	89	112	24

The parameters R_j used in (9.11) for each area are $R_j = [1, 6, 15, 20, 15, 6, 1]$. The corresponding polynomial is $(x + 1)^6$, where all the roots are -1, satisfying the requirement.

9.6.2 Self-Generated Data

In the first case, the varying power generation in each area is shown in Figure 9.4. Note that the functions of the four curves in Figure 9.4 are unknown to the controllers. In the controller design, we choose $s_j = 3$ in (9.3). Note that this does not mean the actual power variation (curves in Figure 9.4) is the superposition of only three sinusoidal functions.

In this subsection, $\overline{P}_j^{in}$ in each area are [15, 15, 15, 15] MW, which are the prediction of aggregated load. It should be pointed out that the prediction is not accurate. The offset errors are [1, 1, 1, 5] MW, which are relatively small but unknown. We compare the performances using controller (9.10) and (9.11). Both the two controllers are applied at $t = 20$ s. The system frequencies are compared in Figure 9.5.

The dotted line stands for the frequency dynamics using (9.10). The frequency oscillation is fierce and nadir is quite low. The black line stands for frequency dynamics using (9.11). In this situation, the nominal frequency is recovered fast without oscillations. The frequency nadir is much less than that using (9.10). This result confirms that our controller can still work well when $\overline{P}_j^{in} \neq 0$.

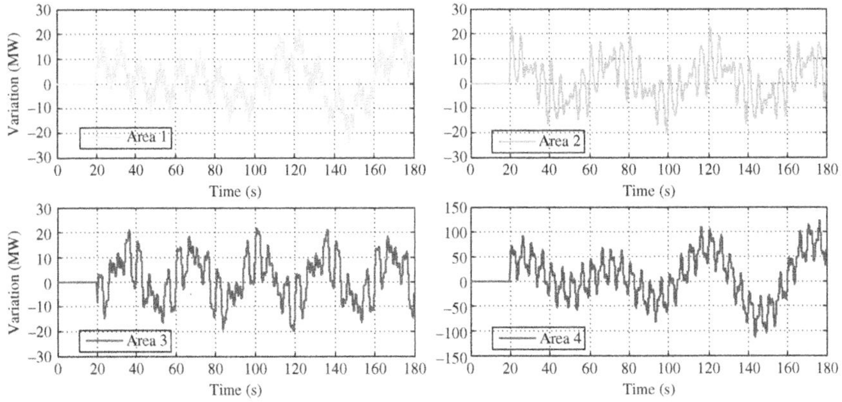

Figure 9.4 Power variation of renewable resources in each area.

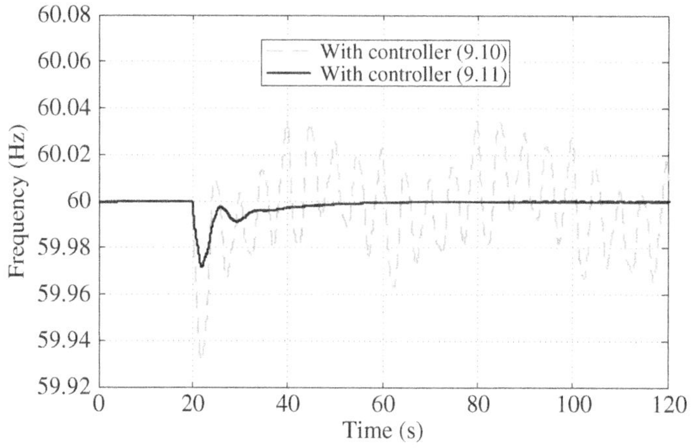

Figure 9.5 System frequencies under two controls.

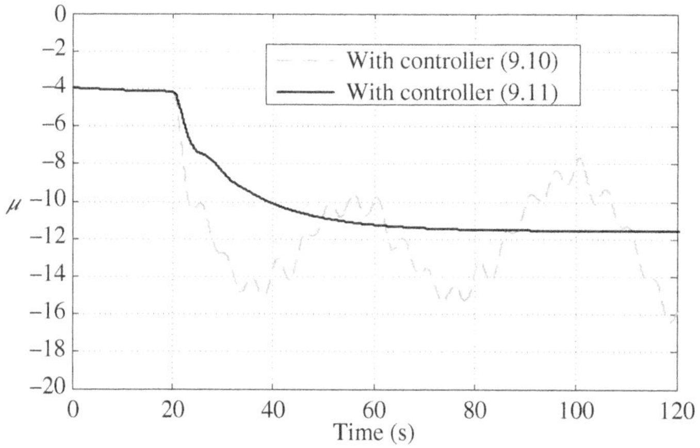

Figure 9.6 Dynamics of μ.

The dynamics of μ_j are given in Figure 9.6. The dotted line stands for μ_j using (9.10) while the black line stands for that using (9.11). μ_j of each area converges to the same value, which implies the optimality is achieved, i.e. $\overline{P}_j^{in}$ is balanced economically.

In this scenario, the controllable load in each area is also composed of two parts: a steady part to balance $\overline{P}_j^{in}$ and a variation part to mitigate the effects of $\tilde{q}_j(t)$. The steady part of controllable load is given in Figure 9.7. The controllable loads in the steady state are [63.8, 39.8, 39.8, 79.6] MW. The result is the same as that obtained using CVX to solve the optimization counterpart (i.e. OLC problem (9.9)).

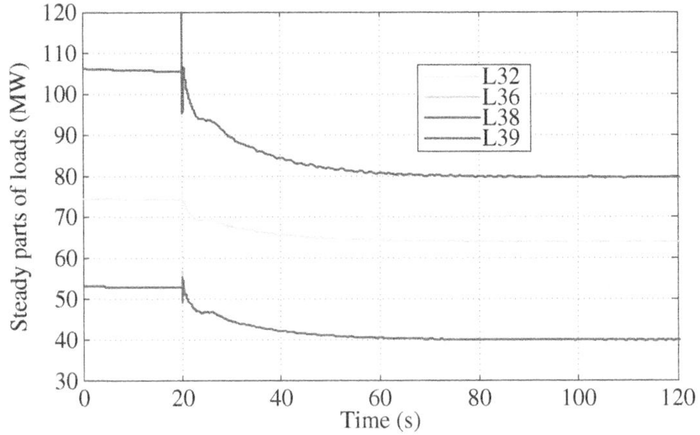

Figure 9.7 Dynamics of steady parts of controllable loads.

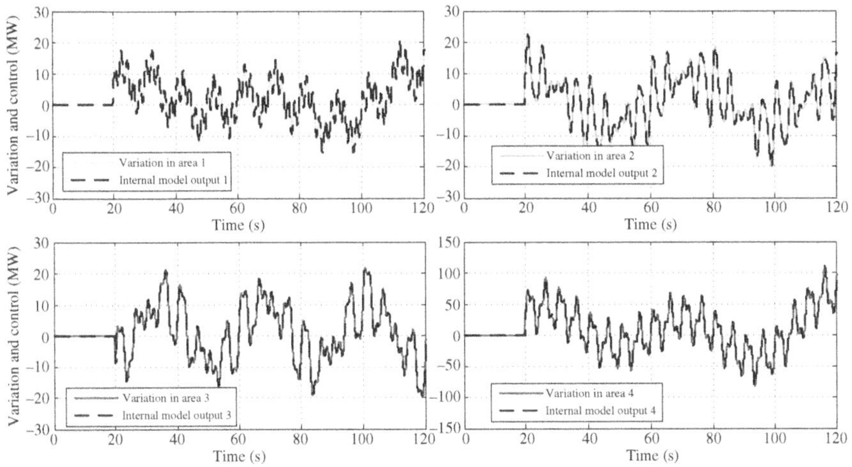

Figure 9.8 Variation and internal model output with load increases.

To demonstrate it more clearly, we define an error index Err_j as below:

$$Err_j := \frac{\int_{t_0}^{t_1} \sqrt{\left(\tilde{A}_j(\hat{\alpha}_j)\zeta_j - \tilde{q}_j(t)\right)^2} \mathrm{d}t}{\int_{t_0}^{t_1} \sqrt{\left(\tilde{q}_j(t)\right)^2} \mathrm{d}t} \tag{9.57}$$

The performance of controllable load tracking power variation in each area is given in Figure 9.8. We can find that the controllable loads coincide to the power variations with high accuracy. Again, the error index Err_j with $t_0 = 20$ and $t_1 = 120$ in this situation is [0.0084, 0.0026, 0.0057, 0.0019], which are also very small.

9.6.3 Performance Under Unknown Disturbances

To test the performance of our controller under high-frequency unknown disturbances, we add random noise $\tilde{w}(t)$ on $\tilde{q}(t)$ into the testing system, which takes the form of $\tilde{w}(t) = [20, 20, 20, 100] \times \text{rand}(t)$ MW, with rand(t) as a function generating a random number between [0, 1] at time t. In the simulation, a random number is generated every 0.01 s. The load control command and the power variations are given Figure 9.9. As the frequency of external disturbance is quite high, the internal model control is not able to follow it accurately. As a consequence, there exist obvious tracking errors. The system frequency is shown in Figure 9.10. The inset zooms into the frequency dynamics between 140 and 160 s when the system converges to the steady state. The maximal frequency deviation is smaller than 0.003 Hz, demonstrating that the unknown disturbance is well attenuated by the proposed controller.

9.6.4 Simulation with Real Data

In this subsection, we use 300 s data points for each area (one data point per second) to illustrate the effectiveness of our controller, which come from a real offshore wind farm. The data is available via the link [22]. Due to agreement with the

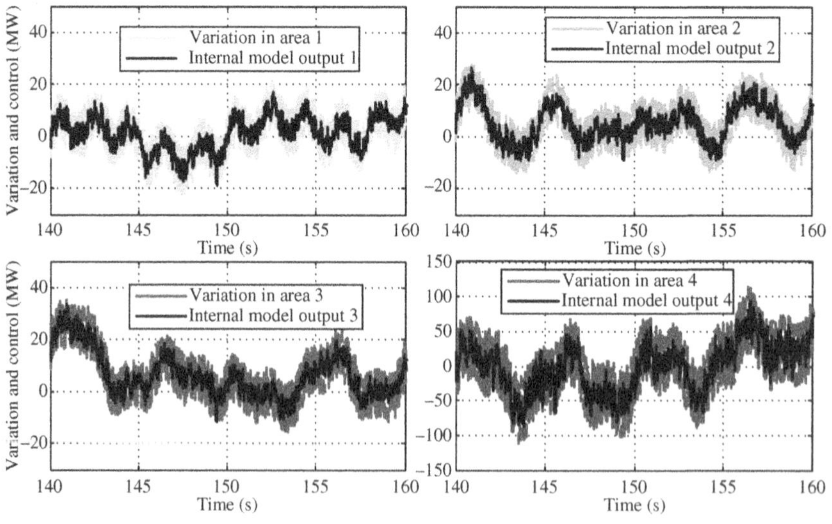

Figure 9.9 Variation and internal model output with noise.

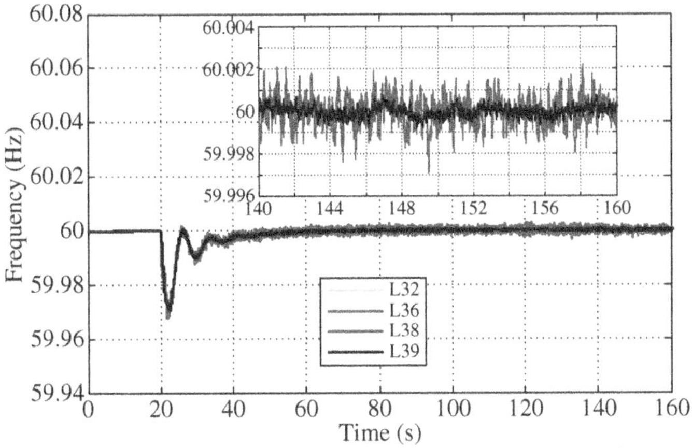

Figure 9.10 Frequency under external noise.

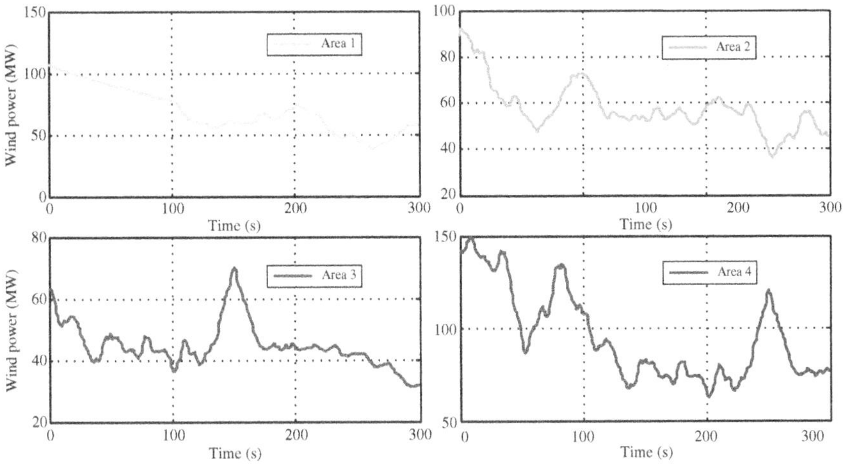

Figure 9.11 Wind power in each area.

data provider, it is for personal use only. The wind generations in the four areas are shown in Figure 9.11, which are added in the simulation at $t = 10$ s. The values of power prediction, $\overline{P}_j^{in}$, are set as [72, 60, 49, 120] MW, respectively. The frequency dynamics using the controllers (9.10) and (9.11) with the real data are given in Figure 9.12. Similar to that in Figure 9.5, the frequency under the controller (9.10) varies and cannot recover to the nominal value due to the variation of wind power. On the contrary, the frequency is very smooth when the controller (9.11) is used. The performance of controllable load tracking wind power variation in each area

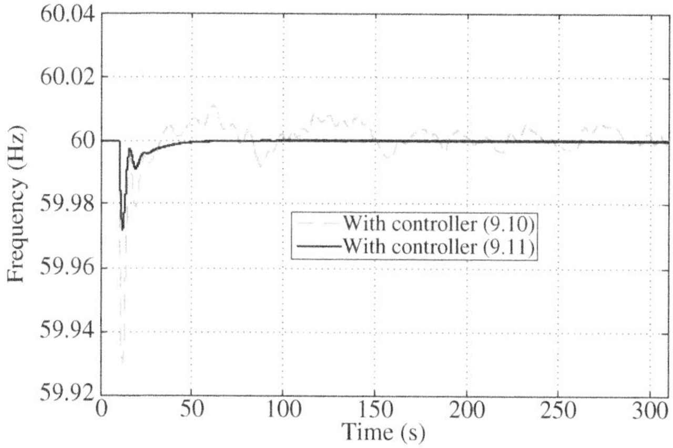

Figure 9.12 Frequency dynamics with real data.

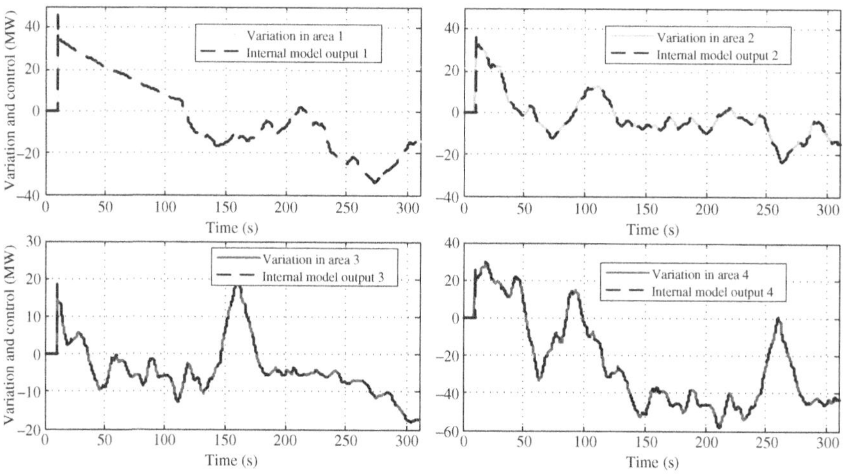

Figure 9.13 Variation and internal model output with real wind data.

is given in Figure 9.13. We can observe that the controllable loads still coincide with the generation variations accurately under the proposed controller.

9.6.5 Comparison with Existing Control Methods

First, we compare the proposed method (9.11) with droop control, and the control demand for the controllable load takes the form of $k_j^{\omega}\omega_j - k_j^P(P_j^l - P_j^{l*})$. Controller's parameters are set as $k_j^{\omega} = [5000, 5000, 5000, 8000]$, $k_j^P = [0.01, 0.01, 0.01, 0.01]$, and $P_j^{l*} = [70, 60, 50, 120]$ MW. The frequency dynamics are given in Figure 9.14a.

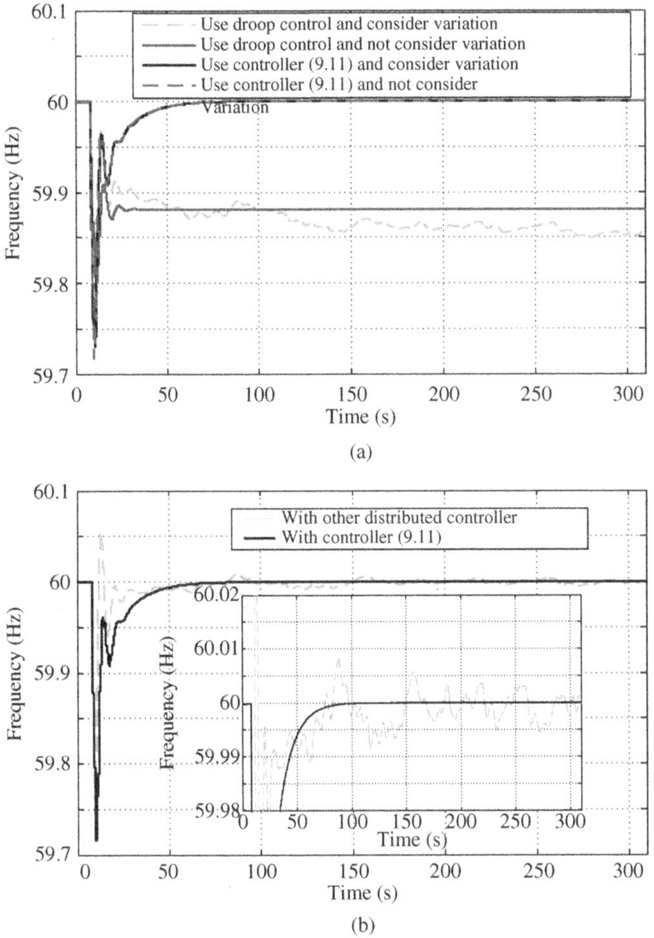

Figure 9.14 Frequency dynamics compared with (a) the droop control and (b) the other distributed controller. (a) Comparison with droop control. (b) Comparison with distributed control.

It is shown that the frequency under the droop control converges to a constant different from the nominal frequency when power imbalance variation is not considered. Therefore, the droop control cannot guarantee to restore the nominal frequency, which is in accordance with the practice. When the power imbalance variation is considered, the frequency under the droop control varies accordingly. In contrast, the nominal frequency can be restored by using the proposed controller whether the variation is considered or not. This verifies the advantage of the proposed method.

We also compare the proposed method (9.11) with the distributed controller in [2]. To make a valid comparison, we do not consider line constraints when using the controller in [2], and the objective function is the same as the proposed controller. The frequency dynamics are given in Figure 9.14b. Similarly, the frequency variation is not eliminated, demonstrating the superiority of our controller in coping with an unknown and time-varying power imbalance.

9.7 Conclusion and Notes

In this chapter, we have addressed the distributed frequency control problem of power systems in the presence of unknown and time-varying power imbalance. We have decomposed the power imbalance into three parts associated with different timescales: the known steady part, the unknown low-frequency variation, and the unknown high-frequency residual. Then the corresponding subproblems are solved under a unified control framework with three timescales:

- **The slow timescale:** designing a consensus-based distributed control to allocate the steady part of power imbalance economically.
- **The medium timescale:** devising an internal model control to accurately track and compensate for unknown time-varying power imbalance locally.
- **The fast timescale:** using the L_2-gain inequality to show the robustness of the controller against uncertain disturbances and parameters.

We have conducted numerical experiments using the data sets of the New England system and real-world wind farms. The empirical results show that the proposed distributed controller can mitigate frequency fluctuations caused by volatile load and renewable generation. The test results also confirm that the distributed optimal frequency controller outperforms existing ones.

The methodology developed in this chapter provides a systematic way to deal with unknown and time-varying power imbalance. It is interesting to combine different types of controllers in a compositional manner. Here, the multi-timescale nature of power systems plays an important role. Except for volatile renewable generation and load, power oscillations and malicious attacks on controllers can also lead to unknown and time-varying power variation in power system operations. Moreover, the proposed methodology could be extended to cope with such problems.

Bibliography

1 C. Zhao, U. Topcu, N. Li, and S. H. Low, "Design and stability of load-side primary frequency control in power systems," *IEEE Transactions on Automatic Control*, vol. 59, no. 5, pp. 1177–1189, 2014.

2 E. Mallada, C. Zhao, and S. Low, "Optimal load-side control for frequency regulation in smart grids," *IEEE Transactions on Automatic Control*, vol. 62, no. 12, pp. 6294–6309, 2017.

3 Z. Wang, F. Liu, J. Z. Pang, S. H. Low, and S. Mei, "Distributed optimal frequency control considering a nonlinear network-preserving model," *IEEE Transactions on Power Systems*, vol. 34, no. 1, pp. 76–86, 2019.

4 A. Kasis, E. Devane, C. Spanias, and I. Lestas, "Primary frequency regulation with load-side participation–Part I: stability and optimality," *IEEE Transactions on Power Systems*, vol. 32, no. 5, pp. 3505–3518, 2017.

5 Z. Wang, F. Liu, S. H. Low, C. Zhao, and S. Mei, "Distributed frequency control with operational constraints, Part I: per-node power balance," *IEEE Transactions on Smart Grid*, vol. 10, no. 1, pp. 40–52, 2019.

6 Z. Wang, F. Liu, S. H. Low, C. Zhao, and S. Mei, "Distributed frequency control with operational constraints, Part II: network power balance," *IEEE Transactions on Smart Grid*, vol. 10, no. 1, pp. 53–64, 2019.

7 L. Y. Lu, H. J. Liu, and H. Zhu, "Distributed secondary control for isolated microgrids under malicious attacks," in *2016 North American Power Symposium (NAPS)*, (Denver, USA), pp. 1–6, IEEE, Sept 2016.

8 S. Trip, M. Bürger, and C. De Persis, "An internal model approach to (optimal) frequency regulation in power grids with time-varying voltages," *Automatica*, vol. 64, pp. 240–253, 2016.

9 K. Xi, H. X. Lin, and J. H. van Schuppen, "Power-imbalance allocation control of power systems - a frequency bound for time-varying loads," in *2017 36th Chinese Control Conference (CCC)*, (Dalian, China), pp. 10528–10533, IEEE, July 2017.

10 E. Weitenberg, Y. Jiang, C. Zhao, E. Mallada, C. De Persis, and F. Dörfler, "Robust decentralized secondary frequency control in power systems: merits and tradeoffs," *IEEE Transactions on Automatic Control*, vol. 64, no. 10, pp. 3967–3982, 2018.

11 N. Li, C. Zhao, and L. Chen, "Connecting automatic generation control and economic dispatch from an optimization view,"*IEEE Transactions on Control of Network Systems*, vol. 3, no. 3, pp. 254–264, 2016.

12 P. Milan, M. Wächter, and J. Peinke, "Turbulent character of wind energy," *Physical Review Letters*, vol. 110, p. 138701, 2013.

13 A. Bušiā and S. Meyn, "Distributed randomized control for demand dispatch," in *Decision and Control (CDC), 2016 IEEE 55th Conference on*, (Las Vegas, USA), pp. 6964–6971, IEEE, 2016.

14 P. Barooah, A. Buic, and S. Meyn, "Spectral decomposition of demand-side flexibility for reliable ancillary services in a smart grid," in *System Sciences (HICSS), 2015 48th Hawaii International Conference on*, pp. 2700–2709, IEEE, 2015.

15 L. A. Aguirre, D. D. Rodrigues, S. T. Lima, and C. B. Martinez, "Dynamical prediction and pattern mapping in short-term load forecasting,"*International Journal of Electrical Power & Energy Systems*, vol. 30, no. 1, pp. 73–82, 2008.

16 G. Obregon-Pulido, B. Castillo-Toledo, and A. Loukianov, "A globally convergent estimator for N-frequencies," *IEEE Transactions on Automatic Control*, vol. 47, no. 5, pp. 857–863, 2002.

17 X. Wang, Y. Hong, P. Yi, H. Ji, and Y. Kang, "Distributed optimization design of continuous-time multiagent systems with unknown-frequency disturbances," *IEEE Transactions on Cybernetics*, vol. 47, no. 8, pp. 2058–2066, 2017.

18 D. Xu, X. Wang, and Z. Chen, "Output regulation of nonlinear output feedback systems with exponential parameter convergence," *Systems & Control Letters*, vol. 88, pp. 81–90, 2016.

19 H. K. Khalil, *Nonlinear Systems*, Upper Saddle River, NJ:Prentice Hall, 1996.

20 K. Zhou, J. C. Doyle, and K. Glover, *Robust and Optimal Control*, Upper Saddle River, NJ:Prentice hall, 1996.

21 B. Qin, X. Zhang, J. Ma, S. Deng, S. Mei, and D. J. Hill, "Input-to-state stability based control of doubly fed wind generator," *IEEE Transactions on Power Systems*, vol. 33, no. 3, pp. 2949–2961, 2017.

22 Z. Wang, "Wind power in each area," https://drive.google.com/drive/folders/1vFXvVp3-mLocxlW4jRXMnhcodOCD8S4q, September 2018.

10

Robustness and Adaptability: Partial Control Coverage in Transient Frequency Control

In the context of distributed optimal control research, one usually, for simplicity by default, assumes a complete control coverage. That means all agents can be taken under control. Nevertheless, unfortunately, this is not the case in practice as many devices in power systems are not controlled by the control center or are even not controllable. This chapter, therefore, shows how to devise distributed optimal controllers that are adaptable to such circumstances. Particularly, this chapter considers a more realistic situation where the distributed optimal controller controls only a part of the agents while others are controlled by their own local controllers, referred to as partial control coverage. In addition, nonlinear power flows and excitation voltage dynamics are included in controller design to consider large disturbances. First we design a distributed controller for the controllable generators by leveraging the primal–dual decomposition technique. Then we utilize the frequency feedback to estimate the virtual load demand of each controllable generator, circumventing the difficulty of load measuring. Finally, we establish incremental passivity conditions for the uncontrollable generators to guarantee the system-level stability under large disturbances. We prove that the closed-loop system is asymptotically stable, and its equilibrium attains the optimal solution to the associated economic dispatch (ED) problem.

10.1 Background

When designing optimal frequency controllers, the choice of power flow models, either the linear ones (usually associated with DC power flow, e.g. [1–10]) or the nonlinear ones (usually associated with AC power flow, e.g. [11–16]), is crucially important. Specifically, the closed-loop system composed of the optimal power flow (OPF) and frequency dynamics can be interpreted in a linear model as carrying out a primal–dual algorithm for solving the associated ED problem.

Merging Optimization and Control in Power Systems: Physical and Cyber Restrictions in Distributed Frequency Control and Beyond, First Edition. Feng Liu, Zhaojian Wang, Changhong Zhao, and Peng Yang.

Here, the idea of forward and reverse engineering is utilized again to facilitate the design of distributed optimal frequency controllers. However, unfortunately, it is not this case for a nonlinear model because it is sophisticated to establish the connection between optimization and control due to the nonlinearity. In addition to nonlinear nature of power flows, excitation voltage dynamics are considered in [11–13], making the model more complicated albeit more realistic. In the literature, the works of distributed optimal frequency control can be roughly divided into two categories up to different power system models: network-reduced models, e.g. [1, 7–11, 13], and network-preserving models, e.g. [2–6, 12].

In network-reduced models, generators and/or loads are aggregated and treated as one bus or control area, which are connected to each other through tie lines. In [1, 8], aggregate generators in each area are driven by automatic generation control (AGC) to restore system frequency. The literatures [7, 9–11, 13] further consider both the aggregate generators and load demands in frequency control. Despite the reduction of network greatly simplifies the analysis and design, it omits the impacts of topology on the dynamics and is practically inconvenient in applications to some extent.

In network-preserving models, generator and load buses are separately handled with different dynamic models and coupled by power flows, rendering a set of differential algebraic equations (DAEs). In [2], an optimal load control (OLC) problem is formulated, and a primary load-side control is derived as a partial primal–dual gradient algorithm for solving the OLC problem. The designing approach is extended to the scope of secondary frequency control (SFC) that restores nominal frequency in [3]. It is further generalized in [4], where sufficient passivity conditions of bus dynamics guaranteeing the system-level stability are proposed for individual local buses. Furthermore, a unified framework combining load and generator control is suggested in [5]. A similar model is considered in [6], where only limited control coverage is available. Similar to [13], the Hamiltonian method is used to analyze the network-preserving model in [12]. Compared with network-reduced models, network-preserving models describe power systems more precisely and are more suitable for analyzing the interaction among control areas. Therefore, this chapter specifically considers network-preserving model.

Most of the aforementioned works assume that all buses are controllable, and load demands at all buses are accurately measurable, especially for SFC. Moreover, it is usually assumed that the communication network has the same topology as the power network. These assumptions might be too strong to be realistic for practice. First, in a practical power system, it is common that only a part of generators and controllable load buses can participate in frequency control. Second, the communication network is usually quite different from the power grid's topology. Third, it is practical challenging to accurately measure the load injections at all buses across the overall power system. In some extreme cases, even the exact number of load buses is unknown. These issues might highlight a part of the reasons why there are so many elegant theories in this area but seldom are put into operation in practice.

In this regard, this chapter aims to devise a novel distributed frequency recovery control that allows partial control coverage. To this end, we adopt a network-preserving power system model with excitation voltage dynamics and nonlinear power flows. This model, unlike traditional works built on linearized frequency dynamics, can result in a valid controller even under large disturbances. Moreover, the controller does not need to measure loads directly, which greatly simplifies the implementation in practice. By using LaSalle's invariance principle, we can prove that the closed-loop system asymptotically converges to an equilibrium point that solves the ED problem associated with the optimal frequency control.

The rest of this chapter is organized as follows. In Section 10.2, we introduce the power system model. Section 10.3 formulates the optimal ED problem. The distributed controller is proposed in Section 10.4, and we further prove the optimality and stability of the corresponding equilibrium point in Section 10.5. The load estimation method is proposed in Section 10.6. We confirm the performance of controllers via simulations on a detailed power system in Section 10.7. Finally, Section 10.8 concludes this chapter with notes.

10.2 Structure-Preserving Model of Nonlinear Power System Dynamics

10.2.1 Power Network

A power system is composed of a number of generators and loads, which are integrated at different buses and interconnected by tie lines. In a power system without load-side control, the buses can be divided into three types, controllable generator buses, uncontrollable generator buses, and load buses.[1] Denote controllable generator buses by $\mathcal{N}_{CG} = \{1,2,\dots,n_{CG}\}$, uncontrollable generator buses by $\mathcal{N}_{UG} = \{n_{CG}+1, n_{CG}+2,\dots,n_{CG}+n_{UG}\}$, and load buses by $\mathcal{N}_L = \{n_{CG}+n_{UG}+1,\dots,n_{CG}+n_{UG}+n_L\}$. Then the set of generator buses is $\mathcal{N}_G = \mathcal{N}_{CG} \cup \mathcal{N}_{UG}$, and the set of all buses is $\mathcal{N} = \mathcal{N}_G \cup \mathcal{N}_L$. Note that both controllable and uncontrollable generator buses may have local load injections. Hence, a load is not necessarily connected at a pure load bus but a bus of any type.

Let $\mathcal{E} \subseteq \mathcal{N} \times \mathcal{N}$ be the set of tie lines, where $(i,j) \in \mathcal{E}$ if buses i and j are connected through a tie line. Then the overall system can be modeled as a connected

1 Here, we omit the pure connecting bus without generation and load for simplicity, as they can be eliminated in the network.

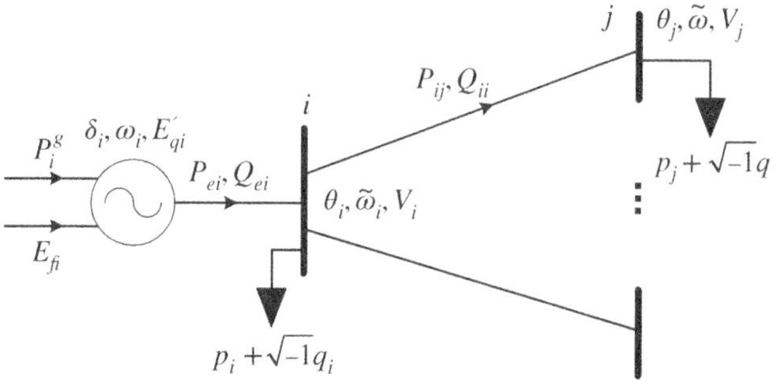

Figure 10.1 Summary of notations.

graph $\mathcal{G} = (\mathcal{N}, \mathcal{E})$. The admittance of each line is $Y_{ij} := G_{ij} + \sqrt{-1}B_{ij}$. It is common to assume $G_{ij} = 0$ in a network-preserving power system model as the resistance of a line is much smaller than its reactance. Denote the bus voltage at bus j by $V_i \angle \theta_i$, where V_i is the amplitude and θ_i is the voltage phase angle. The active and reactive powers P_{ij}, Q_{ij} from bus i to bus j are given by the following power balance equations:

$$P_{ij} = V_i V_j B_{ij} \sin\left(\theta_i - \theta_j\right) \tag{10.1a}$$

$$Q_{ij} = B_{ij} V_i^2 - V_i V_j B_{ij} \cos\left(\theta_i - \theta_j\right) \tag{10.1b}$$

For convenience, most of the notations and their relations are summarized in Figure 10.1.

10.2.2 Synchronous Generators

For the generator buses $i \in \mathcal{N}_G$, the standard third-order generator model is adopted [13, 17, 18], as shown in (10.2a)–(10.2c). Moreover, a simplified governor–turbine model and an excitation voltage control model are included as (10.2d) and (10.2e), respectively:

$$\dot{\delta}_i = \omega_i \tag{10.2a}$$

$$\dot{\omega}_i = \frac{1}{M_i}(P_i^g - D_i\omega_i - P_{ei}) \tag{10.2b}$$

$$\dot{E}'_{qi} = -\frac{E_{qi}}{T'_{d0i}} + \frac{E_{fi}}{T'_{d0i}} \tag{10.2c}$$

$$\dot{P}_i^g = -\frac{P_i^g}{T_i} + u_i^g \tag{10.2d}$$

$$\dot{E}_{fi} = h(E_{fi}, E_{qi}) \tag{10.2e}$$

In this model, M_i is the moment of inertia; D_i is the damping coefficient; T'_{d0i} is the d-axis transient time constant; T_i is the time constant of turbine dynamics; δ_i is the power angle of generator i; ω_i is the generator frequency deviation compared with an steady-state value; P_i^g is the mechanical power input; P_{ei} is the active power injected to network; E'_{qi} is the q-axis transient internal voltage; E_{qi} is the q-axis internal voltage; and E_{fi} is the excitation voltage. E_{qi} is given by

$$E_{qi} = \frac{x_{di}}{x'_{di}} E'_{qi} - \frac{x_{di} - x'_{di}}{x'_{di}} V_i \cos(\delta_i - \theta_i) \tag{10.3}$$

where x_{di} is the d-axis synchronous reactance and x'_{di} is the d-axis transient reactance.

The active and reactive powers (denoted by Q_{ei}) injected to the network are given by

$$P_{ei} = \frac{E'_{qi} V_i}{x'_{di}} \sin(\delta_i - \theta_i) \tag{10.4a}$$

$$Q_{ei} = \frac{V_i^2}{x'_{di}} - \frac{E'_{qi} V_i}{x'_{di}} \cos(\delta_i - \theta_i) \tag{10.4b}$$

For the controllable generators $i \in \mathcal{N}_{CG}$, the following capacity limits should be satisfied:

$$\underline{P}_i^g \leq P_i^g \leq \overline{P}_i^g \tag{10.5}$$

where $\underline{P}_i^g$ and $\overline{P}_i^g$ are the lower and upper limits of P_i^g, respectively.

10.2.3 Dynamics of Voltage Phase Angles

To build the network-preserving power system model, one needs to explicitly establish the relation between generators and the power network. In this chapter, the load connected at bus $i \in \mathcal{N}$ is simply modeled as a constant injection of active and reactive powers. Then the following equations are used to depict the active and reactive power balances at each bus:

$$\dot{\theta}_i = \tilde{\omega}_i, \quad i \in \mathcal{N} \tag{10.6a}$$

$$0 = P_{ei} - \tilde{D}_i \tilde{\omega}_i - p_i - \sum_{j \in N_i} P_{ij}, \quad i \in \mathcal{N}_G \tag{10.6b}$$

$$0 = -\tilde{D}_i \tilde{\omega}_i - p_i - \sum_{j \in N_i} P_{ij}, \quad i \in \mathcal{N}_L \tag{10.6c}$$

$$0 = Q_{ei} - q_i - \sum_{j \in N_i} Q_{ij}, \quad i \in \mathcal{N}_G \tag{10.6d}$$

$$0 = -q_i - \sum_{j \in N_i} Q_{ij}, \quad i \in \mathcal{N}_L \tag{10.6e}$$

where p_i, q_i are active and reactive load demands, respectively; $\tilde{\omega}_i$ is the frequency deviation at bus i, N_i is the set of buses directly connected with bus i, $\tilde{D}_i$ is the damping coefficient, and $\tilde{D}_i\tilde{\omega}_i$ is the change of frequency-dependent load [2].

In a power system, the line flow between buses i and j is mainly determined by the power angle difference between the two buses rather than the power angles independently. Therefore, we define new variables to denote angle differences as $\eta_{ii} := \delta_i - \theta_i, i \in \mathcal{N}_G$ and $\eta_{ij} := \theta_i - \theta_j, i,j \in \mathcal{N}$. The time derivative of η_{ii} and η_{ij} are given by

$$\dot{\eta}_{ii} = \omega_i - \tilde{\omega}_i, \ i \in \mathcal{N}_G \tag{10.7a}$$

$$\dot{\eta}_{ij} = \tilde{\omega}_i - \tilde{\omega}_j, \ i,j \in \mathcal{N}, i \neq j \tag{10.7b}$$

respectively. In the following analysis, we use η_{ii} and η_{ij} as the state variables instead of δ_i and θ_i.

To summarize, (10.1)–(10.4), (10.6b)–(10.6e), (10.7a), and (10.7b) constitute the network-preserving model of power systems with nonlinear dynamics, which is in a form of DAEs.

10.2.4 Communication Network

In this chapter, we consider a communication graph among the buses of controllable generators only. Denote $E \subseteq \mathcal{N}_{CG} \times \mathcal{N}_{CG}$ as the set of communication links. If generators i and j can communicate directly to each other, we say $(i,j) \in E$. Note that the communication graph E is not necessarily the same as the power network topology $\mathcal{E}$. Throughout this chapter we make the following assumption on the communication network.

Assumption 10.1 The communication graph E is undirected and connected.

10.3 Formulation of Optimal Frequency Control

10.3.1 Optimal Power-Sharing Among Controllable Generators

The purpose of optimal frequency control is to have all the controllable generators share power mismatch economically to restore the system frequency. Therefore, we have the following optimization formulation, denoted by SFC.

$$\textbf{SFC:} \quad \min_{P_i^g, i\in\mathcal{N}_{CG}} \sum_{i\in\mathcal{N}_{CG}} f_i(P_i^g) \tag{10.8a}$$

$$\text{s.t.} \quad \sum_{i\in\mathcal{N}_{CG}} P_i^g = \sum_{i\in\mathcal{N}} p_i - \sum_{i\in\mathcal{N}_{UG}} P_i^{g*} \tag{10.8b}$$

$$\underline{P}_i^g \le P_i^g \le \overline{P}_i^g, \quad i \in \mathcal{N}_{CG} \tag{10.8c}$$

where P_i^{g*} is the mechanical power of uncontrollable generator in a steady state. In (10.8a), $f_i(P_i^g)$ is the disutility subject to the controllable generation P_i^g, satisfying the following assumption.

Assumption 10.2 The disutility $f_i(P_i^g)$ is second-order continuously differentiable and strongly convex, and $f_i'(P_i^g)$ is Lipschitz continuous with a Lipschitz constant $l_i > 0$. That is, $\exists\ \alpha_i > 0$ and $l_i > 0$, such that $\alpha_i \leq f_i'(P_i^g) \leq l_i$.

To ensure the feasibility of the optimization problem, we make an additional assumption.

Assumption 10.3 The system satisfies

$$\sum_{i\in\mathcal{N}_{CG}} \underline{P}_i^g \leq \sum_{i\in\mathcal{N}} p_i - \sum_{i\in\mathcal{N}_{UG}} P_i^{g*} \leq \sum_{i\in\mathcal{N}_{CG}} \overline{P}_i^g \tag{10.9}$$

Specifically, we say Assumption A10.3 is *strictly satisfied* if all the inequalities in (10.9) *strictly* hold.

10.3.2 Equivalent Model With Virtual Load

In (10.8b), load demands at buses sometimes cannot be measured accurately, if at all. As a consequence, the values of p_i may be unknown to both the controllable generators $i, i \in \mathcal{N}_{CG}$ and the uncontrollable generators $i, i \in \mathcal{N}_{UG}$. To circumvent such an obstacle, we introduce a set of new variables, $\hat{p}_i$, to reformulate SFC as the following equivalent problem:

$$\textbf{ESFC}: \quad \min_{P_i^g, i\in\mathcal{N}_{CG}} \sum_{i\in\mathcal{N}_{CG}} f_i(P_i^g) \tag{10.10a}$$

$$\text{s.t.} \quad \sum_{i\in\mathcal{N}_{CG}} P_i^g = \sum_{i\in\mathcal{N}_{CG}} \hat{p}_i \tag{10.10b}$$

$$\underline{P}_i^g \leq P_i^g \leq \overline{P}_i^g, \quad i \in \mathcal{N}_{CG} \tag{10.10c}$$

where $\hat{p}_i$ is the *virtual load demand* supplied by generator i in the steady state, which is a constant such that

$$\sum_{i\in\mathcal{N}_{CG}} \hat{p}_i = \sum_{i\in\mathcal{N}} p_i - \sum_{i\in\mathcal{N}_{UG}} P_i^{g*} \tag{10.10d}$$

Obviously, the number of virtual loads should be equal to that of the controllable generators.

Note that the power balance constraint (10.8b) only requires that all the generators supply all the loads while it is not necessary to figure out which loads are supplied exactly by which generators. Hence we treat virtual load demands $\hat{p}_i$ as the effective demands supplied by generator i for dealing with the issue of partial control coverage.

Considering (10.10d), we immediately have the following lemma:

Lemma 10.1 *The problems SFC (10.8) and ESFC (10.10) have the same optimal solutions.*

10.4 Control Design

10.4.1 Controller for Controllable Generators

1) Controller Design Based on Primal–Dual Gradient Algorithm Invoking the primal–dual gradient algorithm, the Lagrangian of the ESFC problem (10.10) is given by

$$\begin{aligned} L = &\sum_{i\in\mathcal{N}_{CG}} f_i(P_i^g) + \mu\left(\sum_{i\in\mathcal{N}_{CG}} P_i^g - \sum_{i\in\mathcal{N}_{CG}} \hat{p}_i\right) \\ &+ \gamma_i^-(\underline{P}_i^g - P_i^g) + \gamma_i^+(P_i^g - \overline{P}_i^g), \qquad i \in \mathcal{N}_{CG} \end{aligned} \tag{10.11}$$

where $\mu, \gamma_i^-, \gamma_i^+$ are Lagrangian multipliers. Based on primal–dual update, the controller for $i \in \mathcal{N}_{CG}$ is designed as

$$u_i^g = \frac{P_i^g}{T_i} - k_{P_i^g}\left(\omega_i + (f_i'(P_i^g) + \mu - \gamma_i^- + \gamma_i^+)\right) \tag{10.12a}$$

$$\dot{\mu} = k_\mu\left(\sum_{i\in\mathcal{N}_{CG}} P_i^g - \sum_{i\in\mathcal{N}_{CG}} \hat{p}_i\right) \tag{10.12b}$$

$$\dot{\gamma}_i^- = k_{\gamma_i}\left[\underline{P}_i^g - P_i^g\right]_{\gamma_i^-}^+ \tag{10.12c}$$

$$\dot{\gamma}_i^+ = k_{\gamma_i}\left[P_i^g - \overline{P}_i^g\right]_{\gamma_i^+}^+ \tag{10.12d}$$

where $k_{P_i^g}, k_\mu, k_{\gamma_i}$ are positive constants, u_i^g is the control input, and $f_i'(P_i^g)$ is the marginal cost at P_i^g. For any $x_i, a_i \in \mathbb{R}$, the operator is defined as $[x_i]_{a_i}^+ = x_i$ if $a_i > 0$ or $x_i > 0$, and $[x_i]_{a_i}^+ = 0$ otherwise.

2) Estimating μ by a Second-Order Consensus In (10.12b), μ serves as a global coordination variable, which is a function of the mechanical powers and loads across the entire system. Here we use a second-order consensus algorithm to estimate μ locally by using neighboring information only. Specifically, for $i \in \mathcal{N}_{CG}$, the controller is revised to

$$u_i^g = \frac{P_i^g}{T_i} - k_{P_i^g}\left(\omega_i + f_i'(P_i^g) + \mu_i - \gamma_i^- + \gamma_i^+\right) \tag{10.13a}$$

$$\dot{\mu}_i = k_{\mu_i}\left(P_i^g - \hat{p}_i - \sum_{j\in N_{ci}}(\mu_i - \mu_j) - \sum_{j\in N_{ci}} z_{ij}\right) \tag{10.13b}$$

$$\dot{z}_{ij} = k_{z_i}(\mu_i - \mu_j) \tag{10.13c}$$

$$\dot{\gamma}_i^- = k_{\gamma_i}\left[\underline{P}_i^g - P_i^g\right]_{\gamma_i^-}^+ \tag{10.13d}$$

$$\dot{\gamma}_i^+ = k_{\gamma_i}\left[P_i^g - \overline{P}_i^g\right]_{\gamma_i^+}^+ \tag{10.13e}$$

where k_{μ_i}, k_{z_i} are positive constants, N_{ci} is the set of neighbors of bus i over the communication graph, and μ_i is the local estimation of μ. Here, (10.13b) and (10.13c) are used to estimate μ locally, where only neighboring information is needed. z_{ij} is an auxiliary variable to guarantee the consistency of all μ_i.

For the Lagrangian multiplier μ, $-\mu$ is often regarded as the marginal cost of generation. Theoretically, $-\mu_i$ should reach consensus for all the generators in the steady state. Since $\dot{\mu}_i = 0$ holds in the steady state, we have $P_i^g - \hat{p}_i - \sum_{j\in N_{ci}} z_{ij} = 0$. Hence, z_{ij} can be regarded as the virtual line power flow of edge (i,j) in the communication graph.

10.4.2 Active Power Dynamics of Uncontrollable Generators

To guarantee the system stability, a sufficient condition is given for calibrating the active power dynamics of uncontrollable generators as follows.

Condition 10.1 The active power dynamics of uncontrollable generators are strictly incrementally output passive with respect to the input $-\omega_i$ and the output P_i^g, i.e. there exists a continuously differentiable and positive semidefinite function S_{ω_i} such that

$$\dot{S}_{\omega_i} \le \left(-\omega_i - (-\omega_i^*)\right)\left(P_i^g - P_i^{g*}\right) - \phi_{\omega_i}(P_i^g - P_i^{g*})$$

where ϕ_{ω_i} is a positive definite function and $\phi_{\omega_i} = 0$ holds only when $P_i^g = P_i^{g*}$.

Condition 10.1 on the active power dynamics of uncontrollable generators is closely related to the incremental passivity condition, which is easy to verify. As an example, it can be verified that the commonly used primary frequency controller

$$u_i^g = -\omega_i + \omega_i^* - k_{\omega_i}(P_i^g - P_i^{g*}) + \frac{P_i^g}{T_i} \tag{10.14}$$

satisfies Condition 10.1 whenever $k_{\omega_i} > 0$. In this case, we have $S_{\omega_i} = \frac{k_1}{2}(P_i^g - P_i^{g*})^2$ with $k_1 > 0$ and $\phi_{\omega_i} = k_2(P_i^g - P_i^{g*})^2$ with $0 < k_2 \leq k_{\omega_i} \cdot k_1$.

10.4.3 Excitation Voltage Dynamics of Generators

Similar to the uncontrollable generators, the following sufficient condition on excitation voltage dynamics of all generators is given to guarantee the system stability, since we do not design specific excitation voltage controllers here.

Condition 10.2 The excitation voltage dynamics are strictly incrementally output passive in terms of the input $-E_{qi}$ and output E_{fi}, i.e. there exists continuously differentiable and positive semidefinite function S_{E_i} such that

$$\dot{S}_{E_i} \leq \left(-E_{qi} - (-E_{qi}^*)\right)\left(E_{fi} - E_{fi}^*\right) - \phi_{E_i}(E_{fi} - E_{fi}^*)$$

where ϕ_{E_i} is a positive definite function and $\phi_{E_i} = 0$ holds only when $E_{fi} = E_{fi}^*$.

Similar to Condition 10.1, Condition 10.2 is also easy to satisfy. As an example, it can be verified that the controller given in [18]

$$h(E_{fi}, E_{qi}) = -E_{fi} + E_{fi}^* - k_{E_i}(E_{qi} - E_{qi}^*) \tag{10.15}$$

with $k_{E_i} > 0$ satisfies Condition 10.2. In this case, $S_{E_i} = \frac{k_3}{2}(E_{fi} - E_{fi}^*)^2$ with $k_3 > 0$ and $\phi_{E_i} = k_4(E_{fi} - E_{fi}^*)^2$ with $0 < k_4 \leq k_{E_i} \cdot k_3$.

10.5 Optimality and Stability

After implementing the controller on the physical power system, the closed-loop system becomes

$$\begin{cases} (10.1)\text{–}(10.4), (10.6b)\text{–}(10.6e), (10.7a), (10.7b) \\ (10.13a)\text{–}(10.13e) \end{cases} \tag{10.16}$$

In this section, we prove the optimality and stability of the closed-loop dynamics (10.16).

10.5.1 Optimality

Denote the trajectory of closed-loop system by

$$v(t) := (\eta(t), \omega(t), \tilde{\omega}(t), P^g(t), \mu(t), z(t), \gamma^-(t), \gamma^+(t), E_q'(t), V(t)).$$

Define the equilibrium set of (10.16) as

$$\mathcal{V} := \{v^* | v^* \text{ is an equilibrium of (10.16)}\} \tag{10.17}$$

We first present the following Theorem.

Theorem 10.1 *Suppose Assumptions 10.1–10.3 hold. In the equilibrium of* (10.16), *the following assertions are true:*

1) *The mechanical power P_i^{g*} is feasible, i.e. it satisfies $\underline{P}_i^g \le P_i^{g*} \le \overline{P}_i^g$, $\forall i \in \mathcal{N}_{CG}$.*
2) *System frequency recovers to the nominal value, i.e. $\omega_i^* = 0, \forall i \in \mathcal{N}_{CG} \cup \mathcal{N}_{UG}$ and $\tilde{\omega}_i^* = 0, \forall i \in \mathcal{N}$.*
3) *The marginal generation costs satisfy $f_i'(P_i^{g*}) - \gamma_i^{-*} + \gamma_i^{+*} = f_j'(P_j^{g*}) - \gamma_j^{-*} + \gamma_j^{+*}, i, j \in \mathcal{N}_{CG}$.*
4) *P_i^{g*} is the unique optimal solution of SFC problem* (10.8).
5) *μ_i^* is unique if Assumption 10.3 is strictly satisfied.*

Proof: From $\dot{\gamma}_i^- = \dot{\gamma}_i^+ = 0$ in (10.13d), it follows that $\underline{P}_i^g \le P_i^{g*} \le \overline{P}_i^g$, which is the first assertion.

From $\dot{z}_{ij} = 0$ in (10.13c), we get $\mu_i^* = \mu_j^* = \mu_0$. Set $\dot{\mu}_i = 0$, and add (10.13b) for all $i \in \mathcal{N}_{CG}$. Recalling (10.10d), we have

$$\sum_{i \in \mathcal{N}_{CG}} P_i^{g*} - \sum_{i \in \mathcal{N}} p_i + \sum_{i \in \mathcal{N}_{UG}} P_i^{g*} = 0 \tag{10.18}$$

The right sides of (10.7a) and (10.7b) vanish in the equilibrium, which implies $\omega_i^* = \tilde{\omega}_i^* = \tilde{\omega}_j^* = \omega_0$.

Setting $\dot{\omega}_i = 0$ and adding (10.2b) and (10.6b) into (10.6c) yield

$$\omega_0 \sum_{i \in \mathcal{N}} \tilde{D}_i = \sum_{i \in \mathcal{N}_{CG}} P_i^{g*} - \sum_{i \in \mathcal{N}} p_i + \sum_{i \in \mathcal{N}_{UG}} P_i^{g*} = 0 \tag{10.19}$$

It implies that $\omega_0 = 0$ by noting $\tilde{D}_i > 0$, which is the second assertion.

Combining (10.13a) and (10.2d) and $\dot{P}_i^g = 0$ gives

$$f_i'(P_i^{g*}) - \gamma_i^{-*} + \gamma_i^{+*} + \omega_i^* + \mu_i^* = 0$$

Hence the third assertion is true by noting $\omega_0 = 0$, $\mu_i^* = \mu_j^* = \mu_0$.

Next we prove the fourth assertion. Since all the constraints of SFC are linear, Assumption 10.3 implies that the Slater's condition holds [[19], Chapter 5.2.3]. Moreover, the objective function is strictly convex. We only need to show that $(P_i^{g*}, \mu_0, \gamma_i^{-*}, \gamma_i^{+*})$ satisfies the KKT condition of SFC for proving the fourth assertion.

The KKT conditions of SFC problem (10.8) are

$$f_i'(P_i^{g*}) - \gamma_i^{-*} + \gamma_i^{+*} + \mu_0 = 0 \tag{10.20a}$$

$$\sum_{i\in\mathcal{N}_{CG}} P_i^{g*} - \sum_{i\in\mathcal{N}} p_i + \sum_{i\in\mathcal{N}_{UG}} P_i^{g*} = 0 \tag{10.20b}$$

$$\underline{P}_i^g \leq P_i^{g*} \leq \overline{P}_i^g \tag{10.20c}$$

$$\gamma_i^{-*} \geq 0, \gamma_i^{+*} \geq 0 \tag{10.20d}$$

$$\gamma_i^{-*}(\underline{P}_i^g - P_i^{g*}) = 0, \gamma_i^{+*}(P_i^{g*} - \overline{P}_i^g) = 0 \tag{10.20e}$$

From $\dot{\gamma}_i^- = \dot{\gamma}_i^+ = 0$, we have (10.20c), (10.20d), and (10.20e). From the third assertion, we have (10.20a). From $\dot{\omega} = 0$ and the second assertion, we have (10.20b). Therefore, the equilibrium points of the closed-loop system (10.16) satisfy the KKT conditions (10.20). It follows the fourth assertion.

If Assumption 10.3 is strictly satisfied, we know $\exists\ i \in \mathcal{N}_{CG}$ that $\gamma_i^{-*} = \gamma_i^{+*} = 0$. Then, $\mu_i^* = -f_i'(P_i^{g*})$ is uniquely determined by P_i^{g*}, implying the last assertion. The proof completes. ■

This theorem shows that the nominal frequency is recovered in the equilibrium point, and the marginal generation costs of all controllable generators are identical, implying the optimality of the equilibrium point.

10.5.2 Stability

This subsection analyzes the stability of the closed-loop dynamics (10.16). First we define a Lagrangian-like function as follows:

$$\begin{aligned}\hat{L} := &\sum_{i\in\mathcal{N}_{CG}} f_i(P_i^{g*}) + \sum_{i\in\mathcal{N}_{CG}} \mu_i(P_i^g - \hat{p}_i)\\ &- \sum_{i\in\mathcal{N}_{CG}} \mu_i z_{ij} - \frac{1}{2}\sum_{i\in\mathcal{N}_{CG}}\left(\mu_i \sum_{j\in N_i}(\mu_i - \mu_j)\right)\\ &+ \sum_{i\in\mathcal{N}_{CG}} \gamma_i^-(\underline{P}_i^g - P_i^g) + \sum_{i\in\mathcal{N}_{CG}} \gamma_i^+(P_i^g - \overline{P}_i^g)\end{aligned} \tag{10.21}$$

Let $x_1 := (P^g)$, $x_2 := (\mu, z, \gamma_i^-, \gamma_i^+)$, and $x := (x_1, x_2)$ for simplicity. It is easy to see $\hat{L}(x_1, x_2)$ is convex in x_1 and concave in x_2.

For the sake of deriving the main result, we construct a Lyapunov candidate function composed of four parts: the quadratic part, the potential energy part, and Conditions 10.1 and 10.2-related parts, as we explain.

For $i \in \mathcal{N}_G$, the quadratic part is given by

$$W_k(\omega, x) = \sum_{i\in\mathcal{N}_G} \frac{1}{2} M_i(\omega_i - \omega_i^*)^2 + \frac{1}{2}(x - x^*)^T K^{-1}(x - x^*) \tag{10.22}$$

where $K = \text{diag}(k_{P_i^g}, k_{\mu_i}, k_{z_i}, k_{\gamma_i})$ is a diagonal positive definite matrix.

Letting $x_p := (E'_{qi}, V_i, \delta_i, \theta_i)$, the potential energy part is

$$W_p(x_p) = \tilde{W}_p(x_p) - (x_p - x_p^*)^{\mathrm{T}} \nabla_{x_p} \tilde{W}_p(x_p^*) - \tilde{W}_p(x_p^*) \tag{10.23}$$

where

$$\begin{aligned} \tilde{W}_p(E'_{qi}, E_i, V_i, \delta_i, \theta_i) &= \sum_{i \in \mathcal{N}} \frac{1}{2} B_{ii} V_i^2 + \sum_{i \in \mathcal{N}} p_i \theta_i \\ &- \sum_{i \in \mathcal{N}} q_i \ln V_i - \frac{1}{2} \sum_{i \in \mathcal{N}} \sum_{j \in N_i} V_i V_j B_{ij} \cos(\theta_i - \theta_j) \\ &- \sum_{i \in \mathcal{N}_G} \frac{E'_{qi} V_i}{x'_{di}} \cos(\delta_i - \theta_i) + \sum_{i \in \mathcal{N}_G} \frac{x_{di}}{2x'_{di}(x_{di} - x'_{di})} \left(E'_{qi}\right)^2 \end{aligned} \tag{10.24}$$

The terms $\sum_{i \in \mathcal{N}_{UG}} S_{\omega_i}$ and $\sum_{i \in \mathcal{N}_G} \frac{1}{T'_{d0i}(x_{di} - x'_{di})} S_{E_i}$ directly come from Conditions 10.1 and 10.2, respectively.

Thereby a Lyapunov function can be constructed as

$$W = W_k + W_p + \sum_{i \in \mathcal{N}_{UG}} S_{\omega_i} + \sum_{i \in \mathcal{N}_G} \frac{S_{E_i}}{T'_{d0i}(x_{di} - x'_{di})} \tag{10.25}$$

To facilitate the analysis, we make the following assumption.

Assumption 10.4 The Hessian of W_p satisfies $\nabla_v^2 W_p(v) > 0$ at equilibrium.

Since the voltage phase deviation between two neighboring buses is usually fairly small in practice, Assumption 10.4 is generally satisfied in power systems. We will further explain it in the end of this subsection.

Now we get ready to present the stability result.

Theorem 10.2 *Suppose Assumptions 10.1–10.4 and Conditions 10.1–10.2 hold. Then for every equilibrium point v^*, there exists a neighborhood $\mathcal{S}$ of v^*, where all trajectories $v(t)$ satisfying* (10.16) *starting from $\mathcal{S}$ asymptotically converge to the set $\mathcal{V}$. In addition, each trajectory converges to an equilibrium point.*

Proof: Recalling (10.21), the dynamics of (10.13a)–(10.13e) can be rewritten as

$$\dot{P}_i^g = -k_{P_i^g} \cdot \left(\omega_i + \frac{\partial \hat{L}(x_1, x_2)}{\partial P_i^g} \right) \tag{10.26a}$$

$$\dot{\mu}_i = k_{\mu_i} \cdot \frac{\partial \hat{L}(x_1, x_2)}{\partial \mu_i} \tag{10.26b}$$

$$\dot{z}_{ij} = k_{z_i} \cdot \frac{\partial \hat{L}(x_1, x_2)}{\partial z_{ij}} \tag{10.26c}$$

$$\dot{\gamma}_i^- = k_{\gamma_i} \cdot \left[\frac{\partial \hat{L}(x_1, x_2)}{\partial \gamma_i^-} \right]^+_{\gamma_i^-} \tag{10.26d}$$

$$\dot{\gamma}_i^+ = k_{\gamma_i} \cdot \left[\frac{\partial \hat{L}(x_1, x_2)}{\partial \gamma_i^+} \right]^+_{\gamma_i^+} \tag{10.26e}$$

Consider the closed-loop dynamics. Following the line of reference [20], we first define two sets, σ^+ and σ^-, as follows:

$$\sigma^+ := \{ i \in \mathcal{N}_{CG} \mid \gamma_i^+ = 0,\ P_i^g - \overline{P}_i^g < 0 \} \tag{10.27a}$$

$$\sigma^- := \{ i \in \mathcal{N}_{CG} \mid \gamma_i^- = 0,\ \underline{P}_i^g - P_i^g < 0 \} \tag{10.27b}$$

Then (10.13d) and (10.13e) are equivalent to

$$\dot{\gamma}_i^+ = \begin{cases} k_{\gamma_i}(P_i^g - \overline{P}_i^g), & \text{if } i \notin \sigma^+; \\ 0, & \text{if } i \in \sigma^+. \end{cases} \tag{10.28a}$$

$$\dot{\gamma}_i^- = \begin{cases} k_{\gamma_i}(\underline{P}_i^g - P_i^g), & \text{if } i \notin \sigma^-; \\ 0, & \text{if } i \in \sigma^-. \end{cases} \tag{10.28b}$$

The derivative of W_k is

$$\begin{aligned}
\dot{W}_k &= \sum_{i \in \mathcal{N}_G} M_i(\omega_i - \omega_i^*)\dot{\omega}_i + (x - x^*)^{\mathrm{T}} \cdot K^{-1}\dot{x} \\
&\le \sum_{i \in \mathcal{N}_G} (\omega_i - \omega_i^*)(P_i^g - D_i\omega_i - P_{ei}) - \sum_{i \in \mathcal{N}_{CG}} (P_i^g - P_i^{g*})\omega_i \\
&\quad -(x_1 - x_1^*)^{\mathrm{T}} \cdot \nabla_{x_1}\hat{L} + (x_2 - x_2^*)^{\mathrm{T}} \cdot \nabla_{x_2}\hat{L} \\
&= \sum_{i \in \mathcal{N}_G} (\omega_i - \omega_i^*)\left(P_i^g - P_i^{g*} - D_i(\omega_i - \omega_i^*) - (P_{ei} - P_{ei}^*)\right) \\
&\quad - \sum_{i \in \mathcal{N}_{CG}} (P_i^g - P_i^{g*}) \cdot (\omega_i - \omega_i^*) \\
&\quad -(x_1 - x_1^*)^{\mathrm{T}} \cdot \nabla_{x_1}\hat{L} + (x_2 - x_2^*)^{\mathrm{T}} \cdot \nabla_{x_2}\hat{L} \\
&= -\sum_{i \in \mathcal{N}_G} D_i(\omega_i - \omega_i^*)^2 - \sum_{i \in \mathcal{N}_G} (\omega_i - \omega_i^*)(P_{ei} - P_{ei}^*) \\
&\quad + \sum_{i \in \mathcal{N}_{UG}} (P_i^g - P_i^{g*})(\omega_i - \omega_i^*) \\
&\quad -(x_1 - x_1^*)^{\mathrm{T}} \cdot \nabla_{x_1}\hat{L} + (x_2 - x_2^*)^{\mathrm{T}} \cdot \nabla_{x_2}\hat{L}
\end{aligned} \tag{10.29}$$

where the inequality comes from

$$\begin{aligned}
(\gamma^- - \gamma^{-*})^{\mathrm{T}}[\underline{P}^g - P^g]^+_{\gamma^-} &\le (\gamma^- - \gamma^{-*})^{\mathrm{T}}(\underline{P}^g - P^g) \\
&= (\gamma^- - \gamma^{-*})^{\mathrm{T}}\nabla_{\gamma^-}\hat{L}
\end{aligned}$$

Here the inequality holds since $\gamma_i^- = 0 \le \gamma_i^{-*}$ and $\underline{P}_i^g - P_i^g < 0$ for $i \in \sigma^-$, i.e. $(\gamma_i^- - \gamma_i^{-*}) \cdot (\underline{P}_i^g - P_i^g) \ge 0$. Similarly for $i \in \sigma^+$, we have

$$\begin{aligned}(\gamma^+ - \gamma^{+*})^{\mathrm{T}}[P^g - \overline{P}^g]^+_{\gamma^+} &\le (\gamma^+ - \gamma^{+*})^{\mathrm{T}}(P^g - \overline{P}^g)\\ &= (\gamma^+ - \gamma^{+*})^{\mathrm{T}}\nabla_{\gamma^+}\hat{L}\end{aligned}$$

From (10.6b) and (10.6c), we have

$$\begin{aligned}0 = &\sum_{i\in\mathcal{N}_G}(\tilde{\omega}_i - \tilde{\omega}_i^*)\left((P_{ei} - P_{ei}^*) - \sum_{j\in N_i}(P_{ij} - P_{ij}^*)\right)\\ &- \sum_{i\in\mathcal{N}}\tilde{D}_i(\tilde{\omega}_i - \tilde{\omega}_i^*)^2 - \sum_{i\in\mathcal{N}_L}(\tilde{\omega}_i - \tilde{\omega}_i^*)\sum_{j\in N_i}(P_{ij} - P_{ij}^*)\end{aligned} \tag{10.30}$$

Adding (10.30)–(10.29) yields

$$\begin{aligned}\dot{W}_k \le &- \sum_{i\in\mathcal{N}_G} D_i(\omega_i - \omega_i^*)^2 - \sum_{i\in\mathcal{N}}\tilde{D}_i(\tilde{\omega}_i - \tilde{\omega}_i^*)^2 + \sum_{i\in\mathcal{N}_G}(\tilde{\omega}_i - \omega_i)(P_{ei} - P_{ei}^*)\\ &- \sum_{(i,j)\in\mathcal{E}}(\tilde{\omega}_i - \tilde{\omega}_j)(P_{ij} - P_{ij}^*) + \sum_{i\in\mathcal{N}_{UG}}(P_i^g - P_i^{g*})(\omega_i - \omega_i^*)\\ &-(x_1 - x_1^*)^{\mathrm{T}} \cdot \nabla_{x_1}\hat{L} + (x_2 - x_2^*)^{\mathrm{T}} \cdot \nabla_{x_2}\hat{L}\end{aligned} \tag{10.31}$$

Since $\hat{L}$ is convex in x_1 and a concave in x_2, we have

$$\begin{aligned}&-(x_1 - x_1^*)^{\mathrm{T}} \cdot \nabla_{x_1}\hat{L}(x_1, x_2) + (x_2 - x_2^*)^{\mathrm{T}} \cdot \nabla_{x_2}\hat{L}(x_1, x_2)\\ &\le \hat{L}(x_1^*, x_2) - \hat{L}(x_1, x_2) + \hat{L}(x_1, x_2) - \hat{L}(x_1, x_2^*)\\ &= \hat{L}(x_1^*, x_2) - \hat{L}(x_1^*, x_2^*) + \hat{L}(x_1^*, x_2^*) - \hat{L}(x_1, x_2^*)\\ &\le 0\end{aligned} \tag{10.32}$$

where the first inequality holds because $\hat{L}$ is convex in x_1 and concave in x_2 while the second inequality holds because (x_1^*, x_2^*) is a saddle point. Therefore, we have

$$\begin{aligned}\dot{W}_k \le &- \sum_{i\in\mathcal{N}_G} D_i(\omega_i - \omega_i^*)^2 - \sum_{i\in\mathcal{N}}\tilde{D}_i(\tilde{\omega}_i - \tilde{\omega}_i^*)^2\\ &- \sum_{i\in\mathcal{N}_G}(\omega_i - \tilde{\omega}_i)(P_{ei} - P_{ei}^*) - \sum_{(i,j)\in\mathcal{E}}(\tilde{\omega}_i - \tilde{\omega}_j)(P_{ij} - P_{ij}^*)\\ &+ \sum_{i\in\mathcal{N}_{UG}}(P_i^g - P_i^{g*})(\omega_i - \omega_i^*)\end{aligned} \tag{10.33}$$

The partial derivatives of $W_p(x_p)$ are

$$\nabla_{E'_{qi}} W_p = \frac{1}{x_{di} - x'_{di}}(E_{qi} - E_{qi}^*) \tag{10.34a}$$

$$\nabla_{V_i} W_p = 0 \tag{10.34b}$$

$$\nabla_{\delta_i} W_p = P_{ei} - P_{ei}^* \tag{10.34c}$$

$$\nabla_{\theta_i} W_p = \sum_{(i,j)\in\mathcal{E}} (P_{ij} - P_{ij}^*) - \sum_{i\in\mathcal{N}_G} (P_{ei} - P_{ei}^*) \tag{10.34d}$$

The derivative of W_p is

$$\begin{aligned}\dot{W}_p = &\sum_{i\in\mathcal{N}_G} \frac{(E_{qi} - E_{qi}^*)(E_{fi} - E_{fi}^*)}{T'_{d0i}(x_{di} - x'_{di})} - \sum_{i\in\mathcal{N}_G} \frac{(E_{qi} - E_{qi}^*)^2}{T'_{d0i}(x_{di} - x'_{di})} \\ &+ \sum_{i\in\mathcal{N}_G} (\omega_i - \tilde{\omega}_i)(P_{ei} - P_{ei}^*) + \sum_{(i,j)\in\mathcal{E}} (\tilde{\omega}_i - \tilde{\omega}_j)(P_{ij} - P_{ij}^*)\end{aligned} \tag{10.35}$$

We further consider the Lyapunov function defined in (10.25), which is

$$W = W_k + W_p + \sum_{i\in\mathcal{N}_{UG}} S_{\omega_i} + \sum_{i\in\mathcal{N}_G} \frac{1}{T'_{d0i}(x_{di} - x'_{di})} S_{E_i} \tag{10.36}$$

Calculating its derivative gives

$$\begin{aligned}\dot{W} &= \dot{W}_k + \dot{W}_p + \sum_{i\in\mathcal{N}_{UG}} \dot{S}_{\omega_i} + \sum_{i\in\mathcal{N}_G} \frac{\dot{S}_{E_i}}{T'_{d0i}(x_{di} - x'_{di})} \\ &\leq -\sum_{i\in\mathcal{N}_G} D_i(\omega_i - \omega_i^*)^2 - \sum_{i\in\mathcal{N}} \tilde{D}_i(\tilde{\omega}_i - \tilde{\omega}_i^*)^2 \\ &\quad - \sum_{i\in\mathcal{N}_G} \frac{(E_{qi} - E_{qi}^*)^2}{T'_{d0i}(x_{di} - x'_{di})} \\ &\quad + \sum_{i\in\mathcal{N}_{UG}} (P_i^g - P_i^{g*})(\omega_i - \omega_i^*) + \sum_{i\in\mathcal{N}_{UG}} \dot{S}_{\omega_i} \\ &\quad + \sum_{i\in\mathcal{N}_G} \frac{(E_{qi} - E_{qi}^*)(E_{fi} - E_{fi}^*)}{T'_{d0i}(x_{di} - x'_{di})} + \sum_{i\in\mathcal{N}_G} \frac{\dot{S}_{E_i}}{T'_{d0i}(x_{di} - x'_{di})} \\ &\leq -\sum_{i\in\mathcal{N}_G} D_i(\omega_i - \omega_i^*)^2 - \sum_{i\in\mathcal{N}} \tilde{D}_i(\tilde{\omega}_i - \tilde{\omega}_i^*)^2 \\ &\quad - \sum_{i\in\mathcal{N}_G} \frac{(E_{qi} - E_{qi}^*)^2}{T'_{d0i}(x_{di} - x'_{di})} - \sum_{i\in\mathcal{N}_{UG}} \phi_{\omega_i} - \sum_{i\in\mathcal{N}_G} \phi_{E_i} \\ &\leq 0\end{aligned} \tag{10.37}$$

The last two inequalities are due to Assumption 10.4.

To prove the locally asymptotic stability of the closed-loop system, we need to show $W > 0$ for all $v \in S \setminus v^*$. It is equivalent to prove $\nabla_v^2 W > 0$ for all $v \in S \setminus v^*$, which is exactly indicated by Assumption 10.4.

Consequently, there exists a neighborhood of v^* $\{\, v | W(v) \leq \epsilon \,\}$ for a sufficiently small $\epsilon > 0$ so that $\nabla_v^2 W(v) > 0$. Hence, there exists a compact set S around v^* contained in this neighborhood, which is positively invariant. Let $Z_1 := \{\, v | \dot{W}(v) = 0 \,\}$. By the LaSalle's invariance principle, each of the trajectories $v(t)$ starting from

$\mathcal{S}$ converges to the largest invariant set Z^+ contained in $\mathcal{S} \cap Z_1$. From the analysis above, if $\dot{W}(v) = 0$, v is an equilibrium point of the closed-loop dynamics (10.16). Hence, v converges to $Z^+ \in \mathcal{V}$.

Finally, we prove that each trajectory $v(t)$ starting from $\mathcal{V}$ converges to a point by following the line of proof of Theorem 1 in [13]. Since $v(t)$ is bounded, its ω-limit set $\Omega(v) \neq \emptyset$. For the sake of contradiction, suppose there exist two different points in $\Omega(v)$, i.e. $v_1^*, v_2^* \in \Omega(v), v_1^* \neq v_2^*$. Since the Hessian of $W_1(v), W_2(v)$ are both positive definite in $\mathcal{S}$, there exist $W_1(v), W_2(v)$ defined by (10.25) with respect to v_1^*, v_2^* and scalars $c_1 > 0, c_2 > 0$ such that the two sets satisfying $W_1^{-1}(\leq c_1) := \{v | W_1(v) \leq c_1\}, W_2^{-1}(\leq c_2) := \{v | W_2(v) \leq c_2\}$ are disjoint (i.e. $W_1^{-1}(\leq c_1) \cap W_2^{-1}(\leq c_2) = \emptyset$) and compact. In addition, $W_1^{-1}(\leq c_1), W_2^{-1}(\leq c_2)$ are positively invariant. By (10.37), there exists a finite time $t_1 > 0$ such that $v(t) \in W_1^{-1}(\leq c_1)$ for $\forall t \geq t_1$. Similarly, there exists a finite time $t_2 > 0$ such that $v(t) \in W_2^{-1}(\leq c_2)$ for $\forall t \geq t_2$. This implies that $v(t) \in W_1^{-1}(\leq c_1) \cap W_2^{-1}(\leq c_2)$ for $\forall t \geq \max(t_1, t_2)$, which is a contradiction. This completes the proof. ■

With the argument above, we further explain the rationale of Assumption 10.4 by referring to [[21], Lemma 3]. Reference [21] investigates the control of inverter-based microgrids based on a network-preserving model while we extend some results to more complicated synchronous-generator based bulk power systems. For simplicity, we first present some notations following [21]. Comparing (10.1) with (10.4), P_{ei}, Q_{ei} have the same structure as P_{ij}, Q_{ij}, respectively. Therefore, one can treat the reactance of a generator as a line with the admittance of $1/x'_{di}$, which connects $i \in \mathcal{N}_G$ and inner node of the generator. We denote the inner nodes of generators as $\mathcal{N}'_G$. Then, an augmented power network is constructed, where its incidence matrix is denoted by $\hat{C}$. The set of nodes in the augmented power network is denoted by $\mathcal{N}' = \mathcal{N} \cup \mathcal{N}'_G$. Let $\hat{V} = (E'_q, V)$ and $\hat{\theta} = (\delta, \theta)$, and denote by $|\hat{C}|$ the matrix obtained from $\hat{C}$ by replacing all the elements c_{ij} with $|c_{ij}|$. Define a matrix as $\Gamma(\hat{V}) := \text{diag}(|B_{ij}|V_iV_j), i, j \in \mathcal{N}'$. Also, define A as

$$A_{ij} = \begin{cases} -|B_{ij}| \cos(\theta_i - \theta_j), & i \neq j, i, j \in \mathcal{N}; \\ \text{diag}(|B_{ii}|), & i = j, i, j \in \mathcal{N}; \\ -\cos(\delta_i - \theta_j)/x'_{di}, & i \in \mathcal{N}'_G, j \in \mathcal{N}_G \end{cases} \tag{10.38}$$

For simplicity, we use the following notation. For an n-dimensional vector $r := \{r_1, r_2, \ldots, r_n\}$, the diagonal matrix $\text{diag}(r_1, r_2, \ldots, r_n)$ is denoted in short by $[r]_D$. Moreover, $\mathbf{cos}(\cdot), \mathbf{sin}(\cdot)$ are defined component-wise.

From the definition of W in (10.25), $\nabla_v^2 W(v) > 0$ if and only if $\nabla_v^2 W_p(v) > 0$, i.e. the matrix

$$\begin{bmatrix} \Gamma(\hat{V})[\mathbf{cos}(\hat{C}^{\mathrm{T}}\hat{\theta})]_D & [\mathbf{sin}(\hat{C}^{\mathrm{T}}\hat{\theta})]_D \Gamma(\hat{V})|\hat{C}|^{\mathrm{T}}[\hat{V}]_D^{-1} \\ [\hat{V}]_D^{-1}|\hat{C}|\Gamma(\hat{V})[\mathbf{sin}(\hat{C}^{\mathrm{T}}\hat{\theta})]_D & A + H(\hat{V}) \end{bmatrix} \tag{10.39}$$

is positive definite, where

$$H(\hat{V}) = \begin{bmatrix} \left[\frac{x_{di}}{2x'_{di}(x_{di}-x'_{di})}\right]_D & 0 \\ 0 & [q_i/V_i^2]_D \end{bmatrix} \tag{10.40}$$

Now we can explain the rationale of Assumption 10.4 more clearly. Note that, the phase-angle difference between two neighboring nodes is usually small at a steady state of power systems. In addition, the difference between δ_i and θ_i is also small. This observation implies that the matrix in (10.39) is diagonal dominant, rendering the positive definiteness of the Hessian matrix. In this context, it is easy to understand that Assumption 10.4 is usually satisfied and makes sense.

10.6 Implementation With Frequency Measurement

10.6.1 Estimating μ_i Using Frequency Feedback

Note that the virtual load demands $\hat{p}_i$ used in the controller (10.13) are difficult to directly measure or estimate in practice. Lemma 10.1 indicates that any profile of $\hat{p}_i$ is valid as long as the power balance equation

$$\sum_{i\in\mathcal{N}_{CG}} \hat{p}_i = \sum_{i\in\mathcal{N}} p_i - \sum_{i\in\mathcal{N}_{UG}} P_i^{g*} \tag{10.41}$$

holds. Noticing that the power imbalance is very small in a normal operation, we have

$$\sum_{i\in\mathcal{N}_{CG}} P_{ei} \approx \sum_{i\in\mathcal{N}} \hat{p}_i$$

In fact, they are identical in steady state. Hence, we approximately specify $\hat{p}_i = P_{ei}$, which implies

$$P_i^g - \hat{p}_i = P_i^g - P_{ei} = M_i\dot{\omega}_i + D_i\omega_i.$$

That leads to an estimation algorithm of μ_i as follows.

$$\begin{aligned}\dot{\mu}_i = k_{\mu_i}\Bigg(&-\sum_{j\in N_i}(\mu_i-\mu_j) - \sum_{j\in N_i} z_{ij} + M_i\dot{\omega}_i + D_i\omega_i \\ &+\tau_i(-\mu_i - f_i'(P_i^g) + \gamma_i^- - \gamma_i^+)\Bigg)\end{aligned} \tag{10.42}$$

where $0 < \tau_i < 4/l_i$.

In this way, we only need to measure local frequencies ω_i at each bus, rather than directly measuring the load demands of all buses across the entire system. Since the controller only requires the information of μ_i from the neighboring buses over the communication graph, it becomes much easier to implement.

Now, we modify the closed-loop dynamics by replacing (10.13b) with (10.42) in (10.16), which yields

$$\begin{cases} (10.1)\text{–}(10.4), (10.6b)\text{–}(10.6e), (10.7a), (10.7b) \\ (10.13a), (10.13c)\text{–}(10.13e), (10.42) \end{cases} \tag{10.43}$$

Regarding the closed-loop dynamics (10.43), we have the following lemma to show the optimality of its equilibrium point.

Lemma 10.2 *Assertions 1)–5) in Theorem 10.1 hold for the equilibrium of the closed-loop dynamics* (10.43).

Proof: From $\dot{z}_{ij} = 0$, we have $\mu_i^* = \mu_j^* = \mu_0$. Setting $\dot{\mu}_i = 0$ and adding (10.42) for all $i \in \mathcal{N}_{CG}$ yield

$$\sum_{i \in \mathcal{N}_{CG}} M_i \dot{\omega}_i + \sum_{i \in \mathcal{N}_{CG}} (D_i + \tau_i)\omega_i = 0 \tag{10.44}$$

The right sides of (10.7a) and (10.7b) vanish at the equilibrium point, implying $\omega_i^* = \tilde{\omega}_i^* = \tilde{\omega}_j^* = \omega_0$. When $\dot{\omega} = 0$ in (10.44), we have

$$\omega_0 \sum_{i \in \mathcal{N}_{CG}} (D_i + \tau_i) = 0 \tag{10.45}$$

This implies that $\omega_0 = 0$ due to $D_i + \tau_i > 0$.

The rest of the proof is the same as that in Theorem 10.1, which is omitted here. ∎

10.6.2 Stability Analysis

Recalling (10.2b), then (10.42) can be converted into

$$\begin{aligned} \dot{\mu}_i = k_{\mu_i} \Bigg(& P_i^g - \hat{p}_i + \hat{p}_i - P_{ei} - \sum_{j \in N_i} (\mu_i - \mu_j) - \sum_{j \in N_i} z_{ij} \\ & + \tau_i(-\mu_i - f_i'(P_i^g) + \gamma_i^- - \gamma_i^+) \Bigg) \end{aligned} \tag{10.46}$$

Define $\rho_i := \hat{p}_i - P_{ei} = P_{ei}^* - P_{ei}$, which stands for the difference between the electric power and its value in the steady state. We further make the following assumption:

Assumption 10.5 The disturbance can be written as $\rho_i = \beta_i(t)\omega_i$, where $|\beta_i(t)| \leq \overline{\beta}_i$ and $\overline{\beta}_i$ is a positive constant. In addition, the set $\{ t < \infty \mid \omega_i(t) = \omega_i^* \}$ is of measure zero except the equilibrium point.

Whenever $\omega_i \neq \omega_i^*$, there exists such $\beta_i(t)$. Assumption 10.5 argues that $\omega_i(t) = \omega_i^*$ only happens at isolated points except the equilibrium point. Generally, this assumption is reasonable in power systems.

Denote the state variable of (10.43) and its equilibrium set by $\tilde{v}$ and $\tilde{\mathcal{V}}$, respectively. Then we have the following theorem to characterize the closed-loop stability.

Theorem 10.3 *Suppose Assumptions 10.1–10.5 and Conditions 10.1–10.2 hold and the constraint (10.9) is not binding. Then for every $\tilde{v}^*$, there exists a neighborhood, S, of $\tilde{v}^*$ where all trajectories $\tilde{v}(t)$ satisfying (10.43) starting from S converge to the set $\tilde{\mathcal{V}}$ if*

$$\bar{\beta}_i < \sqrt{\tau_i D_i (4 - \tau_i l_i)} \tag{10.47}$$

Moreover, the convergence of such trajectories is to an equilibrium point.

Proof: We still use the Lyapunov function (10.25) to analyze the stability of the closed-loop system (10.43). Denote $y = (\omega_i^{\mathrm{T}}, x_1^{\mathrm{T}}, x_2^{\mathrm{T}})^{\mathrm{T}}, i \in \mathcal{N}_{CG}$, and define the following function:

$$\tilde{f}(y) = \begin{bmatrix} -D_i \omega_i \\ -\left(f_i'(P_i^g) + \mu_i - \gamma_i^- + \gamma_i^+\right) \\ f_{\mu_i} \\ \mu_i - \mu_j \\ \underline{P}_i^g - P_i^g \\ P_i^g - \overline{P}_i^g \end{bmatrix}, \quad i \in \mathcal{N}_{CG} \tag{10.48}$$

where[2]

$$f_{\mu_i} = P_i^g - \hat{p}_i - \sum_{j \in N_{ci}} (\mu_i - \mu_j) - \sum_{j \in N_{ci}} z_{ij} - \tau_i \mu_i - \tau_i f_i'(P_i^g) + \beta_i \omega_i$$

The derivative of W_k is

$$\begin{aligned} \dot{W}_k &= \sum_{i \in \mathcal{N}_G} M_i (\omega_i - \omega_i^*) \dot{\omega}_i + (x - x^*)^{\mathrm{T}} \cdot K^{-1} \dot{x} \\ &= \sum_{i \in \mathcal{N}_G} (\omega_i - \omega_i^*)(P_i^g - P_i^{g*} - D_i(\omega_i - \omega_i^*) - (P_{ei} - P_{ei}^*)) \\ &\quad + (x - x^*)^{\mathrm{T}} \cdot K^{-1} \dot{x} \\ &\leq \sum_{i \in \mathcal{N}_G} (\omega_i - \omega_i^*)(P_i^g - P_i^{g*} - (P_{ei} - P_{ei}^*)) \\ &\quad - \sum_{i \in \mathcal{N}_{UG}} D_i (\omega_i - \omega_i^*)^2 + (y - y^*)^{\mathrm{T}} \tilde{f}(y) \end{aligned} \tag{10.49}$$

where the inequality is due to the same reason for that in (10.29).

2 Without confusion, we abuse β_i instead of $\beta_i(t)$ for simplification. In addition, if (10.9) is not binding, we can omit γ_i^-, γ_i^+ in the neighborhood of equilibrium point.

Divide $\dot{W}_k$ into two parts: $\dot{W}_{k1} = (y - y^*)^{\mathrm{T}}\tilde{f}(y)$ and $\dot{W}_{k2} = \dot{W}_k - \dot{W}_{k1}$. Then we first analyze the sign of $\dot{W}_{k1}$:

$$\begin{aligned}
\dot{W}_{k1} &= (y - y^*)^{\mathrm{T}}\tilde{f}(y) \\
&= \int_0^1 (y - y^*)^{\mathrm{T}} \frac{\partial}{\partial \tilde{z}} \tilde{f}(\tilde{z}(s))(y - y^*)\mathrm{d}s + (y - y^*)^{\mathrm{T}}\tilde{f}(y^*) \\
&\leq \frac{1}{2}\int_0^1 (y - y^*)^{\mathrm{T}} \left[\frac{\partial^{\mathrm{T}}}{\partial \tilde{z}}\tilde{f}(\tilde{z}(s)) + \frac{\partial}{\partial \tilde{z}}\tilde{f}(\tilde{z}(s))\right](y - y^*)\mathrm{d}s \\
&= \int_0^1 (y - y^*)^{\mathrm{T}} \left[H(\tilde{z}(s))\right](y - y^*)\mathrm{d}s
\end{aligned} \tag{10.50}$$

where $\tilde{z}(s) = y^* + s(y - y^*)$. The second equation is from the fact that

$$\tilde{f}(y) - \tilde{f}(y^*) = \int_0^1 \frac{\partial}{\partial \tilde{z}}\tilde{f}(\tilde{z}(s))(y - y^*)\mathrm{d}s.$$

The inequality is due to either $\tilde{f}(y^*) = 0$ or $\tilde{f}(y^*) < 0, y_i \geq 0$, i.e. $(y - y^*)^{\mathrm{T}}\tilde{f}(y^*) \leq 0$:

$$\frac{\partial \tilde{f}(y)}{\partial y} = -\begin{bmatrix}
D & 0 & 0 & 0 & 0 & 0 \\
0 & \nabla^2_{Pg} f(P^g) & I & 0 & -I & I \\
\beta & \tau\nabla^2_{Pg} f(P^g) - I & \tau + L_c & C & 0 & 0 \\
0 & 0 & -C^{\mathrm{T}} & 0 & 0 & 0 \\
0 & I & 0 & 0 & 0 & 0 \\
0 & -I & 0 & 0 & 0 & 0
\end{bmatrix} \tag{10.51}$$

where $D = \mathrm{diag}(D_i)$, $\beta = \mathrm{diag}(\beta_i)$, and $\tau = \mathrm{diag}(\tau_i)$, I is the identity matrix with dimension n_{CG}, C is the incidence matrix of the communication graph, and L_c is the Laplacian matrix of the communication graph.

Finally, H in (10.50) is

$$\begin{aligned}
H &= \frac{1}{2}\left[\frac{\partial^{\mathrm{T}}}{\partial y}\tilde{f}(y) + \frac{\partial}{\partial y}\tilde{f}(y)\right] \\
&= \begin{bmatrix}
-D & 0 & -\frac{1}{2}\beta & 0 & 0 & 0 \\
0 & -\nabla^2_{Pg} f(P^g) & -\frac{\tau}{2}\nabla^2_{Pg} f(P^g) & 0 & 0 & 0 \\
-\frac{1}{2}\beta & -\frac{\tau}{2}\nabla^2_{Pg} f(P^g) & -\tau - L_c & 0 & 0 & 0 \\
0 & 0 & 0 & 0 & 0 & 0 \\
0 & 0 & 0 & 0 & 0 & 0 \\
0 & 0 & 0 & 0 & 0 & 0
\end{bmatrix}
\end{aligned} \tag{10.52}$$

$H \leq 0$ holds if and only if

$$\begin{bmatrix}
-D & 0 & -\frac{1}{2}\beta \\
0 & -\nabla^2_{Pg} f(P^g) & -\frac{\tau}{2}\nabla^2_{Pg} f(P^g) \\
-\frac{1}{2}\beta & -\frac{\tau}{2}\nabla^2_{Pg} f(P^g) & -\tau
\end{bmatrix} < 0 \tag{10.53}$$

By Schur complement [22], we only need

$$-\tau_i - \begin{bmatrix} -\frac{1}{2}\beta_i & -\frac{\tau_i}{2}c_i \end{bmatrix} \begin{bmatrix} -D_i & 0 \\ 0 & -c_i \end{bmatrix}^{-1} \begin{bmatrix} -\frac{1}{2}\beta_i \\ -\frac{\tau_i}{2}c_i \end{bmatrix} < 0 \tag{10.54}$$

where $c_i = \nabla^2_{P_i^g} f(P_i^g)$. Solving (10.54) results in

$$-\sqrt{\tau_i D_i(4 - \tau_i c_i)} < \beta_i < \sqrt{\tau_i D_i(4 - \tau_i c_i)} \tag{10.55}$$

By Assumption 10.2, we know $c_i \leq l_i$. Thus we have

$$\sqrt{\tau_i D_i(4 - \tau_i l_i)} \leq \sqrt{\tau_i D_i(4 - \tau_i c_i)}$$

and

$$-\sqrt{\tau_i D_i(4 - \tau_i l_i)} \geq -\sqrt{\tau_i D_i(4 - \tau_i c_i)}$$

provided that $\tau_i > 0,\ 4 - \tau_i l_i > 0$, i.e. $0 < \tau_i < 4/l_i$. It yields

$$-\sqrt{\tau_i D_i(4 - \tau_i l_i)} < \beta_i < \sqrt{\tau_i D_i(4 - \tau_i l_i)} \tag{10.56}$$

i.e. $\overline{\beta}_i < \sqrt{\tau_i D_i(4 - \tau_i l_i)}$, implying (10.47).

The analysis of $\dot{W}_{k2}$ and $\dot{W}_p$ and the convergence to a point are similar to the proof of Theorem 10.2, which are omitted here. ■

It is worth of noting that the range of $\beta(t)$ can be large as long as l_i is small enough. For example, set $\tau_i = \frac{3}{l_i}$; then $\overline{\beta}_i(t) < \sqrt{\frac{3D_i}{l_i}}$. It is well known that the objective function can be scaled up or down by a positive scaling number k, i.e. $k \cdot \sum_{i \in \mathcal{N}_{CG}} f_i(P_i^g), k > 0$, without influencing the optimal solution. Therefore, l_i can be very small by merely choosing a small enough k. In this sense, (10.47) is not conservative.

10.7 Case Studies

10.7.1 Test System and Data

To test the proposed controller, we conduct numerical tests using the New England 39-bus system with 10 generators, as shown in Figure 10.2. In the tests, we apply (10.42) to estimate the virtual load demands. All simulations are implemented in the commercial power system simulation software power systems computer aided design (PSCAD) and are carried on a laptop with 8 GB memory and 2.39 GHz CPU.

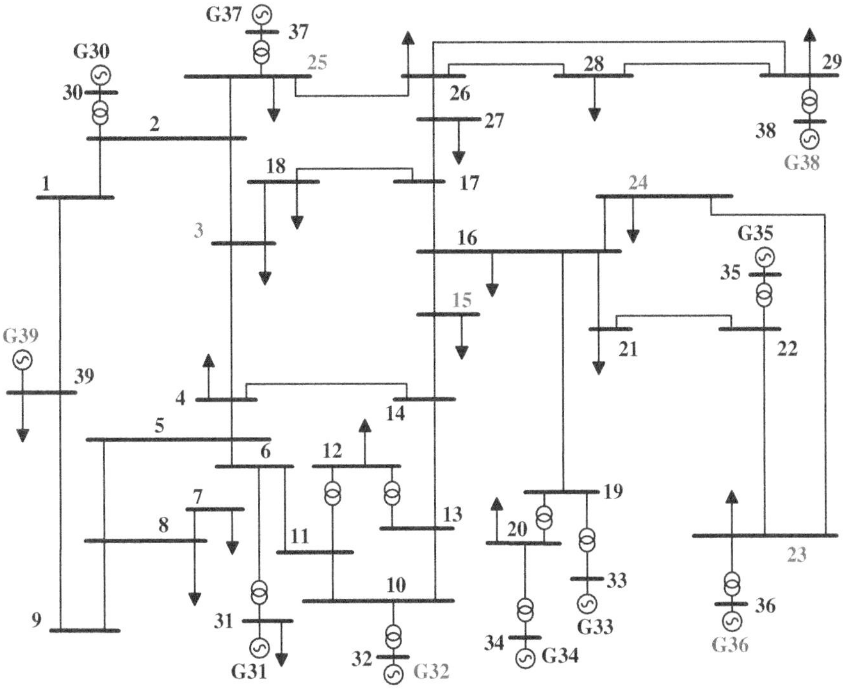

Figure 10.2 The New England 39-bus system.

We control only a subset of these generators, namely, G32, G36, G38, and G39, while the remaining are equipped with the primary frequency control given in (10.14). In particular, we apply the controller (10.13) derived based on a simple model to the much more realistic and complicated model embedded in PSCAD. The detailed electromagnetic transient model of three-phase synchronous machines (sixth-order model) is adopted to simulate the dynamics of synchronous generators with both governors and exciters. All the transmission lines and transformers take typical values of resistance and reactance. The loads are modeled as fixed values given by PSCAD. The communication graph is undirected and connected, which is set as $G32 \leftrightarrow G36 \leftrightarrow G38 \leftrightarrow G39 \leftrightarrow G32$.

The individual disutility functions of generator i are set as

$$f_i = \frac{1}{2}a_i(P_i^g)^2 + b_iP_i^g$$

which are the generation costs of generator i. Capacity limits of P_i^g and parameters a_i, b_i are given in Table 10.1.

Table 10.1 Capacity limits of generators.

Gi	$[\underline{P}_i^g, \overline{P}_i^g]$ (MW)	a_i	b_i
32	[0, 1000]	0.00009	0.032
36	[0, 1000]	0.00014	0.030
38	[0, 850]	0.00010	0.032
39	[0, 1080]	0.00008	0.032

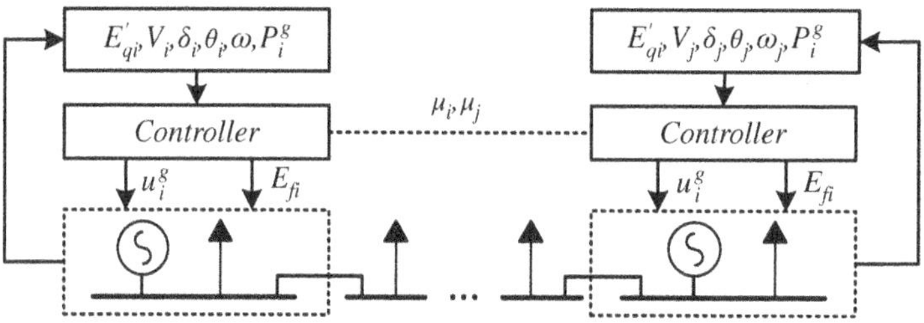

Figure 10.3 Diagram of the closed-loop system.

The closed-loop system is shown in Figure 10.3, where each generator only needs to measure the local frequency, the mechanical power, the terminal voltage, and the phase angle to compute its control command. Note that there is no load measurement and only the information of μ_i is exchanged between neighboring controllable generators.

10.7.2 Performance Under Small Disturbances

We consider the following scenario: (i) at $t = 10$ s, there is a step change of 60 MW load demands at each of buses 3, 15, 23, 24, 25, and (ii) at $t = 70$ s, there is another step change of 120 MW load at bus 23. Neither the original load demands nor their changes are known to all generators.

10.7.2.1 Equilibrium and its Optimality

It is observed that the nominal frequency is well recovered when a steady state is reached. The optimal mechanical powers are given in Table 10.2, which are identical to the optimal solutions to (10.8) being computed by centralized optimization. Stage 1 is for the period from 10 to 70 s, and Stage 2 from 70 to 130 s. The values in Table 10.2 are the generations at the end of each stage. In Stage 1,

Table 10.2 Equilibrium points.

	P^g_{32} (MW)	P^g_{36} (MW)	P^g_{38} (MW)	P^g_{39} (MW)
Stage 1	927	610	834	1043
Stage 2	968	652	850	1080

no generation reaches its limit, while in Stage 2 both G38 and G39 reach their upper limits. At the end of each stage, the marginal generation cost $-\mu_i$ of generator i converges to an identical value (see Figure 10.9), indicating the optimality of the results. The test results confirm the theoretical analyses and demonstrate that the devised controller can automatically attain optimal working points even for the more realistic and sophisticated model.

10.7.2.2 Performance of Frequency Dynamics

In this subsection, we analyze the dynamic performance of the closed-loop system. For comparison, a typical AGC is tested in the same scenario. In the AGC implementation, the signal of area control error (ACE) is given by $ACE = K_f\omega + P_{ij}$ [[23], Chapter 11.6], where K_f is the frequency response coefficient, ω is the frequency deviation, and P_{ij} is the deviation of tie line power. In the case studies, we can treat the whole system as one control area, implying $P_{ij} = 0$. Hence, the control center computes $ACE = K_f\omega$ and allocates it to the generators participating in AGC, i.e. G32, G36, G38, and G39. In this situation, the control command of each controllable generator is $-r_i \cdot ACE$, where $\sum_i r_i = 1$. In this chapter, we set $r_i = 0.25$ for $i = 1,2,3,4$.

The trajectories of frequencies are given in Figure 10.4, where the upper part stands for the devised distributed optimal controller and the lower one for the AGC. It is observed that the frequencies are recovered to the nominal value under both controls. The frequency drops under the two controls are very similar while the recovery time under the proposed control is fairly less than that under the typical AGC.

Although many studies have been devoted to the distributed frequency control of power systems, most of them assume that all the nodes are controllable except [6]. To make a fair comparison, the controller proposed in [6] is adopted as a rival in the tests. As shown in Eq. (8) of [6], each controller needs to predict the load that it should supply in the steady state. However, it is hard to acquire an accurate prediction in practice, which could lead to steady-state frequency error.

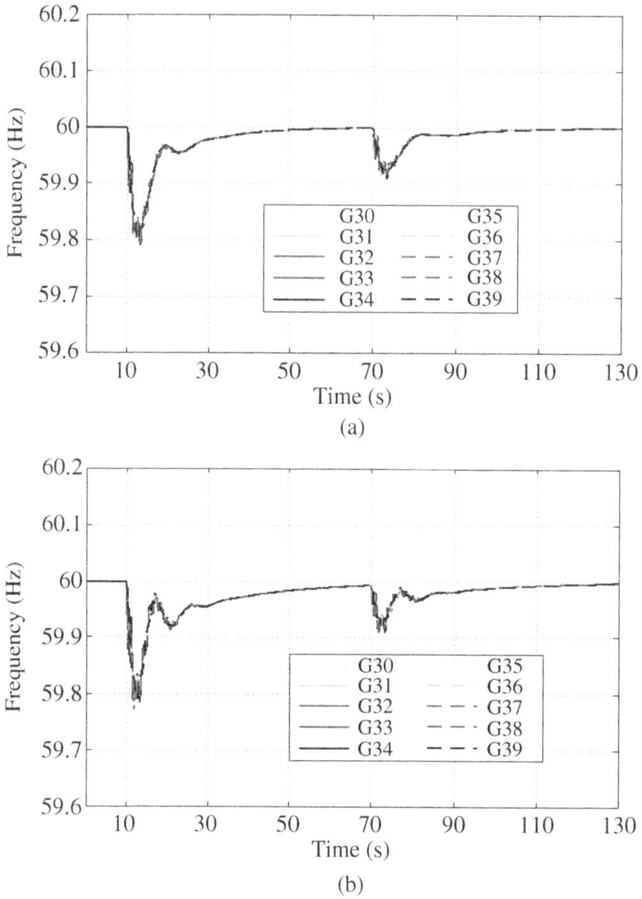

Figure 10.4 Frequency dynamics with (a) distributed controller and (b) AGC.

In the next case, we compare the two controls in the same scenario as that in the Section 10.7.2. The dynamics of frequencies with the controller given in [6] are shown in Figure 10.5. It is observed that there is a frequency deviation in steady state, as the prediction is inaccurate. Although the deviation is usually quite small, it is difficult to completely eliminate. In contrast, when the proposed method is adopted, there is no frequency deviation in steady state, as shown in Figure 10.4. This result is perfectly in coincidence with the indication given by Theorems 10.2, and 10.3 and Lemma 10.2 in this chapter.

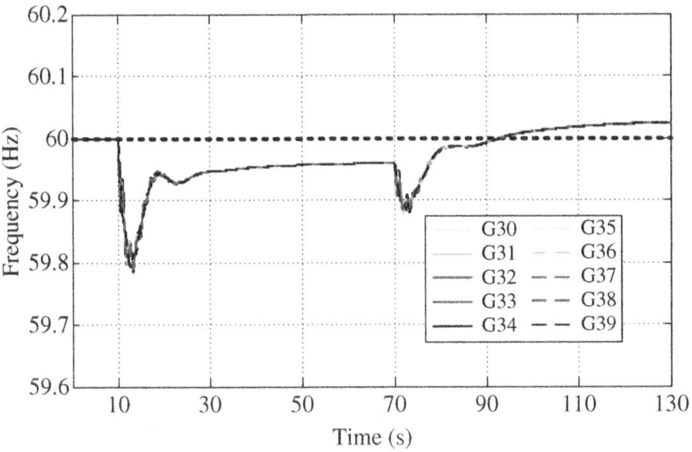

Figure 10.5 Dynamics of frequencies with the controller in [6]. Source: Adapted from [6].

Mechanical power dynamics under the AGC and the proposed controller are shown in Figures 10.6 and 10.7, respectively. The upper parts show the mechanical powers of G32, G36, G38, and G39, while the right parts show the mechanical powers of other generators adopting conventional controllers (10.14). With both controls, the mechanical powers of the generators adopting (10.14) remain identical in the steady state. However, there are two problems when adopting the AGC: (i) mechanical powers are not optimal, and (ii) mechanical power of G39 violates the capacity limit. In contrast, the proposed control can avoid these problems. In Stage 1 of Figure 10.7, no generator reaches capacity limits. In Stage 2, both G38 and G39 reach their upper limits. Then, G38 and G39 stop increasing their outputs while G32 and G36 raise their outputs to balance the load demands. In addition, the steady states of both stages are optimal, which are the same as those in Table 10.2.

We also illustrate in Figure 10.8 the dynamics of voltage at buses 3, 15, 23, 24, and 25. The voltages converge rapidly and only experience small drops when the load changes occur. This result validates the effectiveness of Condition 10.2 for excitation dynamics.

The marginal generation costs of generator i, $-\mu_i$ are shown in Figure 10.9. They converge in both stages, and the steady-state values in Stage 2 are slightly bigger than those in Stage 1, as the load changes lead to an increase in the marginal generation cost. Dynamics of z_{ij}, $(i,j) \in E$ are illustrated in Figure 10.10, which demon-

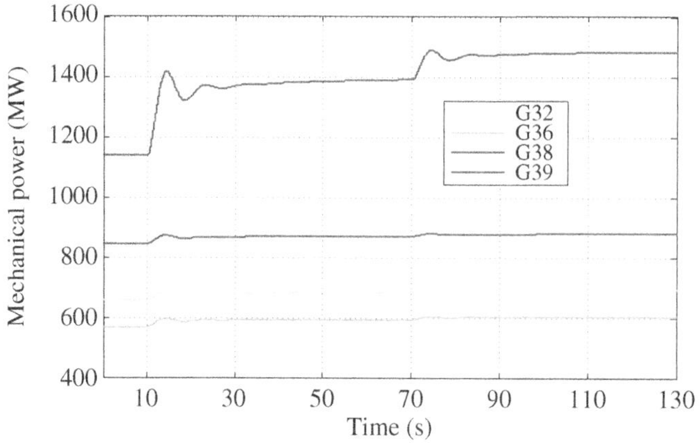

(a)

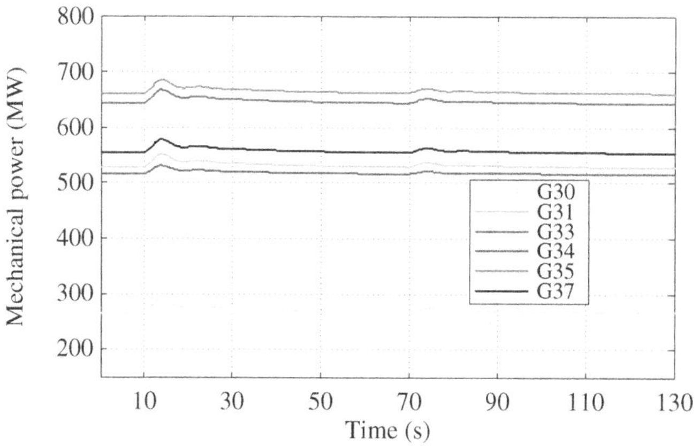

(b)

Figure 10.6 Dynamics of mechanical powers under AGC. (a) Power of controllable generator. (b) power of uncontrollable generator.

strates that the steady-state values do not change in the two stages. In addition, the variation of z in transient is very small as the deviation of μ_i is very small.

10.7.3 Performance Under Large Disturbances

In this subsection, two scenarios of large disturbances are considered. One is a generator tripping, and the other is a short-circuit fault followed by a line tripping.

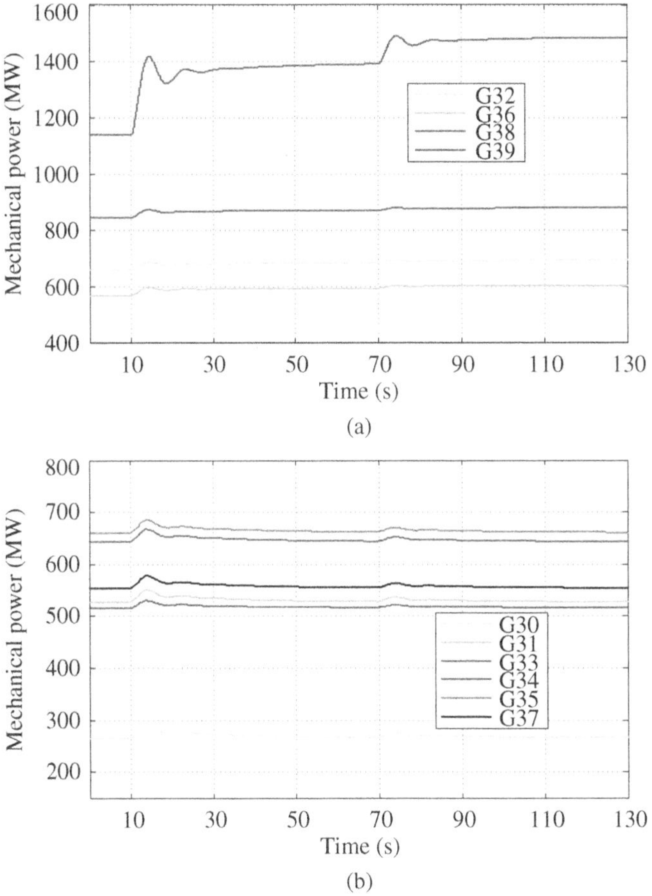

Figure 10.7 Dynamics of mechanical powers under the proposed control. (a) Power of controllable generator. (b) Power of uncontrollable generator.

10.7.3.1 Generator Tripping

At $t = 10s$, the controllable generator G32 is tripped, followed by the occurrence of a large power imbalance. Note that the communication graph remains connected. The output of G32 drops to zero. The frequencies and mechanical powers change accordingly. Frequency dynamics are shown in Figure 10.11, which experience a big drop at first and then recover to the nominal value quickly as other controllable generators increase outputs to rebalance the power mismatch. The mechanical power dynamics of controllable generators are shown in Figure 10.12.

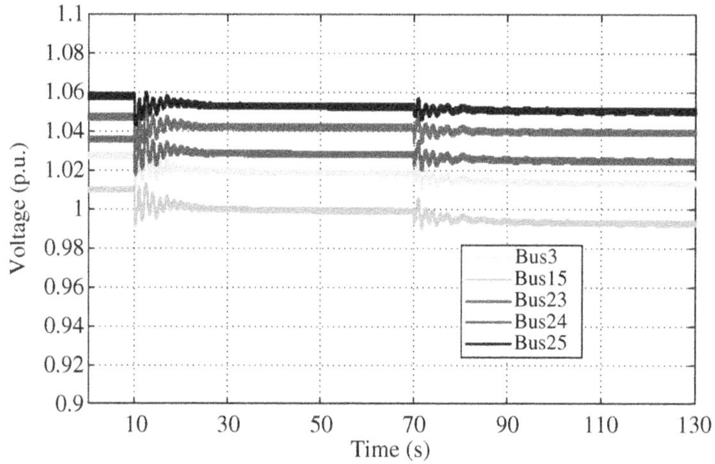

Figure 10.8 Dynamics of bus voltages under load changes.

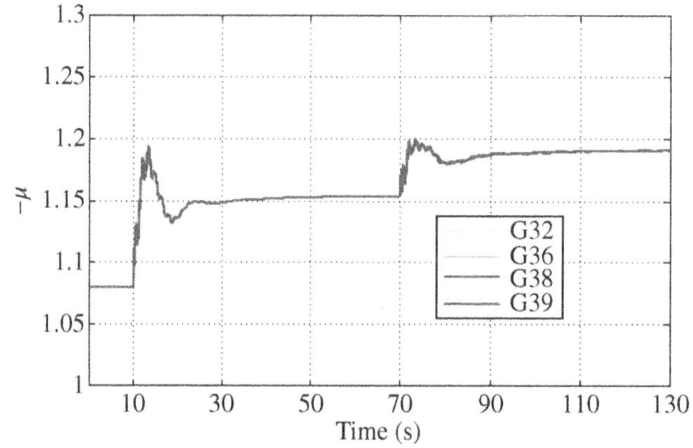

Figure 10.9 Dynamics of $-\mu$ under load changes.

After the output of G32 drops to zero, other generators increase their generations to rebalance the power and the load. These results confirm that the devised controller can adapt to generator tripping autonomously even if the tripped generator is participating in the distributed frequency control.

10.7.3.2 Short-Circuit Fault

At $t = 10$ s, there occurs a three-phase short-circuit fault on the line (4,14). At $t = 10.05$ s, this line is tripped by breakers. At $t = 60$ s, the fault is removed and the line (4,14) is reclosed. Frequency dynamics and voltage dynamics of buses 4 and 14

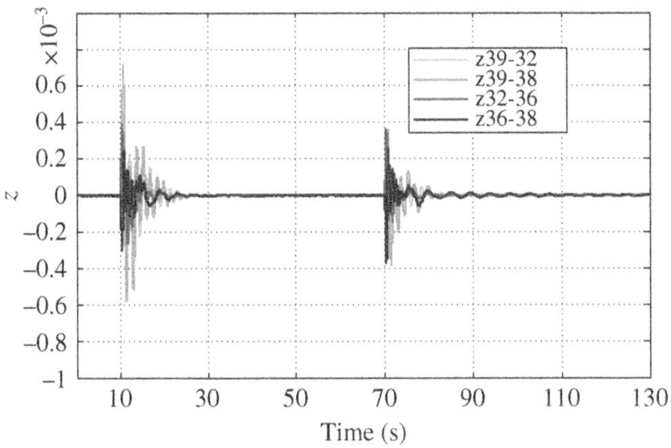

Figure 10.10 Dynamics of z under load changes.

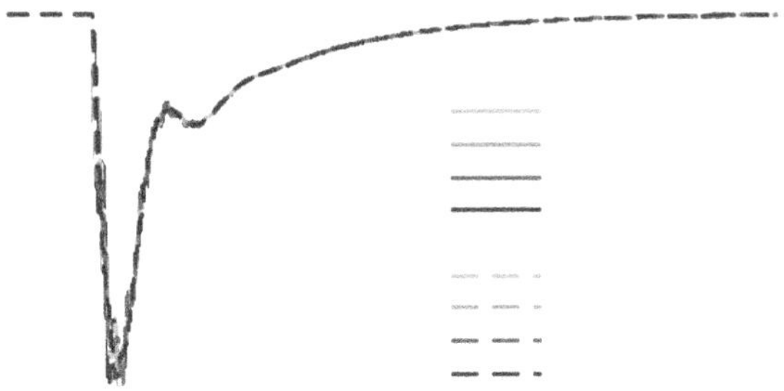

Figure 10.11 Dynamics of frequencies under a generator tripping.

are given in Figures 10.13 and 10.14, respectively. It can be seen that the frequency experiences violent oscillations after the fault happens. Then the system is stabilized quickly once the line is tripped. Small-frequency oscillation occurs when the line is reclosed. At the same time, the voltages of buses 4 and 14 drop to nearly zero when the fault happens and then recover to a new steady state in after about 10 s. They are slightly different from the initial values because the system's operating point has changed due to line tripping. When the tripped line is reclosed, the voltages recover to the initial values quickly.

When the line (4,14) is tripped, the power flow across the power network varies accordingly. As a consequence, the line loss also changes, causing slight variations of mechanical powers, as shown in Figure 10.15. In Figure 10.16, the inset

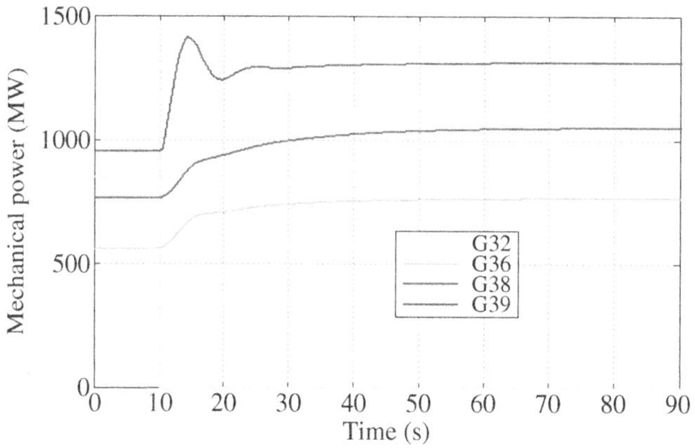

Figure 10.12 Dynamics of mechanical powers under a generator tripping.

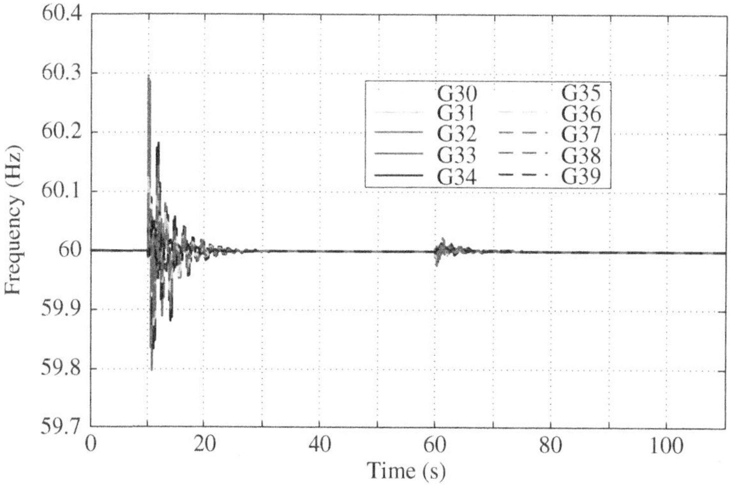

Figure 10.13 Dynamics of frequencies under a line tripping.

is the dynamics of mechanical power of G39 from 30 to 110 s. Similarly, the inset in the right part is the dynamics of $-\mu$ of all generators. Mechanical powers and their marginal costs all increase when the line is tripped. However, the proposed controller compensates for the loss change quickly.

These simulation results demonstrate that the devised distributed optimal frequency controller can cope with large disturbances such as generator tripping and short-circuit fault.

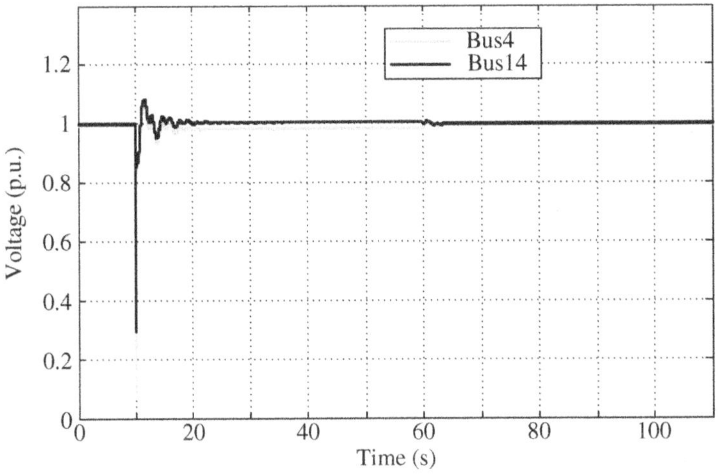

Figure 10.14 Dynamics of voltage under a line tripping.

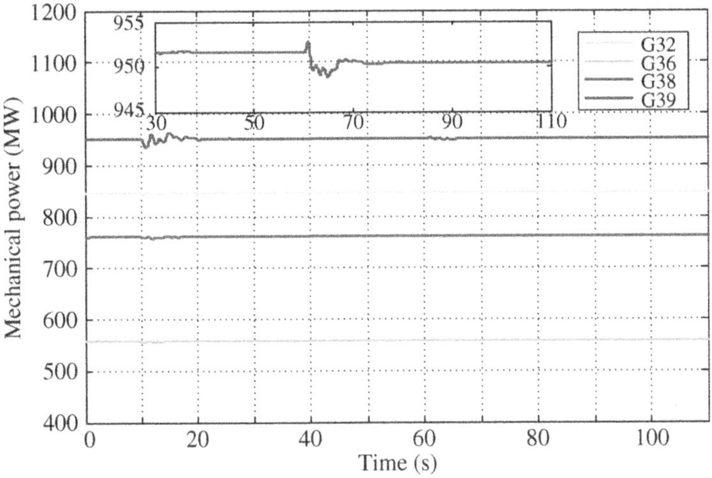

Figure 10.15 Dynamics of $-\mu$ under a line tripping.

10.8 Conclusion and Notes

In this chapter, we have designed a distributed optimal frequency control using a nonlinear network-preserving model, where only a subset of generator buses is controlled. We have also simplified the implementation by relaxing the

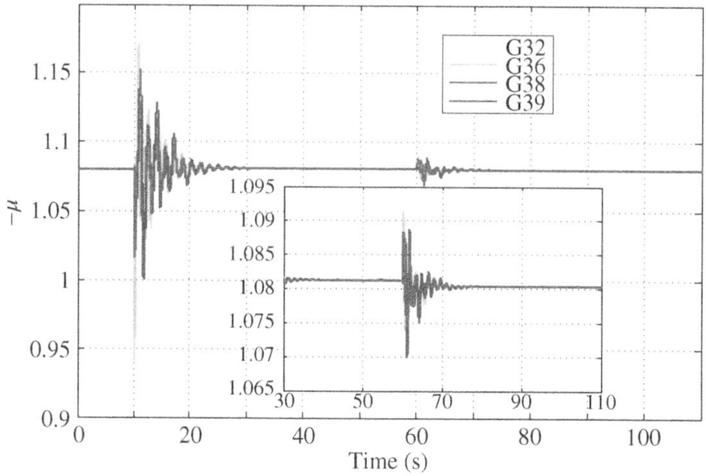

Figure 10.16 Dynamics of mechanical powers under a line tripping.

requirements of load measurements and communication topology. Since the nonlinearity due to the AC power flow model and the excitation voltage dynamics have been taken into account, our controllers can cope with large disturbances such as generator/line tripping. We have proved that the closed-loop system asymptotically converges to the optimal solution to the associated ED problem. We have also carried out substantial simulations on a detailed dynamic power system model to verify the good performance of our controller under both small and large disturbances.

Partial control coverage is an important issue for the implementation of distributed control in practice. This chapter demonstrates how to adapt the distributed control to such a circumstance. The most interesting part might be the use of passivity conditions to restrict the dynamic behaviors of uncontrollable agents such that the system-wide stability can be guaranteed, even when nonlinearity is considered in the model. This idea is further extended to establish distributed stability conditions for power systems with heterogeneous nonlinear bus dynamics [24].

Bibliography

1 N. Li, C. Zhao, and L. Chen, "Connecting automatic generation control and economic dispatch from an optimization view," *IEEE Transactions on Control of Network Systems*, vol. 3, no. 3, pp. 254–264, 2016.

2 C. Zhao, U. Topcu, N. Li, and S. H. Low, "Design and stability of load-side primary frequency control in power systems," *IEEE Transactions on Automatic Control*, vol. 59, no. 5, pp. 1177–1189, 2014.

3 E. Mallada, C. Zhao, and S. Low, "Optimal load-side control for frequency regulation in smart grids," *IEEE Transactions on Automatic Control*, vol. 62, no. 12, pp. 6294–6309, 2017.

4 A. Kasis, E. Devane, C. Spanias, and I. Lestas, "Primary frequency regulation with load-side participation–Part I: stability and optimality," *IEEE Transactions on Power Systems*, vol. 32, no. 5, pp. 3505–3518, 2017.

5 C. Zhao, E. Mallada, S. Low, and J. Bialek, "A unified framework for frequency control and congestion management," in *Power Systems Computation Conference (PSCC)*, 2016, (Genoa, Italy), pp. 1–7, IEEE, 2016.

6 J. Z. F. Pang, L. Guo, and S. H. Low, "Load-side frequency regulation with limited control coverage,"*SIGMETRICS Performance Evaluation Review*, vol. 45, pp. 94–96, 2017.

7 Z. Wang, F. Liu, S. H. Low, C. Zhao, and S. Mei, "Decentralized optimal frequency control of interconnected power systems with transient constraints," in *2016 IEEE 55th Conference on Decision and Control (CDC)*, pp. 664–671, Dec 2016.

8 X. Zhang, N. Li, and A. Papachristodoulou, "Achieving real-time economic dispatch in power networks via a saddle point design approach," in *2015 IEEE PES General Meeting*, (Denver, USA), pp. 1–5, IEEE, July 2015.

9 Z. Wang, F. Liu, S. H. Low, C. Zhao, and S. Mei, "Distributed frequency control with operational constraints, Part I: per-node power balance," *IEEE Transactions on Smart Grid*, vol. 10, no. 1, pp. 40–52, 2019.

10 Z. Wang, F. Liu, S. H. Low, C. Zhao, and S. Mei, "Distributed frequency control with operational constraints, Part II: network power balance," *IEEE Transactions on Smart Grid*, vol. 10, no. 1, pp. 53–64, 2019.

11 T. Stegink, C. D. Persis, and A. van der Schaft, "A port-Hamiltonian approach to optimal frequency regulation in power grids," in *Proceedings of 54th IEEE Conference on Decision and Control (CDC)*, (Osaka, Japan), pp. 3224–3229, IEEE, Dec 2015.

12 T. W. Stegink, C. De Persis, and A. J. van der Schaft, "Stabilization of structure-preserving power networks with market dynamics," IFAC-PapersOnLine 50, no. 1 (2017): 6737–6742.

13 T. Stegink, C. De Persis, and A. van der Schaft, "A unifying energy-based approach to stability of power grids with market dynamics," *IEEE Transactions Automatic Control*, vol. 62, no. 6, pp. 2612–2622, 2017.

14 X. Zhang and A. Papachristodoulou, "A real-time control framework for smart power networks: design methodology and stability," *Automatica*, vol. 58, pp. 43–50, 2015.

15 C. Zhao and S. Low, "Optimal decentralized primary frequency control in power networks," in *Proceedings of the 53rd IEEE Conference on Decision and Control (CDC)*, (Los Angeles, USA), pp. 2467–2473, IEEE, Dec 2014.

16 S. Trip and C. De Persis, "Optimal generation in structure-preserving power networks with second-order turbine-governor dynamics," in *Control Conference (ECC), 2016 European*, (Aalborg, Denmark), pp. 916–921, IEEE, 2016.

17 P. Kundur, *Power System Stability and Control*, New York, USA:McGraw-Hill, 1994.

18 L. Yan-Hong, L. Chun-Wen, and W. Yu-Zhen, "Decentralized excitation control of multi-machine multi-load power systems using hamiltonian function method," *Acta Automatica Sinica*, vol. 35, no. 7, pp. 919–925, 2009.

19 S. Boyd and L. Vandenberghe, *Convex Optimization*. Cambridge, UK: Cambridge University Press, 2004.

20 D. Feijer and F. Paganini, "Stability of primal-dual gradient dynamics and applications to network optimization," *Automatica*, vol. 46, no. 12, pp. 1974–1981, 2010.

21 C. D. Persis, N. Monshizadeh, J. Schiffer, and F. Dörfler, "A Lyapunov approach to control of microgrids with a served differential-algebraic model," in *2016 IEEE 55th Conference on Decision and Control (CDC)*, (Las Vegas, USA), pp. 2595–2600, IEEE, Dec 2016.

22 P. Lin, Y. Jia, and L. Li, "Distributed robust h∞ consensus control in directed networks of agents with time-delay," *Systems & Control Letters*, vol. 57, no. 8, pp. 643–653, 2008.

23 V. V. Arthur and R. Bergen, *Power System Analysis*.Upper Saddle River, NJ: Pretice Hall, 2000.

24 P. Yang, F. Liu, Z. Wang, and C. Shen, "Distributed stability conditions for power systems with heterogeneous nonlinear bus dynamics," *IEEE Transactions on Power Systems*, vol. 35, no. 3, pp. 2313–2324, 2020.

11

Robustness and Adaptability: Heterogeneity in Power Controls of DC Microgrids

When implementing distributed control in practice, there are usually different types of local controllers getting involved. Therefore, it is desired that the distributed control is adaptable to heterogeneous control configurations. This chapter takes direct current (DC) microgrids as an example to show how to design such a distributed optimal control to break the restriction induced by heterogeneity. Microgrids usually belong to different owners and adopt various control strategies. It brings great challenges to optimal operation of microgrids since it is difficult to coordinate diverse local controllers. In this regard, this chapter aims to design a distributed optimal control of stand-alone microgrids with heterogeneous local controllers. Firstly, the optimal power flow (OPF) problem of stand-alone DC microgrids is formulated as an exact second-order cone program (SOCP). It is proved that the optimal solution is unique. Then a dynamic solution algorithm based on the primal–dual decomposition method is constructed. The generation limits and voltage limits in both transient processes and steady states are satisfied. The optimality and convergence are proved theoretically. It is shown that the algorithm allows to unify the following three typical controls of microgrids: (i) power control, (ii) voltage control, and (iii) droop control. Therefore, each microgrid does not need to change its original control strategy in implementation, which is attractive for practical deployment. The designed controller also exhibits a strong capability to support plug-n-play operations.

11.1 Background

Microgrids are clusters of distributed generators (DGs), energy storage systems (ESSs), and loads. They are generally categorized into two types: alternating current (AC) and DC microgrids [1, 2]. In the past years, research has been concentrated on enhancing the performance of AC microgrids. However, some generations and loads are intrinsically of DC type, such as photovoltaic (PV),

Merging Optimization and Control in Power Systems: Physical and Cyber Restrictions in Distributed Frequency Control and Beyond, First Edition. Feng Liu, Zhaojian Wang, Changhong Zhao, and Peng Yang.

battery, computer, and electrical vehicle (EV) [3–5]. DC microgrids can more naturally integrate them and eliminate unnecessary conversion processes, which improves system efficiency and reliability. In addition, DC systems do not face such problems as reactive power compensation, frequency instability, and desynchronization [3], which makes them more and more popular in power systems. In DC microgrids, a hierarchical control architecture is often utilized [6, 7], composed of primary control, secondary control, and tertiary control. It can be implemented in either a centralized manner or a distributed manner. In a centralized manner, a control center is necessary to accumulate information from microgrids, compute control commands, and send them back. With the increasing number of microgrids and uncertainties arising from renewable generations and load demands, centralized control has been facing great challenges, e.g. single-point failures, heavy communication burden of the control center, and slow response to environment changing [8]. These challenges highlight the need for a distributed control paradigm that requires no control center and less communication.

In the hierarchy of power system control, the primary control is usually decentralized without communication. The most popular one might be droop control [6, 9], where load sharing is mainly determined by preset droop coefficients. As pointed out in [10, 11], droop control could lead to unexpected load sharing sometimes, especially in a power system with diverse line parameters and different operation modes. In this regard, many improvements have been done [10–14]. Taking into consideration the effect of different line impedances, Khorsandi et al. [10] proposes a decentralized control strategy to achieve satisfactory load sharing. In [11], a mode-adaptive decentralized control strategy is proposed for the power management in a DC microgrid, which enhances the control flexibility compared with conventional droop controls. In [14], a decentralized method is proposed to adaptively update the droop coefficients according to the state-of-charge (SOC) of energy storage, which can achieve equal load sharing. However, similar to the AC power system, the primary control in DC system suffers from steady-state voltage deviations.

To eliminate the steady-state deviation, distributed secondary control is developed. The most widely used method is consensus based control, where there is usually a global coordination variable, e.g. global voltage deviation, while each agent only has its local estimation. In a DC system, each agent may represent a DG or a DC microgrid. By exchanging information with neighbors, the values of the global coordination variable for all agents will eventually converge [7, 15]–[17]. In [7], each microgrid uses dynamic consensus protocol to estimate the global averaged voltage by communicating with neighboring agents. Then, the estimated voltage is compared with the reference value, and the error is fed to a proportional–integral (PI) controller to eliminate the voltage deviation.

This method is further improved in [16] by adding a current consensus regulator, where the control goal is to achieve globally identical current sharing ratio. By doing so, the equal load sharing can be achieved. A discrete consensus method is also developed in [17] to restore average voltage with accurate load sharing. Although the consensus-based secondary control can realize equality among agents, the results may not be optimal.

Tertiary control is to achieve the optimal operation by controlling the power flow among DC microgrids or among DGs within a microgrid [18–20]. Conventionally, tertiary control provides reference operation points for each subsystem (DC microgrids or DGs). Since its timescale is much slower than real-time control, the fast fluctuation of renewable generations and loads may undermine its performance. One natural idea is to break the hierarchy and combine real-time control and steady-state optimization together. In this circumstance, the optimization solution should be computed in real time and sent to controllers of the system. This idea has been applied in both AC system [21–23] and DC system [24–26]. The point in [24–26] is that, when the working point is optimal, the incremental generation costs of all microgrids should be identical in steady state. In [24], an economic dispatch problem is formulated, and the incremental generation cost is regarded as a consensus variable. Then a consensus-based method is developed to achieve optimality. A similar method is used in [26], where the subgradient is employed to accelerate the convergence of consensus. Despite the inspiring contributions to merging optimization and control, these works still have some restrictions regarding practical implementation. For example, the original controllers of microgrids or DGs have to be removed or modified revised to adapt to the proposed controller. It is practically difficult as the microgrids or DGs may belong to different owners and adopt diverse controllers. This challenge gives rise to the demand to develop a nondisruptive distributed controller that can adapt to diverse local controllers.

In this chapter, we investigate the distributed OPF control among stand-alone DC microgrids, considering different original control configurations of agents. Theoretically, we construct an OPF model for stand-alone microgrids with an exact second-order cone (SOC) relaxation and further prove the uniqueness of its optimal solution. By using the primal–dual decomposition method, a distributed control algorithm is proposed, where we also prove the convergence and optimality of its equilibrium point. From the application point of view, the control algorithm generates control commands for different local controllers such as power control, voltage control, and droop control. There is no need to change the original control configurations/strategies of microgrids. In addition, the constraints arising from generation capacity limits and voltage limits are enforced even in transient process, guaranteeing the feasibility and security of the control commands.

The reset of this chapter is organized as follows. In Section 11.2, the network model of DC microgrids is introduced. In Section 11.3, the OPF model for stand-alone DC microgrids is formulated. In Section 11.4, the distributed control algorithm is proposed with proofs of optimality and convergence. The implementation approach is presented and discussed in Section 11.5. Case studies are given in Section 11.6. Finally, Section 11.7 concludes the chapter with notes.

11.2 Network Model

This chapter considers a stand-alone DC system that is composed of a cluster of microgrids interconnected through tie lines. Each microgrid is treated as a bus with certain generation and load. Then the overall system is modeled as a connected graph $\mathcal{G} := (\mathcal{N}, \mathcal{E})$, where $\mathcal{N} = \{1,2, \dots n\}$ is the set of microgrids and $\mathcal{E} \subseteq \mathcal{N} \times \mathcal{N}$ is the set of lines. If two microgrids i and k are connected by a tie line directly, we denote $(i, k) \in \mathcal{E}$, abbreviated by $i \sim k$. The resistance of line (i, k) is r_{ik}. The power flow from microgrid i to k is P_{ik}, and the current from microgrids i to k is I_{ik}. Let $m := |\mathcal{E}|$ be the number of lines.

For each microgrid $i \in \mathcal{N}$, let $p_i^g(t)$ be the generation at time t and p_i^d be its constant load demand. Denote the voltage at bus i by V_i. DGs in the DC microgrids may have different control configurations or strategies, such as power control, voltage control, and droop control. Power control and voltage control only require the reference values, which are not introduced here in detail. Droop control usually takes the following form:

$$v_i - v_i^* = -k_i(p_i^g - \hat{p}_i) \tag{11.1}$$

where $v_i = V_i^2$, $k_i > 0$ is the droop coefficient, v_i^* is the voltage square reference, and k_i, v_i^* are constants. $\hat{p}_i$ is the power when $v_i = v_i^*$, which is a variable in the rest of the chapter.

Denote by I_{ik} the current in line (i, k) from i to k, which is defined as

$$I_{ik} = \frac{V_i - V_k}{r_{ik}} \tag{11.2}$$

Then the power P_{ik} from i to k is

$$P_{ik} = V_i I_{ik} = \frac{V_i - V_k}{r_{ik}} V_i \tag{11.3}$$

Consequently, the power balance equation for bus i is given by

$$p_i^g - p_i^d = V_i \sum_{k:k \in N_i} \frac{V_i - V_k}{r_{ik}} \tag{11.4a}$$

$$= \sum_{k:k \in N_i} P_{ik}. \tag{11.4b}$$

where N_i is the set of the neighboring microgrids connected to microgrid i directly. Our goal is to generate optimal control commands for microgrids equipped with different local controllers in a distributed manner while satisfying the following operational constraints:

$$0 \leq p_i^g \leq \overline{p}_i^g \tag{11.5a}$$

$$\underline{V}_i \leq V_i \leq \overline{V}_i \tag{11.5b}$$

where $\overline{p}_i^g$ is the upper limit of generation of microgrid/DG i and $\underline{V}_i, \overline{V}_i$ are lower and upper limits of voltage. For a power-controlled microgrid or DG, (11.5a) is a hard limit, which must be satisfied even in transient. For a voltage-controlled microgrid or DG, (11.5b) is not a hard limit. However, it also should be satisfied in transient regarding the security of inverters.

11.3 Optimal Power Flow of DC Networks

11.3.1 OPF Model

Existing OPF models are mainly for grid-connected DC systems [27]. However, DC microgrids can also operate in an isolated mode in many situations such as in remote areas or islands. Since the grid-connected DC system can be regarded as a special case of the stand-alone one (by simply adding an infinite substation bus), here we formulate a generic OPF model for stand-alone DC networks:

$$\mathbf{OPF}: \min_{p^g, v, W} \quad f = \sum_{i \in \mathcal{N}} f_i(p_i^g) \tag{11.6a}$$

$$\text{s. t.} \quad (11.5\text{a})$$

$$\underline{V}_i^2 \leq v_i \leq \overline{V}_i^2, \quad i \in \mathcal{N} \tag{11.6b}$$

$$p_i^g - p_i^d = \sum_{k:k\sim i} \frac{v_i - W_{ik}}{r_{ik}}, \quad i \in \mathcal{N} \tag{11.6c}$$

$$W_{ik} \geq 0, \quad i \sim k \tag{11.6d}$$

$$W_{ik} = W_{ki}, \quad i \sim k \tag{11.6e}$$

$$R_{ik} \succcurlyeq 0, \quad i \sim k \tag{11.6f}$$

$$\text{rank}(R_{ik}) = 1, \quad i \sim k \tag{11.6g}$$

where $R_{ik} = \begin{bmatrix} v_i & W_{ik} \\ W_{ki} & v_j \end{bmatrix}$. If $\text{rank}(R_{ik}) = 1$ holds, W_{ik} can be simplified to be $W_{ik} = V_i V_k$. The cost function (11.6a) is a function of generation in each bus, which should satisfy the following assumption.

Assumption 11.1 The cost function $f_i(p_i^g)$ is strictly increasing when $p_i^g \geq p_i^d$ for all $i \in \mathcal{N}$, second-order continuously differentiable, and strongly convex, i.e. $f_i''(p_i^g) \geq \alpha > 0$.

Constraint (11.6b) is derived from (11.5b) and (11.6c) is from (11.4a). Constraint (11.6f) implies R_{ik} is a positive semidefinite matrix and (11.6g) guarantees R_{ik} is of rank one. The difference between (11.6) and OPF′ model in [27] is that there is no a substation bus with constant voltage in (11.6). Obviously, the model (11.6) is nonconvex due to the constraint (11.6g). Therefore, removing (11.6g) results in a SOC relaxation of (11.6) as follows (SOCP for short):

$$\textbf{SOCP}: \quad \min_{p^g, v, W} \sum_{i \in \mathcal{N}} f_i(p_i^g) \tag{11.7a}$$

$$\text{s. t.} \quad (11.5a), (11.6b) - (11.6f) \tag{11.7b}$$

We claim that the SOC relaxation is exact provided the following conditions are satisfied:

1) $\overline{V}_1 = \overline{V}_2 = \cdots = \overline{V}_n$.
2) $p_i^d > 0$ for $i \in \mathcal{N}$.
3) $\sum_{i \in \mathcal{N}} (p_i^g - p_i^d) > 0$.
4) $f_i(p_i^g)$ is strictly increasing when $p_i^g \geq p_i^d$ for $i \in \mathcal{N}$.

The proofs can be found in Appendix B. In DC power networks, the conditions 1), 2), and 3) are naturally satisfied. Under Assumption 11.1, the condition 4) holds as well.

To improve the numerical stability of the SOCP, we reformulate it into the following stable SOCP model (SSOCP):

$$\textbf{SSOCP}: \quad \min_{p^g, P, l, v} \sum_{i \in \mathcal{N}} f_i(p_i^g) \tag{11.8a}$$

$$\text{s. t.} \quad (11.4b), (11.5a)$$

$$P_{ik} + P_{ki} = r_{ik} l_{ik}, \quad i \sim k \tag{11.8b}$$

$$v_i - v_k = r_{ik} \left(P_{ik} - P_{ki} \right), \quad i \sim k \tag{11.8c}$$

$$l_{ik} \geq \frac{P_{ik}^2}{v_i}, \quad i \sim k \tag{11.8d}$$

$$\underline{V}_i^2 \leq v_i \leq \overline{V}_i^2, \quad i \in \mathcal{N} \tag{11.8e}$$

where $l_{ik} = |I_{ik}|^2$ are squared line currents and $l_{ik} = l_{ki}$. Constraint (11.8d) is the SOC-relaxed form. The detailed explanation of (11.8b)–(11.8d) can be found in reference [27], which is omitted here.

According to Theorem 5 in [27], the SOCP (11.7) and the SSOCP (11.8) are equivalent, i.e. there exists a one-to-one map between the feasible set of SOCP and that of SSOCP, which is

$$
\begin{aligned}
P_{ik} &= \frac{v_i - W_{ik}}{r_{ik}}, \quad i \sim k; \\
l_{ik} &= \frac{v_i - W_{ik} - W_{ki} + v_k}{r_{ik}^2}, \quad i \sim k
\end{aligned} \tag{11.9}
$$

In microgrids, the droop control is extensively utilized. However, the solution to (11.8) may not guarantee that v_i and p_i^g satisfy (11.1), which implies that the optimal solution may not be realized in operation. In this regard, we further incorporate the droop control into the constraints, leading to the following convex problem (DSOCP for short):

$$
\begin{aligned}
\textbf{DSOCP}: \quad & \min_{p^g, P, l, v, \hat{p}} \sum_{i \in \mathcal{N}} f_i(p_i^g) \qquad (11.10) \\
\text{s. t.} \quad & (11.1), (11.4b), (11.5a), (11.8b) - -(11.8e)
\end{aligned}
$$

In the DSOCP, $\hat{p}$ is the decision variable.

To help design the algorithm, we equivalently consider the following extended optimization problem (ESOCP for short):

$$
\begin{aligned}
\textbf{ESOCP}: \quad & \min_{p^g, P, l, v, \hat{p}} \sum_{i \in \mathcal{N}} f_i(p_i^g) + \sum_{i \in \mathcal{N}} \frac{1}{2} y_i^2 + \sum_{i \in \mathcal{N}} \frac{1}{2} z_i^2 \qquad (11.11) \\
\text{s. t.} \quad & (11.1), (11.4b), (11.5a), (11.8b) - -(11.8e)
\end{aligned}
$$

where

$$
\begin{aligned}
y_i &= v_i + k_i p_i^g - v_i^* - k_i \hat{p}_i, \\
z_i &= p_i^g - p_i^d - \sum_{k: k \in N_i} P_{ik}
\end{aligned}
$$

In this formulation, y_i and z_i are added only for facilitating the convergence [29], following the idea of reverse engineering. The equivalence between the ESOCP model and the DSOCP model is straightforward by noting that $y_i = z_i = 0$ holds at the minimum of the ESOCP model and for all feasible solutions to the DSOCP model.

11.3.2 Uniqueness of Optimal Solution

Throughout this chapter, the following assumption is assumed.

Assumption 11.2 The OPF problem (11.6) has at least one feasible solution.

The following theorems characterize the optimal solutions to the aforementioned optimization problems. We start from the SSOCP model (11.8).

Theorem 11.1 *Suppose Assumptions 11.1–11.2 hold. The optimal solution to the SSOCP problem (11.8) is unique.*

Proof: As Assumption 11.2 holds, problem (11.8) is also feasible due to the one-to-one map (11.9). It suffices to prove the uniqueness of the optimal solution to SOCP (11.7) as they are equivalent. Let $x^{1*} = (p^{g1*}, v^{1*}, W^{1*})$ and $x^{2*} = (p^{g2*}, v^{2*}, W^{2*})$ be two optimal solutions to the SOCP; then we prove they are equal. First we have

$$\sum_{i\in\mathcal{N}} f_i(p_i^{g1*} - p_i^d) = \sum_{i\in\mathcal{N}} f_i(p_i^{g2*} - p_i^d) \tag{11.12}$$

From the proof of Theorem 3 in [27], we know

$$\frac{v_i^{1*}}{v_i^{2*}} = \frac{v_k^{1*}}{v_k^{2*}} = \eta, \quad i \sim k$$

$$W_{ik}^{1*} = \sqrt{v_i^{1*} v_k^{1*}} = \eta\sqrt{v_i^{2*} v_k^{2*}} = \eta W_{ik}^{2*}$$

From (11.4a), we have

$$\begin{aligned} p_i^{g1*} - p_i^d &= \sum_{k:k\sim i} \frac{v_i^{1*} - W_{ik}^{1*}}{r_{ik}} \\ &= \sum_{k:k\sim i} \frac{\eta v_i^{2*} - \eta W_{ik}^{2*}}{r_{ik}} \\ &= \eta(p_i^{g2*} - p_i^d) \end{aligned}$$

Since $f_i(p_i^g - p_i^d)$ is strictly increasing, there must be $\eta = 1$; otherwise it contradicts (11.12). Therefore, we have $x^{1*} = x^{2*}$, justifying the uniqueness of the solution to the SOCP. According to the one-to-one map (11.9), the solution to the SSOCP (11.8) is also unique, which completes the proof. ∎

Theorem 11.2 *Denote the optimal solution to the SSOCP model (11.8) by $x^{1*} = (p^{g1*}, P^{1*}, l^{1*}, v^{1*})$ and the optimal solution to the DSOCP (11.10) by $x^{2*} = (p^{g2*}, P^{2*}, l^{2*}, v^{2*}, \hat{p}^{2*})$. Then, see the following:*

1) *There exists a unique $\hat{p}^{2*}$ such that $(x^{1*}, \hat{p}^{2*})$ is the optimal solution to the DSOCP.*
2) *The optimal solution to the DSOCP is unique.*
3) *$(p^{g2*}, P^{2*}, l^{2*}, v^{2*}) = (p^{g1*}, P^{1*}, l^{1*}, v^{1*})$.*

Proof:

Assertion 1). Supposing $x^{1*} = (p^{g1*}, P^{1*}, l^{1*}, v^{1*})$ is the optimal solution to the SSOCP (11.8), there exists a unique $\hat{p}^*$ satisfying (11.1). Noting that models (11.8) and (11.10) have the same objective functions and constraints except the constraint (11.1), we know $(x^{1*}, \hat{p}^{2*})$ must be the optimal solution to the DSOCP (11.10).

Assertion 2). By Theorem 11.1 and assertion 1) of Theorem 11.2, this assertion is trivial.

Assertion 3). Since $(p^{g2*}, P^{2*}, l^{2*}, v^{2*})$ is the optimal solution to the DSOCP, it also satisfies all the constraints of the SSOCP. Moreover, the DSOCP and the SSOCP have identical objective functions; hence $(p^{g2*}, P^{2*}, l^{2*}, v^{2*})$ is also the optimal solution to the SSOCP. Due to the uniqueness of the optimal solution to SSOCP, we have $(p^{g2*}, P^{2*}, l^{2*}, v^{2*}) = (p^{g1*}, P^{1*}, l^{1*}, v^{1*})$. This completes the proof. ■

Suppose the optimal solution to the DSOCP is $x^{2*} = (p^{g2*}, P^{2*}, l^{2*}, v^{2*}, \hat{p}^{2*})$ with the droop coefficient k_{i1}. From the proof of the Assertion 1) of Theorem 11.2, if k_{i1} changes to k_{i2}, there exists a unique $\hat{p}^{2*}(k_{i2})$ such that $(p^{g2*}, P^{2*}, l^{2*}, v^{2*}, \hat{p}^{2*}(k_{i2}))$ is the optimal solution to the DSOCP. It indicates that the droop coefficients do not influence the optimal solution to the SSOCP.

Lemma 11.1 *The optimization problems ESOCP and DSOCP have identical feasible solutions.*

It is easy to prove Lemma 11.1 as $y_i = 0$ and $z_i = 0$ for any feasible solution.

Remark 11.1 Theorem 11.2 shows that the unique optimal solution to the SOCP problem still exists even if a droop controller is incorporated, and we can obtain the corresponding optimal $\hat{p}_2^*$. Moreover, for those microgrids that do not adopt droop controllers, the ESOCP can provide the optimal reference values of output power and voltages. In addition, for different droop coefficient k_i, $(p_2^{g*}, P_2^*, l_2^*, v_2^*)$ in the optimal solution does not change. Thus, for microgrids that do not adopt droop control, we can simply assign an imaginary droop control to them when formulating the ESOCP, i.e. assuming all the microgrids adopt droop control therein. This does not influence the optimal solutions to those microgrids adopting power control and voltage control. In this regard, our method adapts to all the three different controllers.

11.4 Control Design

11.4.1 Distributed Optimization Algorithm

Based on the primal–dual dynamics, we propose the following distributed algorithm to solve the ESOCP, which is

$$\dot{p}_i^g = \left[p_i^g - \left(G_i(p_i^g) - \mu_i + k_i \epsilon_i + z_i + k_i y_i\right)\right]_0^{\overline{p}_i^g} - p_i^g \tag{11.13a}$$

$$\dot{v}_i = \left[v_i - \left(y_i + \sum_{k \in N_i} \gamma_{ik} + \epsilon_i - \sum_{k \in N_i} \rho_{ik} \frac{P_{ik}^2}{v_i^2}\right)\right]_{\underline{V}_i^2}^{\overline{V}_i^2} - v_i \tag{11.13b}$$

$$\dot{P}_{ik} = -\left(\mu_i + \lambda_{ik} - \gamma_{ik} r_{ik} + 2\rho_{ik} P_{ik}/v_i - z_i\right) \tag{11.13c}$$

$$\dot{l}_{ik} = -\left(-\lambda_{ik} r_{ik} - \rho_{ik} - \rho_{ki}\right) \tag{11.13d}$$

$$\dot{\hat{p}}_i = k_i \epsilon_i + k_i y_i \tag{11.13e}$$

$$\dot{\mu}_i = -\left(p_i^g - p_i^d - \sum_{k:k \in N_i} P_{ik}\right) \tag{11.13f}$$

$$\dot{\epsilon}_i = v_i + k_i p_i^g - v_i^* - k_i \hat{p}_i \tag{11.6g}$$

$$\dot{\lambda}_{ik} = P_{ik} + P_{ki} - r_{ik} l_{ik} \tag{11.13h}$$

$$\dot{\gamma}_{ik} = v_i - v_k - r_{ik}\left(P_{ik} - P_{ki}\right) \tag{11.13i}$$

$$\dot{\rho}_{ik} = \left[\frac{P_{ik}^2}{v_i} - l_{ik}\right]_{\rho_{ik}}^{+} \tag{11.13j}$$

where $G_i(p_i^g)$ is defined as

$$G_i(p_i^g) := \frac{\partial}{\partial p_i^g} f_i(p_i^g) \tag{11.14}$$

For any x_i, a_i, b_i with $a_i \leq b_i$, we define

$$[x_i]_{a_i}^{b_i} := \min\{b_i, \max\{a_i, x_i\}\}$$

Moreover, $[x_i]_{a_i}^{+}$ is the projection operator defined as

$$[x_i]_{a_i}^{+} = \begin{cases} x_i, & \text{if } a_i > 0 \text{ or } x_i > 0; \\ 0, & \text{otherwise} \end{cases}$$

The algorithm (11.13) is fully distributed where each MG updates its internal states p_i^g, P_{ik}, l_{ik}, v_i, $\hat{p}_i$, μ_i, ϵ_i, λ_{ik}, γ_{ik}, and ρ_{ik} relying only on local information and neighboring communication. The neighboring information only appears in the variables $P_{ki}, v_k, \rho_{ki}, k \in N_i$. For a microgrid not adopting a droop control, $\hat{p}_i$ in (11.13e) is only a variable in the cyber system computed to help controller design. Hence it is feasible to use the hypothesis in practical microgrid operations.

Next, we will investigate the boundedness of $(p_i^g(t), v_i(t))$. Firstly we introduce the following assumption.

Assumption 11.3 The initial states of the dynamic system (11.13), $(p_i^g(0), v_i(0))$, are finite and satisfy constraint (11.5).

Define a set X as

$$X := \{(p_i^g, v_i) | 0 \le p_i^g \le \overline{p}_i^g, \underline{v}_i \le v_i \le \overline{v}_i\} \tag{11.15}$$

Then we prove the boundedness of $(p_i^g(t), v_i(t))$ for all $tge0$.

Lemma 11.2 *Suppose Assumption 11.3 holds. Then constraint (11.5) is satisfied for all $t \ge 0$, i.e. $(p^g(t), v(t)) \in X$ for all $t \ge 0$ where X is defined in (11.15).*

Proof: Note that (11.13a) is a first-order inertia dynamic with the input

$$u_i^g = [p_i^g - (G_i(p_i^g) - \mu_i + k_i\epsilon_i + z_i + k_iy_i)]_0^{\overline{p}_i^g}$$

According to the feature of the first-order inertia dynamic, $p^g(t) \in X$ holds for all $t \ge 0$ as long as $u_i^g(t) \in X$ for all $t \ge 0$. Thus, we know $p^g(t) \in X$ for all $t \ge 0$ always holds. Similarly, $v(t) \in X$ for all $t \ge 0$ always holds. This completes the proof. ∎

Lemma 11.2 implies that the system trajectories $p_i^g(t)$ and $v_i(t)$ satisfy the local inequality constraints for all time even in transient.

11.4.2 Optimality of Equilibrium

In this subsection, we prove that the equilibrium points of (11.13) are primal–dual optimal for the ESOCP problem and its dual and vice versa.

Given $x_p := (p^g, P, l, v, \hat{p})$, $x_d := (\mu, \epsilon, \lambda, \gamma, \rho)$, the following two definitions are introduced.

Definition 11.1 A point $(x_p^*, x_d^*) := (p^{g*}, P^*, l^*, v^*, \hat{p}^*, \mu^*, \epsilon^*, \lambda^*, \gamma^*, \rho^*)$ is an equilibrium point of the ESOCP model (11.13) if the right-hand side of (11.13) vanishes at (x_p^*, x_d^*).

Definition 11.2 A point (x_p^*, x_d^*) is primal–dual optimal if x_p^* is optimal for ESOCP and x_d^* is optimal for its dual problem.

To prove the optimality of (x_p^*, x_d^*), we make the following assumption.

Assumption 11.4 The Slater's condition holds for the ESOCP model (11.11).

We first show that the control saturation does not influence the optimal solution to the ESOCP, which is justified by Lemma 11.3.

Lemma 11.3 *Suppose Assumptions 11.1–11.3 hold. If (x_p^*, x_d^*) is primal–dual optimal, then there must be*

$$p_i^{g*} = \left[p_i^{g*} - \left(G_i(p_i^{g*}) - \mu_i^* + k_i\epsilon_i^* + z_i^* + k_iy_i^*\right)\right]_0^{\overline{p}_i^g}$$

$$v_i^* = \left[v_i^* - \left(y_i^* + \sum_{k\in N_i}\gamma_{ik}^* + \epsilon_i^* - \sum_{k\in N_i}\rho_{ik}^*\frac{(P_{ik}^*)^2}{(v_i^*)^2}\right)\right]_{\underline{V}_i^2}^{\overline{V}_i^2}$$

Proof: With Assumptions 11.1, 11.2, and 11.3, the strong duality holds. (x_p^*, x_d^*) is primal–dual optimal if and only if it satisfies the KKT conditions.

The Lagrangian of the ESOCP is given in (11.16):

$$\begin{aligned} L = & \sum_{i\in\mathcal{N}} f_i(p_i^g - p_i^d) + \sum_{i\in\mathcal{N}}\frac{1}{2}z_i^2 + \sum_{i\in\mathcal{N}}\frac{1}{2}y_i^2 \\ & + \sum_{i\in\mathcal{N}}\epsilon_i(v_i + k_ip_i^g - v_i^* - k_i\hat{p}_i) \\ & + \sum_{(i,k)\in\mathcal{E}}\lambda_{ik}\left(P_{ik} + P_{ki} - r_{ik}l_{ik}\right) \\ & + \sum_{i\in\mathcal{N}}\gamma_{ik}\left(v_i - v_k - r_{ik}\left(P_{ik} - P_{ki}\right)\right) \\ & + \sum_{(i,k)\in\mathcal{E}}\rho_{ik}\left(\frac{P_{ik}^2}{v_i} - l_{ik}\right) \end{aligned} \tag{11.16}$$

Based on (11.16) we can obtain the KKT conditions

$$G_i(p_i^{g*}) - \mu_i^* + k_i\epsilon_i^* + z_i^* + k_iy_i^* \begin{cases} \geq 0, & p_j^{g*} = 0 \\ = 0, & 0 < p_j^{g*} < \overline{p}_j^g \\ \leq 0, & p_j^{g*} = \overline{p}_j^g \end{cases} \tag{11.17a}$$

$$y_i^* + \sum_{k\in N_i}\gamma_{ik}^* + \epsilon_i^* - \sum_{k\in N_i}\rho_{ik}^*\frac{(P_{ik}^*)^2}{(v_i^*)^2} \begin{cases} \geq 0, & p_j^{g*} = 0 \\ = 0, & 0 < p_j^{g*} < \overline{p}_j^g \\ \leq 0, & p_j^{g*} = \overline{p}_j^g \end{cases} \tag{11.17b}$$

$$0 = -\left(\mu_i^* + \lambda_{ik}^* - \gamma_{ik}^*r_{ik} + 2\rho_{ik}^*\frac{P_{ik}^*}{v_i^*} - z_i^*\right) \tag{11.17c}$$

$$0 = -\left(-\lambda_{ik}^*r_{ik} - \rho_{ik}^* - \rho_{ki}^*\right) \tag{11.17d}$$

$$0 = k_i\epsilon_i^* + k_iy_i^* \tag{11.17e}$$

$$0 = -\left(p_i^{*g} - p_i^d - \sum_{k:k\in N_i}P_{ik}^*\right) \tag{11.17f}$$

$$0 = v_i^* + k_i p_i^{g*} - v_i^* - k_i \hat{p}_i^* \tag{11.17g}$$

$$0 = P_{ik}^* + P_{ki}^* - r_{ik} l_{ik}^* \tag{11.17h}$$

$$0 = v_i^* - v_k^* - r_{ik}^* \left(P_{ik}^* - P_{ki}^*\right) \tag{11.17i}$$

$$0 = \left(\frac{(P_{ik}^*)^2}{v_i^*} - l_{ik}^* \right) \rho_{ik}^*, \ \rho_{ik}^* \geq 0 \tag{11.17j}$$

(x_p^*, x_d^*) is primal–dual optimal if and only if it satisfies the KKT conditions. It can be checked that (11.17a) and (11.17b) are equivalent to

$$p_i^{g*} = [p_i^{g*} - (G_i(p_i^{g*}) - \mu_i^* + k_i \epsilon_i^* + z_i^* + k_i y_i^*)]_0^{\overline{p}_i^g}$$

$$v_i^* = \left[v_i^* - \left(y_i^* + \sum_{k \in N_i} \gamma_{ik}^* + \epsilon_i^* - \sum_{k \in N_i} \rho_{ik}^* \frac{(P_{ik}^*)^2}{(v_i^*)^2} \right) \right]_{\underline{V}_i^2}^{\overline{V}_i^2}$$

This completes the proof. ∎

Based on Lemma 11.3, we have the following Theorem.

Theorem 11.3 *Suppose Assumptions 11.1–11.4 hold. A point* (x_p^*, x_d^*) *is primal–dual optimal for the ESOCP model (11.11) if and only if it is an equilibrium point of the dynamic system (11.13).*

Proof: **Sufficiency.** Suppose (x_p^*, x_d^*) is primal–dual optimal. Therefore, it satisfies the KKT conditions. It can be obtained directly from (11.17c)–(11.17i) that the right sides of dynamics (11.13c)–(11.13i) vanish. The right sides of (11.13a) and (11.13b) also vanish by invoking Lemma 11.3. From (11.17j) and the exactness of the convex relaxation,[1] we know

$$\frac{(P_{ik}^*)^2}{v_i^*} - l_{ik}^* = 0,$$

and

$$\rho_{ik}^* \left(\frac{(P_{ik}^*)^2}{v_i^*} - l_{ik}^* \right) = 0$$

Then, the right side of (11.13j) vanishes, which indicates that (x_p^*, x_d^*) is an equilibrium of (11.13).

Necessity. Supposing that (x_p^*, x_d^*) is an equilibrium of (11.13), then all the right sides of (11.13) vanish. (11.13a)–(11.17i) are exactly the KKT conditions (11.17a)–(11.13i). $\dot{\rho}_{ik} = 0$ implies $\left(\frac{(P_{ik}^*)^2}{v_i^*} - l_{ik}^* \right) \rho_{ik}^* = 0$ and $\rho_{ik}^* \geq 0$, which is identical to (11.17j). Thus, (x_p^*, x_d^*) is primal–dual optimal for the ESOCP model (11.11). This completes the proof. ∎

1 The detailed proofs can be found Appendix B.

11.4.3 Convergence Analysis

In this subsection, we justify the convergence of the algorithm (11.13) by leveraging the projection gradient theory combined with the LaSalle's invariance principle for switched systems.

Define the sets σ_ρ as

$$\sigma_\rho := \left\{(i,k) \in \mathcal{E} \mid \rho_{ik} = 0,\ P_{ik}^2/v_i - l_{ik} < 0\right\}$$

Then (11.13j) is equivalent to

$$\dot{\rho}_{ik} = \begin{cases} \rho_{ik}\left(\frac{P_{ik}^2}{v_i} - l_{ik}\right), & \text{if } (i,k) \notin \sigma_\rho; \\ 0, & \text{if } (i,k) \in \sigma_\rho. \end{cases} \tag{11.18}$$

From (11.18), it is easy to know $\rho_{ik}(t) \geq 0, \forall t$.

Let $x := (p_i^g, v_i, P_{ik}, l_{ik}, \hat{p}_i, \mu_i, \epsilon_i, \lambda_{ik}, \gamma_{ik}, \rho_{ik})$, and define $F(x)$ in a fixed σ_ρ as

$$F(x) = \begin{bmatrix} G_i(p_i^g) + k_i\epsilon_i - \mu_i + z_i + k_i y_i \\ y_i + \sum\limits_{k:k\in N_i} \gamma_{ik} + \epsilon_i - \varphi_i^- + \varphi_i^+ - \sum\limits_{k\in N_i} \rho_{ik}\frac{P_{ik}^2}{v_i^2} \\ \mu_i + \lambda_{ik} - \gamma_{ik} r_{ik} + 2\rho_{ik}\frac{P_{ik}}{v_i} - z_i \\ -\lambda_{ik} r_{ik} - \rho_{ik} - \rho_{ki} \\ -k_i\epsilon_i - k_i y_i \\ p_i^g - p_i^d - \sum\limits_{k:k\in N_i} P_{ik} \\ -(v_i + k_i p_i^g - v_i^* - k_i\hat{p}_i) \\ -(P_{ik} + P_{ki} - r_{ik} l_{ik}) \\ -(v_i - v_k - r_{ik}\left(P_{ik} - P_{ki}\right)) \\ -\left[\frac{P_{ik}^2}{v_i} - l_{ik}\right]_{\rho_{ik}}^+ \end{bmatrix} \tag{11.19}$$

Obviously, $F(x)$ is continuously differentiable in a fixed σ_ρ.

We further define the set S as

$$S := X \times R^{7m+3n}$$

where X is given in (11.15). For any x, the projection of $(x - F(x))$ onto S is

$$H(x) := \mathcal{P}_S(x - F(x)) := \arg\min_{y\in S} \|y - (x - F(x))\|_2$$

where $\|\cdot\|_2$ is the Euclidean norm. Then, the algorithm (11.13) can be rewritten as

$$\dot{x}(t) = H(x(t)) - x(t) \tag{11.20}$$

A point $x^* \in S$ is an equilibrium of (11.20) if and only if it is a fixed point of the projection $H(s)$, i.e.

$$H(x^*) = x^*.$$

Define E as the set of the equilibrium points of (11.20). That is

$$E := \{ x \in S \mid H(x(t)) - x(t) = 0 \} \tag{11.21}$$

Then we have the following theorem.

Theorem 11.4 *Suppose Assumptions 11.1–11.4 hold. Then every trajectory $x(t)$ of (11.20) starting from a finite initial state asymptotically converges to some equilibrium $x^* \in E$ as $t \to +\infty$ that is optimal for the problem ESOCP.*

Before proving Theorem 11.4, we introduce some definitions and notations for clarity and convenience. Similar to Chapter 5, we define the following function:

$$\tilde{U}(x) := -F(x)^{\mathrm{T}} \cdot (H(x) - x) - \frac{1}{2}||H(x) - x||_2^2 + \frac{1}{2}||x - x^*||_2^2 \tag{11.22}$$

From [30], we know that $\tilde{U}(x) \geq 0$ and $\tilde{U}(x) = 0$ hold only at an equilibrium point x^*.

For a fixed σ_p, $F(x)$ is continuously differentiable, so is $\tilde{U}$. Moreover, $\tilde{U}$ is non-increasing for a fixed σ_p, as we prove in Lemma 11.4. It is worthy of noting that the index set σ_p may sometimes change, resulting in discontinuity of $\tilde{U}$ [29]. To circumvent this issue, we slightly modify the definition of $\tilde{U}$ at the discontinuous points as follows:

1) $U(x) := \tilde{U}(x)$, if $\tilde{U}(x)$ is continuous in x.
2) $U(x) := \limsup_{w \to x} \tilde{U}(w)$, if $\tilde{U}(x)$ is discontinuous in x.

The modified function, $U(x)$, is upper semicontinuous with $U(x) \geq 0$ on S. It is obvious that $U(x) = 0$ holds only at an equilibrium satisfying $x^* = H(x^*)$.

Note that U is continuous almost everywhere except at the switching points. Hence $U(x)$ is *nonpathological* [31, Definition 3 and 4].

Now we are ready to prove Theorem 11.4. We start with the following key lemma.

Lemma 11.4 *Suppose Assumptions 11.1–11.3 hold. Then see the following:*

1) *$U(x)$ is decreasing along the trajectory, $x(t)$, of system (11.20).*
2) *The trajectory $x(t)$ starting from a finite initial state is bounded.*
3) *Every trajectory $x(t)$ starting from a finite initial state ultimately converges to the largest weakly invariant subset Z^* of $Z^+ := \{ x \mid \dot{U}(x) = 0 \}$.*
4) *Every $x^* \in Z^*$ is an equilibrium point of (11.13).*

Proof:

Assertion 1). In light of Theorem 3.2 in [30], $U(x)$ is continuously differentiable if $F(x)$ is continuously differentiable. Its gradient is

$$\nabla_x U(x) = F(x) - (\nabla_x F(x) - I)(H(x) - x) + x - x^* \tag{11.23}$$

Then the derivative of $U(x)$ is

$$\dot{U}(x) = \nabla_x^{\mathrm{T}} U(x) \cdot \dot{x} = \nabla_x^{\mathrm{T}} U(x) \cdot (H(x) - x) \tag{11.24}$$

Combining (11.23) and (11.24), we have

$$\begin{aligned}
\dot{U}(x) &= -(H(x) - x)^{\mathrm{T}} \nabla_x F(x)(H(x) - x) \\
&\quad + \langle F(x) + H(x) - x,\ (H(x) - x) \rangle \\
&\quad + \langle x - x^*,\ (H(x) - x) \rangle \\
&= \langle F(x) + H(x) - x,\ H(x) - x^* + x^* - x \rangle \\
&\quad + \langle x - x^*,\ H(x) - x \rangle \\
&\quad - (H(x) - x)^{\mathrm{T}} \nabla_x F(x)(H(x) - x) \\
&= \langle F(x) + H(x) - x,\ H(x) - x^* \rangle && (11.25\text{a}) \\
&\quad + \langle x - x^*, -F(x) \rangle && (11.25\text{b}) \\
&\quad - (H(x) - x)^{\mathrm{T}} \nabla_x F(x)(H(x) - x) && (11.25\text{c})
\end{aligned}$$

Next, we will prove that (11.25a), (11.25b), and (11.25c) are all nonpositive. For ξ and χ, the projection has the following property [30]:

$$\left\langle \xi - \mathcal{P}(\xi)_S,\ \chi - \mathcal{P}(\xi)_S \right\rangle \leq 0 \quad \forall \chi \in S$$

Set $\xi = x - F(x)$, $\chi = x^*$; then we have

$$\langle F(x) + H(x) - x,\ H(x) - x^* \rangle \leq 0 \tag{11.26}$$

It implies that (11.25a) is nonpositive.

Let $x_1 := (p_i^g, v_i, P_{ik}, l_{ik}, \hat{p}_i)$ and $x_2 := (\mu_i, \epsilon_i, \lambda_{ik}, \gamma_{ik}, \rho_{ik})$; then L defined in (11.16) is convex in x_1 and concave in x_2. It can be verified that

$$\begin{aligned}
\langle x - x^*, -F(x) \rangle &= -(x_1 - x_1^*)^{\mathrm{T}} \nabla_{x_1}^{\mathrm{T}} L + (x_2 - x_2^*)^{\mathrm{T}} \nabla_{x_2}^{\mathrm{T}} L \\
&\leq L(x_1^*, x_2) - L(x_1, x_2) + L(x_1, x_2) - L(x_1, x_2^*) \\
&= \underbrace{L(x_1^*, x_2) - L(x_1^*, x_2^*)}_{\leq 0} + \underbrace{L(x_1^*, x_2^*) - L(x_1, x_2^*)}_{\leq 0} \\
&\leq 0
\end{aligned} \tag{11.27}$$

which indicates that (11.25b) is nonpositive.

For (11.25c), we have

$$\begin{aligned}
&- (H(x) - x)^{\mathrm{T}} \nabla_x F(x)(H(x) - x) \\
&= - \dot{x}^{\mathrm{T}} \nabla_x F(x) \dot{x} \\
&= - \dot{x}^{\mathrm{T}} Q \dot{x} \\
&\leq 0
\end{aligned} \tag{11.28}$$

where Q is given in (11.29) with

$$M_n = \mathrm{diag}\left(\sum_{k \in N_i} \frac{2\rho_{ik} P_{ik}^2}{v_i^3} \right);$$

$$
Q = \begin{bmatrix}
\nabla_{p_i^g} G + I_n + K^2 & -T & 0 & K & -K^2 & -I_n & K & 0 & 0 & 0 & 0 & 0 \\
-T^{\mathrm{T}} & \left[\frac{2\rho_{ik}}{v_i}\right]_d + I_{2m} & 0 & D^{\mathrm{T}} & -T^{\mathrm{T}} & 0 & (\tilde{I}^1)^{\mathrm{T}} & (\tilde{I}^1)^{\mathrm{T}} & 0 & 0 & \left[\frac{2P_{ik}}{v_i}\right]_d & 0 \\
0 & 0 & 0 & 0 & 0 & 0 & -[r_{ik}]_d & 0 & 0 & 0 & -\tilde{I}^1 & 0 \\
K & D & 0 & M_n + I_n & -K & I_n & 0 & \tilde{I}^2 & -I_n & I_n & R & 0 \\
-K^2 & 0 & 0 & -K & K^2 & 0 & -K & 0 & 0 & 0 & 0 & 0 \\
I_n & T & 0 & 0 & 0 & 0 & 0 & 0 & 0 & 0 & 0 & 0 \\
-K & 0 & 0 & -I_n & 0 & 0 & K & 0 & 0 & 0 & 0 & 0 \\
0 & -\tilde{I}^1 & [r_{ik}]_d & 0 & 0 & 0 & 0 & 0 & 0 & 0 & 0 & 0 \\
0 & -\tilde{I}^1 & 0 & -(\tilde{I}^2)^{\mathrm{T}} & 0 & 0 & 0 & 0 & 0 & 0 & 0 & 0 \\
0 & 0 & 0 & I_n & 0 & 0 & 0 & 0 & 0 & 0 & 0 & 0 \\
0 & 0 & 0 & -I_n & 0 & 0 & 0 & 0 & 0 & 0 & 0 & 0 \\
0 & \left[-\frac{2P_{ik}}{v_i}\right]_d & (\tilde{I}^1)^{\mathrm{T}} & -R^{\mathrm{T}} & 0 & 0 & 0 & 0 & 0 & 0 & 0 & 0
\end{bmatrix}
\tag{11.29}
$$

$$T_{n\times 2m}(s,t) = \begin{cases} 1, & \text{if } (s,t+1) \in \mathcal{E} \text{ or } (s,t-m+1) \in \mathcal{E} \\ 0, & \text{otherwise} \end{cases};$$

$$D_{n\times 2m}(s,t) = \begin{cases} -\frac{2\rho_{s,t+1}P_{s,t+1}}{v_s^2}, & \text{if } (s,t+1) \in \mathcal{E} \\ & \text{or } (s,t-m+1) \in \mathcal{E} \\ 0, & \text{otherwise} \end{cases};$$

$$R_{n\times 2m}(s,t) = \begin{cases} -\frac{P_{s,t+1}^2}{v_s^2}, & \text{if } (s,t+1) \in \mathcal{E} \\ & \text{or } (s,t-m+1) \in \mathcal{E} \\ 0, & \text{otherwise} \end{cases};$$

$$\tilde{I}^1{}_{m\times 2m}(s,t) = \begin{cases} 1, & \text{if } (s,t+1) \in \mathcal{E} \text{ or } (s,t-m+1) \in \mathcal{E} \\ 0, & \text{otherwise} \end{cases};$$

$$\tilde{I}^2{}_{m\times 2m}(s,t) = \begin{cases} 1, & \text{if } (s,t+1) \in \mathcal{E} \text{ and } s \le t \\ -1, & \text{if } (s,t-m+1) \in \mathcal{E} \text{ and } s > t-m \\ 0, & \text{otherwise} \end{cases}.$$

Here, Q is a semi-definite positive matrix. I is an identity matrix, where the subscript stands for its dimension. $[c_i]_d$ is the diagonal matrix composed of c_i with proper dimensions. Moreover, Q is the sum of two matrices, one of which is skew symmetric and the other is positive symmetric.

Note that the index set σ_ρ may change during the decreasing of U. We have the following two kinds of observations:

1) The set σ_ρ is reduced, which only happens when $\frac{P_{ik}^2}{v_i} - l_{ik}$ changes its sign from negative to positive. Hence an extra term will be added to U. As this term is initially zero, there is no discontinuity of U in this case.
2) The set σ_ρ is enlarged when ρ_{ik} goes to zero from positive while $\frac{P_{ik}^2}{v_i} - l_{ik} \le 0$.

   Here U will lose the positive term $\left(\frac{P_{ik}^2}{v_i} - l_{ik}\right)^2/2$, causing discontinuity.

Therefore, it can be concluded that U is decreasing even when σ_ρ changes, which justifies Assertion 1).

Assertion 2). Note that [[30], Theorem 3.1] has proved that

$$-F(x)^{\mathrm{T}}(H(x)-x) - \frac{1}{2}||H(x)-x||_2^2 \ge 0.$$

Therefore, we have

$$\frac{1}{2}||x-x^*||_2^2 \le U(t) \le U(0)$$

It indicates that $x(t)$ is bounded, which verifies Assertion 2).

Assertion 3). Given an initial point $x(0)$, there is a compact set $\Omega_0 := \Omega(x(0)) \subset S$ such that $x(t) \in \Omega_0$ for $t \ge 0$ and $\dot{U}(x) \le 0$ in Ω_0.

In addition, U is radially unbounded and positive definite except at equilibrium. As U and $\dot{U}$ are nonpathological, we conclude that any trajectory $x(t)$ starting from Ω_0 eventually converges to the largest weakly invariant subset Z^* contained in $Z^+ = \{ x \in \Omega_0 \mid \dot{U}(x) = 0 \}$ [[31], Proposition 3]. Therefore, the third assertion holds.

Assertion 4). To satisfy $\dot{U}(x) = 0$, both terms in (11.27) have to be zero, implying that $L(x_1^*, x_2) \equiv L(x_1^*, x_2^*)$ must hold in Z^+. Differentiating it with respect to t gives

$$\left(\frac{\partial}{\partial x_2} L(x_1^*, x_2(t))\right)^{\mathrm{T}} \cdot \dot{x}_2(t) = 0 = \dot{x}_2(t)^{\mathrm{T}} \dot{x}_2(t) \tag{11.30}$$

The second equality holds due to (11.13f)–(11.13j). Then, we conclude $\dot{x}_2(t) = 0$ due to the boundedness of $x(t)$. It directly follows that $\mu_i, \epsilon_i, \lambda_{ik}, \gamma_{ik}, \rho_{ik}$ are constants and $y_i = z_i = 0$ in Z^+. Then we further know $\dot{l}_{ik} = \dot{\hat{p}}_i = 0$ from (11.13d), (11.13e), and the boundedness of $x(t)$.

Combining (11.28) and (11.29) yields

$$\begin{aligned}
\dot{U}(x) &\leq -\dot{x}^{\mathrm{T}} Q \dot{x} \\
&= -\sum_{i\in\mathcal{N}} (\dot{p}_i^g)^{\mathrm{T}} \cdot \nabla^2_{p^g} f \cdot \dot{p}_i^g - \sum_{i\in\mathcal{N}} \left(\dot{p}_i^g - \sum_{k\in N_i} \dot{P}_{ik}\right)^2 \\
&\quad - \sum_{i\in\mathcal{N}} \frac{2\rho_{ik}}{v_i} \left(\dot{P}_{ik} - \sum_{k\in N_i} \frac{|P_{ik}|}{v_i} \dot{v}_i\right)^2 \\
&\quad - \sum_{i\in\mathcal{N}} (\dot{v}_i + k_i \dot{p}_i^g - k_i \dot{\hat{p}}_i)^2
\end{aligned} \tag{11.31}$$

We can directly get $\dot{p}_i^g = 0$ by noting Assumption 11.1. From $\dot{v}_i + k_i \dot{p}_i^g - k_i \dot{\hat{p}}_i = 0$ and $\dot{p}_i^g = \dot{\hat{p}}_i = 0$, we have $\dot{v}_i = 0$. If $\rho_{ik} = 0$, then $\dot{P}_{ik}$ is a constant, implying $\dot{P}_{ik} = 0$. If $\rho_{ik} > 0$, then $\frac{P_{ik}^2}{v_i} = l_{ik}$, implying P_{ik} a constant. Therefore, we conclude that $\dot{P}_{ik} = 0$ always holds. Consequently, we have that $\dot{x}(t) = 0$ in Z^*, which proves the last assertion of Lemma 11.4. ■

With Lemma 11.4, we can further prove Theorem 11.4.

Proof: [Proof of Theorem 11.4]

Consider a trajectory $(x(t), t \geq 0)$ of (11.20) starting from an arbitrary initial state $x(0)$. As mentioned in the proof of Lemma 11.4, $x(t)$ stays entirely in a compact set Ω_0. Hence there exists an infinite sequence of time instants t_k such that $x(t_k) \to \hat{x}^*$ as $t_k \to \infty$, for some $\hat{x}^* \in Z^*$. The Assertion 4) in Lemma 11.4 guarantees that $\hat{x}^*$ is an equilibrium point of the system (11.20), and hence we know $\hat{x}^* = H(\hat{x}^*)$. Then, using this specific equilibrium point $\hat{x}^*$ in the definition of U, we have

$$\begin{aligned}
U^* &= \lim_{t\to\infty} U(x(t)) = \lim_{t_k\to\infty} U(x(t_k)) \\
&= \lim_{x(t_k)\to\hat{x}^*} U\left(x(t_k)\right) = U(\hat{x}^*) \\
&= 0
\end{aligned}$$

Here, the first equality uses the fact that $U(t)$ is nonincreasing in t, the second equality uses the fact that t_k is the infinite sequence of t, the third equality uses the fact that $x(t)$ is absolutely continuous in t, the fourth equality is due to the upper semicontinuity of $U(x)$, and the last equality holds as $\hat{x}^*$ is an equilibrium point of $U(x)$.

The quadratic term $(x - \hat{x}^*)^T(x - \hat{x}^*)$ in $U(x)$ then guarantees that $x(t) \to \hat{x}^*$ as $t \to \infty$, which completes the proof. ∎

11.5 Implementation

We consider heterogeneous DC microgrids, where each microgrid may adopt one of the three types of controls: voltage control, power control, and droop control. The required control commands are p_i^g, v_i, and $\hat{p}_i$, respectively. To merge the optimization and the real-time control, the values of p_i^g, v_i, and $\hat{p}_i$ during transient are sent to the controllable DGs as control commands. The control commands are denoted by p_{gi}^{ref}, v_i^{ref}, and $\hat{p}_i^{\text{ref}}$, respectively, to distinguish them from the state variables p_i^g, v_i, and $\hat{p}_i$.

For voltage control and power control, the algorithm (11.13) can generate optimal p_{gi}^{ref} and v_i^{ref} that are feasible even during transient. For droop control, $\hat{p}_i^{\text{ref}}$ is generated to ensure the optimal working point in steady state. The distributed control diagram unifying the three types of microgrid controls is shown in Figure 11.1.

In Figure 11.1, the left part is the proposed distributed algorithm. Its inputs are the local information of p_i^g, P_{ik}, l_{ik}, v_i, $\hat{p}_i$ μ_i, ϵ_i, λ_{ik}, γ_{ik}, ρ_{ik} and the neighboring information of P_{ki}, v_k, ρ_{ki}, $k \in N_i$. The outputs are p_{gi}^{ref}, v_i^{ref}, and $\hat{p}_i^{\text{ref}}$. The right part is the diagrams of the three controls. For a power-controlled DG, it has two control loops, the power loop and the current loop, where I_{ref} is the current reference for the current control loop and I_i is the measured current. For a voltage controlled DG, it also has two control loops, the power control loop and the current control loop, where v_i is the voltage measurement. For a droop-controlled DG, it has three control loops, the droop control loop, the power control loop, and the current control loop, where both the voltage and current need to be measured.

Figure 11.1 shows that the proposed distributed optimal control unifies three commonly used controls, to a large extent, breaking the restrictions arising from the heterogeneity of microgrid controls.

Remark 11.2 For microgrid i, its neighbors' information P_{ki} and v_k can be estimated locally using the following equations:

$$P_{ki} = P_{ik} - r_{ik} I_{ik}^2 \tag{11.32}$$

$$v_k = (\sqrt{v_i} - r_{ik} I_{ik})^2 \tag{11.33}$$

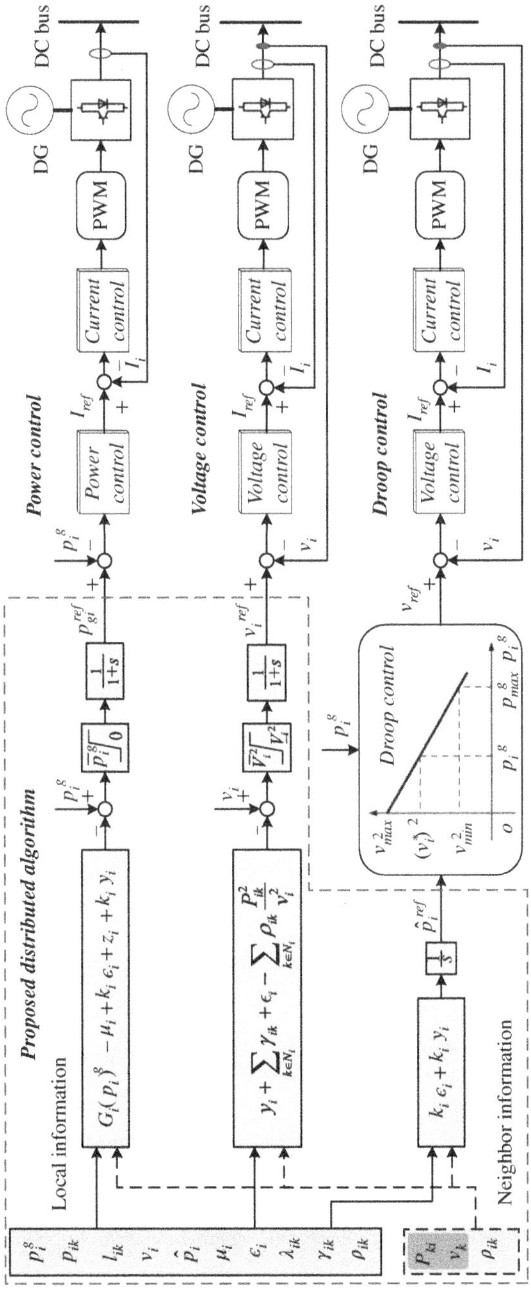

Figure 11.1 The proposed distributed optimal control diagram that unifies three control modes.

where the line current I_{ik} from microgrid i to k can be measured locally. Then, only $\rho_{ki}, k \in N_i$ need to be exchanged between neighbors, significantly reducing the communication burden.

11.6 Case Studies

11.6.1 Test System and Data

To verify the effectiveness of the devised approach, we carry out numerical tests on a multi-microgrid DC system, whose topology is based on the low-voltage microgrid benchmark presented in [32]. The system includes two feeders with six dispatchable DGs, which are divided into six microgrids (Figure 11.2). The Breaker 1 is open; thus the system operates in a stand-alone mode. All simulations are performed in PSCAD with Matlab on a laptop with 8 GB memory and 2.39 GHz CPU. The DC microgrids system is modeled in PSCAD, whereas the distributed optimal control algorithm (11.13) runs in Matlab. They are combined through a user-defined interface. The control commands obtained in Matlab are sent to

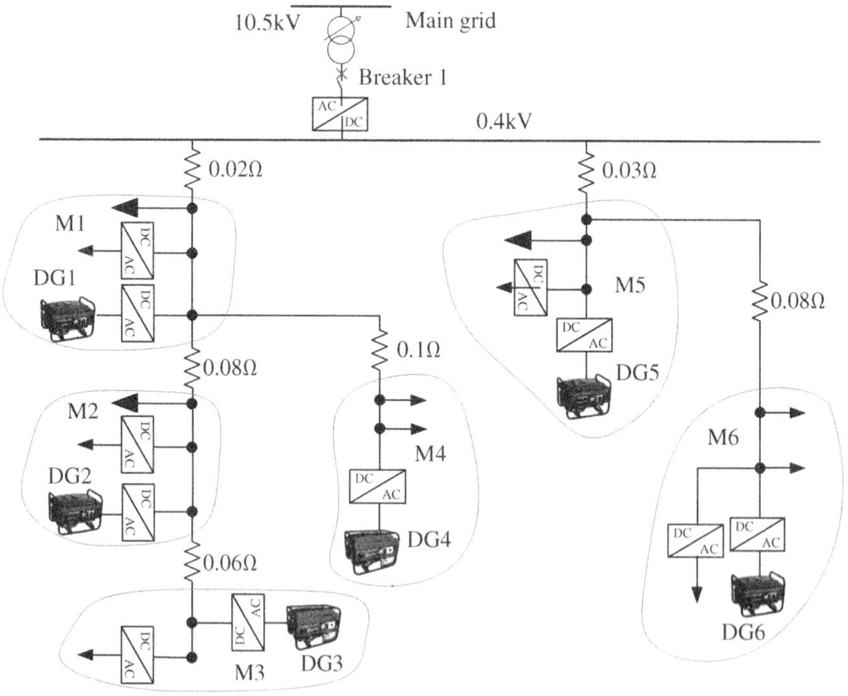

Figure 11.2 Six-microgrid system.

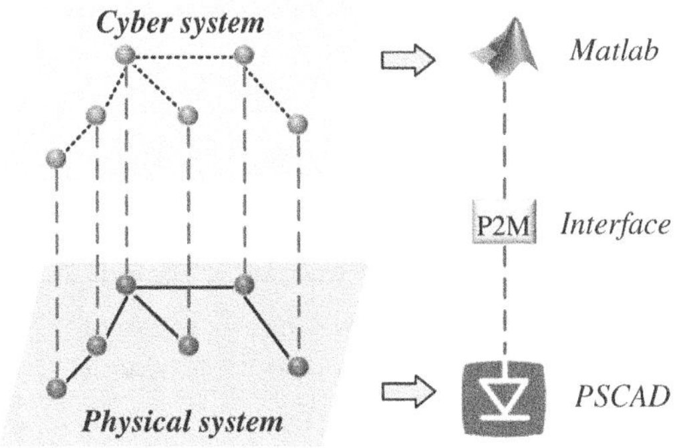

Figure 11.3 Simulation design combined with PSCAD and Matlab.

corresponding microgrids in PSCAD via the interface. Conversely, MATLAB can also collect data from PSCAD. In this regard, the PSCAD represents the physical subsystem, while MATLAB represents the cyber subsystem. Communication processes in the cyber subsystems to obtain information from neighboring microgrids. The joint simulation flowchart is illustrated in Figure 11.3. The control algorithm (11.13) in a continuous form in the analysis is discretized in realization. At each control step, the individual local controller only needs to compute a few multiplication and addition operations based on the latest information. In the case studies, it takes only 0.1 ms per step in average, which is fast enough for the purpose of real-time control.

The objective function is set as $f_i(p_i^g) = \frac{a_i}{2}\left(p_i^g\right)^2 + b_i p_i^g$ [26, 33]. It satisfies Assumption 11.1 when $b_i > 0$. Key parameters for the microgrids are given in Table 11.1. The initial load demands in each microgrid is [41, 40, 42, 39, 42, 40] kW. Then they jump to [51, 50, 52, 49, 52, 50] kW at the moment of 1s. All the three regular local controls are utilized in the DC microgrid system, where DG1 and DG6 adopt the droop control, DG2 and DG5 adopt the power control, and DG3 and DG4 adopt the voltage control.

Table 11.1 System parameters.

	M_1	M_2	M_3	M_4	M_5	M_6
a_i	0.036	0.03	0.035	0.03	0.035	0.042
b_i	1	1	1	1	1	1
p_i^d(kW)	51	50	52	49	52	50
$\overline{p}_i^g$(kW)	50	60	55	60	55	45
$\overline{V}_i$ (V)	420	420	420	420	420	420
$\underline{V}_i$ (V)	380	380	380	380	380	380
k_i	0.12	0.125	0.164	0.131	0.156	0.131

Table 11.2 Comparision with centralized optimization.

	Generation (kW)			Power reference (kW)		
DG_i	p_g^d	p_g^c	e (%)	$\hat{p}^d$	$\hat{p}^c$	e (%)
1	48.1701	48.1823	−0.0253	46.7315	46.8543	−0.2621
2	57.4041	57.4227	−0.0324	56.3536	56.7590	−0.5381
3	49.3516	49.3602	−0.0174	48.2000	48.6611	−0.3311
4	56.9853	56.9861	−0.0014	56.9264	56.9861	−0.1048
5	49.9761	50.0053	−0.0584	48.8660	48.3499	0.4470
6	42.1217	42.1495	−0.0660	39.2465	39.2176	0.0737

11.6.2 Accuracy Analysis

In this subsection, we use the convex optimization (CVX) tool built in Matlab to solve the ESOCP problem. The results after load changes are utilized as the baseline to validate the accuracy of the proposed distributed control algorithm. Numerical results of these two methods are listed in Table 11.2 for comparison.

In Table 11.2, p_g^d and $\hat{p}^d$ are given by the proposed distributed algorithm, while p_g^c and $\hat{p}^c$ are obtained using the CVX tool. e is the relative error between p^d (or p^c) and $\hat{p}^d$ (or $\hat{p}^c$). The results in Table 11.2 indicate that the errors between p_g^d and p_g^c in all MGs are smaller than 0.07%, while the errors between $\hat{p}^d$ and $\hat{p}^c$ are smaller than 0.6%. The results well validate the accuracy of the proposed distributed optimal control.

11.6.3 Dynamic Performance Verification

In this subsection, we analyze the impacts of generation limits on the dynamic property. To do so, we compare dynamic responses of the inverter outputs in M2 and M5 in two scenarios: with and without generation constraints. The trajectories in the two cases are given in Figure 11.4. In both cases, the same steady-state generations are achieved. However, with the proposed controller, the generations of DG2 and DG5 without constraints remain within the limits in both transient and steady states. On the contrary, the generations of DG2 and DG5 violate their upper limits in transient, which is practically infeasible.

Similarly, we compare the voltage dynamics of DG3 and DG4 in the following two scenarios: with and without capacity constraints. The trajectories in the two cases are given in Figure 11.5. In both cases, the same steady-state voltages are achieved. However, with the saturated controller, the voltages of DG3 and DG4 remain within the limits in both transient and steady states. On the contrary, voltages of DG3 and DG4 violate their upper limits in the transient process if

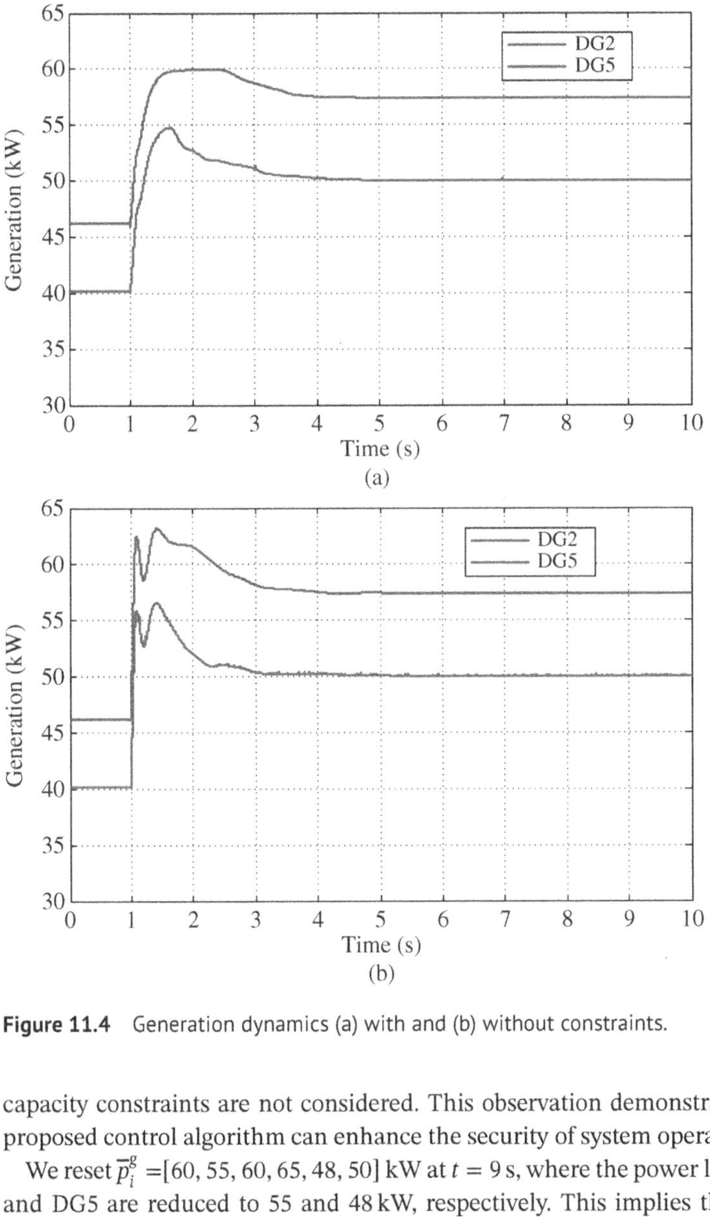

Figure 11.4 Generation dynamics (a) with and (b) without constraints.

capacity constraints are not considered. This observation demonstrates that the proposed control algorithm can enhance the security of system operation.

We reset $\overline{p}_i^g$ =[60, 55, 60, 65, 48, 50] kW at $t = 9$ s, where the power limits of DG2 and DG5 are reduced to 55 and 48 kW, respectively. This implies that they can be strictly reached in the steady state. This scenario often happens in microgrids since the generation limits of renewable resources such as wind turbines and PVs may vary rapidly. The generations of all DGs with different capacity constraints are given in Figure 11.6.

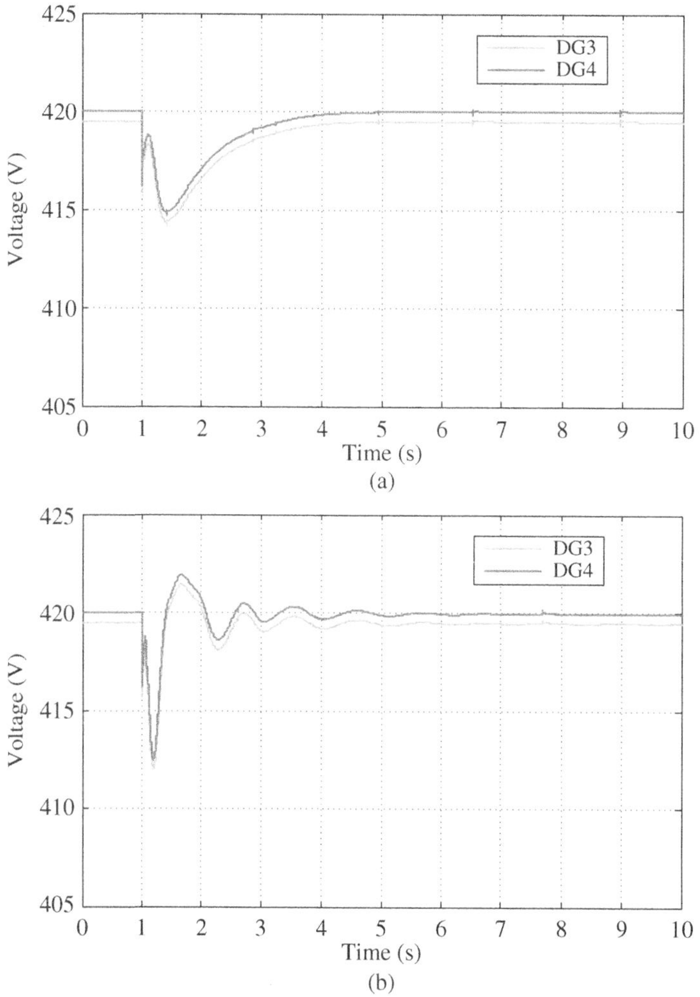

Figure 11.5 Voltage dynamics (a) with and (b) without constraints.

The steady-state generations in Figure 11.6 are identical to the results obtained by CVX. In Figure 11.6, it is shown that generations of DG2 and DG5 reduce rapidly to the capacity limits after load changes. Other DGs will change their generations to rebalance the power mismatch across the overall system. It indicates that the proposed distributed optimal control can adapt to unknown variation caused by load change or generation volatility.

In power systems, time delay always exists in communication regardless of whether the control is centralized or distributed. Figures 11.7 and 11.8 shows the dynamics of generation of DG5 and voltage of DG3 under time delays of 5, 10, 15,

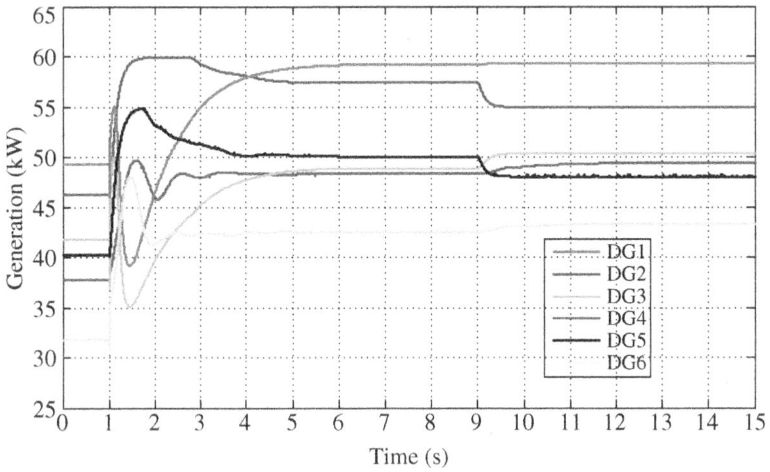

Figure 11.6 Generation dynamics of different constraints.

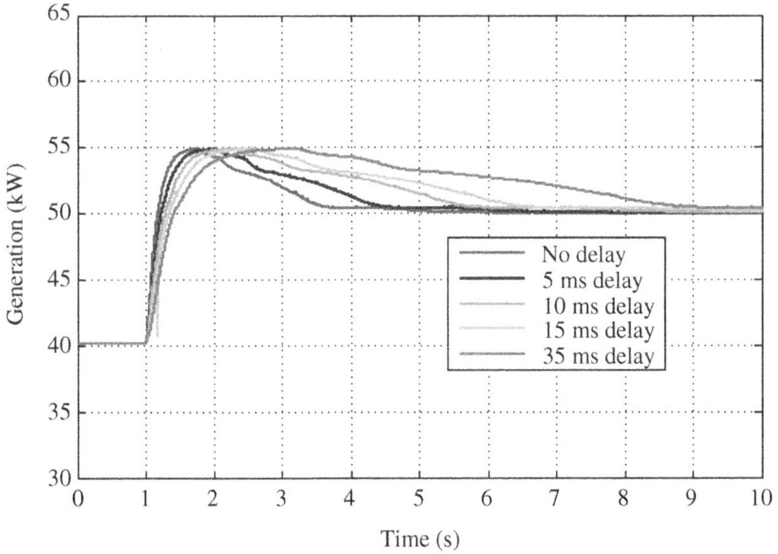

Figure 11.7 Generation dynamics under different time delays.

and 35 ms, respectively. It is shown that the steady-state values of both generation and voltage under different time delays are identical, which validates that the proposed distributed optimal control algorithm can adapt to different time delays and meet real-time requirements. Large delays, however, can lead to a slower convergence rate and a severer voltage nadir.

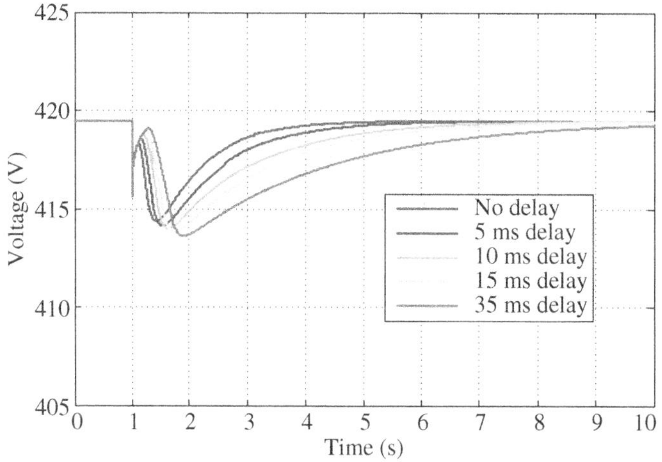

Figure 11.8 Voltage dynamics under different time delays.

11.6.4 Performance in Plug-n-play Operations

In this case, M6 is switched off at 1 s; then it is switched on at 9 s. When M6 is switched off, it has to supply the load demand itself while microgrids M1–M5 remain connected. The generation and voltage dynamics in this process are illustrated in Figures 11.9 and 11.10, respectively.

It is observed that the output of DG6 increases to 50 kW to supply the load in M6 after being disconnected from the main system. At the same time, the voltage

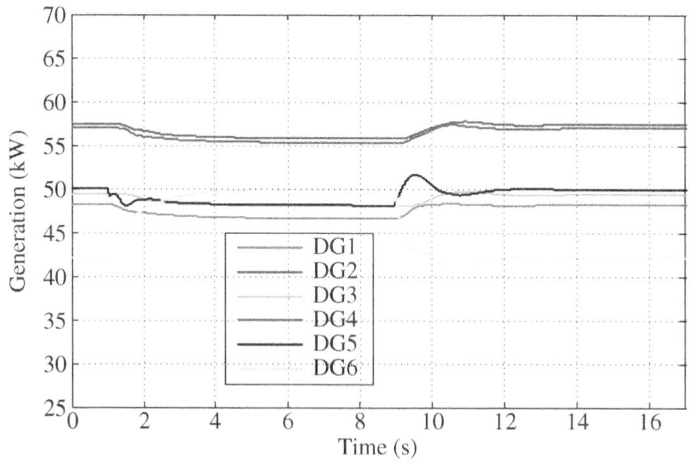

Figure 11.9 Generation dynamics of plug and play.

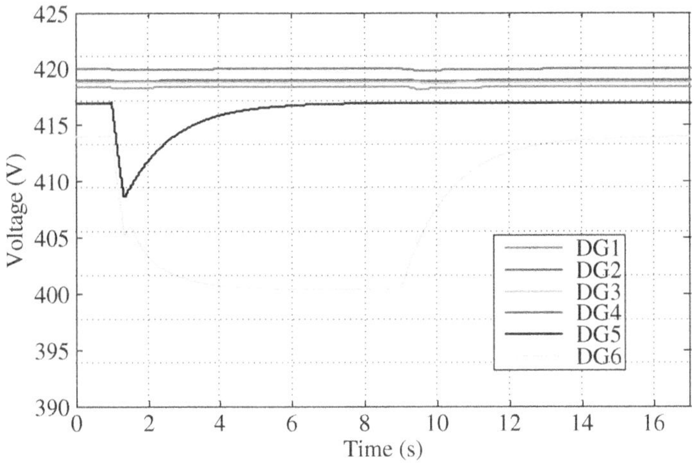

Figure 11.10 Voltage dynamics of plug and play.

of DG6 drops to 400 V. The result is identical to that obtained by CVX. In addition, after M6 is reconnected, the generations and voltages of all DGs recover to the nominal values. Comparing the voltages of different MGs in transient, it is shown that only the two MGs connected directly with the breaking point are influenced remarkably, while other MGs experience a moderate transient process. This result validates that the proposed distributed optimal control algorithm can admit plug-n-play operations.

11.7 Conclusion and Notes

This chapter addresses the distributed optimal control of stand-alone DC microgrids, where each microgrid may adopt one of the three different control strategies, such as power control, voltage control, and droop control. The controller can provide commands for all these strategies, which implies it breaks the restriction of various control strategies to achieve an optimal operation point. A six-microgrid system based on the microgrid benchmark is utilized to demonstrate the efficacy of our designs. The error of results between the proposed method and the CVX tool is smaller than 0.6%, which validates the accuracy of the proposed approach. Moreover, the commands for power-controlled and voltage-controlled microgrids satisfy generation limits and voltage limits in both transient processes and steady states. This enhances the security of the DC system. In addition, our controller can adapt to the uncertainties of renewable generations. Finally, the proposed approach can support the plug-n-play operation of DC microgrids.

It is common that diverse control configurations are involved in distributed controller design. This chapter suggests a methodology to systematically design a nondisruptive distributed optimal controller that can unify and coordinate heterogeneous local controllers. Again, the idea of forward and reverse engineering plays a crucially important role.

It should be noted that, in this chapter, we use the SOC relaxation to convexify the model of DC microgrids. However, the exactness of the convexification has not been justified here. We leave it to Appendix B as it is beyond the focus of this chapter to some extent.

Bibliography

1 J. J. Justo, F. Mwasilu, J. Lee, and J.-W. Jung, "AC-microgrids versus DC-microgrids with distributed energy resources: a review," *Renewable and Sustainable Energy Reviews*, vol. 24, pp. 387–405, 2013.

2 E. Planas, J. Andreu, J. I. Gárate, I. M. de Alegría, and E. Ibarra, "AC and DC technology in microgrids: a review," *Renewable and Sustainable Energy Reviews*, vol. 43, pp. 726–749, 2015.

3 T. Dragicevic, X. Lu, J. C. Vasquez, and J. M. Guerrero, "DC microgrids - Part I: a review of control strategies and stabilization techniques," *IEEE Transactions on Power Electronics*, vol. 31, no. 7, pp. 4876–4891, 2016.

4 T. Dragicevic, X. Lu, J. C. Vasquez, and J. M. Guerrero, "DC microgrids - Part II: a review of power architectures, applications, and standardization issues," *IEEE Transactions on Power Electronics*, vol. 31, no. 7, pp. 3528–3549, 2016.

5 A. T. Elsayed, A. A. Mohamed, and O. A. Mohammed,"DC microgrids and distribution systems: an overview," *Electric Power Systems Research*, vol. 119, pp. 407–417, 2015.

6 J. M. Guerrero, J. C. Vasquez, J. Matas, L. G. de Vicuna, and M. Castilla, "Hierarchical control of droop-controlled AC and DC microgrids - a general approach toward standardization," *IEEE Transactions on Industrial Electronics*, vol. 58, no. 1, pp. 158–172, 2011.

7 Q. Shafiee, T. Dragicevic, J. C. Vasquez, and J. M. Guerrero, "Hierarchical control for multiple DC-microgrids clusters," *IEEE Transactions on Energy Conversion*, vol. 29, no. 4, pp. 922–933, 2014.

8 M. Yazdanian and A. Mehrizi-Sani, "Distributed control techniques in microgrids," *IEEE Transactions on Smart Grid*, vol. 5, no. 6, pp. 2901–2909, 2014.

9 A. Maknouninejad, Z. Qu, F. L. Lewis, and A. Davoudi, "Optimal, nonlinear, and distributed designs of droop controls for DC microgrids," *IEEE Transactions on Smart Grid*, vol. 5, no. 5, pp. 2508–2516, 2014.

10 A. Khorsandi, M. Ashourloo, and H. Mokhtari, "A decentralized control method for a low-voltage DC microgrid," *IEEE Transactions on Energy Conversion*, vol. 29, no. 4, pp. 793–801, 2014.
11 Y. Gu, X. Xiang, W. Li, and X. He, "Mode-adaptive decentralized control for renewable DC microgrid with enhanced reliability and flexibility," *IEEE Transactions on Power Electronics*, vol. 29, no. 9, pp. 5072–5080, 2014.
12 D. Chen and L. Xu, "Autonomous dc voltage control of a DC microgrid with multiple slack terminals," *IEEE Transactions on Power Systems*, vol. 27, no. 4, pp. 1897–1905, 2012.
13 M. D. Cook, G. G. Parker, R. D. Robinett, and W. W. Weaver, "Decentralized mode-adaptive guidance and control for DC microgrid," *IEEE Transactions on Power Delivery*, vol. 32, no. 1, p. 263–2711, 2017.
14 X. Lu, K. Sun, J. M. Guerrero, J. C. Vasquez, and L. Huang, "State-of-charge balance using adaptive droop control for distributed energy storage systems in DC microgrid applications," *IEEE Transactions on Industrial Electronics*, vol. 61, no. 6, pp. 2804–2815, 2014.
15 Q. Shafiee, T. Dragicevic, F. Andrade, J. C. Vasquez, and J. M. Guerrero, "Distributed consensus-based control of multiple DC-microgrids clusters," in *IECON 2014 - 40th Annual Conference of the IEEE Industrial Electronics Society*, (Dallas, USA), pp. 2056–2062, IEEE, Oct 2014.
16 V. Nasirian, S. Moayedi, A. Davoudi, and F. L. Lewis, "Distributed cooperative control of DC microgrids," *IEEE Transactions on Power Electronics*, vol. 30, no. 4, pp. 2288–2303, 2015.
17 Z. Lv, Z. Wu, X. Dou, and M. Hu, "Discrete consensus-based distributed secondary control scheme with considering time-delays for DC microgrid," in *IECON 2015 - 41st Annual Conference of the IEEE Industrial Electronics Society*, (Yokohama, Japan), pp. 002898–002903, IEEE, Nov 2015.
18 L. Che and M. Shahidehpour, "DC microgrids: economic operation and enhancement of resilience by hierarchical control," *IEEE Transactions on Smart Grid*, vol. 5, no. 5, pp. 2517–2526, 2014.
19 J. Xiao, P. Wang, and L. Setyawan, "Hierarchical control of hybrid energy storage system in DC microgrids," *IEEE Transactions on Industrial Electronics*, vol. 62, no. 8, pp. 4915–4924, 2015.
20 S. Moayedi and A. Davoudi, "Distributed tertiary control of DC microgrid clusters," *IEEE Transactions on Power Electronics*, vol. 31, no. 2, pp. 1717–1733, 2016.
21 C. Zhao, U. Topcu, N. Li, and S. H. Low, "Design and stability of load-side primary frequency control in power systems," *IEEE Transactions on Automatic Control*, vol. 59, no. 5, pp. 1177–1189, 2014.
22 N. Li, C. Zhao, and L. Chen, "Connecting automatic generation control and economic dispatch from an optimization view,"*IEEE Transactions on Control of Network Systems*, vol. 3, no. 3, pp. 254–264, 2016.

23 F. Dorfler, J. W. Simpson-Porco, and F. Bullo, "Breaking the hierarchy: distributed control and economic optimality in microgrids," *IEEE Transactions on Control of Network Systems*, vol. 3, no. 3, pp. 241–253, 2016.

24 S. Moayedi and A. Davoudi, "Unifying distributed dynamic optimization and control of islanded DC microgrids," *IEEE Transactions on Power Electronics*, vol. 32, no. 3, pp. 2329–2346, 2017.

25 A. A. Hamad, M. A. Azzouz, and E. F. El-Saadany, "Multiagent supervisory control for power management in DC microgrids," *IEEE Transactions on Smart Grid*, vol. 7, no. 2, pp. 1057–1068, 2016.

26 Z. Wang, W. Wu, and B. Zhang, "A distributed control method with minimum generation cost for DC microgrids," *IEEE Transactions on Energy Conversion*, vol. 31, no. 4, pp. 1462–1470, 2016.

27 L. Gan and S. H. Low, "Optimal power flow in direct current networks," *IEEE Transactions on Power Systems*, vol. 29, no. 6, pp. 2892–2904, 2014.

28 J. Li, F. Liu, Z. Wang, S. Low, and S. Mei, "Optimal power flow in stand-alone DC microgrids," *IEEE Transactions on Power Systems*, vol. 33, no. 5, pp. 5496–5506, 2018.

29 D. Feijer and F. Paganini, "Stability of primal-dual gradient dynamics and applications to network optimization," *Automatica*, vol. 46, pp. 1974–1981, 2010.

30 M. Fukushima, "Equivalent differentiable optimization problems and descent methods for asymmetric variational inequality problems," *Mathematical Programming*, vol. 53, no. 1, pp. 99–110, 1992.

31 A. Bacciotti and F. Ceragioli, "Nonpathological Lyapunov functions and discontinuous carathéodory systems," *Automatica*, vol. 42, no. 3, pp. 453–458, 2006.

32 S. Papathanassiou, N. Hatziargyriou, K. Strunz, *et al.*, "A benchmark low voltage microgrid network," in *Proceedings of the CIGRE Symposium: Power Systems with Dispersed Generation*, (Athens, Greece), pp. 1–8, CIGRE, 2005.

33 Y. Xu and Z. Li, "Distributed optimal resource management based on the consensus algorithm in a microgrid," *IEEE Transactions on Industrial Electronics*, vol. 62, no. 4, pp. 2584–2592, 2015.

Appendix A

Typical Distributed Optimization Algorithms

In the distributed framework, an optimization problem is usually decomposed into a set of subproblems. Individual agents exchange necessary information with each other over a communication network and iteratively solve the subproblems till the convergence. In the past decades, a number of distributed optimization methods have been developed to deal with a wide variety of optimization problems. This appendix provides readers with a short summary of typical distributed methods, including consensus-based algorithms, first-order (sub)gradient-based algorithms, second-order Newton-based algorithms, and zeroth-order online algorithms.

We adopt the following notations in this appendix. We use $\mathbb{R}^n$ ($\mathbb{R}^n_+$) to denote the n-dimensional (nonnegative) Euclidean space. For a column vector $x \in \mathbb{R}^n$ (a matrix $A \in \mathbb{R}^{m \times n}$), x^{T} (A^{T}) represents its transpose. For $x, y \in \mathbb{R}^n$, we denote the inner product by $\langle x, y \rangle = x^{\mathrm{T}} y$. For $x \in \mathbb{R}^n$, $\operatorname{diag}(x) \in \mathbb{R}^{n \times n}$ represents the diagonal matrix whose diagonal entries are the entries of x one for one. Denote by $\mathbf{1}$ and I the all-ones vector and the identity matrix, respectively. For a set Ω, $|\Omega|$ stands for its cardinality. Let $[x]_+$ denote the projection of x onto $\mathbb{R}^n_+$. For a nonempty set Ω, define the indicator function ι_Ω as

$$\iota_\Omega(x) := \begin{cases} 0, & x \in \Omega \\ +\infty, & x \notin \Omega \end{cases} \tag{A.1}$$

A.1 Consensus-Based Algorithms

Consensus-based algorithms are usually applied to multi-agent decision-making systems, where the agents can choose some particular variable as a consensus indicator to coordinate individual behaviors. In this circumstance, one can decompose the original optimization problem into a set of individual subproblems that

Merging Optimization and Control in Power Systems: Physical and Cyber Restrictions in Distributed Frequency Control and Beyond, First Edition. Feng Liu, Zhaojian Wang, Changhong Zhao, and Peng Yang.

are solved locally. As long as all the subproblems reach the consensus, the global optimum is obtained. Due to its simplicity, this approach has been widely deployed in various kinds of optimization problems.

A.1.1 Consensus Algorithms

In many multi-agent systems, a fundamental task is to achieve an agreement among the agents regarding a certain output with only neighboring communications. The consensus algorithm is proposed to deal with such an agreement problem. In this scheme, all agents repeat exchanging values with their neighbors through the communication network and modifying their own outputs until reaching an agreement. Consensus algorithms are simple but powerful, which have exhibited an excellent performance on coping with both the time-varying topology and communication delay [1–4]. To cope with complicated problems, consensus methods are usually combined with other distributed algorithms, e.g. the dual decomposition (DD) method [5], the alternating direction method of multipliers (ADMM) [6], and the primal–dual gradient (PDG) method [7].

Consider a multi-agent system with a set $\mathcal{N}$ of agents and a set $\mathcal{L}$ of communication links. Assume that the graph $\mathcal{G} = \{\mathcal{N}, \mathcal{L}\}$ is an undirected connected graph. $a_{ij} \geq 0$ is the weight of the link $(i,j) \in \mathcal{L}$, where $a_{ij} = 0$ if $(i,j) \notin \mathcal{L}$. For agent $i \in \mathcal{N}$, the set of its neighbors are defined as $\mathcal{N}_i = \{j \in \mathcal{N} \mid (i,j) \in \mathcal{L}\}$. The variable for consensus of the agent i is denoted by x_i. Then the standard consensus algorithm is presented as [2]

$$x_i(k+1) = x_i(k) + \alpha_k \sum_{j\in\mathcal{N}_i} a_{ij}\left(x_j(k) - x_i(k)\right) \tag{A.2}$$

where $x_i(k)$ is the consensus output of the agent i at the kth iteration and $\alpha_k > 0$ is the step size. When the algorithm converges, all the consensus variables are identical, i.e. $x_i^* = x_j^*, \forall i,j \in \mathcal{N}$.

Besides the discrete form, the consensus algorithm is also applied to the continuous systems [2]

$$\dot{x}_i = \alpha \sum_{j\in\mathcal{N}_i} a_{ij}\left(x_j - x_i\right) \tag{A.3}$$

When reaching a steady state, $\dot{x}_i = 0$ indicates that $x_i^* = x_j^*, \forall i,j \in \mathcal{N}$.

Consider the classic consensus-based method, i.e. the average consensus algorithm

$$x_i(k+1) = \sum_{j\in\mathcal{N}} a_{ij} x_j(k) \tag{A.4a}$$

$$= x_i(k) + \sum_{j\neq i,\ j\in\mathcal{N}} a_{ij}\left(x_j(k) - x_i(k)\right) \tag{A.4b}$$

where the weight a_{ij} satisfies the following:

1) $a_{ij} \geq 0, \ \forall i,j \in \mathcal{N}$.
2) $\sum_{j \in \mathcal{N}} a_{ij} = 1$ and $\sum_{i \in \mathcal{N}} a_{ij} = 1$.

The average consensus method has the following convergence result [8, Theorem 3.1].

Theorem A.1 *Given the initial* $x_i(0), i \in \mathcal{N}$, *the average consensus algorithm reaches the agreement*

$$\lim_{k \to \infty} x_i(k) \to \frac{1}{|\mathcal{N}|} \sum_{j \in \mathcal{N}} x_j(0), \quad \forall i \in \mathcal{N} \tag{A.5}$$

The consensus algorithm can be utilized directly for simple problems, while the combinations of the consensus algorithms and other distributed methods can handle more complicate problems. For instance, reference [9] deals with the following optimization problem:

$$\min_{x} \sum_{i \in \mathcal{N}} f_i(x) \tag{A.6a}$$

$$\text{s.t. } x \in \mathcal{X} \tag{A.6b}$$

where $x \in \mathbb{R}^n$ is the decision variable; $\mathcal{X} \subseteq \mathbb{R}^n$ is nonempty, closed, and convex; and $f_i : \mathcal{X} \to \mathbb{R}$ is convex. Objective functions $f_i(x), i \in \mathcal{N}$ are assumed to be non-differentiable, whose set of subgradient is denoted by $\partial f_i(x)$. To solve this problem, a subgradient-based consensus algorithm is proposed as

$$v_i(k+1) = x_i(k) - \beta_k g_i(x_i(k)), \tag{A.7a}$$

$$x_i(k+1) = \mathcal{P}_{\mathcal{X}}\left(v_i(k) + \alpha_k \sum_{j \in \mathcal{N}_i} a_{ij} \left(v_j(k) - v_i(k) \right) \right), \tag{A.7b}$$

where $x_i(k) \in \mathbb{R}^n$ and $v_i \in \mathbb{R}^n$ are the real value and auxiliary value of the agent i, $\alpha_k > 0$ and $\beta_k > 0$ are step sizes, and $g_i(x_i(k)) \in \partial f_i(x_i(k))$ is a subgradient of $f_i(x_i(k))$.

A.1.2 Cutting-Plane Consensus Algorithm

There is an alternative consensus-based problem, where the consensus of agents is the feasible region of the optimization problem, instead of certain variables. In this context, a cutting plane consensus algorithm is proposed to solve such problems. Specifically, in a convex optimization problem, each agent solves its

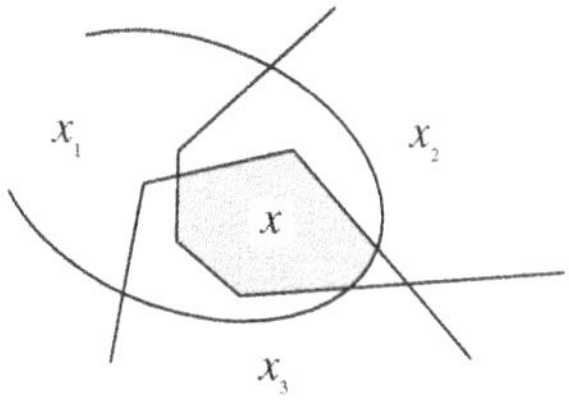

Figure A.1 An illustration of the individual and global feasible regions.

own local subproblem and uses a polyhedral outer approximation to estimate the global feasible region. By iteratively generating cutting planes and exchanging them with neighbors, all agents may finally reach an agreement on the global feasible region and then obtain the optimal solution to the original problem. Although the communication burden is high in the presence of many global constraints, the cutting-plane consensus algorithm is fairly robust to various kinds of communication issues, e.g. time delays, packet drops, and change of the communication topology [10–12].

Consider the following optimization problem: [13].

$$\min_{x} \; c^{\mathrm{T}} x \tag{A.8a}$$

$$\text{s.t. } x \in \mathcal{X} = \bigcap_{i \in \mathcal{N}} \mathcal{X}_i \tag{A.8b}$$

The individual feasible region $\mathcal{X}_i \subseteq \mathbb{R}^n$ is nonempty, convex, and closed. The global feasible region $\mathcal{X} \subseteq \mathbb{R}^n$ is the intersection of all individual feasible regions, which is assumed to be nonempty. Obviously, for a finite number of agents, the global feasible region is also convex and closed. Figure A.1 illustrates the individual and global feasible regions.

Here, each agent tries to reach a consensus on the global feasible region. To this end, the cutting-plane consensus algorithm is an appealing solution. To explain, we begin with some necessary definitions. Define the half space as $h := \{x \mid a^{\mathrm{T}} x - b \leq 0\}$. Given a nonempty, closed, and convex set $\mathcal{X}$ and a query point x_q, a half space $h\left(x_q\right)$ is called a cutting plane, if it satisfies

$$a\left(x_q\right)^{\mathrm{T}} x \leq b\left(x_q\right), \quad \forall x \in \mathcal{X}, \tag{A.9a}$$

$$a\left(x_q\right)^{\mathrm{T}} x_q > b\left(x_q\right) \tag{A.9b}$$

Let **ORC** $\left(x_q, \mathcal{X}\right)$ denote a cutting-plane oracle, which is defined as follows:

1) If $x_q \notin \mathcal{X}$, return a cutting plane, which is denoted by $h\left(x_q\right)$.
2) If $x_q \in \mathcal{X}$, return an empty h.

Hereafter, we use a specific cutting-plane oracle for a set $\mathcal{X} = \{x \mid f(x) \leq 0\}$. If $x_q \notin \mathcal{X}$, let

$$h\left(x_q\right) := \left\{x \mid g^{\mathrm{T}}\left(x_q\right)\left(x - x_q\right) + f\left(x_q\right) \leq 0\right\} \tag{A.10}$$

where $g\left(x_q\right) \in \partial f\left(x_q\right)$ is a subgradient. Otherwise, i.e. $x_q \in \mathcal{X}$, return an empty h.

Given a collection of cutting planes $H = \{h_1, h_2, \dots, h_m\}$, define the corresponding set $\mathcal{H} := \bigcap_{i=1}^{m} h_i$. Directly, if $B \subset H$, we have $\mathcal{H} \subseteq \mathcal{B}$. Given H, for the following problem

$$\min_{x} \; c^{\mathrm{T}} x \tag{A.11a}$$

$$\text{s.t. } a_i^{\mathrm{T}} x - b_i \leq 0, \quad \forall h_i \in H, \tag{A.11b}$$

the constraint (A.11b) is equivalent to $x \in \mathcal{H}$.

Denote by Γ_H the set of optimal solutions to the problem (A.11). We are interested in the unique optimal solution in the sense of the minimal 2-norm, which is defined as

$$x_H^* := \arg \min_{x \in \Gamma_H} \|x\|_2 \tag{A.12}$$

Given a collection of cutting planes H, a subset $B \subset H$ is called a basis of H, if $x_H^* = x_B^*$ and $x_{B'}^* \neq x_B^*, \forall B' \subset B$. Define the active constraints of the solution x_H^* as the collection of cutting planes $h_i \in H$ satisfying $a_i^{\mathrm{T}} x_H^* - b_i = 0$. Directly, the active constraints of x_H^* are identical to those of x_B^*.

Then we have the cutting-plane consensus algorithm to solve the problem (A.8). In this algorithm, agents need to exchange information with their neighbors. Assume that the communication network is an undirected connected graph. Denote by $\mathcal{N}_i$ the set of the neighbors of the agent i. Each agent repeats the following steps [13, 14]:

1) Exchange its basis $B_i(k)$ with all its neighbor $\mathcal{N}_i$, and formulate a temporary collection of cutting planes

$$H_{i,tmp}(k) = B_i(k) \bigcup \left[\bigcup_{j \in \mathcal{N}_i} B_j(k) \right]; \tag{A.13}$$

2) Computes a query point $x_i(k)$ as

$$x_i(k) = \arg \min_{x \in \Gamma_{H_{i,tmp}(k)}} \|x\|_2 \tag{A.14}$$

and selects a minimal collection of active constraints $B_{i,tmp}(k)$

3) Generate a cutting-plane oracle for the individual feasible region $\mathcal{X}_i$ at the query point $x_i(k)$ as

$$h(x_i(k)) = \mathbf{ORC}(x_i(k), \mathcal{X}_i) \tag{A.15}$$

4) Update its basis $B_i(k+1)$: if $x_i(k) \notin \mathcal{X}_i$, $B_i(k+1)$ is the set of the minimal basis of $B_{i,tmp}(k) \cup h(x_i(k))$; otherwise, $B_i(k+1) = B_{i,tmp}(k)$.

The convergence of the cutting-plane consensus algorithm is guaranteed by the following theorem [13].

Theorem A.2 *Assume that the communication network, as an undirected (directed) graph, is (strongly) connected and the* **ORC** $(x_q, \mathcal{X})$ *satisfies the following:*

1) There exists a finite scalar $L > 0$ *such that* $\left\|a\left(x_q\right)\right\|_2 < L$ *for all* x_q.

2) For any sequence $\left\{x_q(k)\right\}$, $x_q(k) \rightarrow \bar{z}$ *and* $a\left(x_q\right)^{\mathrm{T}} x_q - b\left(x_q\right) \rightarrow 0$ *imply* $\bar{z} \in \mathcal{X}$.

Then the cutting-plane consensus algorithm approaches to the agreement as

$$\lim_{k \rightarrow \infty} x_i(k) \rightarrow x^*, \quad \forall i \in \mathcal{N} \tag{A.16}$$

where x^* *is the unique optimal solution to* (A.8).

In the power system community, the cutting-plane consensus method has been successfully utilized in microgrid control [13], dynamic economic dispatch [15], electric vehicle charging [12], etc. In [13], the cutting-plane consensus algorithm is applied to solving the robust optimization problem considering microgrid control with uncertain constraints. Reference [15] utilizes a modified cutting-plane consensus method to successfully solve the multi-area dynamic economic dispatch problem, where the transmission and elimination of the cutting planes are customized to speed up the convergence. In [12], the cutting-plane consensus algorithm is applied to solve the electric vehicles charging problem, and a completely localized stopping criteria is proposed.

A.2 First-Order Gradient-Based Algorithms

The first-order gradient-based algorithms have been extensively used in a broad variety of fields. To explain how they work, we begin with the definitions of subgradient and gradient.

Given a set $\mathcal{X} \subseteq \mathbb{R}^n$ and a function $f: \mathcal{X} \rightarrow \mathbb{R}$, a vector $g \in \mathbb{R}^n$ is called a subgradient of f at $x \in \mathcal{X}$, if

$$f(z) - f(x) \geq g^{\mathrm{T}}(z - x), \quad \forall z \in \mathcal{X} \tag{A.17}$$

The set of all subgradients at x is called the subdifferential at x and is denoted by $\partial f(x)$.

If f is differentiable at x, define the gradient at x as

$$\nabla f(x) := \left[\frac{\partial f}{\partial x_1}(x), \frac{\partial f}{\partial x_2}(x), \ldots, \frac{\partial f}{\partial x_n}(x)\right]^{\mathrm{T}} \tag{A.18}$$

where $\frac{\partial f}{\partial x_i}(x)$ is the partial derivatives of f over x_i at x.

If f is convex and differentiable at x, then the subdifferential $\partial f(x) = \{\nabla f(x)\}$.

Typical first-order gradient-based algorithms include DD methods, ADMMs, PDG methods, and proximal gradient methods, all of which are based on the first-order gradient or subgradient. Generally, it suffices to use gradient if only differentiable functions are involved in the problem; otherwise, one likely needs to use subgradient.

A.2.1 Dual Decomposition

The main idea of the DD algorithm is to use the Lagrangian and dual theories to handle coupled constraints in convex optimization problems. The coupled constraints are multiplied by the associated dual variables and added into the objective function, resulting in a separable problem that can be solved in a distributed fashion. In this framework, agents repeatedly solve their own subproblems and update the dual variables through the communication network with their neighbors until the algorithm converges.

Formally, DD is applicable to the constrained convex optimization problems in the form of [16, Section 8]

$$\min_{x} \sum_{i \in \mathcal{N}} f_i\left(x_i\right) \tag{A.19a}$$

$$\text{s.t.} \sum_{i \in \mathcal{N}} h_{ij}\left(x_i\right) \leq 0,\ j = 1,2, \dots, m \tag{A.19b}$$

$$x_i \in \mathcal{X}_i, \dots \forall i \in \mathcal{N} \tag{A.19c}$$

where $x_i \in \mathbb{R}^{n_i}$ is the decision variable of the agent i and $\mathcal{X}_i \subseteq \mathbb{R}^{n_i}$ is the individual feasible region. The functions $f_i : \mathcal{X}_i \to \mathbb{R}$ and $h_{ij} : \mathcal{X}_i \to \mathbb{R}$ are convex. (A.19b) describes m coupled global constraints.

The coupled global constraints (A.19b) usually reflect the network topology of the agents, which is sparse and distributed in our problems. For instance, assume that the jth constraint represents the physical relation between the agents i and i'. Then we have $h_{ij}\left(x_i\right) + h_{i'j}\left(x_i'\right) \leq 0$ and $h_{kj}\left(x_k\right) \equiv 0$, for any other $k \in \mathcal{N} / \left\{i, i'\right\}$. Hence, the steps related to the jth coupled constraint in distributed algorithms can be implemented locally by the agents i and i'.

Note that the constraint (A.19b) are coupled, while the objective function and the constraint (A.19c) are separable. The Lagrangian and the dual theory are utilized to transform (A.19) into subproblems.

For simplicity, let $x := \left[x_1^{\mathrm{T}}, x_2^{\mathrm{T}}, \dots, x_{|\mathcal{N}|}^{\mathrm{T}}\right]^{\mathrm{T}}$ and $\mathcal{X} := \Pi_{i \in \mathcal{N}} \mathcal{X}_i$. Define the Lagrangian multipliers (dual variables) of (A.19b) as $\lambda_1, \lambda_2, \dots, \lambda_m$. The Lagrangian of (A.19) is defined as

$$L(x, \lambda) = \sum_{i \in \mathcal{N}} f_i\left(x_i\right) + \sum_{j=1}^{m} \lambda_j \sum_{i \in \mathcal{N}} h_{ij}\left(x_i\right)$$

$$= \sum_{i \in \mathcal{N}} \left(f_i(x_i) + \sum_{j=1}^{m} \lambda_j h_{ij}(x_i) \right)$$
$$= \sum_{i \in \mathcal{N}} L_i(x_i, \lambda),$$

where $L_i(x_i, \lambda) = f_i(x_i) + \sum_{j=1}^{m} \lambda_j h_{ij}(x_i)$.

Then we have the following DD algorithm:

$$x_i(k+1) \in \arg \min_{x_i \in \mathcal{X}_i} L_i(x_i, \lambda(k)), \tag{A.20a}$$

$$\lambda_j(k+1) = \left[\lambda_j(k) + \alpha_k \left(\sum_{i \in \mathcal{N}} h_{ij}(x_i(k)) \right) \right]_+, \tag{A.20b}$$

where $[\cdot]_+$ is the projection operator onto $\mathbb{R}_+$ and $\alpha_k > 0$ is the step size.

Mathematically, DD can be regarded as a subgradient ascent method of the dual problem defined as

$$\max_{\lambda > 0} D(\lambda) \tag{A.21}$$

where the dual function $D(\lambda)$ is defined as

$$D(\lambda) := \inf_{x \in \mathcal{X}} L(x, \lambda) = \sum_{i \in \mathcal{N}} \inf_{x_i \in \mathcal{X}_i} L_i(x_i, \lambda) \tag{A.22}$$

Assume that the Slater's condition of (A.19) holds, i.e. there exist a point $\hat{x} \in \mathcal{X}$ such that $\sum_{i \in \mathcal{N}} h_{ij}(\hat{x}_i) < 0$, $j = 1,2, \ldots, m$. Then the optimal value of (A.19) equals that of (A.21), i.e. the strong duality holds.

Invoking the definition of the subgradient, we have

$$g(\lambda) = [g_1(\lambda), g_2(\lambda), \ldots, g_m(\lambda)]^{\mathrm{T}}, \tag{A.23a}$$

$$g_j(\lambda) = \sum_{i \in \mathcal{N}} h_{ij}(x_i(\lambda)), \tag{A.23b}$$

$$x_i(\lambda) \in \arg \min_{x_i \in \mathcal{X}_i} L_i(x_i, \lambda) \tag{A.23c}$$

Under some mild assumptions, the subgradient (A.23) is equivalent to the gradient of $D(\lambda)$, which indicates that $D(\lambda)$ is differentiable therein. As proved in [17, Prop. 6.1.1], if $\mathcal{X}$ is nonempty and compact, f_i and h_{ij} are continuous in $\mathcal{X}$ for all $i \in \mathcal{N}, j = 1,2, \ldots, m$, and $L_i(x_i, \lambda)$ is minimized over $x_i \in \mathcal{X}_i$ at a unique point $x_i(\lambda)$ for any $i \in \mathcal{N}$, then $D(\lambda)$ is differentiable. Moreover, its gradient is defined as (A.23).

The convergence analysis of DD methods largely depends on whether (A.23) is a gradient of $D(\lambda)$ or not. If (A.23) is a subgradient, a convergence result is presented in [16, Prop. 8.2.4], which reads the following.

Theorem A.3 *If the step size α_k satisfies*

$$\sum_{k=0}^{\infty} \alpha_k \to \infty, \quad \sum_{k=0}^{\infty} \alpha_k^2 < \infty \tag{A.24}$$

then it follows that

$$\lim_{k \to \infty} \inf D(\lambda(k)) = D^* \tag{A.25}$$

where D^* is the optimal value of the problem (A.21).

To better understand this convergence result, we provide an intuitive explanation. According to [16, Prop. 8.2.2], if a constant step size α is used, the value of the objective will converge to a neighborhood of the optimal value, where the maximal range of the neighborhood is proportional to α. As α_k vanishes to zero, the value of the objective will converge to the optimal value. Typically, to satisfy the requirements of a shrinking sequence of step size, we can simply set α_k as $\frac{1}{k}$ or $\frac{r}{k+r}$ with $r > 0$.

When (A.23) is a gradient, a classic convergence result is given by [18, Prop. 3.4], which reads the following.

Theorem A.4 *Assume that the gradient defined in* (A.23) *is L-Lipschitz continuous. If the step size α satisfies*

$$0 < \alpha < \frac{2}{L} \tag{A.26}$$

then it follows that

$$\lim_{k \to \infty} \left[x_1(k)^{\mathrm{T}}, x_2(k)^{\mathrm{T}}, \ldots, x_{|\mathcal{N}|}(k)^{\mathrm{T}} \right]^{\mathrm{T}} \to x^* \tag{A.27}$$

where x^* is the optimal solution to the primal problem (A.19).

DD has been widely utilized in power systems, e.g. demand response [19–21], distribution network voltage control [22, 23], multi-energy system operation [24], energy dispatch [25, 26], and optimal power flow (OPF) [27–29]. In [19], a DD-based method is applied to schedule residential loads, where the authors prove the convergence of the algorithm in the presence of metering message loss. A multiseller–multibuyer market clearing scheme considering demand response is built on the DD method in [20], while aggregators are introduced to integrate end users and to participate in the electricity market in [21]. Reference [22] applies the DD method to voltage control based on a linearized model of distribution networks. Thanks to the radial structure of distribution networks, reference [23] proposes a fully distributed DD algorithm for distribution network voltage control. Reference [24] formulates an asynchronous DD method to manage the multi-energy system consisting of gas storages, fuel cells, microcombined heat power systems, and heat buffers. In reference [25], the L-BFGS-B method is introduced to accelerate the DD method for the stochastic economic dispatch problem. In [26], a hierarchical distributed method is formulated as a double-loop iteration of DD, corresponding to the hierarchical architecture of the distribution network operator, aggregators, and end users. DD-based methods are also applied to OPF problems with nonconvex power flows [27], convex power flow by semidefinite relaxation [28], and convex power flow by the second-order cone relaxation [29].

A.2.2 Alternating Direction Method of Multipliers

The ADMM is similar to the DD algorithm, but it uses an augmented Lagrangian and mainly focuses on the coupled constraints of linear equations. ADMM is commonly applied to solve the following optimization problem:

$$\min_{x,z} f(x) + h(z) \tag{A.28a}$$

$$\text{s.t. } Ax + Bz = c, \tag{A.28b}$$

where $x \in \mathbb{R}^n$ and $z \in \mathbb{R}^m$ are decision variables; $A \in \mathbb{R}^{p\times n}$, $B \in \mathbb{R}^{p\times m}$, and $c \in \mathbb{R}^p$ are constant; $f : \mathbb{R}^n \to \mathbb{R}$ and $h : \mathbb{R}^m \to \mathbb{R}$ are convex functions.

Define the Lagrangian multiplier (dual variable) of (A.28b) as $y \in \mathbb{R}^p$. Then the augmented Lagrangian of (A.28) is in the form of

$$L_\rho (x, z, y) := f(x) + h(z) + y^{\mathrm{T}} (Ax + Bz - c) + \frac{\rho}{2} \|Ax + Bz - c\|_2^2 \tag{A.29}$$

where $\rho > 0$ is a positive constant.

Then ADMM algorithm is given as

$$x(k+1) = \arg \min_{x} L_\rho (x, z(k), y(k)), \tag{A.30a}$$

$$z(k+1) = \arg \min_{z} L_\rho (x(k+1), z, y(k)), \tag{A.30b}$$

$$y(k+1) = y(k) + \rho (Ax(k+1) + Bz(k+1) - c) \tag{A.30c}$$

where ρ can also be regarded as the step size to update y.

ADMM is not inherently distributed. It can be implemented in both centralized and distributed manners. The distributed ADMM is usually realized by two ways: (i) combined with consensus algorithm [30–33] and (ii) utilizing the distributed structure of power systems [34–37].

The convergence performance of ADMM is tightly related to the parameter ρ. A too large or too small ρ will slow down the convergence. In practice, we usually use the following criterion to tune the value of ρ. Define the primal residual and the dual residual as

$$r(k) := \|Ax(k) + Bz(k) - c\|_2, \tag{A.31}$$

$$s(k) := \rho\left\|A^{\mathrm{T}}B(z(k) - z(k-1))\right\|_2 \tag{A.32}$$

During the iteration, if $r(k) \gg s(k)$, then we tend to pick a larger ρ. However, if $r(k) \ll s(k)$, then we tend to pick a smaller ρ; otherwise, the current ρ is suitable.

Reference [6] presents a general convergence result for ADMM as follows.

Theorem A.5 *Suppose the problem* (A.28) *satisfies the following:*

1) *The functions $f(x)$ and $h(z)$ are closed, proper, and convex.*
2) *The unaugmented Lagrangian $L(x, z, y) = f(x) + h(z) + y^{\mathrm{T}}(Ax + Bz - c)$ has a saddle point, i.e. there exists (x^*, z^*, y^*) such that*

$$L(x^*, z^*, y) \leq L(x^*, z^*, y^*) \leq L(x, z, y^*)$$

holds for all x, z, y

Then ADMM (A.30) *has the following convergence performance:*

1) *Primal residual convergence: $\lim_{k\to\infty} r(k) \to 0$.*
2) *Objective convergence: $\lim_{k\to\infty} f(x(k)) + h(z(k)) \to p^*$, where p^* is the optimal value of the problem* (A.28).
3) *Dual variable convergence: $\lim_{k\to\infty} y(k) \to y^*$, where y^* is an optimal dual variable.*

The convergence of ADMM also holds for the following problem:

$$\min_{x,z} \; f(x) + h(z) \tag{A.33a}$$

$$\text{s.t. } Ax + Bz = c \tag{A.33b}$$

$$x \in \mathcal{X}, z \in \mathcal{Z}, \tag{A.33c}$$

where the individual feasible regions $\mathcal{X}$ and $\mathcal{Z}$ are nonempty, closed, and convex.

To this end, we can simply use the indicator functions $\iota_{\mathcal{X}}$ and $\iota_{\mathcal{Z}}$ to convert the original problem into

$$\min_{x,z} \; \left\{f(x) + \iota_{\mathcal{X}}(x)\right\} + \left\{h(z) + \iota_{\mathcal{Z}}(z)\right\} \tag{A.34a}$$

$$\text{s.t. } Ax + Bz = c \tag{A.34b}$$

which immediately follows the form of problem in Theorem A.5.

A.2.3 Primal–Dual Gradient Algorithm

From a dynamic point of view, a PDG algorithm can be regarded as a gradient dynamic to seek the saddle point of the Lagrangian. Different from DD algorithms, the primal and dual variables are both updated along with the gradient directions in PDG algorithms. Hence, this kind of algorithms are sometimes called the primal–dual dynamics. Compared with other gradient-based methods, PDG algorithms are more suitable for system control design, as it allows the combination of physical system dynamics and primal–dual dynamics, which can be comprehensively studied by employing mature theories in control and stability analysis.

Consider the following convex optimization problem:

$$\min_{x} \sum_{i\in\mathcal{N}} f_i\left(x_i\right) \tag{A.35a}$$

$$\text{s.t.} \sum_{i\in\mathcal{N}} h_{ij}\left(x_i\right) \leq 0,\ j = 1, 2, \dots, m \tag{A.35b}$$

where $x_i \in \mathbb{R}^{n_i}$ is the decision variable of the agent i, x is defined as $x = \left[x_1^{\mathrm{T}}, x_2^{\mathrm{T}}, \dots, x_{|\mathcal{N}|}^{\mathrm{T}}\right]^{\mathrm{T}} \in \mathbb{R}^n$ with $n = \sum_{i\in\mathcal{N}} n_i$, and the functions $f_i, h_{ij} : \mathbb{R}^{n_i} \to \mathbb{R}$ are convex and differentiable. The sparse and distributed structure of the coupled global constraints (A.35b) is the same as that of the problem (A.19).

Note that the constraint (A.35b) is coupled, while the objective function is separable. The Lagrangian is utilized to transfer the problem into a set of subproblems.

The Lagrangian of the problem (A.35) is

$$L(x, \lambda) = \sum_{i\in\mathcal{N}} f_i\left(x_i\right) + \sum_{j=1}^{m} \lambda_j \sum_{i\in\mathcal{N}} h_{ij}\left(x_i\right) \tag{A.36}$$

where λ_j is the Lagrangian multiplier of the constraint $\sum_{i\in\mathcal{N}} h_{ij}\left(x_i\right) \leq 0$ and $\lambda = \left[\lambda_1, \lambda_2, \dots, \lambda_m\right]^{\mathrm{T}} \in \mathbb{R}^m$.

Assume that the Slater's condition holds, i.e. there exist a point $\hat{x} \in \mathbb{R}^n$ such that $\sum_{i\in\mathcal{N}} h_{ij}\left(\hat{x}_i\right) < 0,\ j = 1,2, \dots, m$. Then the optimal value of (A.35) is equal to that of the dual problem

$$\max_{\lambda > 0} \inf_{x} L(x, \lambda) \tag{A.37}$$

Then the problem is transferred to seeking the saddle point (x^*, λ^*) of $L(x, \lambda)$, i.e.

$$L(x^*, \lambda) \leq L(x^*, \lambda^*) \leq L(x, \lambda^*) \tag{A.38}$$

holds for all $x \in \mathbb{R}^n, \lambda \in \mathbb{R}_+^m$.

Then we have the discrete form of the PDG algorithm as follows:

$$x_i(k+1) = x_i(k) - \beta_k \nabla_{x_i} L(x(k), \lambda(k)) \tag{A.39a}$$

$$\lambda_j(k+1) = \left[\lambda_j(k) + \alpha_k \nabla_{\lambda_j} L(x(k), \lambda(k))\right]_+ \tag{A.39b}$$

where $\alpha_k > 0$ and $\beta_k > 0$ are the step sizes. The algorithm is similar to the DD method (A.20), but the primal variable is updated along the gradient instead of solving the subproblem.

In addition to the discrete form, we also have a continuous form of the PDG as follows:

$$\dot{x}_i = -\nabla_{x_i} L(x, \lambda) \tag{A.40a}$$

$$\dot{\lambda}_j = \left[\nabla_{\lambda_j} L(x, \lambda)\right]^+_{\lambda_j} \tag{A.40b}$$

where the operator $[x]_y^+$ is defined as

$$[x]_y^+ := \begin{cases} x, & x > 0 \text{ or } y > 0 \\ 0, & \text{otherwise} \end{cases} \tag{A.41}$$

By [38], the PDG method (A.40) of the problem (A.35) has the following convergence result.

Theorem A.6 *The set of primal–dual solutions to* (A.35) *is globally asymptotically stable on* $\mathbb{R}^n \times \mathbb{R}^m_+$ *under the primal–dual dynamics* (A.40), *and the convergence of each solution is to a point.*

Based on the standard PDG method, the partial PDG method is proposed to deal with optimization problems with the following form:

$$\min_{x,y} f(x) + h(z) \tag{A.42a}$$

$$\text{s.t. } Ax + Bz = c, \tag{A.42b}$$

where $x \in \mathbb{R}^n$ and $z \in \mathbb{R}^m$ are decision variables; $A \in \mathbb{R}^{p\times n}$, $B \in \mathbb{R}^{p\times m}$, and $c \in \mathbb{R}^p$ are constant; and the function $f : \mathbb{R}^n \to \mathbb{R}$ is strictly convex and twice differentiable and the function $h : \mathbb{R}^m \to \mathbb{R}$ is convex and differentiable.

Define the Lagrangian as

$$L(x, z, y) := f(x) + h(z) + y^{\mathrm{T}}(Ax + Bz - c) \tag{A.43}$$

where $y \in \mathbb{R}^p_+$ is the Lagrangian multiplier of (A.42b). Then we have the partial PDG algorithm [39] as

$$x = \arg\min_x L(x, z, y), \tag{A.44a}$$

$$\dot{z} = -\beta \nabla_z L(x, z, y), \tag{A.44b}$$

$$\dot{y} = \alpha(Ax + Bz - c) \tag{A.44c}$$

where $\alpha > 0$ and $\beta > 0$ are the step sizes.

A revised version of PDG algorithm is the so-called **partial PDG algorithm**. Consider the following optimization problem:

$$\min_{x\in X, y\in\Omega} \quad f(x) + h(y) \tag{A.45a}$$

$$\text{s.t.} \quad Ax + By = C \tag{A.45b}$$

where $X \subset \mathbb{R}^n$ and $\Omega \subset \mathbb{R}^m$ are closed convex sets and $A \in \mathbb{R}^{p\times n}$, $B \in \mathbb{R}^{p\times m}$, $C \in \mathbb{R}^p$ are constant matrices. We make following assumptions.

Assumption A.1 The Slater's condition of (A.45) holds.

Assumption A.2 For some $\alpha > 0$, $f(x)$ is α strongly convex and twice differentiable, i.e. $\nabla^2 f(x) \succeq \alpha I$.

Assumption A.3 The function $h(y)$ is Lipschitz continuous and β-strongly convex on Ω for some $\beta > 0$, that is, $\langle y_1 - y_2, g_h(y_1) - g_h(y_2)\rangle \geq \beta\|y_1 - y_2\|^2$, $\forall y_1, y_2 \in \Omega$ where $g_h(y_1) \in \partial h(y_1)$ and $g_h(y_2) \in \partial h(y_2)$.

Assumption A.4 The matrix A has full row rank and $\kappa_1 I \preceq AA^{\mathrm{T}}$ for some $\kappa_1 > 0$.

The Lagrangian of (A.45) is

$$\underset{x\in X, y\in\Omega}{L(x, y, \lambda)} = f(x) + h(y) + \lambda^{\mathrm{T}}(Ax + By - C) \tag{A.46}$$

where $\lambda \in \mathbb{R}^p$ is the Lagrangian multiplier vector.

Define functions

$$\varphi(\lambda) \triangleq \min_{x\in X} \left\{f(x) + \lambda^{\mathrm{T}} Ax\right\}, \tag{A.47}$$

$$\underset{y\in\Omega}{\hat{L}(y, \lambda)} \triangleq \varphi(\lambda) + h(y) + \lambda^{\mathrm{T}}(By - C) \tag{A.48}$$

Then, the partial primal–dual gradient dynamics is

P-PDGD:

$$x(t) = \arg\min_{q\in X} \left\{f(q) + \lambda^{\mathrm{T}}(t)Aq\right\}, \tag{A.49a}$$

$$\dot{y}(t) \in P_{T_\Omega(y(t))}\left(-\partial h\left(y(t)\right) - B^{\mathrm{T}}\lambda(t)\right), \tag{A.49b}$$

$$\dot{\lambda}(t) = \nabla\varphi(\lambda(t)) + By(t) - C \tag{A.49c}$$

The (partial) PDG methods are widely applied to the power system operations in the discrete form [40–42] and the continuous form [43–45]. In [40], an asynchronous distributed PDG method is proposed, where the augmented Lagrange function is used to improve the convergence performance. Reference

[41] formulates a PDG method and designs a hierarchical communication protocol to alleviate the communicational burden in the large-scale distribution systems. In [42], an asynchronous distributed PDG algorithm is designed to solve the economic dispatch problem of microgrids and is proved to converge under asynchronous communication environments. In [43], the primary frequency regulation in the power system is combined with the PDG method, achieving the frequency regulation and the optimal cost. Reference [44] proposes a distributed optimal frequency control of multi-area power systems, where the system frequency is measured and used in the gradient method to adapt to unknown load disturbance. Reference [45] further extends [44] to limit tie-line powers within secure ranges.

A.2.4 Proximal Gradient Method

The proximal gradient method is commonly utilized in the optimization problems with nonsmooth objectives. We begin with the definition of the proximal operator

$$\text{prox}_{\alpha f}(x) = \arg\min_{z} f(z) + \frac{1}{2\alpha}\|z - x\|_2^2 \tag{A.50}$$

where $x \in \mathbb{R}^n$ is a variable vector, $\alpha > 0$ is a positive constant, and $f : \mathbb{R}^n \to \mathbb{R}$ is a real-value proper, closed, and convex function. Specially, if $\alpha = 1$, the proximal operator is reduced to $\text{prox}_f(x)$.

The proximal operator has the following properties:

1) The proximal operator is always single value, since the term $\|z - x\|_2^2$ is strongly convex.
2) A point $x^* \in \mathbb{R}^n$ minimizes $f(x)$, if and only if x^* is a fixed point of $\text{prox}_{\alpha f}(x)$, i.e.
$$x^* = \text{prox}_{\alpha f}(x^*) \tag{A.51}$$
3) The proximal operator is nonexpansive, i.e. for any $x, y \in \mathbb{R}^n$, it holds that
$$\left\|\text{prox}_{\alpha f}(x) - \text{prox}_{\alpha f}(y)\right\|_2 \le \|x - y\|_2 \tag{A.52}$$
4) For any nonempty closed convex set $\Omega \subseteq \mathbb{R}^n$, the projection operator $\mathcal{P}_\Omega(\cdot)$ is the special proximal operator $\text{prox}_{\iota_\Omega}(\cdot)$, since
$$\begin{aligned}\mathcal{P}_\Omega(x) &= \arg\min_{z\in\Omega} \|z - x\|_2^2 \\ &= \arg\min_{z} \iota_\Omega(z) + \frac{1}{2}\|z - x\|_2^2 \\ &= \text{prox}_{\iota_\Omega}(x)\end{aligned}$$

Using the proximal operator, the proximal gradient method can be derived to solve the problems in the form of

$$\min_{x} f(x) + \theta(x) \tag{A.53a}$$

where $x \in \mathbb{R}^n$ is the decision value, $f : \mathbb{R}^n \to \mathbb{R}$ is convex and differentiable, and $\theta : \mathbb{R}^n \to \mathbb{R}$ is convex but not necessarily differentiable.

The proximal gradient method to solve (A.53) reads

$$z(k+1) = x(k) - \alpha_k \nabla f(x(k)), \tag{A.54a}$$

$$y(k+1) = \mathrm{prox}_{\alpha_k \theta}(z(k+1)), \tag{A.54b}$$

$$x(k+1) = y(k+1) + \beta_k (y(k+1) - y(k)) \tag{A.54c}$$

where α_k is the step size and β_k is the parameter to accelerate the convergence. If $\beta_k = 0, \forall k$, then algorithm (A.54) is reduced to the standard proximal gradient method. An ingenious assignments of β_k can remarkably accelerate the convergence. In practice, β_k is commonly set to $\frac{k-1}{k+2}$ in the accelerated proximal gradient methods [46–48].

The standard proximal gradient method, i.e. $\beta_k = 0$, has the following convergence result as presented in [49, Thm. 10.21]

Theorem A.7 *Assume that θ is proper closed and convex, f is proper closed and differentiable, and ∇f is L-Lipschitz continuous. Set $\alpha_k = 1/L, \forall k$; then it follows that*

$$f(x(k)) + \theta(x(k)) - p^* \leq \frac{L\,\|x(0) - x^*\|_2^2}{2k} \tag{A.55}$$

where x^* is the optimal solution to (A.53) *and p^* is the optimal value of* (A.53).

As a type of distributed optimization methods, the proximal gradient method is commonly applied to cope with problems with the following form:

$$\min_x \sum_{i\in\mathcal{N}} f_i(x) + \theta_i(x) \tag{A.56a}$$

where $f_i(x) + \theta_i(x)$ is the objective function of the agent i, $x \in \mathbb{R}^n$ is the consensus decision value, $f_i : \mathbb{R}^n \to \mathbb{R}$ is convex and differentiable, and $\theta_i : \mathbb{R}^n \to \mathbb{R}$ is convex but generally nondifferentiable.

Assume that the agents communicate with each other through a undirected connected graph $\mathcal{G} = \{\mathcal{N}, \mathcal{L}\}$, where $\mathcal{N}$ is the set of agents and $\mathcal{L}$ is the set of communication links. Then a plenty of distributed proximal gradient-based methods have been derived to solve the problem (A.56). Here we only introduce an influential work [50] as follows:

1) Initialization. Given arbitrary initial $x_i(0) \in \mathbb{R}^n$ for any agent $i \in \mathcal{N}$, it follows that

$$y_i(1) = \sum_{j\in\mathcal{N}} w_{ij} x_j(0) - \alpha \nabla f_i(x_i(0)), \tag{A.57a}$$

$$x_i(1) = \mathrm{prox}_{\alpha\theta_i}(y_i(1)) \tag{A.57b}$$

2) Iteration. For $k = 1, 2, \ldots$, each agent $i \in \mathcal{N}$ does

$$y_i(k+1) = y_i(k) + \sum_{j \in \mathcal{N}} \left[w_{ij} x_j(k) - \tilde{w}_{ij} x_j(k-1) \right] - \alpha \left[\nabla f_i\left(x_i(k)\right) - \nabla f_i\left(x_i(k-1)\right) \right], \tag{A.58a}$$

$$x_i(k+1) = \operatorname{prox}_{\alpha\theta_i}\left(y_i(k+1)\right) \tag{A.58b}$$

where $\alpha > 0$ is the step size and w_{ij} and $\tilde{w}_{ij}$ are the i-row j-column entries of the matrices W and $\tilde{W}$.

The matrices W and $\tilde{W}$ are supposed to satisfy the following:

1) If $i \neq j$ and $(i,j) \notin \mathcal{L}$, then $w_{ij} = \tilde{w}_{ij} = 0$.
2) $W = W^{\mathrm{T}}$ and $\tilde{W} = \tilde{W}^{\mathrm{T}}$.
3) $\mathrm{null}\{W - \tilde{W}\} = \mathrm{span}\{1\}$ and $\mathrm{span}\{1\} \subseteq \mathrm{null}\{I - \tilde{W}\}$,
4) $\tilde{W} > 0$ and $\frac{I+W}{2} \geq \tilde{W} \geq W$.

Here, $\mathrm{null}\{W\} := \{x \in \mathbb{R}^n \mid Wx = 0\}$ is the null space of the matrix W, and $\mathrm{span}\{W\} := \{x \in \mathbb{R}^n \mid x = Wy, \forall y \in \mathbb{R}^m\}$ is the subspace spanned by the columns of W.

Distributed proximal gradient methods have been widely applied to unconstrained convex optimization problems. Since the constraints, e.g. the power balance constraints and the power flow equations, are commonly considered in the power system operations, the appliance of the distributed proximal gradient methods are limited. Nevertheless, the proximal gradient-based methods are tightly related to ADMM, which is also extensively used in distributed optimization of power systems.

Consider the problem (A.28), which can be solved using ADMM. From an alternative perspective, the dual problem of (A.28) is

$$\min_y d_1(y) + d_2(y) \tag{A.59}$$

where the dual functions $d_1(y)$ and $d_2(y)$ are defined as

$$d_1(y) = -\min_x f(x) + y^{\mathrm{T}} A x, \tag{A.60a}$$

$$d_2(y) = -\min_z h(z) + y^{\mathrm{T}}(Bz - c) \tag{A.60b}$$

Note that the dual problem (A.59) is unconstrained. We can solve this problem using a proximal based method, known as the Douglas–Rachford splitting method

$$v(k+1) = v(k) + \operatorname{prox}_{\rho d_1}(2y(k) - v(k)) - y(k), \tag{A.61a}$$

$$y(k+1) = \operatorname{prox}_{\rho d_2}(v(k+1)) \tag{A.61b}$$

where the parameter ρ is identical to that in the ADMM (A.30). By [51], the ADMM (A.30) of the primal problem (A.28) is equivalent to the Douglas–Rachford splitting method (A.61) of the dual problem (A.60), i.e. the variable $y(k+1)$ is the same as in (A.30) and (A.61) and the variable $v(k+1)$ in (A.61) equals to $y(k) + \rho Ax(k+1)$ in (A.30).

A.3 Second-Order Newton-Based Algorithms

The first-order (sub)gradient-based methods usually have sublinear or linear convergence rates, whereas the second-order Newton-based methods may achieve superlinear convergence rates. Here we will first introduce the centralized second-order Newton-based method and then extend to distributed methods.

Consider the following problem:

$$\min_{x} f(x) \tag{A.62a}$$

$$\text{s.t. } h_j(x) \leq 0,\ j = 1, 2, \ldots, m \tag{A.62b}$$

$$Ax = b, \tag{A.62c}$$

where $x \in \mathbb{R}^n$ is the decision variable, $f, h_1, \ldots, h_m : \mathbb{R}^n \to \mathbb{R}$ are convex and twice continuously differentiable, and $A \in \mathbb{R}^{p \times n}$, $b \in \mathbb{R}^p$. Assume that the Slater's condition holds, i.e. there exists $\tilde{x} \in \mathbb{R}^n$ such that $A\tilde{x} = b$ and $h_j(\tilde{x}) < 0,\ j = 1, 2, \ldots, m$.

There are mainly two types of Newton-based methods to solve the problem (A.62): the barrier method and the primal–dual interior-point method.

A.3.1 Barrier Method

The barrier method uses the barrier function to eliminate the inequality constraint (A.62b) and then solves the modified equation constrained optimization problem by employing the Newton methods, e.g. the infeasible start Newton method.

Using the indicator function as the barrier, the original problem (A.62) can be equivalently converted into the following problem:

$$\min_{x} f(x) + \sum_{j=1}^{m} \iota_{\mathcal{X}_j}(x) \tag{A.63a}$$

$$\text{s.t. } Ax = b \tag{A.63b}$$

where the set $\mathcal{X}_j = \{x \in \mathbb{R}^n \mid h_j(x) \leq 0\}$.

Note that the indicator function $\iota_{\mathcal{X}_j}$ is nondifferentiable. Therefore, the logarithmic barrier function

$$\phi(x) = -\sum_{j=1}^{m} \log\left(-h_j(x)\right) \tag{A.64}$$

is often applied instead, yielding the approximate problem

$$\min_{x} f(x) + \frac{1}{t}\phi(x) \tag{A.65a}$$

$$\text{s.t. } Ax = b \tag{A.65b}$$

where $t > 0$ is a constant.

Let $x^*(t)$ be the optimal solution to (A.65). By [52, Sec. 11.2], $x^*(t)$ is proved m/t–suboptimal, i.e.

$$f(x^*(t)) - F^* \leq \frac{m}{t},$$

where F^* is the optimal value of the problem (A.62). It indicates that the approximation with the logarithmic barrier is accurate if $t \to \infty$, and we can obtain an ϵ-suboptimal solution by setting $t = m/\epsilon$.

Given a strictly feasible x, $t > 0$, $\mu > 1$, and the tolerance $\epsilon_{\text{out}} > 0$, the barrier method repeats the following steps [52, Sec. 11.3]:

1) Solve the approximate problem (A.65) with the initial point x. Then compute the optimal solution $x^*(t)$, and set $x \leftarrow x^*(t)$.
2) Stop the iteration and return $x^* \leftarrow x$ if $\frac{m}{t} < \epsilon_{\text{out}}$; otherwise, set $t \leftarrow \mu t$.

The main step of the barrier method is to solve (A.65) with the Newton-based methods, e.g. the infeasible start Newton method. Define the objective of (A.65) as $f_t(x) := f(x) + \frac{1}{t}\phi(x)$. The primal and dual residuals of the infeasible start Newton method are defined as

$$r_{\text{dual}}(x, \nu) := \nabla f_t(x) + A^{\mathrm{T}}\nu, \tag{A.66a}$$

$$r_{\text{pri}}(x, \nu) := Ax - b \tag{A.66b}$$

where $\nu \in \mathbb{R}^p$ is the dual variable associated with the equation constraint (A.65b). Let the residual vector $r(x, \nu) := \left[r_{\text{dual}}^{\mathrm{T}}(x, \nu), r_{\text{pri}}^{\mathrm{T}}(x, \nu)\right]^{\mathrm{T}}$.

Given arbitrary x, ν, $0 < \alpha < 0.5$, $0 < \beta < 1$, and the tolerance $\epsilon_{in} > 0$, the infeasible start Newton method repeats the following steps [52, Sec. 10.3]:

1) Compute the primal and dual Newton increments Δx_{nt}, $\Delta \nu_{nt}$ by

$$\begin{bmatrix} \nabla^2 f_t(x) & A^{\mathrm{T}} \\ A & 0 \end{bmatrix} \begin{bmatrix} \Delta x_{nt} \\ \Delta \nu_{nt} \end{bmatrix} = -r(x, \nu) \tag{A.67}$$

2) Determine the step size $s > 0$ by the backtracking line search approach.
3) Update the primal and dual variables by $x \leftarrow x + s\Delta x_{nt}$, $\nu \leftarrow \nu + s\Delta \nu_{nt}$.
4) Stop the iteration and return $x^*(t) \leftarrow x$, if $Ax = b$ and $\|r(x, \nu)\|_2 \leq \epsilon_{in}$.

The algorithm consists of the barrier method as the outer iteration and the infeasible start Newton method as the inner iteration. More details can be found in [52, Sec. 10.3, 11.3].

A.3.2 Primal–Dual Interior-Point Method

Different from the barrier method, the primal–dual interior-point method is a single-loop iterative method. In practice, the primal–dual interior-point method usually performs more efficiently than the barrier method and hence is more popular.

The primal–dual interior-point method directly deals with the optimization problems with inequality constraints. To this end, the modified

Karush–Kuhn–Tucker (KKT) equations are formulated and are solved by using a Newton-based approach.

Since Newton-based approaches can only deal with equations, we formulate the modified KKT equations of the problem (A.62) as

$$\nabla f(x) + \sum_{j=1}^{m} \lambda_j \nabla h_j(x) + A^{\mathrm{T}} \nu = 0, \tag{A.68a}$$

$$-\lambda_j h_j(x) = \frac{1}{t}, j = 1, 2, \ldots, m, \tag{A.68b}$$

$$Ax = b \tag{A.68c}$$

where $\lambda_j \in \mathbb{R}_+$ is the dual variable associated with the inequality constraint $h_j(x) \leq 0$ and $\nu \in \mathbb{R}^p$ is the dual variable associated with the equation constraint $Ax = b$.

Define the primal, dual, and centrality residuals as

$$r_{\text{dual}}(x, \lambda, \nu) := \nabla f(x) + Dh(x)^{\mathrm{T}} \lambda + A^{\mathrm{T}} \nu, \tag{A.69a}$$

$$r_{\text{cent}}(x, \lambda, \nu) := -\operatorname{diag}(\lambda) h(x) - \frac{1}{t}\mathbf{1}, \tag{A.69b}$$

$$r_{\text{pri}}(x, \lambda, \nu) := Ax - b \tag{A.69c}$$

where $h : \mathbb{R}^n \to \mathbb{R}^m$ and $Dh : \mathbb{R}^n \to \mathbb{R}^{m \times n}$ are defined as

$$h(x) := \begin{bmatrix} h_1(x) \\ \vdots \\ h_m(x) \end{bmatrix}, \quad Dh(x) := \begin{bmatrix} \nabla h_1(x)^{\mathrm{T}} \\ \vdots \\ \nabla h_m(x)^{\mathrm{T}} \end{bmatrix} \tag{A.70}$$

Given x satisfying $h_j(x) < 0, j = 1, 2, \ldots, m$, $\lambda > 0$, $\mu > 1$, and the tolerances $\epsilon_1, \epsilon_2 > 0$, the primal–dual interior-point method repeats the following steps [52, Sec. 11.7]:

1) Compute the surrogate duality gap

$$\hat{\eta}(x, \lambda) = -h(x)^{\mathrm{T}} \lambda \tag{A.71}$$

2) Set $t \leftarrow \mu m / \hat{\eta}(x, \lambda)$;
3) Compute the Newton increments $\Delta x_{pd}, \Delta \lambda_{pd}, \Delta \nu_{pd}$ by

$$H(x, \lambda, \nu) \begin{bmatrix} \Delta x_{pd} \\ \Delta \lambda_{pd} \\ \Delta \nu_{pd} \end{bmatrix} = - \begin{bmatrix} r_{\text{dual}}(x, \lambda, \nu) \\ r_{\text{cent}}(x, \lambda, \nu) \\ r_{\text{pri}}(x, \lambda, \nu) \end{bmatrix} \tag{A.72}$$

with

$$H(x, \lambda, \nu) = \begin{bmatrix} \nabla^2 f(x) + \sum_{j=1}^{m} \lambda_j \nabla^2 h_j(x) & Dh(x)^{\mathrm{T}} & A^{\mathrm{T}} \\ -diag(\lambda) Dh(x) & -diag(h(x)) & 0 \\ A & 0 & 0 \end{bmatrix}$$

4) Determine the step size $s > 0$ by the backtracking line search approach.
5) Update the primal and dual variables by $x \leftarrow x + s\Delta x_{pd}$, $\lambda \leftarrow \lambda + s\Delta\lambda_{pd}$, $\nu \leftarrow \nu + s\Delta\nu_{nt}$.
6) Stop the iteration and return $x^* (t) \leftarrow x$, if

$$\left\| r_{\text{pri}} (x, \lambda, \nu) \right\|_2 \leq \epsilon_1, \ \left\| r_{\text{dual}} (x, \lambda, \nu) \right\|_2 \leq \epsilon_1, \ \hat{\eta} (x, \lambda) \leq \epsilon_2$$

To implement the primal–dual interior-point algorithm in a distributed manner, three steps need to be modified, including 3) computing the Newton increments, 4) determining the step size, and 6) checking the termination. More details about the distributed primal–dual interior-point method can be found in [53].

The distributed primal–dual interior-point method has various applications in power systems, especially OPF problems [53–57], due to its superlinear convergence rate. In [53], the primal–dual interior-point method is applied to the DC OPF problem, while [54] proposes a modified predictor–corrector interior point for the AC OPF problem. Reference [55] applies the primal–dual interior-point method to multi-area OPF control, where the Newton-based method is extended to deal with overlapping areas. Reference [56] explores how the partition of areas affects the convergence performance. In [57], a fully decentralized primal–dual interior-point method is formulated for the multi-area OPF problem, and its convergence is proved by referring to a centralized interior-point method. Distributed primal–dual interior-point methods are also utilized in other problems, such as optimal electric vehicle charging [58] and distributed generation sitting and sizing in planning [59].

A.4 Zeroth-Order Online Algorithms

The zeroth-order (or gradient-free, derivative-free, model-free) method is usually used in the case that the gradient is difficult to compute and traditional first-order gradient-based methods are not applicable therein.

Consider a convex optimization problem with a convex objective function $f : \mathbb{R}^n \to \mathbb{R}$, which is nondifferentiable at $x \in \mathbb{R}^n$. In this situation, the first-order gradient-based methods can not apply directly. However, we may alternatively pursuit an approximate gradient from the first-order difference equation as

$$Df_{u,\xi} (x) = \frac{1}{2u} \left[f (x + u\xi) - f (x - u\xi) \right] \xi \tag{A.73}$$

where $Df_{u,\xi} (x)$ is the approximate gradient of $f (x)$ at x, $u > 0$ is a small enough scalar, and $\xi \in \mathbb{R}^n$ is the perturbation vector.

Equation (A.73) is called a zeroth-order gradient estimator, or simply a gradient estimator. The performance of estimator largely depends on the selection of ξ. Unfortunately, there is no systematic method to determine it. In practice, ξ is commonly randomly generated, e.g. by sampling a sinusoidal signal

$$\xi_i(t) = \sqrt{2}\sin\left(\omega_i t\right) \quad i = 1, 2, \dots, n, \tag{A.74}$$

or by picking random numbers that obey a normal distribution

$$\xi_i(t) \sim N\left(0, \sigma_i\right), \quad i = 1, 2, \dots, n. \tag{A.75}$$

Sometimes, the function f may be an implicit function, and its value may be difficult to compute. However, if it can be directly observed or measured from the physical system, we can use the measurements to estimate the gradient, other than directly computing the value. For example, we may apply $x + u\xi$ and $x - u\xi$ to the real system and observe the values $f(x + u\xi)$ and $f(x - u\xi)$, respectively. In this regard, the zeroth-order methods are often deployed in online algorithms. Figure A.2 illustrates the framework of zeroth-order online algorithms. If the gradient is available, the zeroth-order online algorithm degenerates into the gradient-based online algorithm.

Different gradient-based methods can deal with different optimization problems. The combinations of the gradient estimator with the gradient-based methods form various zeroth-order algorithms. Next we take the zeroth-order algorithm in [60] as an example to explain how it works. Consider the following problem:

$$\min_x f(x) := \frac{1}{n} f_i\left(x_1, x_2, \dots, x_n\right) \tag{A.76}$$

where $x \in \mathbb{R}^n$ is the compact decision variable vector and $x_i \in \mathbb{R}$ is the decision variable of the agent i. f_i is the cost function of the agent i. The derivative of f_i is assumed to be unavailable.

Assume that agents communicate with each other through the set of communication link $\mathcal{L}$. Agents i and j are neighbors if and only if $(i,j) \in \mathcal{L}$. Given initial $x(0)$, $u > 0$, $D_j^i(-1) = 0$, $\tau_j^i(-1)$, and the tolerance $\epsilon > 0$, the distributed zeroth-order online algorithm to solve (A.76) repeats the following steps [60]:

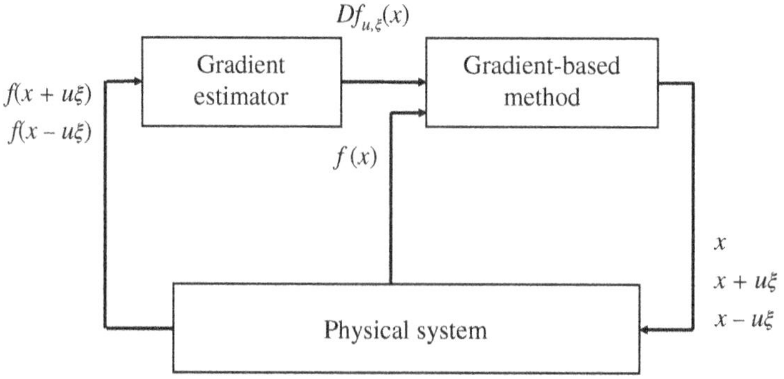

Figure A.2 The framework of the zeroth-order online algorithm.

1) Each agent i generates $\xi^i(t) \sim N(0, \sigma_i)$.
2) Each agent i takes action $x^i(t) + u\xi^i(t)$.
3) Each agent i observes $f_i(x^i(t) + u\xi^i(t))$.
4) Each agent i takes action $x^i(t) - u\xi^i(t)$.
5) Each agent i observes $f_i(x^i(t) - u\xi^i(t))$.
6) Each agent i estimates the individual approximate gradient by

$$Df_i^i(t) = \frac{f_i(x^i(t) + u\xi^i(t)) - f_i(x^i(t) - u\xi^i(t))}{2u}$$

   and set $\tau_i^i(t) = t$.

7) Each agent i receives $\left\{Df_j^{k\to i}(t), \tau_j^{k\to i}(t)\right\}_{j=1}^n$ from its neighbor k and set

$$k_j^i(t) = \arg \max_{k:(k,i)\in\mathcal{L}} \tau_j^{k\to i}(t)$$

   and

$$\tau_j^i(t) = \tau_j^{k_j^i(t)\to i}(t), \quad Df_j^i(t) = Df_j^{k_j^i(t)\to i}(t)$$

   for each $j \neq i$.

8) Each agent i sends $\left\{Df_j^i(t), \tau_j^i(t)\right\}_{j=1}^n$ to his neighbor k.
9) Each agent i updates

$$G^i(t) = \frac{1}{n} \sum_{j:\tau_j^i(t)\geq 0} Df_j^i(t) z^i\left(\tau_j^i(t)\right) \tag{A.77}$$

$$x^i(t+1) = x^i(t) - \eta G^i(t) \tag{A.78}$$

10) If $\max_{i=1,2,\dots,n} |G^i(t)| < \epsilon$, then stop the iteration and returns $x^i(t+1)$, $i = 1, 2, \dots, n$; otherwise, set $t \leftarrow t + 1$.

The zeroth-order online method is an emerging algorithm in distributed optimization. The associated theoretical works include [61–65] while there are a few studies in the context of power systems operation [60, 66]. In [60], the authors formulate a zeroth-order method to cope with the output power maximization problem of wind farms and prove that the convergence rate is in the order of $\mathcal{O}(1/\sqrt{T})$. Reference [66] considers the distribution network market (as a game) with electric vehicles charging involved and applies the distributed zeroth-order method to seek the equilibrium individually.

Bibliography

1 Y. G. Sun, L. Wang, and G. Xie, "Average consensus in networks of dynamic agents with switching topologies and multiple time-varying delays," *System & Control Letters*, vol. 57, no. 2, pp. 175–183, 2008.

2 R. Olfati-Saber and R. M. Murray, "Consensus problems in networks of agents with switching topology and time-delays," *IEEE Transactions on Automatic Control*, vol. 49, no. 9, pp. 1520–1533, 2004.
3 F. Xiao and L. Wang, "State consensus for multi-agent systems with switching topologies and time-varying delays," *International Journal of Control*, vol. 79, no. 10, pp. 1277–1284, 2006.
4 F. Xiao and L. Wang, "Asynchronous consensus in continuous-time multi-agent systems with switching topology and time-varying delays," *IEEE Transactions on Automatic Control*, vol. 53, no. 8, pp. 1804–1816, 2008.
5 A. Simonetto and H. Jamali-Rad, "Primal recovery from consensus-based dual decomposition for distributed convex optimization," *Journal of Optimization Theory and Applications*, vol. 168, no. 1, pp. 172–197, 2016.
6 S. Boyd, N. Parikh, and E. Chu, *Distributed Optimization and Statistical Learning via the Alternating Direction Method of Multipliers*. Now Publishers Inc., 2011.
7 D. Yuan, S. Xu, and H. Zhao, "Distributed primal–dual subgradient method for multiagent optimization via consensus algorithms," *IEEE Transactions on Systems, Man, and Cybernetics, Part B (Cybernetics)*, vol. 41, no. 6, pp. 1715–1724, 2011.
8 A. Bemporad, M. Heemels, M. Johansson *et al.*, *Networked Control Systems*. Springer, 2010.
9 B. Johansson, T. Keviczky, M. Johansson, and K. H. Johansson, "Subgradient methods and consensus algorithms for solving convex optimization problems," in *2008 47th IEEE Conference on Decision and Control*. IEEE, 2008, pp. 4185–4190.
10 M. Bürger, G. Notarstefano, and F. Allgöwer, "Distributed robust optimization via cutting-plane consensus," in *2012 IEEE 51st IEEE Conference on Decision and Control (CDC)*. IEEE, 2012, pp. 7457–7463.
11 R. Carli, G. Notarstefano, L. Schenato, and D. Varagnolo, "Analysis of Newton-Raphson consensus for multi-agent convex optimization under asynchronous and lossy communications," in *2015 54th IEEE Conference on Decision and Control (CDC)*. IEEE, 2015, pp. 418–424.
12 Y. Zhang, F. Liu, Z. Wang, J. Wang, Y. Su, Y. Chen, C. Wang, and Q. Wu, "Asynchrony-resilient and privacy-preserving charging protocol for plug-in electric vehicles," *arXiv preprint arXiv:2005.12519*, 2020.
13 M. Bürger, G. Notarstefano, and F. Allgöwer, "A polyhedral approximation framework for convex and robust distributed optimization," *IEEE Transactions on Automatic Control*, vol. 59, no. 2, pp. 384–395, 2013.
14 B. C. Eaves and W. Zangwill, "Generalized cutting plane algorithms," *SIAM Journal on Control*, vol. 9, no. 4, pp. 529–542, 1971.

15 W. Zhao, M. Liu, J. Zhu, and L. Li, "Fully decentralised multi-area dynamic economic dispatch for large-scale power systems via cutting plane consensus," *IET Generation, Transmission and Distribution*, vol. 10, no. 10, pp. 2486–2495, 2016.

16 D. Bertsekas, A. Nedic, and A. Ozdaglar, *Convex Analysis and Optimization*. Athena Scientific, 2003.

17 D. P. Bertsekas, *Nonlinear Programming*, 2nd ed. Athena Scientific, 1999.

18 D. Bertsekas and J. Tsitsiklis, *Parallel and Distributed Computation*. Old Tappan, NJ, USA: Prentice Hall Inc., 1989.

19 N. Gatsis and G. B. Giannakis, "Residential load control: distributed scheduling and convergence with lost AMI messages," *IEEE Transactions on Smart Grid*, vol. 3, no. 2, pp. 770–786, 2012.

20 R. Deng, Z. Yang, F. Hou, M.-Y. Chow, and J. Chen, "Distributed real-time demand response in multiseller–multibuyer smart distribution grid," *IEEE Transactions on Power Systems*, vol. 30, no. 5, pp. 2364–2374, 2014.

21 N. Gatsis and G. B. Giannakis, "Decomposition algorithms for market clearing with large-scale demand response," *IEEE Transactions on Smart Grid*, vol. 4, no. 4, pp. 1976–1987, 2013.

22 H. Almasalma, J. Engels, and G. Deconinck, "Dual-decomposition-based peer-to-peer voltage control for distribution networks," *CIRED-Open Access Proceedings Journal*, vol. 2017, no. 1, pp. 1718–1721, 2017.

23 S. Magnússon, G. Qu, and N. Li, "Distributed optimal voltage control with asynchronous and delayed communication," *IEEE Transactions on Smart Grid*, vol. 11, no. 4, pp. 3469–3482, 2020.

24 D. Alkano, J. M. Scherpen, and Y. Chorfi, "Asynchronous distributed control of biogas supply and multienergy demand," *IEEE Transactions on Automation Science and Engineering*, vol. 14, no. 2, pp. 558–572, 2017.

25 S. Huang, Y. Sun, and Q. Wu, "Stochastic economic dispatch with wind using versatile probability distribution and L-BFGS-B based dual decomposition," *IEEE Transactions on Power Systems*, vol. 33, no. 6, pp. 6254–6263, 2018.

26 Y. Su, Z. Wang, F. Liu, P. Yang, Y. Zhang, and M. Jia, "*Hierarchical decomposition based distributed energy management of distribution networks*," 8th Renewable Power Generation Conference (RPG 2019), pp. 1–7, 2019.

27 S. Mhanna, A. C. Chapman, and G. Verbič, "Component-based dual decomposition methods for the OPF problem," *Sustainable Energy, Grids and Networks*, vol. 16, pp. 91–110, 2018.

28 A. Y. Lam, B. Zhang, and N. T. David, "Distributed algorithms for optimal power flow problem," in *2012 IEEE 51st IEEE Conference on Decision and Control (CDC)*. IEEE, 2012, pp. 430–437.

29 Z. Cheng and M.-Y. Chow, "Collaborative distributed AC optimal power flow: a dual decomposition based algorithm," *Journal of Modern Power Systems and Clean Energy*, vol. 9, no. 6, pp. 1414–1423, 2021.

30 T.-H. Chang, M. Hong, and X. Wang, "Multi-agent distributed optimization via inexact consensus ADMM," *IEEE Transactions on Signal Processing*, vol. 63, no. 2, pp. 482–497, 2014.

31 K. Huang and N. D. Sidiropoulos, "Consensus-ADMM for general quadratically constrained quadratic programming," *IEEE Transactions on Signal Processing*, vol. 64, no. 20, pp. 5297–5310, 2016.

32 T.-H. Chang, "A proximal dual consensus ADMM method for multi-agent constrained optimization," *IEEE Transactions on Signal Processing*, vol. 64, no. 14, pp. 3719–3734, 2016.

33 L. Yang, J. Luo, Y. Xu, Z. Zhang, and Z. Dong, "A distributed dual consensus ADMM based on partition for DC-DOPF with carbon emission trading," *IEEE Transactions on Industrial Informatics*, vol. 16, no. 3, pp. 1858–1872, 2019.

34 W. Zheng, W. Wu, B. Zhang, H. Sun, and Y. Liu, "A fully distributed reactive power optimization and control method for active distribution networks," *IEEE Transactions on Smart Grid*, vol. 7, no. 2, pp. 1021–1033, 2015.

35 Y. Han, L. Chen, Z. Wang, S. Mei, and W. Liu, "Fully distributed optimal power flow for unbalanced distribution networks based on ADMM," in *2016 IEEE International Conference on Power System Technology (POWERCON)*. IEEE, 2016, pp. 1–6.

36 M. Wang, Y. Su, L. Chen, Z. Li, and S. Mei, "Distributed optimal power flow of dc microgrids: a penalty based ADMM approach," *CSEE Journal of Power and Energy Systems*, vol. 7, no. 2, pp. 339–347, 2019.

37 A. Maneesha and K. S. Swarup, "A survey on applications of alternating direction method of multipliers in smart power grids," *Renewable and Sustainable Energy Reviews*, vol. 152, 111687, 2021.

38 A. Cherukuri, E. Mallada, and J. Cortés, "Asymptotic convergence of constrained primal–dual dynamics," *System & Control Letters*, vol. 87, pp. 10–15, 2016.

39 Z. Wang, W. Wei, C. Zhao, Z. Ma, Z. Zheng, Y. Zhang, and F. Liu, "Exponential stability of partial primal–dual gradient dynamics with nonsmooth objective functions," *Automatica*, vol. 129, 109585, 2021.

40 B. Millar and D. Jiang, "Smart grid optimization through asynchronous, distributed primal dual iterations," *IEEE Transactions on Smart Grid*, vol. 8, no. 5, pp. 2324–2331, 2016.

41 X. Zhou, Z. Liu, C. Zhao, and L. Chen, "Accelerated voltage regulation in multi-phase distribution networks based on hierarchical distributed algorithm," *IEEE Transactions on Power Systems*, vol. 35, no. 3, pp. 2047–2058, 2019.

42 Z. Wang, L. Chen, F. Liu, P. Yi, M. Cao, S. Deng, and S. Mei, "Asynchronous distributed power control of multimicrogrid systems," *IEEE Transactions on Control of Network Systems*, vol. 7, no. 4, pp. 1960–1973, 2020.

43 C. Zhao, U. Topcu, N. Li, and S. Low, "Design and stability of load-side primary frequency control in power systems," *IEEE Transactions on Automatic Control*, vol. 59, no. 5, pp. 1177–1189, 2014.

44 Z. Wang, F. Liu, S. H. Low, C. Zhao, and S. Mei, "Distributed frequency control with operational constraints, Part I: per-node power balance," *IEEE Transactions on Smart Grid*, vol. 10, no. 1, pp. 40–52, 2019.

45 Z. Wang, F. Liu, S. H. Low, C. Zhao, and S. Mei, "Distributed frequency control with operational constraints, Part II: network power balance," *IEEE Transactions on Smart Grid*, vol. 10, no. 1, pp. 53–64, 2019.

46 M. Schmidt, N. L. Roux, and F. Bach, "Convergence rates of inexact proximal-gradient methods for convex optimization," *arXiv preprint arXiv:1109.2415*, 2011.

47 A. I. Chen and A. Ozdaglar, "A fast distributed proximal-gradient method," in *2012 50th Annual Allerton Conference on Communication, Control, and Computing (Allerton)*. IEEE, 2012, pp. 601–608.

48 G. Silva and P. Rodriguez, "Efficient consensus model based on proximal gradient method applied to convolutional sparse problems," *arXiv preprint arXiv:2011.10100*, 2020.

49 A. Beck, *First-Order Methods in Optimization*. SIAM, 2017.

50 W. Shi, Q. Ling, G. Wu, and W. Yin, "A proximal gradient algorithm for decentralized composite optimization," *IEEE Transactions on Signal Processing*, vol. 63, no. 22, pp. 6013–6023, 2015.

51 D. O'Connor and L. Vandenberghe, "Primal-dual decomposition by operator splitting and applications to image deblurring," *SIAM Journal on Imaging Sciences*, vol. 7, no. 3, pp. 1724–1754, 2014.

52 S. Boyd, S. P. Boyd, and L. Vandenberghe, *Convex Optimization*. Cambridge University Press, 2004.

53 A. Minot, Y. M. Lu, and N. Li, "A parallel primal-dual interior-point method for DC optimal power flow," in *2016 Power Systems Computation Conference (PSCC)*. IEEE, 2016, pp. 1–7.

54 R. A. Jabr, A. H. Coonick, and B. J. Cory, "A primal-dual interior point method for optimal power flow dispatching," *IEEE Transactions on Power Systems*, vol. 17, no. 3, pp. 654–662, 2002.

55 G. Hug-Glanzmann and G. Andersson, "Decentralized optimal power flow control for overlapping areas in power systems," *IEEE Transactions on Power Systems*, vol. 24, no. 1, pp. 327–336, 2009.

56 J. Guo, G. Hug, and O. K. Tonguz, "Intelligent partitioning in distributed optimization of electric power systems," *IEEE Transactions on Smart Grid*, vol. 7, no. 3, pp. 1249–1258, 2015.

57 W. Lu, M. Liu, S. Lin, and L. Li, "Fully decentralized optimal power flow of multi-area interconnected power systems based on distributed interior point method," *IEEE Transactions on Power Systems*, vol. 33, no. 1, pp. 901–910, 2017.

58 J. Zhang, Y. He, M. Cui, and Y. Lu, "Primal dual interior point dynamic programming for coordinated charging of electric vehicles," *Journal of Modern Power Systems and Clean Energy*, vol. 5, no. 6, pp. 1004–1015, 2017.

59 Z. Liu, F. Wen, G. Ledwich, and X. Ji, "Optimal sitting and sizing of distributed generators based on a modified primal-dual interior point algorithm," in *2011 4th International Conference on Electric Utility Deregulation and Restructuring and Power Technologies (DRPT)*. IEEE, 2011, pp. 1360–1365.

60 Y. Tang, Z. Ren, and N. Li, "Zeroth-order feedback optimization for cooperative multi-agent systems," in *2020 59th IEEE Conference on Decision and Control (CDC)*. IEEE, 2020, pp. 3649–3656.

61 Y. Tang, J. Zhang, and N. Li, "Distributed zero-order algorithms for nonconvex multiagent optimization," *IEEE Transactions on Control of Network Systems*, vol. 8, no. 1, pp. 269–281, 2020.

62 D. Hajinezhad, M. Hong, and A. Garcia, "Zone: Zeroth-order nonconvex multiagent optimization over networks," *IEEE Transactions on Automatic Control*, vol. 64, no. 10, pp. 3995–4010, 2019.

63 Y. Li, Y. Tang, R. Zhang, and N. Li, "Distributed reinforcement learning for decentralized linear quadratic control: a derivative-free policy optimization approach," in *Learning for Dynamics and Control*. PMLR, 2020, pp. 814.

64 A. K. Sahu, D. Jakovetic, D. Bajovic, and S. Kar, "Distributed zeroth order optimization over random networks: a Kiefer-Wolfowitz stochastic approximation approach," in *2018 IEEE Conference on Decision and Control (CDC)*. IEEE, 2018, pp. 4951–4958.

65 Z. Yu, D. W. Ho, and D. Yuan, "Distributed randomized gradient-free mirror descent algorithm for constrained optimization," *IEEE Transactions on Automatic Control*, vol. 67, no. 2, pp. 957–964, 2021.

66 Y. Chen, S. Zou, and J. Lygeros, "Game theoretic stochastic energy coordination under a distributed zeroth-order algorithm," *IFAC-PapersOnLine*, vol. 53, no. 2, pp. 4070–4075, 2020.

Appendix B

Optimal Power Flow of Direct Current Networks

In Chapter 11, we have suggested a unified approach to driving the direct current (DC) microgrids to the optimum of the corresponding optimal power flow (OPF) problem subject to heterogeneous controllers. Noting that the OPF problem is generally non-convex, we alternatively use the second-order conic (SOC) relaxation to transfer it into a convex counterpart. However, we have not proved whether the relaxation is exact or not. Here we will fill this theoretic gap. Indeed, Gan and Low [1] has presented the exactness result for grid-connected DC networks by assuming an infinite substation bus, which is not the case in Chapter 11. This appendix extends the theory by removing this key assumption, which admits arbitrary network topologies and operating modes.

B.1 Mathematical Model

B.1.1 Formulation

Consider a graph $\mathcal{G} := (\mathcal{N}, \mathcal{E})$, where $\mathcal{N} := \{1, \dots, n\}$ denotes the set of all buses and $\mathcal{E}$ denotes the set of all lines in the DC network. $\mathcal{G}$ is assumed to be connected. Index the buses by $1, \dots, n$, and abbreviate $\{i,j\} \in \mathcal{E}$ as $i \sim j$. Denote $(i \sim j \;\&\; i < j)$ by $i \to j$. For each bus $i \in \mathcal{N}$, denote V_i as its voltage and p_i as its power injection. For each line $i \sim j$, y_{ij} denotes its conductance. A letter without subscripts denotes a vector of the corresponding quantities, e.g. $V = [V_i]_{i \in \mathcal{N}}$. The notations are summarized in Figure B.1.

Merging Optimization and Control in Power Systems: Physical and Cyber Restrictions in Distributed Frequency Control and Beyond, First Edition. Feng Liu, Zhaojian Wang, Changhong Zhao, and Peng Yang.

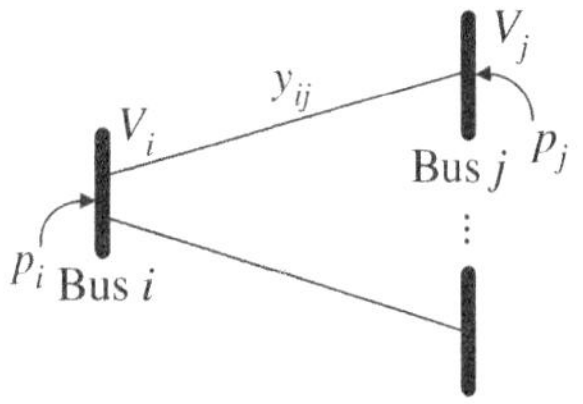

Figure B.1 Summary of notations.

In a (stand-alone) DC network, V_i, p_i, and y_{ij} are real numbers, and $y_{ij} > 0$. Then the corresponding OPF problem reads

$$\textbf{OPF1}: \quad \min_{p,V} \; h(p) = \sum_{i \in \mathcal{N}} f_i(p_i)$$

$$\text{s.t. } p_i = \sum_{j:j\sim i} V_i \left(V_i - V_j\right) y_{ij}, \quad i \in \mathcal{N}; \tag{B.1a}$$

$$\underline{p}_i \leq p_i \leq \overline{p}_i, \quad i \in \mathcal{N}; \tag{B.1b}$$

$$\underline{V}_i \leq V_i \leq \overline{V}_i, \quad i \in \mathcal{N}. \tag{B.1c}$$

Here, $f_i(p_i)$ is strictly increasing in p_i. When $f_i(p_i)$ denotes the generation cost function, $h(p)$ stands for the total generation cost, e.g. $f_i(p_i) = a_i p_i + b_i$ where a_i, b_i, are coefficients. When $f_i(p_i) = p_i$, $h(p)$ stands for the total network loss. Equation (B.1a) is the power injection equation for bus i. The nodal voltages are constrained by (B.1c) with the lower bound $\underline{V}_i > 0$ and upper bound $\overline{V}_i$. The power injections are constrained by (B.1b) with the lower bound $\underline{p}_i$ and upper bound $\overline{p}_i$. It is assumed that $\underline{p}_i \leq 0$, which indicates the following two cases:

1) Bus i is a pure generation bus without load changes. In this case, $\underline{p}_i = 0$ as the generators in a DC network can be turned off;
2) Bus i is a pure load bus without generation or a mixture power injection of both load and generation. In this case, there is $\underline{p}_i < 0$.

The model of a grid-connected DC network [1] can be viewed as a special case of OPF1 including a substation bus with an unconstrained power injection ($\underline{p}_0 = -\infty, \overline{p}_0 = \infty$) and a fixed nodal voltage ($V_0 = V_0^{\text{ref}}$).

The OPF problem only provides the control targets for the devices, and then the devices are controlled to track the targets using appropriate control strategies. The control range of device is reflected in the boundaries of power injections and nodal voltages. That is, we consider the OPF problem in a system level. Other detailed constraints of device are omitted here since they should be considered at a device level.

Due to over-provisioned design, the line capacity is usually large enough to fulfill the maximal load without causing congestions. It means that the current across the lines should always be less than the maximal values in a normal operating condition. In this situation, it is reasonable to ignore line constraint since it is inactive. In this section, we first consider the OPF problem of DC networks without line constraints. Then, we discuss the influence of line constraints on the exactness of SOC relaxation. Note that, throughout this part, we do not assume any specific topology of the power network, since the exactness of SOC relaxation holds for arbitrary topologies of DC networks, as we will explain later on.

B.1.2 Equivalent Transformation

Obviously, the OPF problem (B.1) in consideration is a nonlinear non-convex problem. By introducing slack variables, it can be transformed into an equivalent counterpart, where the non-convex power injection equation (B.1a) is converted into a rank constraint.

Specifically, introduce slack variables to formulate a map f such that

$$f := v_i = V_i^2, i \in \mathcal{N}; \tag{B.2a}$$

$$W_{ij} = V_i V_j, i \sim j; \tag{B.2b}$$

and define a matrix

$$R_{ij} := \begin{bmatrix} v_i & W_{ij} \\ W_{ji} & v_j \end{bmatrix} \tag{B.3}$$

for every $i \to j$. Then OPF1 (B.1) is converted into

$$\textbf{OPF2}: \quad \min_{p,v,W} h(p)$$

$$\text{s.t.} \quad p_i = \sum_{j:j\sim i} (v_i - W_{ij}) y_{ij}, \quad i \in \mathcal{N} \tag{B.4a}$$

$$\underline{p}_i \leq p_i \leq \overline{p}_i, \quad i \in \mathcal{N} \tag{B.4b}$$

$$\underline{V}_i^2 \leq v_i \leq \overline{V}_i^2, \quad i \in \mathcal{N} \tag{B.4c}$$

$$W_{ij} \geq 0, \quad i \to j \tag{B.4d}$$

$$W_{ij} = W_{ji}, \quad i \to j \tag{B.4e}$$

$$R_{ij} \succeq 0, \quad i \to j \tag{B.4f}$$

$$\text{rank}(R_{ij}) = 1, \quad i \to j \tag{B.4g}$$

where the non-convexity in (B.1a) (in OPF1) is converted into the non-convexity in the rank constraint (B.4g) (in OPF2). R_{ij} are positive semidefinite as shown in (B.4f).

Theorem B.1 *OPF1 and OPF2 are equivalent.*

To prove Theorem B.1, we first give the following lemma.

Lemma B.1 *Given $v_i > 0$ for $i \in \mathcal{N}$ and $W_{ij} \geq 0$ for $i \to j$, let $W_{ij} = W_{ji}$ for $i \to j$. If rank$(R_{ij}) = 1$ for $i \to j$, then there exists a unique V satisfying $V_i > 0$ for $i \in \mathcal{N}$ and (B.2). Moreover, V is determined by $V_i = \sqrt{v_i}$ for $i \in \mathcal{N}$.*

It is easy to prove Lemma B.1 by noting that $v_i v_j - W_{ij} W_{ji} = 0$ $(i \to j)$ due to $\text{rank}(R_{ij}) = 1$. Lemma B.1 implies that for each (v, W), there exists a unique V that satisfies the map given by (B.2). It should be noted that, differing from [1], we do not have a substation bus with a fixed voltage. Instead, we assume $V_i > 0$ for all $i \in \mathcal{N}$.

With Lemma B.1, Theorem B.1 can be proved. Let $\mathcal{F}_{OPF1}$ and $\mathcal{F}_{OPF2}$ denote the feasible sets of OPF1 and OPF2, respectively. Since OPF1 and OPF2 have the same objective function, it suffices to show that there exists a one-to-one map between $\mathcal{F}_{OPF1}$ and $\mathcal{F}_{OPF2}$. Specifically, the map $f : (V) \mapsto (v, W)$ given by (B.2) is one to one.

It is worthy of noting that Theorem B.1 does not rely on any specific topology of the network.

B.2 Exactness of SOC Relaxation

B.2.1 SOC Relaxation of OPF in DC Networks

By removing (B.4g), the original non-convex problem OPF2 is transformed to a second-order conic programming (SOCP) problem (named as RL1):

$$\begin{aligned} \textbf{RL1}: \quad & \min_{p,v,W} \; h(p) \\ \text{s.t.} \quad & \text{(B.4a)–(B.4f)} \end{aligned}$$

The only difference between RL1 and OPF2 is that constraint (B.4g) is absent in RL1. Therefore, RL1 is exact if its every optimal solution satisfies (B.4g). To ensure the exactness of the SOC relaxation, additional assumptions are required.

B.2.2 Assumptions

Throughout this section, we make the following assumptions.

Assumption B.1 $\overline{V}_1 = \overline{V}_2 = \cdots \overline{V}_n > 0$.

Assumption B.2 $\sum_{i \in \mathcal{N}} p_i > 0$.

Assumption B.1 requires all the nodal voltages have the same upper bounds in per unit value, while the real values can be different from each other. Assumption B.2 is trivial as it means the total network loss is positive, which is guaranteed by the physical power system in practice. We introduce it just for mathematical rigor. Actually, Assumption B.2 can be guaranteed by simply letting $y_{ij} > 0$ and at least one nonzero power injection p_i. Such assumptions relax those in [1], which only require negative power injection lower bounds and an unconstrained power injection, and admit the unique features of stand-alone DC networks.

B.2.3 Exactness of the SOC Relaxation

The following Theorem claims that RL1 is an exact SOC relaxation of OPF2 (equivalently OPF1) under Assumptions B.1 and B.2.

Theorem B.2 *RL1 is exact if Assumptions B.1 and B.2 hold.*

To prove this theorem, we first introduce the following lemmas.

Lemma B.2 *Assume Assumption B.1 holds. Let (p, v, W) be feasible for RL1 but violate the rank constraint* (B.4g) *on a certain line $(s \to t) \in \mathcal{E}$. If $p_s = \underline{p}_s$, then $v_s < \overline{V}_s^2$. Meanwhile, if $p_t = \underline{p}_t$, then $v_t < \overline{V}_t^2$.*

Proof: For brevity, we only prove the case of bus s as the proof of bus t is the same. With regard to the value of $\underline{p}_s$, we only need to discuss two cases: $p_s = \underline{p}_s < 0$ and $p_s = \underline{p}_s = 0$. In the first case, we have

$$p_s = \sum_{i:i\sim s} \left(v_s - W_{si}\right) y_{si} < 0 \tag{B.5}$$

according to (B.4a).

Therefore, $\left(v_s - W_{si}\right) y_{si} < 0$ for some $i^\flat \in \mathcal{N}$. It follows from (B.4d)–(B.4f) that $W_{si^\flat} \leq \sqrt{v_s v_{i^\flat}}$. According to Assumption B.1, $\overline{V}_s = \overline{V}_{i^\flat}$. Thus, we have

$$v_s < W_{si^\flat} \leq \sqrt{v_s v_{i^\flat}} \leq \sqrt{\overline{V}_s^2 \overline{V}_{i^\flat}^2} = \overline{V}_s^2 \tag{B.6}$$

In the second case, we have

$$p_s = \sum_{i:i\sim s} \left(v_s - W_{si}\right) y_{si} = 0 \tag{B.7}$$

according to (B.4a). Then we only need to discuss the following two cases:

1) When $v_s - W_{si} = 0$ for all $i \sim s$, we have $v_s = W_{st}$ for $i = t$. Since (p, v, W) violates the rank constraint (B.4g) for $s \to t$, R_{st} is non-singular. Hence, we have

$$W_{st} \neq \sqrt{v_s v_t} \Rightarrow W_{st} < \sqrt{v_s v_t} \tag{B.8}$$

due to (B.4d)–(B.4f). According to Assumption B.1, $\overline{V}_s = \overline{V}_t$. It follows that

$$v_s = W_{st} < \sqrt{v_s v_t} \leq \sqrt{\overline{V}_s^2 \overline{V}_t^2} = \overline{V}_s^2 \tag{B.9}$$

2) When $v_s - W_{si} = 0$ does not hold for some of $i \sim s$, there must exist at least one $i^\flat \sim s$ such that $v_s - W_{si^\flat} < 0$ (and accordingly there exists at least another $i^\sharp \sim s$ such that $v_s - W_{si^\sharp} > 0$ to satisfy (B.7)). Hence we have

$$v_s < W_{si^\flat} \leq \sqrt{v_s v_{i^\flat}} \leq \sqrt{\overline{V}_s^2 \overline{V}_{i^\flat}^2} = \overline{V}_s^2 \tag{B.10}$$

due to $\overline{V}_s = \overline{V}_{i^\flat}$. This completes the proof. ■

Lemma B.2 implies that, for each bus, the power injection's lower bound and the nodal voltage's upper bound cannot be binding at the same time. Therefore, for any feasible solution (p, v, W) of RL1, if it violates the rank constraint (B.4g) for a certain line $(s \to t) \in \mathcal{E}$ and the constraint $p_s \geq \underline{p}_s$ (or $p_t \geq \underline{p}_t$) is binding, then $v_s \leq \overline{V}_s^2$ (or $v_t \leq \overline{V}_t^2$) will never bind.

Lemma B.3 *Assume Assumptions B.1 and B.2 hold, and let (p, v, W) be a feasible solution to RL1:*

1) *(p, v, W) violates the rank constraint (B.4g) on a certain line $(s \to t) \in \mathcal{E}$.*
2) *$p_s > \underline{p}_s, p_t > \underline{p}_t$*

Then there exists another feasible solution (p', v, W') that does the following:

1) *Satisfy (B.4a)–(B.4f).*
2) *Satisfy $h(p') < h(p)$.*

Proof: Since (p, v, W) satisfies (B.4d)–(B.4f), we have $0 \leq W_{ij} \leq \sqrt{v_i v_j}$ for $i \to j$. Since (p, v, W) violates the rank constraint (B.4g) for $s \to t$, we have $W_{st} \neq \sqrt{v_s v_t}$ and $W_{st} < \sqrt{v_s v_t}$. We can always choose a small enough number $\epsilon > 0$ such that

$$\epsilon < \min \left\{ \frac{p_s - \underline{p}_s}{y_{st}}, \frac{p_t - \underline{p}_t}{y_{st}}, \sqrt{v_s v_t} - W_{st} \right\}$$

Following the line of the proof of Lemma 7 in [1], we can use ϵ to construct W' as

$$W'_{ij} := \begin{cases} W_{ij} + \epsilon & \text{if } \{i, j\} = \{s, t\}; \\ W_{ij} & \text{otherwise} \end{cases}$$

Also, we construct p' as $p'_i := \sum_{j:j\sim i} \left(v_i - W'_{ij}\right) y_{ij}, \ i \in \mathcal{N}$

The point (p', v, W') satisfies (B.4a) due to the construction of p'. When $i \neq s, t$, we have

$$\begin{aligned} p'_i &= \sum_{j:j\sim i} \left(v_i - W'_{ij}\right) y_{ij} \\ &= \sum_{j:j\sim i} \left(v_i - W_{ij}\right) y_{ij} \\ &= p_i \end{aligned}$$

When $i = s, t$, we have

$$\begin{aligned} p'_i &= \sum_{j:j\sim i} (v_i - W'_{ij}) y_{ij} \\ &= \sum_{j:j\sim i} (v_i - W_{ij}) y_{ij} - y_{st}\epsilon \\ &= p_i - y_{st}\epsilon \\ &\in (\underline{p}_i, \overline{p}_i) \end{aligned}$$

Hence, (p', v, W') satisfies (B.4b) and $p'_i < p_i$ when $i = s, t$. Otherwise, $p'_i = p_i$. It follows that $h(p') < h(p)$ since f is strictly increasing in p_i for $i \in \mathcal{N}$. The point (p', v, W') satisfies (B.4c) since v remains the same as the feasible point (p, v, W). The point (p', v, W') satisfies (B.4d) due to the construction of W'. The point (p', v, W') satisfies (B.4e) since $W'_{ij} - W'_{ji} = W_{ij} - W_{ji} = 0$ for $i \to j$. Additionally, the point (p', v, W') satisfies (B.4f) since $W'_{ij} = W_{ij} \in [0, \sqrt{v_i v_j}]$ when $\{i, j\} \neq \{s, t\}$ and $W'_{ij} = W_{ij} + \epsilon \in [0, \sqrt{v_i v_j})$ when $\{i, j\} = \{s, t\}$. This completes the proof. ■

Lemma B.3 says that if a feasible point (p, v, W) violates the rank constraint (B.4g) for a certain line $s \to t$, while both p_s and p_t are not binding, then we can always find another feasible point (p', v, W') with a better objective value. It implies that if the optimal solution (p^*, v^*, W^*) to RL1 violates the rank constraint (B.4g) for a certain line $s \to t$, then at least one of p^*_s and p^*_t must have reached its lower bound.

Lemma B.4 *Assume Assumptions B.1 and B.2 hold, and let (p, v, W) be a feasible solution to RL1:*

1) *(p, v, W) violates the rank constraint* (B.4g) *on a certain line $(s \to t) \in \mathcal{E}$.*
2) *Either $(p_s = \underline{p}_s \,\&\, p_t > \underline{p}_t)$ or $(p_s > \underline{p}_s \,\&\, p_t = \underline{p}_t)$.*

Then there exists another feasible solution (p', v', W') that does the following:

1) *Satisfy* (B.4a)–(B.4f).
2) *Satisfy $h(p') < h(p)$.*

Proof: We only prove the case $(p_s = \underline{p}_s \,\&\, p_t > \underline{p}_t)$. The proof for the case $(p_s > \underline{p}_s \,\&\, p_t = \underline{p}_t)$ is similar and hence omitted.

Similar to the proof of Lemma B.3, we have $W_{st} < \sqrt{v_s v_t}$. Additionally, it follows from Lemma B.2 that $v_s < \overline{V}_s^2$, so we can always choose a small enough number $\epsilon > 0$ such that

$$\epsilon < \min \left\{ \frac{p_t - \underline{p}_t}{y_{st}}, \sqrt{v_s v_t} - W_{st}, \frac{\sum_{j:j \sim s} y_{sj}}{y_{st}} \left(\overline{V}_s^2 - v_s \right) \right\}$$

Then we have

$$p_t - y_{st}\epsilon > \underline{p}_t$$

$$W_{st} + \epsilon < \sqrt{v_s v_t}$$

$$v_s + \frac{y_{st}}{\sum_{j:j \sim s} y_{sj}} \epsilon < \overline{V}_s^2$$

Following the line of the proof of Lemma 8 in [1], we can use ϵ to construct W' as

$$W'_{ij} := \begin{cases} W_{ij} + \epsilon & \text{if } \{i,j\} = \{s,t\}; \\ W_{ij} & \text{otherwise} \end{cases}$$

Also, we construct v' as

$$v'_i := \begin{cases} v_i + \frac{y_{st}}{\sum_{j:j\sim i} y_{ij}}\epsilon & \text{if } \quad i = s; \\ v_i & \text{otherwise} \end{cases}$$

and p' as $p'_i := \sum_{j:j\sim i} (v'_i - W'_{ij})y_{ij},\ i \in \mathcal{N}$

(p', v', W') satisfies (B.4a) by noting the construction of p'. When $i \neq s, t$, we have

$$\begin{aligned} p'_i &= \sum_{j:j\sim i} \left(v'_i - W'_{ij}\right) y_{ij} \\ &= \sum_{j:j\sim i} \left(v_i - W_{ij}\right) y_{ij} \\ &= p_i \end{aligned}$$

When $i = s$, we have

$$\begin{aligned} p'_s &= \sum_{j:j\sim s} \left(v'_s - W'_{sj}\right) y_{sj} \\ &= \sum_{j:j\sim s} \left(v_s - W_{sj}\right) y_{sj} + (v'_s - v_s) \sum_{j:j\sim s} y_{sj} - y_{st}\epsilon \\ &= \sum_{j:j\sim s} \left(v_s - W_{sj}\right) y_{sj} \\ &= p_s \end{aligned}$$

When $i = t$, we have

$$\begin{aligned} p'_t &= \sum_{j:j\sim t} \left(v'_t - W'_{tj}\right) y_{tj} \\ &= \sum_{j:j\sim t} \left(v_t - W_{tj}\right) y_{tj} - y_{st}\epsilon \\ &= p_t - y_{st}\epsilon \\ &\in \left(\underline{p}_t, \overline{p}_t\right) \end{aligned}$$

Thus, (p', v', W') satisfies (B.4b) and $p'_i < p_i$ if $i = t$. Otherwise, $p'_i = p_i$. It follows that $h(p') < h(p)$. Since $v'_i = v_i$ if $i \neq s$ and $v_i < v'_i < \overline{V}_i^2$ if $i = s$, the point (p', v', W') satisfies (B.4c).

Similar to the proof of Lemma B.3, it is straightforward to check that the point (p', v', W') satisfies (B.4d), (B.4e), and (B.4f). This completes the proof. ■

Lemma B.4 means that if a feasible point (p, v, W) violates the rank constraint (B.4g) for a certain line $s \to t$, while either p_s or p_t is binding, then we can always find another feasible point (p', v', W') with a better objective value. Lemmas B.3 and B.4 imply that if the optimal solution (p^*, v^*, W^*) to RL1 violates the rank constraint (B.4g) for a certain line $s \to t$, then it must satisfy $p_s^* = \underline{p}_s$ and $p_t^* = \underline{p}_t$. Otherwise, we can always find a better solution.

Lemma B.5 *Assume Assumptions B.1 and B.2 hold, and let (p, v, W) be a feasible solution to RL1:*

1) *(p, v, W) violates the rank constraint* (B.4g) *on a certain line $(s \to t) \in \mathcal{E}$.*
2) *$p_s = \underline{p}_s$ and $p_t = \underline{p}_t$.*

Then there always exists another solution (p, v', W') that does the following:

1) *Satisfy* (B.4a)–(B.4f).
2) *Violate rank constraint* (B.4g) *for all the neighboring lines of $s \to t$, i.e. $i \to j$ with $\{\{i\}, \{j\}\} \cap \{\{s\}, \{t\}\} \neq \emptyset$.*

Proof: Similar to the proof of Lemma B.3, we have $W_{st} < \sqrt{v_s v_t}$. According to Lemma B.2, we have $v_s < \overline{V}_s^2, v_t < \overline{V}_t^2$.

Therefore, we can always choose a small enough number $\epsilon > 0$ so that

$$\epsilon < \min \left\{ \sqrt{v_s v_t} - W_{st}, \frac{\sum_{j:j\sim s} y_{sj}}{y_{st}} \left(\overline{V}_s^2 - v_s\right), \frac{\sum_{j:j\sim t} y_{tj}}{y_{st}} \left(\overline{V}_t^2 - v_t\right) \right\}$$

Then

$$W_{st} + \epsilon < \sqrt{v_s v_t}$$
$$v_s + \frac{y_{st}}{\sum_{j:j\sim s} y_{sj}} \epsilon < \overline{V}_s^2$$
$$v_t + \frac{y_{st}}{\sum_{j:j\sim t} y_{tj}} \epsilon < \overline{V}_t^2$$

Following the line of the proof of Lemma 9 in [1], we can use ϵ to construct W' as

$$W'_{ij} := \begin{cases} W_{ij} + \epsilon & \text{if } \{i,j\} = \{s,t\}; \\ W_{ij} & \text{otherwise} \end{cases}$$

and construct v' as

$$v'_i := \begin{cases} v_i + \frac{y_{st}}{\sum_{j:j\sim i} y_{ij}} \epsilon & i = s, t; \\ v_i & \text{otherwise} \end{cases}$$

The point (p, v', W') satisfies (B.4a) since when $i = s, t$:

$$\begin{aligned}&\sum_{j:j\sim i}\left(v_i' - W_{ij}'\right) y_{ij}\\ &= \sum_{j:j\sim i}\left(v_i - W_{ij}\right) y_{ij} + \sum_{j:j\sim i}\frac{y_{st}}{\sum_{j:j\sim i} y_{ij}}\epsilon y_{ij} - \epsilon y_{st}\\ &= \sum_{j:j\sim i}\left(v_i - W_{ij}\right) y_{ij} = p_i\end{aligned}$$

When $i \neq s, t$, we have

$$\sum_{j:j\sim i}\left(v_i' - W_{ij}'\right) y_{ij} = \sum_{j:j\sim i}\left(v_i - W_{ij}\right) y_{ij} = p_i$$

Since p does not change, the point (p, v', W') satisfies (B.4b). It follows from (B.2.1) that $v_i' = v_i$ if $i \neq s, t$ and $v_i < v_i' < \overline{V}_i^2$ if $i = s, t$. Hence, the point (p, v', W') satisfies (B.4c). Similar to the proof of Lemma B.3, it is straightforward to check that the point (p', v', W') satisfies (B.4d), (B.4e), and (B.4f). Since $\mathcal{G}$ is connected, there always exists some $i \to j$ such that $\{i,j\} \neq \{s,t\}$ and $\{\{i\},\{j\}\} \cap \{\{s\},\{t\}\} \neq \emptyset$. Then we have

$$W_{ij}' = W_{ij} \leq \sqrt{v_i v_j} < \sqrt{v_i' v_j'}$$

Particularly, when $\{i,j\} = \{s,t\}$,

$$W_{ij}' = W_{ij} + \epsilon < \sqrt{v_i v_j} < \sqrt{v_i' v_j'}$$

It means (p, v', W') violates the rank constraint (B.4g) for $s \to t$ and all its neighboring lines. This completes the proof. ■

Lemma B.5 says that if a feasible point (p, v, W) violates the rank constraint (B.4g) for a certain line $s \to t$, while the corresponding power injections' lower bounds are binding, we can always find another feasible point (p, v', W') that violates the rank constraint (B.4g) for $s \to t$ and all its neighboring lines. It should be noted that in Lemma B.5, the construction of feasible point (p, v', W') does not change p. Since the network $\mathcal{G}$ is connected, Lemma B.5 implies that we can continue such propagation to obtain a feasible point that violates the rank constraint (B.4g) for all the lines, without changing p.

If the optimal solution (p^*, v^*, W^*) to RL1 violates the rank constraint (B.4g) for a certain line $s \to t$, Lemmas B.3 and B.4 ensure that it must satisfy $p_s^* = \underline{p}_s$ and $p_t^* = \underline{p}_t$. In this case, according to Lemma B.5, we can always find a feasible (p', v', W') that satisfies $p' = p^*$ and violates the rank constraint (B.4g) for all the lines. Following this idea, now we can prove Theorem B.2 based on Lemmas B.2–B.5.

Proof: [Proof of Theorem B.2].

To prove that RL1 is exact, it suffices to show that any optimal solution to RL1 satisfies the rank constraint (B.4g). We first assume RL1 is *not* exact for the sake of contradiction. Then there must exist at least one optimal solution (p^*, v^*, W^*) of RL1, which violates (B.4g) for a certain line $(s \to t) \in \mathcal{E}$. We discuss from the following three cases of the power injections p_s and p_t corresponding to $s \to t$.

Case 1): $(p_s^* > \underline{p}_s)$ & $(p_t^* > \underline{p}_t)$.

In this case, as Lemma B.3 points out, there must exist a feasible (p', v', W') that has a smaller objective value than (p^*, v^*, W^*). Therefore, (p^*, v^*, W^*) cannot be optimal regarding RL1.

Case 2): $(p_s^* = \underline{p}_s$ & $p_t^* > \underline{p}_t)$ or $(p_s^* > \underline{p}_s$ & $p_t^* = \underline{p}_t)$.

In this case, Lemma B.4 states that there must exist a feasible (p', v', W') that has a smaller objective value than (p^*, v^*, W^*). Therefore, (p^*, v^*, W^*) cannot be optimal regarding RL1.

Case 3): $(p_s^* = \underline{p}_s)$ & $(p_t^* = \underline{p}_t)$.

In this case, we show such an optimal solution (p^*, v^*, W^*) does not exist. According to Lemma B.5, there always exists a feasible (p', v', W') that violates the rank constraint (B.4g) for $s \to t$ and all its neighboring lines. Since $\mathcal{G}$ is connected and $p' = p^*$, we can further propagate such construction to obtain another feasible $(p^*, v^\dagger, W^\dagger)$ of RL1, which violates the rank constraint (B.4g) for all the lines. Hence, p^* satisfies $p_i^* = \underline{p}_i$ for all $i \in \mathcal{N}$. It follows that

$$\sum_{i \in \mathcal{N}} p_i^* = \sum_{i \in \mathcal{N}} \underline{p}_i \leq 0 \tag{B.11}$$

Since it contradicts Assumption B.2 that states $\sum_{i \in \mathcal{N}} p_i^* > 0$, such a (p^*, v^*, W^*) does not exist.

Therefore, every optimal solution must satisfy the rank constraint (B.4g), i.e. RL1 is exact. This completes the proof. ■

Remark B.1 In [1], the exactness of SOC relaxation for grid-connected DC network is presented, which motivates us to investigate a more general problem: Can the requirement of infinite substation bus be removed? Compared with [1], the proofs of the first two cases are similar, while the third case are quite different. In this section, since there is no infinite substation bus, the proof in [1] does not apply, and a more general contradiction has to be derived. By deduction, we finally have $\sum_{i \in \mathcal{N}} p_i^* = \sum_{i \in \mathcal{N}} \underline{p}_i \leq 0$. This contradicts Assumption B.2 that states $\sum_{i \in \mathcal{N}} p_i^* > 0$. To this end, Lemma B.2 is proved considering non-positive power injection lower bounds, instead of negative ones as in [1]. Based on Lemma B.2, Lemmas B.4 and B.5 can be proved.

B.2.4 Topological Independence

Recall the SOC relaxation in AC networks [2]. The approach consists of two relaxation steps: angle relaxation and conic relaxation. Similarly, our method also contains two steps: equivalent transformation and conic relaxation. Differing from AC networks, DC networks do not involve voltage angles. The first step in our method (i.e. equivalent transformation) is exact, as Theorem B.1 states. However, in the second step, directly removing the rank constraints may result in inexactness. Thus, additional assumptions (Assumptions B.1 and B.2) are made to ensure the exactness of conic relaxation. By noting that none of the two steps depends on specific network topologies, we directly have the following theorem.

Theorem B.3 *Under Assumptions B.1 and B.2, the exactness of RL1 is independent of network topologies.*

Remark B.2 In terms of a grid-connected DC networks, it has also been demonstrated that the SOC relaxation is topology independent [1]. Actually, when a DC network works in a grid-connected state, one can simply assign a substation bus in RL1 by letting $\underline{p}_0 = -\infty, \overline{p}_0 = +\infty$ and $v_0 = (V_0^{\text{ref}})^2$. Hence, combining the results in [1] and Theorem B.3 immediately concludes that the SOC relaxation of the OPF problem of DC networks does not rely on network topology, no matter the DC network works in a grid-connected or a stand-alone mode.

B.2.5 Uniqueness of the Optimal Solution

Next we show the optimal solution to RL1 is unique.

Theorem B.4 *Suppose Assumptions B.1 and B.2 hold. The optimal solution to RL1 is unique.*

Proof: Let $x^{1*} := (p^{1*}, v^{1*}, W^{1*})$ and $x^{2*} := (p^{2*}, v^{2*}, W^{2*})$ be two optimal solutions of RL1; then we have

$$\sum_{i \in \mathcal{N}} f_i(p_i^{1*}) = \sum_{i \in \mathcal{N}} f_i(p_i^{2*}) \tag{B.12}$$

Following the line of the proof of Theorem 3 in [1], it holds that

$$\frac{v_i^{1*}}{v_i^{2*}} = \frac{v_j^{1*}}{v_j^{2*}} = \eta, \quad i \sim j$$

$$W_{ij}^{1*} = \sqrt{v_i^{1*} v_j^{1*}} = \eta \sqrt{v_i^{2*} v_j^{2*}} = \eta W_{ij}^{2*}$$

It follows from (B.4a) that

$$
\begin{aligned}
p_i^{1*} &= \sum_{j:j\sim i} \left(v_i^{1*} - W_{ij}^{1*} \right) y_{ij} \\
&= \sum_{j:j\sim i} \left(\eta v_i^{2*} - \eta W_{ij}^{2*} \right) y_{ij} \\
&= \eta p_i^{2*}
\end{aligned}
$$

Since $f_i(p_i)$ is strictly increasing, we must have $\eta = 1$. Otherwise, it contradicts (B.12). Thus, $x^{1*} = x^{2*}$, i.e. RL1 has at most one optimal solution. This completes the proof. ■

For a grid-connected DC network, it has also been proved that the SOCP problem has at most one optimal solution [1]. Thus, we can conclude that the uniqueness of optimal solution to the SOCP problem does not depend on the operating modes of DC networks.

B.2.6 Branch Flow Model

In an optimal solution to RL1, v_i and W_{ij} may be numerically very close to each other, since the range of nodal voltage is small (usually 0.95–1.05 p.u.) and $\text{Rank}(R_{ij}) = 1$ is satisfied, which implies $v_i v_j = W_{ij} W_{ji}$. Thus, RL1 is ill-conditioned as (B.4a) requires the subtractions of v_i and W_{ij}. However, such subtractions can be avoided by converting RL1 into a branch flow model (BFM), so that the numerical stability is improved. In light of [1], by defining $z_{ij} := 1/y_{ij}$ and using new variables P, l, RL1 can be converted into a BFM via the map $g : (v, W) \mapsto (v', P, l)$ defined as below:

$$
g := \begin{cases} v_i' = v_i, & i \in \mathcal{N}; \\ P_{ij} = \left(v_i - W_{ij} \right) y_{ij}, & i \to j; \\ l_{ij} = y_{ij}^2 \left(v_i - W_{ij} - W_{ji} + v_j \right), & i \to j \end{cases} \tag{B.13a}
$$

Using the map g (B.13), OPF2 is converted into the following optimization problem with a BFM:

$$
\textbf{OPF3}: \quad \min_{p,P,v,l} h(p)
$$

$$
\text{s.t.} \quad p_i = \sum_{j:j\sim i} P_{ij}, \quad i \in \mathcal{N}; \tag{B.14a}
$$

$$
\underline{p}_i \le p_i \le \overline{p}_i, \quad i \in \mathcal{N}; \tag{B.14b}
$$

$$
\underline{V}_i^2 \le v_i \le \overline{V}_i^2, \quad i \in \mathcal{N}; \tag{B.14c}
$$

$$
P_{ij} + P_{ji} = z_{ij} l_{ij}, \quad i \to j; \tag{B.14d}
$$

$$v_i - v_j = z_{ij}\left(P_{ij} - P_{ji}\right), \quad i \to j; \tag{B.14e}$$

$$l_{ij} = \frac{P_{ij}^2}{v_i}, \quad i \sim j \tag{B.14f}$$

With g, RL1 is converted into the following RLS1, which is a convex relaxation of OPF3.

$$\textbf{RLS1:} \quad \min_{p,P,v,l} \; h(p)$$

$$\text{s.t.} \quad \text{(B.14a)–(B.14e)}$$

$$l_{ij} \geq \frac{P_{ij}^2}{v_i}, \; i \sim j \tag{B.15}$$

where P_{ij} denotes the power flow through line $i \to j$ and l_{ij} denotes the magnitude square of the current through line $i \to j$.

Theorem B.5 *RL1 and RLS1 are equivalent.*

Proof: Let $\mathcal{F}_{RL1}$ and $\mathcal{F}_{RLS1}$ be the feasible sets of RL1 and RLS1, respectively. It suffices to show that the map g (B.13) is one to one between $\mathcal{F}_{RL1}$ and $\mathcal{F}_{RLS1}$, since p is determined by (v, W) in $\mathcal{F}_{RL1}$ and by P in $\mathcal{F}_{RLS1}$. On the one hand, it is straightforward that for any $(v_1, W_1) \neq (v_2, W_2) \in \mathcal{F}_{RL1}, g(v_1, W_1) \neq g(v_2, W_2) \in \mathcal{F}_{RLS1}$. On the other hand, for any $(v', P, l) \in \mathcal{F}_{RLS1}$, it can be verified that there exists $g^{-1}(v', P, l) = (v, W) \in \mathcal{F}_{RL1}$. ∎

Under Assumptions B.1 and B.2, let $\mathcal{F}$ be a feasible set of a certain optimization problem, which is indicated by the subscript. Then the relationship between different feasible sets is

$$\mathcal{F}_{\text{OPF2}} = f(\mathcal{F}_{\text{OPF1}}) \subset \mathcal{F}_{\text{RL1}},$$
$$\mathcal{F}_{\text{RLS1}} = g(\mathcal{F}_{\text{RL1}})$$

where f is the one-to-one map given in Theorem B.1 and g is the one-to-one map given in Theorem B.5. The relationship between different feasible sets is depicted in Figure B.2. $\mathcal{F}_{\text{OPF1}}$ is a non-convex region, and (p^*, V^*) denotes the optimal solution to OPF1. $\mathcal{F}_{\text{OPF1}}$ is transformed equivalently into another non-convex region $\mathcal{F}_{\text{OPF2}}$ by the map f. $\mathcal{F}_{\text{OPF2}}$ is convexified by removing the rank constraint (B.4g), yielding $\mathcal{F}_{\text{RL1}}$, which is larger than $\mathcal{F}_{\text{OPF2}}$. The exactness of RL1 ensures that the optimal solution to RL1, i.e. (p^*, v^*, W^*), is not in the shaded region of $\mathcal{F}_{\text{RL1}}$. Since the objective function of OPF2 and RL1 are the same, (p^*, v^*, W^*) is also the optimal solution to OPF2. Applying the one-to-one map f, (p^*, v^*, W^*) is transformed into (p^*, V^*). The one-to-one map g converts $\mathcal{F}_{\text{RL1}}$ into an equivalent convex region $\mathcal{F}_{\text{RLS1}}$ and converts $\mathcal{F}_{\text{OPF2}}$ into an equivalent non-convex region

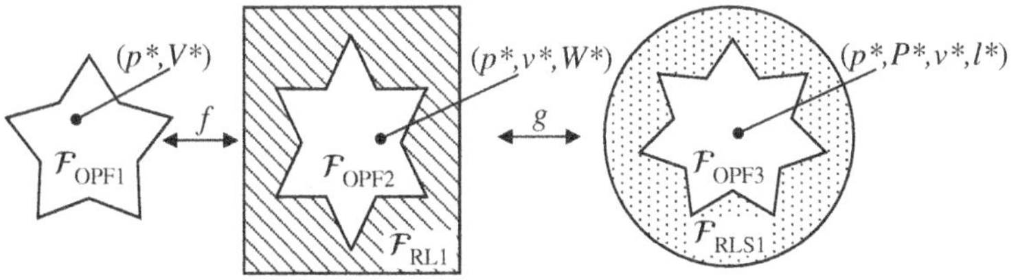

Figure B.2 Relationship between different feasible sets.

$\mathcal{F}_{\text{OPF3}}$. (p^*, P^*, v^*, l^*) is the optimal solution to RLS1, which is also the optimal solution to OPF3, since RLS1 is exact. Moreover, (p^*, P^*, v^*, l^*) can be converted equivalently into (p^*, v^*, W^*) using the map g.

Extending the work in [1], we have shown that under the proposed assumptions, the OPF problem in DC networks can be converted equivalently into a convex SOCP problem, regardless of topologies or operating modes. Thereby solving the relaxed convex SOCP can obtain the global optimum with theoretic guarantee. Furthermore, we have shown that the SOCP problem has at most one optimal solution, whether the DC network is working in grid-connected or stand-alone mode.

Remark B.3 In OPF1–OPF3, line flow constraints are not considered. Whereas we have not found any violation of line constraints in all our tests, we cannot theoretically exclude the possibility of such violation. In such a circumstance, approximate solutions can be heuristically constructed. We will discuss this issue later on in Section B.4.

B.3 Case Studies

B.3.1 16-Bus System

Tests are conducted on the modified 16-bus system [3] to demonstrate the effectiveness of the proposed method when the system works in different operating modes and network topologies. The three-feeder system is shown in Figure B.3 where 6 distributed generators (DGs) are added with the same capacity of 5 MW. The dashed lines represent tie lines with tie breakers. The system is able to work in two modes:

1) **Grid-connected**: the system is connected to a power grid via all the three feeders (Feeders A, B, and C).
2) **Stand-alone**: all the three feeders are disconnected from the grid.

Additionally, the network can switch between tree and mesh topologies by opening or closing the tie breakers.

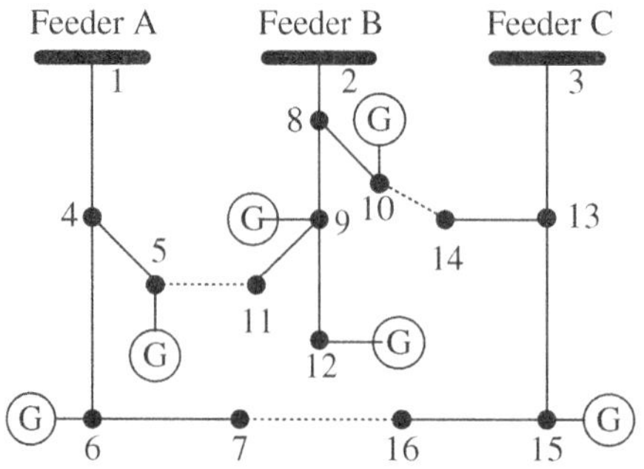

Figure B.3 16-bus system with DGs.

1) **Tree topology**: all the tie breakers are switched off.
2) **Mesh topology**: all the tie breakers are switched on.

Thus, the following four cases will be studied:

1) **GT**: grid-connected mode with a tree topology.
2) **GM**: grid-connected mode with a mesh topology.
3) **ST**: stand-alone mode with a tree topology.
4) **SM**: stand-alone mode with a mesh topology.

The voltage lower and upper bounds are set as 0.95 and 1.05 p.u., respectively. The objective is to minimize total network loss, and the problems are solved using MOSEK. The results are listed in Table B.1.

At a numerical RLS1 solution (p, v, W), R_{ij} can be obtained for each $i \to j$. If RLS1 is exact, then for all $i \to j$, we have $\text{rank}(R_{ij}) = 1$, i.e. the difference $D_{ij} := v_i v_j - W_{ij} W_{ji} = 0$. Hence the smaller D_{ij} indicates the closer R_{ij} is to rank one. The row "Exactness" lists the maximum value of D_{ij} for all $i \to j$, indicating RLS1 is exact for all the test networks. Moreover, it demonstrates that the proposed SOC relaxation does not depend on operating mode or network topology. The objective values indicate that in this system, mesh topology reduces network loss in both grid-connected and stand-alone modes, since it is able to support higher nodal voltages. Additionally, working in the same network topology, grid-connected mode experiences less network loss than stand-alone mode, since more power is injected

Table B.1 Results of 16-bus system.

	GT	GM	ST	SM
Exactness	5.73E−9	1.46E−9	5.59E−10	1.72E−10
Objective value (p.u.)	0.012	0.009	0.017	0.013
Computation time (s)	0.13	0.20	0.12	0.12

Table B.2 Results of power injections and nodal voltages.

	Power injection (p.u.)				Nodal voltage (p.u.)			
Bus	GT	GM	ST	SM	GT	GM	ST	SM
1	0.072	0.072	0	0	1.050	1.050	1.042	1.042
2	0.144	0.131	0	0	1.050	1.050	1.026	1.029
3	0.085	0.041	0	0	1.050	1.050	1.034	1.044
4	−0.200	−0.200	−0.200	−0.200	1.045	1.045	1.042	1.042
5	0.068	0.132	0.107	0.192	1.050	1.050	1.050	1.050
6	0.211	0.226	0.246	0.260	1.050	1.050	1.050	1.050
7	−0.150	−0.150	−0.150	−0.150	1.044	1.044	1.044	1.044
8	−0.400	−0.400	−0.400	−0.400	1.035	1.036	1.026	1.029
9	0.100	0.100	0.100	0.100	1.044	1.047	1.039	1.044
10	0.144	0.235	0.231	0.324	1.050	1.050	1.050	1.050
11	−0.060	−0.060	−0.060	−0.060	1.038	1.048	1.033	1.047
12	0.079	0.038	0.139	0.078	1.050	1.050	1.050	1.050
13	−0.100	−0.100	−0.100	−0.100	1.041	1.046	1.034	1.044
14	−0.100	−0.100	−0.100	−0.100	1.032	1.046	1.026	1.045
15	0.329	0.253	0.415	0.279	1.050	1.050	1.050	1.050
16	−0.210	−0.210	−0.210	−0.210	1.042	1.042	1.042	1.042

from the feeders to support higher nodal voltages. The results of power injections and nodal voltages are listed in Table B.2.

B.3.2 Larger-Scale Systems

More tests are conducted on both mesh and tree networks to check the exactness of the SOC relaxation. In addition, the objective value and computation time of the relaxed model (i.e. RLS1) and the original model (i.e. OPF1) are compared to show the efficacy of the relaxation. The objective of both models is to minimize total network loss. The mesh networks [1] are modified from MATPOWER by ignoring line reactances. All the line resistances are reduced to 10% of the original values to simulate the DC network condition. Particularly, the zero resistances are reset as 10^{-3} p.u. Additionally, the IEEE 118-bus system is applied to show the scalability of the proposed method. Four radial distribution networks in the literature are also used to verify the topological independence of the relaxation. In these systems, all the line resistances are also reduced to 10% of the original values to simulate the

Table B.3 Exactness of RLS1 and comparison of two models.

Topology	System	Exactness of RLS1	Objective value (p.u.)		Time (s)	
			RLS1	OPF1	RLS1	OPF1
Mesh	case6ww	1.24E−10	3.17E−03	3.17E−03	0.14	0.13
	case9	7.17E−12	5.72E−03	5.72E−03	0.11	0.14
	case_ieee30	2.37E−11	1.52E−03	1.52E−03	0.14	0.19
	case39	3.64E−11	1.30E−01	1.30E−01	0.16	0.16
	case118	6.38E−11	7.98E−03	7.98E−03	0.21	0.23
Tree	33-bus [4]	1.28E−11	1.47E−01	1.47E−01	0.13	0.14
	70-bus [5]	5.35E−12	1.10E−01	1.10E−01	0.23	0.21
	94-bus [6]	3.09E−11	3.63E−01	3.63E−01	0.22	0.36

DC network condition. In all the test systems, the voltage lower and upper bounds are set as 0.95 and 1.05 p.u., respectively. RLS1 is solved using MOSEK, while OPF1 is solved using IPOPT. The results are listed in Table B.3.

It is observed that RLS1 is exact for both mesh and tree topologies. The scale of a DC network is small; however, more and more DC micro-grids may be built in the future to integrate distributed renewable energy generation and electric vehicles. The results of computation time imply that the proposed method may be applied in large scale systems, for example, the cluster of DC micro-grids.

We also tested the model with line constraints (i.e. RLS2), on all the above systems. However, we have not found any case where the optimal solution violates the rank constraints (B.4g). In this regard, we conjecture that the RLS2 is almost always exact in practice. Since we cannot theoretically exclude the possibility of inexactness of OPF in this case, we give some discussions and insights in the next section.

B.4 Discussion on Line Constraints

B.4.1 OPF with Line Constraints

Let $\bar{I}_{ij}$ be the threshold of current through line $i \rightarrow j$; then the line constraint can be formulated as

$$y_{ij}^2(v_i - W_{ij} - W_{ji} + v_j) \leq \bar{I}_{ij}^2 \tag{B.16}$$

which is corresponding to

$$y_{ij}^2(V_i - V_j)^2 \leq \overline{I}_{ij}^2 \tag{B.17}$$

in the original problem. By adding (B.16) to RL1, we have the following model.

$$\textbf{RL2:} \quad \min_{p,v,W} h(p)$$
$$\text{s.t.} \quad \text{(B.4a)–(B.4f), (B.16)}$$

Using the transformation in Theorem B.5, the line constraint (B.16) can be added into RLS1, yielding the following model.

$$\textbf{RLS2:} \quad \min_{p,P,v,l} h(p)$$
$$\text{s.t.} \quad \text{(B.14a)– (B.14e), (B.15)}$$

$$l_{ij} \leq \overline{I}_{ij}^2 \tag{B.18}$$

As shown in Figure B.4a, when the optimal solution is exact, i.e. $\text{rank}(R_{ij}) = 1$ holds, the optimal solution (v_i, v_j, W_{ij}) is on the conical surface Ω_C defined by $v_i v_j - W_{ij}^2 = 0$. As shown in Figure B.4b, when the line constraint (B.16) is considered simultaneously, if the optimal solution is exact, it should be on the remaining conical surface Ω_{CP} cut by the plane Ω_P defined by (B.16). Thus, two cases are possible: (i) If the line constraint is *not* binding, which means the optimal solution is *above* the plane Ω_P, then Theorem B.2 still holds, implying that the optimal solution is on the conical surface Ω_{CP}, and (ii) if the line constraint is binding, which means the optimal solution is *on* the plane Ω_P, then Theorem B.2 may not hold. Meanwhile, the exactness requires the optimal solution to be on the conical surface Ω_{CP}. Thus, the optimal solution should be on the intersections of the plane Ω_P and the conical surface Ω_{CP}, i.e. Ω_I.

B.4.2 Exactness Conditions with Line Constraints

We first introduce the following theorem to give the conditions that guarantee the exactness of RL2.

Theorem B.6 *Under Assumptions B.1 and B.2, RL2 is exact under either of the following conditions:*

1) *$\forall (i \to j) \in \mathcal{E}$, constraint* (B.16) *is not binding.*
2) *For each line $(s \to t) \in \mathcal{E}$, if constraint* (B.16) *is binding for this line, then the corresponding lower bounds on p_s and p_t are not binding.*

Proof: When the line constraint (B.16) is not binding, RL2 is equivalent to RL1, which is exact under Assumptions B.1 and B.2 according to Theorem B.2.

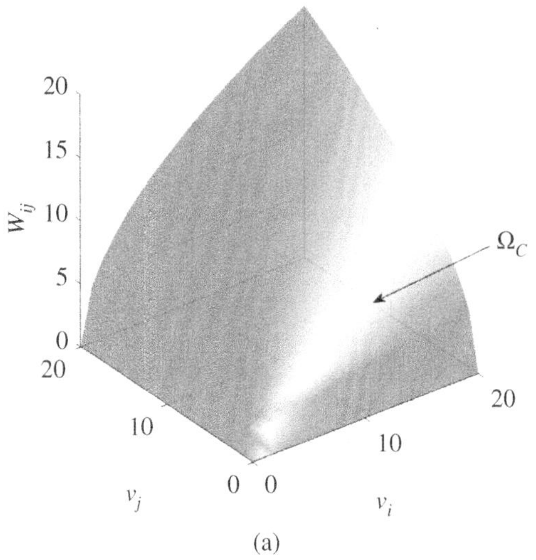

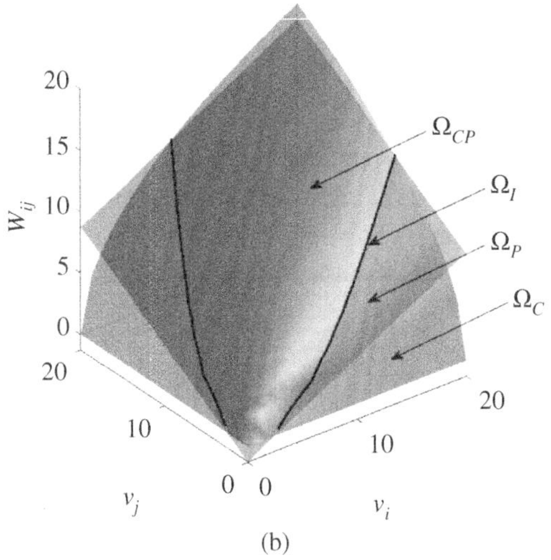

Figure B.4 Geometrical interpretation of rank and line constraints. (a) Rank constraint. (b) Rank constraint and line constraint.

However, when constraint (B.16) is binding, Theorem B.2 does not hold. In this situation, assume (p^*, v^*, W^*) is the optimal solution to RL2. Also, assume (p^*, v^*, W^*) violates the rank constraint (B.4g) for a certain line $(s \to t) \in \mathcal{E}$ where constraint (B.16) is binding and the lower bounds of p_s and p_t are not binding. Since (p^*, v^*, W^*) satisfies (B.4d)–(B.4f), we have

$$0 \le W_{ij}^* \le \sqrt{v_i^* v_j^*}, \quad \forall i \to j$$

Since (p^*, v^*, W^*) violates the rank constraint (B.4g) for $s \to t$, we have $W^*_{st} \neq \sqrt{v^*_s v^*_t}$. Thus, $W^*_{st} < \sqrt{v^*_s v^*_t}$. We can always choose a small enough number $\epsilon > 0$ such that

$$\epsilon < \min\left\{\frac{p^*_s - \underline{p}_s}{y_{st}}, \frac{p^*_t - \underline{p}_t}{y_{st}}, \sqrt{v^*_s v^*_t} - W^*_{st}\right\}$$

Then we can use ϵ to construct (p', v^*, W') where

$$W'_{ij} := \begin{cases} W^*_{ij} + \epsilon & \text{if } \{i,j\} = \{s,t\}; \\ W^*_{ij} & \text{otherwise;} \end{cases}$$

$$p'_i := \sum_{j:j\sim i} \left(v^*_i - W'_{ij}\right) y_{ij}, \quad i \in \mathcal{N}$$

It is easy to verify that (p', v^*, W') satisfies (B.4a)–(B.4f) and $h(p') < h(p^*)$. Additionally, when $\{i,j\} \neq \{s,t\}$, we have

$$y^2_{ij}(v^*_i - W'_{ij} - W'_{ji} + v^*_j) = y^2_{ij}(v^*_i - W^*_{ij} - W^*_{ji} + v^*_j) \leq \bar{I}^2_{ij}$$

When $\{i,j\} = \{s,t\}$, since the line constraint (B.16) is binding,

$$y^2_{ij}(v^*_i - W'_{ij} - W'_{ji} + v^*_j) = y^2_{ij}(v^*_i - W^*_{ij} - W^*_{ji} + v^*_j - 2\epsilon) < \bar{I}^2_{ij}$$

Thus, the point (p', v^*, W') satisfies (B.16). It means that the point (p', v^*, W') is feasible for RL2 and has a smaller objective value than (p^*, v^*, W^*). Thus, (p^*, v^*, W^*) cannot be the optimal solution. In contrast, the optimal solution to RL2 must satisfy the rank constraint (B.4g). This completes the proof. ∎

Furthermore, if the solution to RL2 is not exact, it is possible to find an approximate solution by relaxing the bounds on the power injections, as the following theorem indicates.

Theorem B.7 *Assume Assumptions B.1 and B.2 hold, and let (p, v, W) be a feasible solution to RL2:*

1) *(p, v, W) violates the rank constraint* (B.4g) *for a certain line $(s \to t) \in \mathcal{E}$ where the line constraint* (B.16) *is binding.*
2) *At least one of $p_s = \underline{p}_s$ and $p_t = \underline{p}_t$ is satisfied.*

Then there must exist another solution (p', v, W') that does the following:

1) *Satisfy all the constraints of RL2 except for* (B.4b) *(i.e.* (B.4a), (B.4c)–(B.4f), *and* (B.16)*).*
2) *Satisfy the rank constraint* (B.4g).
3) *Have a lower objective value than (p, v, W).*

Proof: Noting that (p, v, W) satisfies (B.4d)–(B.4f), we have $0 \le W_{ij} \le \sqrt{v_i v_j}$ for $i \to j$. Since (p, v, W) violates the rank constraint (B.4g) for $s \to t$, $W_{st} \neq \sqrt{v_s v_t}$. Hence $W_{st} < \sqrt{v_s v_t}$. Then we can always choose $\epsilon := \sqrt{v_s v_t} - W_{st} > 0$ to construct another solution (p', v, W') as below:

$$W'_{ij} := \begin{cases} W_{ij} + \epsilon & \text{if } \{i,j\} = \{s,t\}; \\ W_{ij} & \text{otherwise;} \end{cases}$$

$$p'_i := \sum_{j:j\sim i} \left(v_i - W'_{ij}\right) y_{ij}, \quad i \in \mathcal{N}$$

It is easy to verify that (p', v, W') satisfies (B.4a), (B.4c)–(B.4f), and (B.16). Hence the first assertion is proved.

In terms of the second assertion, (p', v, W') satisfies (B.4g) by noting $W'_{ij} = W_{ij} = \sqrt{v_i v_j}$ for $\{i,j\} \neq \{s,t\}$ and $W'_{ij} = W_{ij} + \epsilon = \sqrt{v_i v_j}$ for $\{i,j\} = \{s,t\}$.

Next we consider the third assertion. According to the construction of (p', v, W'), for any $i \neq s, t$, p' satisfies

$$\begin{aligned} p'_i &= \sum_{j:j\sim i} \left(v_i - W'_{ij}\right) y_{ij} \\ &= \sum_{j:j\sim i} \left(v_i - W_{ij}\right) y_{ij} \\ &= p_i \end{aligned}$$

and for any $i = s, t$,

$$\begin{aligned} p'_i &= \sum_{j:j\sim i} (v_i - W'_{ij}) y_{ij} \\ &= \sum_{j:j\sim i} (v_i - W_{ij}) y_{ij} - y_{st}\epsilon \\ &= p_i - y_{st}\epsilon \\ &< p_i \end{aligned}$$

Then (p', v, W') has a lower objective value than (p, v, W) as the objective function is strictly increasing in p_i. This completes the proof. ■

Theorem B.7 indicates that if the adjustment of power injection $y_{st}\epsilon$ is allowed at bus s and bus t, then a solution that satisfies the rank constraint (B.4g) can be constructed. In a DC network, such adjustment may be achieved by employing demand response or energy storage.

B.4.3 Constructing Approximate Optimal Solutions

Since RL2 is not always exact, we can check the solution after solving RL2. If the solution satisfies the rank constraint (B.4g), then it is globally optimal for OPF1 (B.1) with line constraint (B.17). Otherwise, inspired by the recovery methods of semidefinite program (SDP) relaxation [7], we propose two heuristic methods to construct approximate solutions.

Theorem B.6 indicates that if (p^*, v^*, W^*) is the optimal solution to RL2 and violates the rank constraint (B.4g) for a certain line $(s \to t) \in \mathcal{E}$ where the line constraint (B.16) is binding, then at least one of p_s and p_t must reach its lower bound. Otherwise, we can always find another feasible point, which has a smaller objective value than (p^*, v^*, W^*). Therefore, we only need to discuss the cases that at least one of p_s and p_t reaches its lower bound.

B.4.3.1 Direct Construction Method

In [7], a direct recovery method is proposed for AC networks, using the first column of the optimal solution matrix W^* to recover a nearly optimal solution. It inspires a direct construction method for DC networks.

Assume (p^*, v^*, W^*) is the optimal solution to RL2, which violates the rank constraint (B.4g) for some lines. Instead of using the first column of W^* (see in [7]), we use v^* to construct an approximate solution (p, V) to the original problem OPF1 with line constraint (B.17). First, $V_i (i \in \mathcal{N})$ is derived by letting $V_i = \sqrt{v_i^*}$. Then, $p_i (i \in \mathcal{N})$ is derived by substituting V_i into the power balance equation (B.1a).

Next, we show the approximate solution (p, V): (i) Satisfy power balance equation (B.1a), voltage constraint (B.1c), and line constraint (B.17). (ii) Possibly violate (B.1b) for some $i \in \mathcal{N}$, but the violations have limited bounds.

According to the construction of (p, V), it is straightforward to check that (p, V) satisfies (B.1a) and (B.1c). Hence we only need to examine constraint (B.17) to justify the first assertion.

For any $i \in \mathcal{N}$, let C_i be the set of buses immediately connected to bus i. Let $\mathcal{B}_i \subseteq C_i$ be the subset such that for any $j \in \mathcal{B}_i$, line $i \sim j$ violates the rank constraint (B.4g). If all the neighboring lines of bus i do not violate the rank constraint (B.4g), we have $\mathcal{B}_i = \emptyset$. It follows that for all $(i \sim j) \in \mathcal{E}$, $\sqrt{v_i^* v_j^*} > W_{ij}^*$ for $j \in \mathcal{B}_i$ while $\sqrt{v_i^* v_j^*} = W_{ij}^*$ for $j \notin \mathcal{B}_i$. Since (p^*, v^*, W^*) satisfies (B.16), for any $i \sim j (j \in \mathcal{B}_i)$, it follows that

$$
\begin{aligned}
y_{ij}^2 (V_i - V_j)^2 &= y_{ij}^2 (V_i^2 - 2V_i V_j + V_j^2) \\
&= y_{ij}^2 (v_i^* - 2\sqrt{v_i^* v_j^*} + v_j^*) \\
&< y_{ij}^2 (v_i^* - 2W_{ij}^* + v_j^*) \\
&\leq \bar{I}_{ij}^2
\end{aligned}
$$

and for any $i \sim j(j \notin \mathcal{B}_i)$

$$\begin{aligned}
y_{ij}^2(V_i - V_j)^2 &= y_{ij}^2(V_i^2 - 2V_iV_j + V_j^2) \\
&= y_{ij}^2(v_i^* - 2\sqrt{v_i^*v_j^*} + v_j^*) \\
&= y_{ij}^2(v_i^* - 2W_{ij}^* + v_j^*) \\
&\leq \bar{I}_{ij}^2
\end{aligned}$$

Therefore, (p, V) satisfies (B.17).

Next, we check constraint (B.1b). It follows from (B.1a) that

$$\begin{aligned}
p_i &= \sum_{j:j\sim i} V_i\left(V_i - V_j\right)y_{ij} \\
&= \sum_{j:j\sim i} \sqrt{v_i^*}\left(\sqrt{v_i^*} - \sqrt{v_j^*}\right)y_{ij} \\
&= \sum_{j:j\sim i} \left(v_i^* - \sqrt{v_i^*v_j^*}\right)y_{ij} \\
&= \sum_{j:j\sim i, j\notin\mathcal{B}_i} \left(v_i^* - \sqrt{v_i^*v_j^*}\right)y_{ij} + \sum_{j:j\sim i, j\in\mathcal{B}_i} \left(v_i^* - \sqrt{v_i^*v_j^*}\right)y_{ij} \\
&\leq \sum_{j:j\sim i, j\notin\mathcal{B}_i} \left(v_i^* - W_{ij}^*\right)y_{ij} + \sum_{j:j\sim i, j\in\mathcal{B}_i} \left(v_i^* - W_{ij}^*\right)y_{ij} \\
&= p_i^*
\end{aligned}$$

Hence, p_i may violate (B.1b), which is bounded by

$$p_i - p_i^* \leq \sum_{j:j\sim i, j\in\mathcal{B}_i} \left(-\sqrt{v_i^*v_j^*} + W_{ij}^*\right)y_{ij}$$

B.4.3.2 Slack Variable Method

According to Theorem B.6, RL2 is exact if the lower bounds of power injections are not binding. Thus, after solving RL2, if the solution violates the rank constraint (B.4g) for a certain line $(s \sim t) \in \mathcal{E}$ and any lower bound of p_s and p_t is binding, for example, $p_s = \underline{p}_s$, then a corresponding slack variable ε_s $(\varepsilon_s > 0)$ can be added into (B.4b) to reformulate the constraint as $\underline{p}_s \leq p_s - \varepsilon_s \leq \overline{p}_s$. This is so that p_s will not reach its lower bound in the next iteration to solve RL2. Additionally, to minimize ε_s, it is also added into the objective function as min $h(p) + \sum_{i\in\hat{\mathcal{N}}}\varepsilon_i$, where $\hat{\mathcal{N}}$ is the set of buses in which rank constraint (B.4g) is violated and the power injection lower bound is binding at the same time.

Bibliography

1 L. Gan and S. H. Low, "Optimal power flow in direct current networks," *IEEE Transactions on Power Systems*, vol. 29, no. 6, pp. 2892–2904, 2014.

2 M. Farivar and S. H. Low, "Branch flow model: relaxations and convexification-Part I," *IEEE Transactions Power Systems*, vol. 28, no. 3, pp. 2554–2564, 2013.

3 S. Civanlar, J. J. Grainger, H. Yin, and S. S. H. Lee, "Distribution feeder reconfiguration for loss reduction," *IEEE Transactions on Power Delivery*, vol. 3, no. 3, pp. 1217–1223, 1988.

4 M. E. Baran and F. F. Wu, "Network reconfiguration in distribution systems for loss reduction and load balancing," *IEEE Transactions on Power Delivery*, vol. 4, no. 2, pp. 1401–1407, 1989.

5 D. Das, "A fuzzy multiobjective approach for network reconfiguration of distribution systems," *IEEE Transactions on Power Delivery*, vol. 21, no. 1, pp. 202–209, 2006.

6 C.-T. Su and C.-S. Lee, "Network reconfiguration of distribution systems using improved mixed-integer hybrid differential evolution," *IEEE Transactions on Power Delivery*, vol. 18, no. 3, pp. 1022–1027, 2003.

7 G. Fazelnia, R. Madani, A. Kalbat, and J. Lavaei,"Convex relaxation for optimal distributed control problems," *IEEE Transactions on Automatic Control*, vol. 62, no. 1, pp. 206–221, 2017.

Index